T0399289

LIDAR PRINCIPLES, PROCESSING AND APPLICATIONS IN FOREST ECOLOGY

LIDAR PRINCIPLES, PROCESSING AND APPLICATIONS IN FOREST ECOLOGY

QINGHUA GUO

Professor at Peking University, and serves as the director of the Institute of Remote Sensing and Geographical Information System of Peking University, China

YANJUN SU

Professor at the Institute of Botany, Chinese Academy of Sciences, China

TIANYU HU

Associate professor at the Institute of Botany, Chinese Academy of Sciences, China

ACADEMIC PRESS

An imprint of Elsevier

Academic Press is an imprint of Elsevier
125 London Wall, London EC2Y 5AS, United Kingdom
525 B Street, Suite 1650, San Diego, CA 92101, United States
50 Hampshire Street, 5th Floor, Cambridge, MA 02139, United States
The Boulevard, Langford Lane, Kidlington, Oxford OX5 1GB, United Kingdom

ISBN: 978-0-12-823894-3

For information on all Academic Press publications visit our website at
https://www.elsevier.com/books-and-journals

Publisher: Nikki P. Levy
Acquisitions Editor: Glyn Jones
Editorial Project Manager: Naomi Robertson
Production Project Manager: Fahmida Sultana
Cover Designer: Miles Hitchen

Typeset by TNQ Technologies

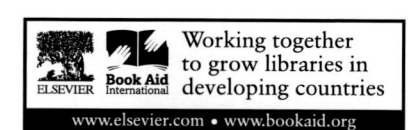

Contents

Preface

In 2005, I started my academic career at the University of California, Merced, USA. The first large-scale interdisciplinary project I was involved in was the "Sierra Nevada Adaptive Management Project" commissioned by the State of California, USA. The University of California, as an independent third party, was invited to conduct a comprehensive evaluation of the impact of the US Federal Forest Service's forest management strategies on California's Sierra Nevada forests (including forest health, endangered species, forest fires, water resources, and public forest awareness). The project involved ecology, hydrology, geology, social sciences, and other disciplines. More than 10 professors from six teams and three campuses of the University of California (Berkeley, Davis, and Merced) participated in the project. I had the privilege of joining the project as a project leader on the spatial analysis team. Since the results of this project would directly affect the United States Forest Service's forest management policy, obtaining high-precision terrain, forest vegetation, and species distribution information were critical to project success. I proposed using light detection and ranging (LiDAR) technology to extract precise forest structure attributes and terrain attributes to evaluate the impact of forest thinning on the environment. In addition, as a member of the Critical Zone Observatory (CZO), funded by the US National Science Foundation, I was responsible for collecting and processing LiDAR data for all CZO sites. These two projects started my LiDAR career and made me aware of the important role of LiDAR in forestry studies and forest management practices. From then on, I completed several projects on this topic with the support from the Strategic Priority Research Program of Chinese Academy of Sciences, the National Key Basic Research and Development Program of China, the National Key R&D Program of China, the National Natural Science Foundation of China, the US Geological Survey, the US National Park Service, the California State Department of Water Resources, the Moore Foundation, and others. These research projects enabled us to push scientific boundaries in underforest terrain extraction, forest structure attribute inversion, biodiversity monitoring, crop phenotyping, urban three-dimensional (3-D) green volume monitoring, and multisensor integration.

While implementing these projects, my team and I broke through a series of LiDAR data acquisition and information extraction bottlenecks, solving the information asymmetry problem between the ecological niche requirements of biological species in 3-D space and the two-dimensional (2-D) habitat information provided by traditional passive remote sensing technology. Meanwhile, we deeply understand that LiDAR technology will provide a new 3-D perspective for ecological research and promote the transition of quantitative ecological studies from "2-D" to "3-D." However, we note the continued lack of a comprehensive introduction to this emerging technology for LiDAR applications in forest ecosystems. This book is intended to systematically introduce readers to LiDAR technology and its application and frontier development in forest ecological research. Its audience includes senior undergraduate and graduate students majoring in ecology and geosciences, as well as scholars who wish to learn more about the field. Since LiDAR technology is still undergoing rapid development, many algorithms have not yet matured. To understand LiDAR principles and applications, readers must often use programming languages to process data. Therefore, the book provides Python code for relevant content when it is introduced. Relevant data and Python code can be freely downloaded from our team website, http://www.3dEcology.org.

This book is written based on nearly 20 years of research results for LiDAR software and hardware development and forest ecological applications carried out by my team and me. I am forever grateful to the team members who have studied with me over the years. Special thanks go to the team members who participated in writing and editing some chapters, including Wenkai Li, Shengli Tao, Qin Ma, Shichao Jin, Chunyue Niu, Hongcan Guan, Kai Cheng, Yu Ren, Zekun Yang, Guangcai Xu, Mengxi Chen, Yi Wang, Xiaoqiang Liu, Xiaoyong Wu, Qiuli Yang, Qin Ma, Zhonghua Liu, Mengqi Cao, Xuejing Wang, Xiaoxia Zhao, Jiatong Wang, Bingwei Yang, and Xiliang Sun.

The original idea to write this book was born while I worked at the University of California, Merced. After I returned to China to start my new academic position, I taught two LiDAR classes and organized seven workshops on the topic of LiDAR applications in forestry, hoping to promote the use of LiDAR in forestry studies and forest management practices. Over 6000 students, researchers, and forest managers participated in these classes and workshops. Their thirst for knowledge about advanced technology has urged me to improve the book's structure and finally finish the writing. In publishing this book, I have received strong support from Yan Guan, an editor of the Higher Education Press of China, and on behalf of all the authors of this book, I would like to express my deepest gratitude.

This book was supported by the Strategic Priority Research Program of Chinese Academy of Sciences (Grant No. XDA19050401), the National Natural Science Foundation of China (Grant Nos. 31971575, 41871332, and 41901358), and China Classics International.

Due to the author's limited knowledge, this book may inevitably have errors. We welcome corrections from readers.

Qinghua Guo

Institute of Remote Sensing and Geographic Information Systems
School of Earth and Space Sciences
Peking University

CHAPTER 1

The Origin and Development of LiDAR Techniques

Contents

Remote sensing, literally interpreted as "remotely sensed," refers to a noncontact, long-range detection technology. Remote sensing can reveal the spatial distribution and spatiotemporal changes in objects on the Earth's surface qualitatively and quantitatively through the transmission, transformation, and processing of electromagnetic radiation information. According to various methods of signal acquisition (i.e., electromagnetic wave energy), remote sensing can be categorized as passive remote sensing and active remote sensing. A passive remote sensing system does not have a radiation source, only receiving and recording the electromagnetic information emitted or reflected by the target. In contrast, an active remote sensing system carries an artificial radiation source. It receives and records information by actively emitting electromagnetic waves toward a target. Synthetic aperture radar, ground penetrating radar, and light detection and ranging (LiDAR) are all active remote sensing systems. In recent decades, active remote sensing technology has developed rapidly because it can collect data day and night and does not depend on solar radiation. It can also select the wavelength of emitted electromagnetic waves and the emission mode to detect different targets. As one of the main branches of active remote sensing technology, LiDAR is widely used in fields such as forestry, agriculture, meteorology, surveying, and archaeology because it

LiDAR Principles, Processing and Applications in Forest Ecology
ISBN 978-0-12-823894-3
https://doi.org/10.1016/B978-0-12-823894-3.00001-3

can provide high-precision three-dimensional information on objects. This chapter introduces the origin and development of LiDAR and illustrates three LiDAR systems mounted on different platforms. Mainstream commercial LiDAR devices and data processing software are briefly introduced. Finally, the importance of applying LiDAR to forest ecology research is emphasized.

1.1 Introduction and History of LiDAR

LiDAR is an acronym for "light detection and ranging." It can quantify three-dimensional (3-D) structural information of a target accurately by measuring the distance between the sensor and the target and analyzing the energy, amplitude, frequency, and phase of the reflected beam (Fig. 1.1).

LiDAR was first developed and mounted to airborne platforms to measure near-shore ocean depth. The first LiDAR system was built in 1968

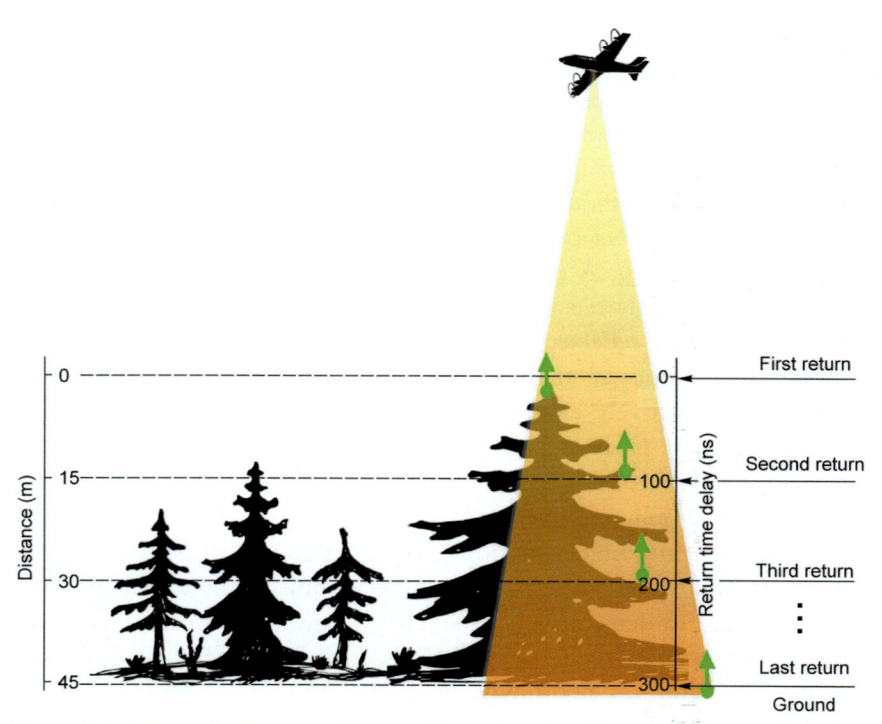

Figure 1.1 Schematic diagram of the working principle of light detection and ranging (LiDAR) systems.

by Hickman and Hogg of Syracuse University in the United States. It measured ocean depth by calculating the time difference between emitted and received echoes. Their system demonstrated, for the first time, the feasibility of using lasers to detect water depth (Hickman & Hogg, 1969). In the late 1970s, the National Aeronautics and Space Administration (NASA) successfully developed Airborne Oceanographic LiDAR (AOL), which could scan and record data quickly. Hoge et al. (1980) measured water depth in the Atlantic Ocean and Chesapeake Bay using AOL and mapped the submarine geomorphology at water depths of less than 10 m. After that, the potential of airborne LiDAR systems began to attract broad attention and was applied to topographic land surveys in the early 1980s (Arp et al., 1982; Krabill et al., 1984). It is worth mentioning that, at that time, vegetation was usually regarded as an obstacle in topographic surveys because the laser was often intercepted in areas with high canopy density. In 1984, however, Nelson et al. (1984) found that the penetration of laser pulses was closely related to canopy density in a study at Blue Mountain, Pennsylvania, USA. They noted that LiDAR could detect vertical forest structure and estimate tree height. At that time, LiDAR systems could not detect and identify trees shorter than 8 m, which hindered their application in forestry to a certain extent (Nelson et al., 1984).

In the late 1980s, the development of global navigation satellite systems (GNSSs) and inertial navigation systems (INSs) made it possible to precisely measure the location and posture of the scanner in real time during laser scanning. The world's first laser cross section measurement system, developed in 1990 by Professor Ackermann from Stuttgart University in Germany, was the most notable achievement during this period. This system successfully combined laser scanning technology with a real-time positioning system to form an airborne LiDAR (Ackermann-19). In 1993, the first commercial airborne LiDAR, TopScan ALTM 1020, was developed in Germany, and in 1995, airborne LiDAR was commercially manufactured. Since then, airborne LiDAR technology has become an important addition to forest resource surveys. It is widely used to acquire a range of forest structure information quickly (e.g., tree height and canopy volume) and provide a quantitative estimation of forest attributes (e.g., vertical stratification structure, carbon, and flammable litter) for forest ecological research and forest management (Pang et al., 2005).

The development and application of spaceborne LiDAR gradually matured through the 1990s. In 2003, NASA proposed a plan to measure the change in the ice surface at the poles using spaceborne LiDAR and

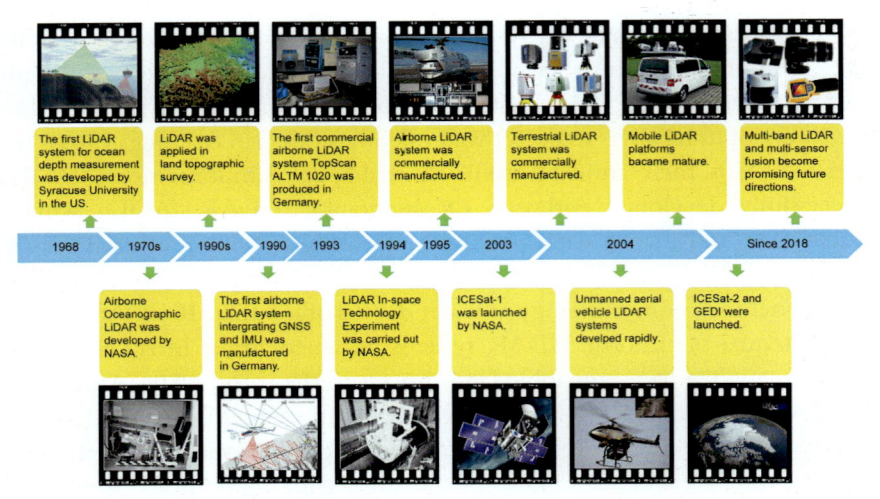

Figure 1.2 A brief history of LiDAR development and key milestones. NASA represents the National Aeronautics and Space Administration; GNSS and IMU represent global Navigation satellite system and inertial measurement unit; and ICESat and GEDI represent the Ice, Cloud, and Land Elevation Satellite and the Global Ecosystem Dynamics Investigation, respectively.

officially included the Geoscience Laser Altimeter System (GLAS) in the Earth Observation System. GLAS was carried on the Ice, Cloud, and land Elevation Satellite (ICESat) launched in January 2003 and was in operation for 7 years. ICESat-2 (a successor to the first ICESat mission) and the Global Ecosystem Dynamics Investigation (GEDI) were launched on September 15, 2018, and December 5, 2018, respectively. They greatly expanded the detection performance of spaceborne LiDAR and brought new opportunities for ecologists to understand forest structural attributes at large scales (Hand, 2014; White et al., 1996). A brief history of the development and key milestones of LiDAR are shown in Fig. 1.2.

1.2 Classification of LiDAR Hardware

With the development of new technology, LiDAR has become more diverse. According to the principles of ranging, laser scanners are mainly divided into pulse-based and phase-based. The pulse-based laser scanner emits a very narrow light pulse to the target. It determines the distance from the laser to the target by measuring the time interval between transmitting the pulse and receiving it in reflected form. In contrast, the phase-based laser scanner emits a modulated laser beam and then compares the phase shifts between the emitted and returned laser beams to determine the distance. This means that the phase-based laser scanner uses consecutive laser beams. Regardless, the phase-based laser scanner is generally used in close-range measurement owing to the limitations of laser energy.

Laser scanners can be mounted on various platforms such as tripods, backpacks, automobiles, unmanned aerial vehicles (UAVs), manned aircraft, and satellites, with capabilities ranging from plot to global scales (Fig. 1.3). Terrestrial LiDAR, also called terrestrial laser scanning (TLS), is commonly used for fine-scale 3-D data acquisition of a single target or small scenes. TLS usually consists of a laser ranging system, laser scanning system, charge-coupled device digital camera, and internal calibration system. The scanning distance of TLS ranges from a few meters to a few thousand meters. Traditionally, TLS is mounted on fixed tripods and uses a stop-and-go mode to acquire data, a method requiring substantial effort to register multiscan LiDAR data. Recent advances in integrated navigation techniques and simultaneous localization and mapping algorithms create new possibilities to mount laser scanners on mobile platforms, such as backpacks and mobile vehicles, greatly improving data acquisition efficiency. A stationary terrestrial LiDAR system measures the whole area using single station scanning and multistation registration. In contrast, mobile LiDAR

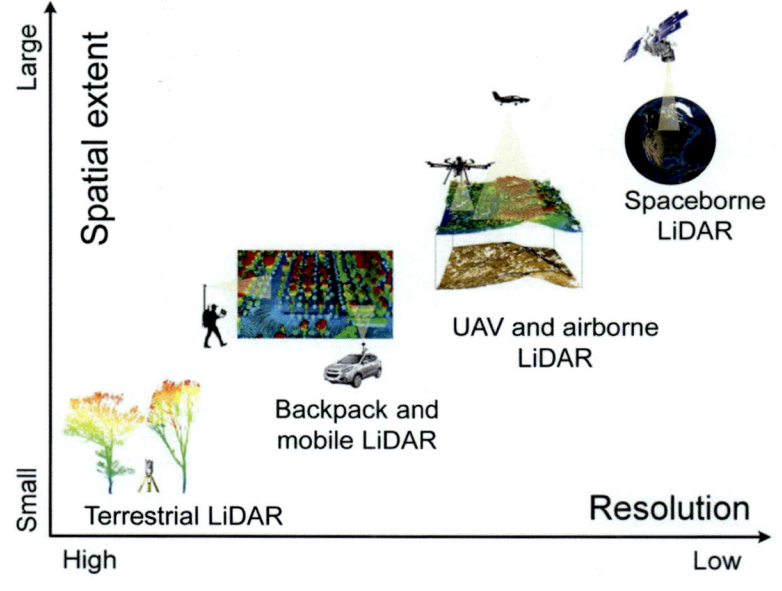

Figure 1.3 Different LiDAR sensors mounted on various platforms.

systems, exemplified by backpack LiDAR and mobile LiDAR, consist of a LiDAR sensor, GNSS, and IMU that can automatically acquire point cloud data of the whole study area. TLS makes up for the shortcomings of current observation methods in forestry and is gradually becoming a powerful tool for forestry inventory because of its high-quality 3-D point cloud data. Moreover, TLS provides a nondestructive approach to acquiring high-accuracy 3-D data on individual trees and forest stands, which makes it possible to automatically extract the structural information of individual trees and reconstruct real 3-D forest scenes.

Airborne LiDAR systems, also known as airborne laser scanning, integrate a laser scanner, GNSS, and IMU and greatly increase the ranging capacity of LiDAR. They are often mounted on aircraft such as planes, airships, powered delta wings, and UAVs, of which planes and UAVs are two typical platforms. ALS provides a new technology for forest inventory and the extraction of forest structural attributes because it can quickly obtain 3-D point cloud data at a regional scale. Nowadays, ALS systems capture forest structure attributes from landscape to regional scales. With the development of manufacturing technology, commercial laser scanners are becoming lighter, smaller, and cheaper. For example, solid-state LiDAR sensors can be lighter than 1 kg in weight and cheaper than USD1000 in price. These sensors have greatly promoted the development of UAV LiDAR systems, which significantly increase flexibility and reduce the cost of collecting point clouds at the landscape scale.

Compared with terrestrial and airborne LiDAR systems, spaceborne LiDAR systems have the greatest ranging capability and are configured with a profiling design including multiple lasers. They can produce tracks or transects of laser pulses. Compared with ALS, spaceborne LiDAR has a higher orbit and a wider observation field of view. It is theoretically capable of providing global data, which gives it unique advantages for topographic mapping, environmental monitoring, forest surveys, and other applications (Lefsky, 2010; Lefsky et al., 1999; Saatchi et al., 2011). Nonetheless, because of its orbital altitude and emission frequency limitations, spaceborne LiDAR usually has low data density. It cannot obtain spatially continuous observations like passive remote sensing imagery. To date, spaceborne LiDAR systems have been mainly carried on six Earth Observation LiDAR missions. These missions were designed for a wide range of applications. The GLAS aboard ICESat (launched in 2003) and the Advanced Topographic Laser Altimetry System aboard ICESat-2 (launched in 2018) were designed for measuring the ice-cap volume. The Cloud-

Aerosol LiDAR with Orthogonal Polarization aboard the Cloud-Aerosol LiDAR and Infrared Pathfinder Satellite Observation satellite and the cloud profiling radar system on the CloudSat satellite were mainly used for probing the vertical structure and properties of thin clouds and aerosols over the globe. GEDI, developed by the University of Maryland in collaboration with NASA's Goddard Space Flight Center, focuses on measuring the 3-D structure and dynamics of the Earth's ecosystems. Aeolus was launched in 2018 by the European Space Agency (ESA) and carried a Doppler wind LiDAR; it is the first satellite mission to acquire profiles of Earth's winds on a global scale.

1.3 Commercial LiDAR Hardware

Currently, commercial LiDAR systems are still mainly based on the development of terrestrial and airborne systems. There are few commercial spaceborne LiDAR systems because of the high cost of development, maintenance, and launch. Therefore, this book focuses on commercial terrestrial and airborne LiDAR systems.

After decades of exploration and research, commercial LiDAR technology has become relatively mature. Many world-renowned manufacturers have emerged, such as Teledyne Optech, RIEGL Laser Measurement Systems, Trimble Inc., Velodyne LiDAR, Faro, Shanghai Hesai Photoelectric Technology Co., Ltd., Radium God Intelligent System Co., Ltd., Shenzhen DJI Sciences and Technologies Ltd., and Beijing Beike Tianhui Technology Co., Ltd. (Fig. 1.4). The current representative commercial LiDAR systems and their parameter information are described in Table 1.1. The typical representatives of stationary terrestrial LiDAR include the RIEGL VZ series, Trimble VX, and Faro Focus3D, characterized by their medium size and weight, ability to measure distances ranging from hundreds of meters to several kilometers, and relatively high ranging accuracy and transmission frequency. The Velodyne Puck VLP16, Velodyne HDL series, Hesai Pandar series, DJI Livox series, and RIEGL VUX series are the most common portable LiDAR scanners and the preferred choice for mobile and UAV LiDAR systems (Fig. 1.5). Although the scan range of such devices is relatively limited, the operating efficiency and coverage of the entire system are greatly improved owing to the platform's mobility. The mainstream equipment types of ALS are the Optech ALTM series,

Phase-based LiDAR

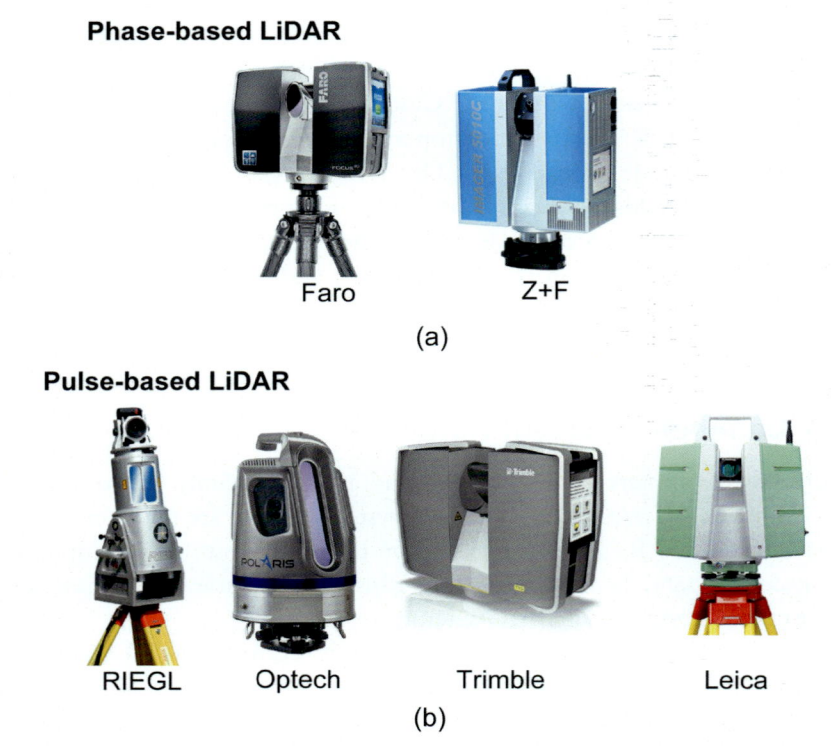

Faro   Z+F

(a)

Pulse-based LiDAR

RIEGL   Optech   Trimble   Leica

(b)

Figure 1.4 (a) Examples of commercial phase-based terrestrial LiDAR systems and (b) examples of commercial pulse-based terrestrial LiDAR systems.

RIEGL LMS series, and Leica ALS series, which are well suited to fast-moving aircraft platforms because such devices have a high long-range detection ability and positioning accuracy, although they are bulky and expensive. In addition, manufacturers are researching LiDAR system integration and have developed commercial systems. Beijing GreenValley Technology Co., Ltd. is typical of this type of manufacturer. It has developed the state-of-the-art LiBackpack backpack LiDAR system, LiAir UAV LiDAR system, and LiEagle airborne LiDAR system. In addition, it develops a supporting data processing software system named LiDAR360, which has been widely used in forestry, the power industry, surveying and mapping, agriculture, ecology, and other fields, along with its hardware systems.

Table 1.1 Mainstream commercial light detection and ranging (LiDAR) scanners.

Manufacturer	Country	Model	Scanning type	Distance (m)	Scan view (degree)	Ranging accuracy	Maximum scanning rate (points/s)	Laser level[a]
Leica	Switzerland	HDS3000	Pulse	2−100	360 × 270	6 mm@50 m	4000	III
		HDS6000	Phase	1−79	360 × 310	6 mm@50 m	500,000	III
		ScanStation2	Pulse	2−300	360 × 270	6 mm@50 m	50,000	III
		HDS C10	Pulse	0.1−300	360 × 270	2 mm@100 m	50,000	III
		ALS70-HP	Pulse	1−3500	75	10 cm@1000 m	1500	III
Trimble	USA	GS 100	Pulse	1−100	360 × 60	6 mm@50 m	5000	II
		GX 3D	Pulse	1−350	360 × 60	12 mm@100 m	5000	III
		GX	Pulse	≤350	360 × 270	7.2 mm@100 m	50,000	3R
		VX	Pulse	>150	Viewfinder window	10 mm (<150 m)	15	−
RIEGL	Austria	LPM-2K	Pulse	2−2000	360 × 190	50 mm	3600	I
		LPM-321	Pulse	10−6000	360 × 150	15−25 mm	1000	I
		LMS-Z620	Pulse	2−2000	360 × 80 degrees	10 mm	11,000	I
		VZ-400	Pulse	1−500	360 × 100	5 mm	12,500	I
		VUX-1	Pulse	3−300	330	10 mm	550,000	I

Continued

Table 1.1 Mainstream commercial light detection and ranging (LiDAR) scanners.—cont'd

Manufacturer	Country	Model	Scanning type	Distance (m)	Scan view (degree)	Ranging accuracy	Maximum scanning rate (points/s)	Laser level[a]
Faro	USA	LS420	Phase	0.6–20	360 × 320	3 mm@20 m	120,000	III
		LS840/880	Phase	0.6–0/70	360 × 320	3 mm@25 m	120,000	III
		Photo 20/80	Phase	0.6–0/76	360 × 320	2 mm@25 m	120,000	III
		FOCUS 3D	Phase	≤153	360 × 305	2 mm@25 m	976,000	3R
Optech	Canada	ILRIS-3D/3D^{VP}	Pulse	3–1500	40 × 40	7 mm@100 m	2500	I
		ILRIS-36D	Pulse	3–1500	360 × 110	7 mm@100 m	2500	I
		ILRIS-3D/3D^{ER}	Pulse	3–2000	40 × 40 50	7 mm@100 m	2500 280	I
		ALTM Gemini	Pulse	150–4000		5–30 cm		I
Z + F	Germany	Imagery 5010	Phase	0.3–187	360 × 320	1 mm@50 m	1,016,700	I
Velodyne	USA	VLP16	Pulse	1–100	360 × 30	3 mm@100 m	300,000	I
		HDL-32E	Pulse	1–70	360 × 40	3 mm@100 m	700,000	I
		HDL-64E	Pulse	1–120	360 × 26.9	2 mm@120 m	>1,330,000	I

Hesai	China	Pandar64	Pulse	0.3−200	360 × 40	2 cm@200 m	2,304,000	I
		Pandar40P	Pulse	0.3−200	360 × 40	2 cm@200 m	1,440,000	I
		Pandar40M	Pulse	0.3−200	360 × 40	2 cm@200 m	1,440,000	I
		QT64	Pulse	0.1−30	360 × 104.2	3 cm	768,000	I
		Pandar128	Pulse	0.3−200	360 × 40	2 cm@200 m	6,912,000	I
		XT32	Pulse	80	360 × 31	1 cm	1,280,000	I
		AT128	Pulse	0.3−200	120 × 25	2 cm@200 m	3,072,000	I
		QT128	Pulse	0.1−20	360 × 105.2	3 cm	1,728,000	I
Livox	China	Mid-100	Pulse	0−260	98.4 × 38.4	2 cm@20 m	300,000	I
		Mid-70	Pulse	0−260	360 × 70.4	2 cm@20 m	200,000	I
		Mid-40	Pulse	0−260	360 × 38.4	2 cm@20 m	100,000	I
		Avia	Pulse	0−450	77.2 × 70.4	2 cm@20 m	720,000	I
		Horizon	Pulse	0−260	81. 7 × 25.1	2 cm@20 m	480,000	I
		Tale-15	Pulse	0−500	16.2 × 14.5	2 cm@20 m	480,000	I

[a]Classified according to the potential damage to humans. Class I lasers have the lowest energy; class II lasers are safe for humans and free of static electricity; class III lasers are continuous and can hurt human eyes.

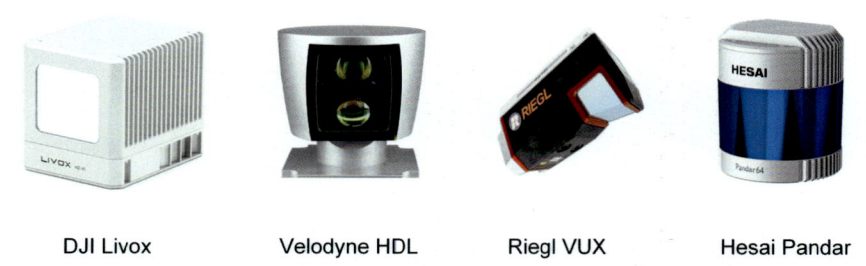

| DJI Livox | Velodyne HDL | Riegl VUX | Hesai Pandar |

Figure 1.5 Examples of portable LiDAR scanners for mobile and unmanned aerial vehicle LiDAR systems.

1.4 LiDAR Software

Professional processing software for LiDAR data is an important factor that promotes applications of LiDAR technology. Compared with traditional optical remote sensing processing software, LiDAR data processing software performs poorly in richness, ease of use, and degree of automation. In recent years, professional LiDAR processing software has appeared, including open-source software like Cloud Compare, SAGA GIS, Fusion, Whitebox GAT, Fugro Viewer, and MCC-LiDAR, and commercial software like Terrasolid, Quick Terrain Modeler, ENVI LiDAR, LP360, LiDAR360, Lastools, and Global Mapper LiDAR. The specific functions of each software are listed and compared in Table 1.2.

Basic functions of LiDAR processing software include display, filtering, terrain processing, and forestry attribute extraction. Displaying point clouds is the most basic function of LiDAR data processing software, which usually includes options to exhibit point clouds in different symbols or colors based on the properties of points and users' requirements. It is also the most convenient method to check the quality of LiDAR data. Currently, the volume of LiDAR data is becoming larger, and the efficiency and quality of displaying enormous amounts of data have become one of the main in-dicators for evaluating LiDAR processing software. Even before considering the limitations of the computer processors, much available software cannot display large volumes of LiDAR data efficiently. Comparatively, LiDAR360 has excellent performance in displaying large amounts of LiDAR data and can open and display over 400 GB of point cloud data. Filtering is a core step for processing LiDAR data, but not all software provides this function. Terrasolid is the most widely used software in this field. However, LiDAR360 is increasingly used because its novel filtering

Table 1.2 Comparison of LiDAR processing software functions.

Software	3-D display function	Filter	Terrain module	Forestry module	Large data processing	Batch processing	Ease of use
Cloud compare	√	X	X	X	X	X	General
SAGA GIS	√ (general)	√	√	√	X	√	Hard
Fusion	X	X	X	√	X	X	Easy
Lastools	√ (general)	√	√	√	X	√	Easier
Terrasolid	√	√ (nice)	√	X	X	√	General
Quick Terrain Modeler	√	X	X	X	X	X	Easy
ENVI LiDAR	√	X	X	X	X	X	Easier
LP360	√	√	√	X	X	√	Easy
LiDAR360	√ (nice)	√ (nice)	√	√	√	√	Easy

algorithm can achieve higher accuracy and robustness in filtering. The filtering function of other software like SAGA GIS, LP360, and Lastools is not always accurate enough. Forestry attribute extraction is also an important professional requirement for LiDAR data processing software. Fusion and Lastools can extract simple forestry attributes, such as the height quantile and signal strength distribution. SAGA GIS can also calculate some forestry attributes, but it requires manual intervention at each step (Lim et al., 2003). The forestry module in LiDAR360 can extract forestry attributes automatically and quickly, as well as execute batch processing.

1.5 The Importance of LiDAR in Forest Ecology Applications

Forests are an important part of the terrestrial ecosystem. On the one hand, forest ecosystems provide many material resources, such as forest products and by-products, and direct economic benefits for human life and economic development (Wang et al., 2001). On the other hand, they maintain the stability of the biosphere, improve the ecological environment, conserve runoff by regulating the latent heat and sensible heat flux on the local land surface, and change the hydrological conditions (Bonan, 2008; Lefsky et al., 2002). Forest ecosystems are also the largest terrestrial carbon sink, which makes them essential for reducing carbon dioxide concentrations and slowing global warming (Fang et al., 2001; Houghton, 2005; Pan et al., 2011). In addition, they play a key role in ecological functions such as maintaining water and soil and preventing wind and sand erosion. Therefore, the accurate and efficient acquisition of forest-related information and timely monitoring of the dynamic changes in forest ecosystems are among the major tasks of forestry departments.

Compared with other ecosystems, forest ecosystems have high heterogeneity along both horizontal and vertical dimensions. Different forest ecosystem structures (both horizontal and vertical) lead to different ecological functions. Therefore, it is important to study the relationship between forest structure and function, species distribution and environment, the growth and development of forests, and the laws of regeneration and succession (Hurtt et al., 2004, 2010; Medvigy et al., 2009; Shugart et al., 2010). In the early stages, surveys of forest structure are usually conducted at the plot level by experienced personnel, which is time-consuming and labor-intensive. Furthermore, most survey results are recorded as text or statistical descriptions, which cannot effectively capture the spatial heterogeneity of forest ecosystems (Fang et al., 1996, 2001; Malhi et al., 2002). With the development of satellite remote sensing, optical imagery has been used to derive forest structural attributes and biomass at local and regional scales (Defries et al., 2000; Lu, 2006; Wessman et al., 1988). However, this kind of optical remote sensing has difficulty penetrating forest canopies and reaching the understory and the ground, making it hard to capture the vertical structural information of forests (Li et al., 2012). Moreover, various metrics derived from optical remote sensing data such as albedo can become saturated when there is high canopy closure or biomass, which reduces the sensitivity of optical imagery to changes in forest attributes (e.g., biomass, leaf area index) and limits its application in estimating forest attributes at the regional scale (Baccini et al., 2008; Feng et al., 2005; Luckman et al., 1997). There is an important and urgent need to achieve real-time 3-D monitoring of forest ecosystems to overcome the limitations of passive remote sensing.

In contrast to traditional passive remote sensing, LiDAR can obtain the 3-D coordinates of objects directly, quickly, and accurately. The wavelengths of pulses emitted by different LiDAR scanners are mostly in the infrared and near-infrared bands, which are not affected by natural illumination conditions. Therefore, LiDAR can acquire data in all weather conditions and at all times. The point cloud of a mangrove forest is shown in Fig. 1.6 as an example.

Figure 1.6 Point cloud of a mangrove forest in Leizhou City, Guangdong Province, China, collected using unmanned aerial vehicle LiDAR.

Fixed–point observations and model simulations are two of the fundamental approaches to understanding forest ecosystems. LiDAR can contribute to both of the approaches and thus promote the development of forest ecology.

Plot surveys are the most widely used method to observe forest ecosystems. Usually, a 1-ha plot is the recommended unit for field surveys (Shugart et al., 2010). Forest surveys in such plots are highly representative and reliable. In practice, however, surveys are usually based on samples instead of covering the whole plot due to the workload, which may influence the representativeness of samples in areas with high heterogeneity, such as mountainous areas. On larger scales, satellite remote sensing observation is a more suitable method, mainly relying on optical imagery. Compared with field observation, regional and global observations using satellite imagery can describe the spatial heterogeneity of forest ecosystems quite accurately at large scales. However, optical remote sensing observations at the regional scale also contain considerable errors (Xue et al., 2015). Establishing a new method that links observations at the plot scale and the regional scale is of great significance for cross–scale studies of forest ecology (Fu, 2010). LiDAR can help achieve this goal because it focuses on the landscape scale (a few square kilometers to tens of square kilometers). LiDAR observations maintain the plot-level detail and can be upscaled using optical and other remote sensing observation data. There has been extensive research in this field. Taking the estimation of biomass at the regional scale as an example, first, the relationship between biomass at the plot scale and canopy attributes extracted from airborne LiDAR data such as tree height can be established; second, the relationship can be used to estimate biomass at a larger scale based on combining airborne LiDAR data and optical remote sensing data (Boudreau et al., 2008; Lefsky et al., 2006; Li et al., 2015; Su et al., 2016).

Model simulation is also a key approach to understanding the Earth system. For forest ecosystems, their complex structures, together with climate change and global warming, bring great uncertainties to simulating and predicting the dynamics of forest ecosystems using models. These uncertainties mostly relate to three aspects: model structure, driving data, and model parameters. The uncertainty from the model structure is mainly embodied in the equations that express the physical and biochemical processes, which can be improved by revising the equations based on a continuously improved understanding of the ecological processes of forest ecosystems (Purves & Pacala, 2008; Sitch et al., 2008). The uncertainty from driving data is due to the uncertain boundary conditions of a model, which can be reduced by increasing ground observations (Barman et al., 2014). The uncertainty from model parameters refers to the fact that some biophysical attributes of the ground and vegetation, such as tree height, are not available at the regional scale or are highly heterogeneous (Yu et al., 2011). Despite this variation, the models used usually adopt a single value to represent the whole area in the simulation process, which leads to uncertainties in the simulation results (Castanho et al., 2013). Traditional optical remote sensing can provide useful data products such as leaf area index for model simulations, especially large-scale simulations. Owing to the limitations of optical remote sensing imagery, however, it cannot provide all required model parameters, e.g., tree height (Chen et al., 2003; Heinsch et al., 2006). At present, many ecological models use multiyear climate data to simulate the succession of vegetation communities from bare land to a balanced ecosystem to obtain the potential vegetation distribution (Cohen & Goward, 2004; Hall et al., 1991). If the initial vegetation distribution provided to the models is quite different from the actual vegetation distribution, considerable uncertainties may be introduced into the simulated results (Kucharik et al., 2000; Sitch et al., 2003). LiDAR can address this issue by obtaining tree height and biomass information over large scales (Fig. 1.7) (Li et al., 2015). Additionally, assimilation methods can be used to optimize the model parameters and improve the prediction accuracy in the model simulation.

Recently, LiDAR has been extensively explored and successfully applied to 3-D agricultural phenotypes, and it has also shed new light in forestry studies (Jin et al., 2021; Tao et al., 2022). LiDAR can efficiently capture the 3-D structure information of forests to understand their states, such as canopy cover and dynamics at a stand and larger levels. This provides an unprecedented opportunity to describe forest phenotyping accurately and quantitatively. This information is crucial for tree breeding and forest management, such as selecting genetic material (e.g., tree volume). In addition to extracting phenotyping information and dynamics modeling for management, recent efforts have been devoted to linking LiDAR-derived tree phenotypes with genetic traits (Dungey et al., 2018). However, owing to the long cycle of forest breeding, LiDAR-based modeling has rarely been investigated. A primary practice is estimating heritability and genetic gain for easily accessible phenotypes, such as height percentiles and canopy metrics (Bombrun et al., 2020). Complex phenotypes such as tree volume and crown architecture need to be explored in the future. These

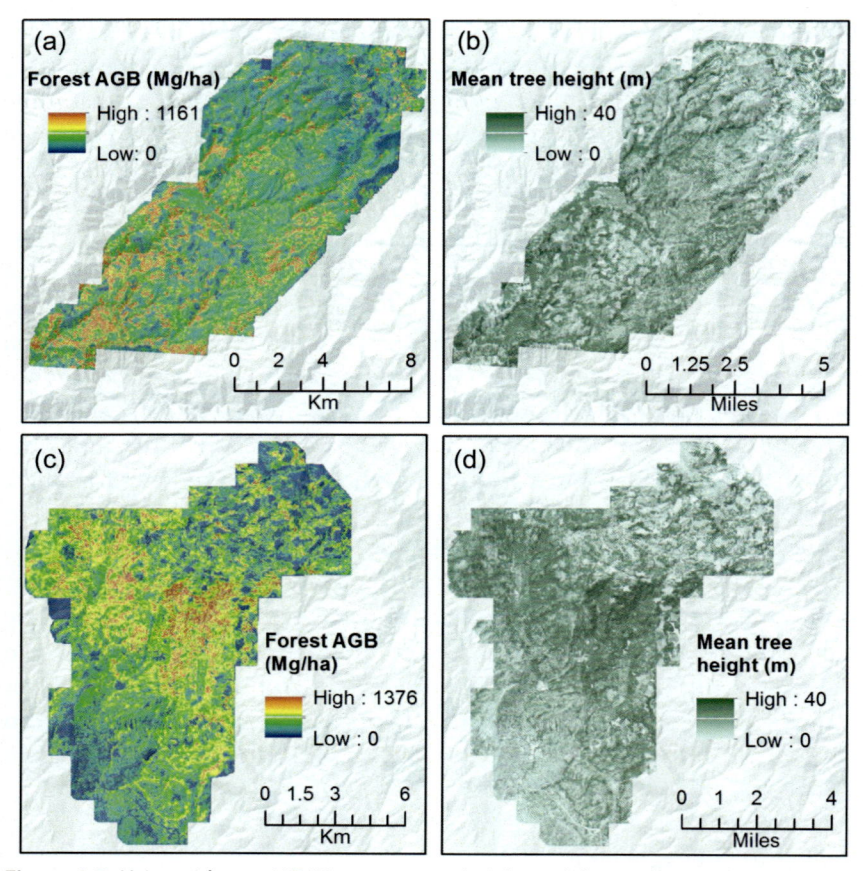

Figure 1.7 Using airborne LiDAR to map tree height and forest aboveground biomass (AGB) at regional scale (data acquisition time: September 2007 and September 2008; location: Sierra Nevada, California, USA) *(From Li, L., Guo, Q., Tao, S., Kelly, M., & Xu, G. (2015). Lidar with multi-temporal MODIS provide a means to upscale predictions of forest biomass.* ISPRS Journal of Photogrammetry and Remote Sensing, 102, *198–208.)*

phenotypes are critical for improving breeding by selecting the best progeny in provenance trials. It is important to note that LiDAR-derived phenotype heritability is influenced by both environmental factors and internal physiological properties. The integration of LiDAR with multi-source remote sensing technologies, GNSS, and geographic information systems should be further explored to phenotype forests and track their genetic performance (Dungey et al., 2018).

1.6 Chapter Summary

As an emerging active remote sensing technology, LiDAR can be used to accurately acquire 3-D structural attributes of forests and boost the evolution of forest ecosystem studies from 2-D to 3-D. However, the promotion and application of LiDAR in forest ecology were limited in the past because of technical reasons and the high costs of LiDAR systems. In recent years, the rapid development of LiDAR hardware, especially the rise of near-surface mobile LiDAR systems on different platforms such as backpacks, mobile vehicles, and UAVs, has greatly improved the accuracy and timeliness of LiDAR data acquisition while reducing the cost. Extracting 3-D structural and functional information from the forest has become a new bottleneck in promoting the application of LiDAR in forest ecology. Hence, in subsequent chapters of this book, we focus on the principles, types, and components of various LiDAR systems; methods and algorithms for LiDAR data processing and forestry attribute extraction; and applications of LiDAR in forest and other ecosystems.

References

Arp, H., Griesbach, J., & Burns, J. (1982). Mapping in tropical forests: A new approach using the laser APR. *Photogrammetric Engineering and Remote Sensing, 45*, 785–792.

Baccini, A., Laporte, N., Goetz, S., Sun, M., & Dong, H. (2008). A first map of tropical Africa's above-ground biomass derived from satellite imagery. *Environmental Research Letters, 3*(4), 045011.

Barman, R., Jain, A. K., & Liang, M. (2014). Climate-driven uncertainties in modeling terrestrial energy and water fluxes: A site-level to global-scale analysis. *Global Change Biology, 20*(6), 1885–1900.

Bombrun, M., Dash, J. P., Pont, D., Watt, M. S., Pearse, G. D., & Dungey, H. S. (2020). Forest-scale phenotyping: Productivity characterisation through machine learning. *Frontiers in Plant Science, 11*, 99.

Bonan, G. B. (2008). Forests and climate change: Forcings, feedbacks, and the climate benefits of forests. *Science, 320*(5882), 1444–1449.

Boudreau, J., Nelson, R. F., Margolis, H. A., Beaudoin, A., Guindon, L., & Kimes, D. S. (2008). Regional aboveground forest biomass using airborne and spaceborne LiDAR in Québec. *Remote Sensing of Environment, 112*(10), 3876–3890.

Castanho, A. D. A., Coe, M. T., Costa, M. H., Malhi, Y., Galbraith, D., & Quesada, C. A. (2013). Improving simulated Amazon forest biomass and productivity by including spatial variation in biophysical parameters. *Biogeosciences, 10*(4), 2255–2272.

Chen, J. M., Liu, J., Leblanc, S. G., Lacaze, R., & Roujean, J.-L. (2003). Multi-angular optical remote sensing for assessing vegetation structure and carbon absorption. *Remote Sensing of Environment, 84*(4), 516–525.

Cohen, W. B., & Goward, S. N. (2004). Landsat's role in ecological applications of remote sensing. *Bioscience, 54*(6), 535–545.

Defries, R. S., Hansen, M. C., Townshend, J. R., Janetos, A., & Loveland, T. R. (2000). A new global 1-km dataset of percentage tree cover derived from remote sensing. *Global Change Biology, 6*(2), 247–254.

Dungey, H. S., Dash, J. P., Pont, D., Clinton, P. W., Watt, M. S., & Telfer, E. J. (2018). Phenotyping whole forests will help to track genetic performance. *Trends in Plant Science, 23*(10), 854–864.

Fang, J., Chen, A., Peng, C., Zhao, S., & Ci, L. (2001). Changes in forest biomass carbon storage in China between 1949 and 1998. *Science, 292*(5525), 2320–2322.

Fang, J., Liu, G., & Xu, S. (1996). Biomass and net production of forest vegetation in China. *Acta Ecologica Sinica, 16*(5), 497–508 (in Chinese).

Feng, Z., Luo, X., & Shi, L. (2005). Some problems and perfect approaches of research on forest biomass. *World Forestry Research, 18*(3), 25.

Fu, B. (2010). Trends and priority areas in ecosystem research of China. *Geogr Res, 29*(3), 383–396.

Hall, F. G., Botkin, D. B., Strebel, D. E., Woods, K. D., & Goetz, S. J. (1991). Large-scale patterns of forest succession as determined by remote sensing. *Ecology, 72*(2), 628–640.

Hand, E. (2014). Carbon-mapping satellite will monitor plants' faint glow. *Science, 344*, 1211–1212.

Heinsch, F. A., Zhao, M., Running, S. W., Kimball, J. S., Nemani, R. R., Davis, K. J., Bolstad, P. V., Cook, B. D., Desai, A. R., & Ricciuto, D. M. (2006). Evaluation of remote sensing based terrestrial productivity from MODIS using regional tower eddy flux network observations. *IEEE Transactions on Geoscience and Remote Sensing, 44*(7), 1908–1925.

Hickman, G. D., & Hogg, J. E. (1969). Application of an airborne pulsed laser for near shore bathymetric measurements. *Remote Sensing of Environment, 1*(1), 47–58.

Hoge, F., Swift, R. N., & Frederick, E. B. (1980). Water depth measurement using an airborne pulsed neon laser system. *Applied Optics, 19*(6), 871–883.

Houghton, R. (2005). Aboveground forest biomass and the global carbon balance. *Global Change Biology, 11*(6), 945–958.

Hurtt, G. C., Dubayah, R., Drake, J., Moorcroft, P. R., Pacala, S. W., Blair, J. B., & Fearon, M. G. (2004). Beyond potential vegetation: Combining lidar data and a height-structured model for carbon studies. *Ecological Applications, 14*(3), 873–883.

Hurtt, G., Fisk, J., Thomas, R., Dubayah, R., Moorcroft, P., & Shugart, H. (2010). Linking models and data on vegetation structure. *Journal of Geophysical Research: Biogeosciences, 115*(G2).

Jin, S., Sun, X., Wu, F., Su, Y., Li, Y., Song, S., Xu, K., Ma, Q., Baret, F., & Jiang, D. (2021). Lidar sheds new light on plant phenomics for plant breeding and management: Recent advances and future prospects. *ISPRS Journal of Photogrammetry and Remote Sensing of Environment, 171*, 202–223.

Krabill, W., Collins, J., Link, L., Swift, R., & Butler, M. (1984). Airborne laser topographic mapping results. *Photogrammetric Engineering and Remote Sensing, 50*(6), 685–694.

Kucharik, C. J., Foley, J. A., Delire, C., Fisher, V. A., Coe, M. T., Lenters, J. D., Young-Molling, C., Ramankutty, N., Norman, J. M., & Gower, S. T. (2000). Testing the performance of a dynamic global ecosystem model: Water balance, carbon balance, and vegetation structure. *Global Biogeochemical Cycles, 14*(3), 795–825.

Lefsky, M. A. (2010). A global forest canopy height map from the moderate resolution imaging spectroradiometer and the geoscience laser altimeter system. *Geophysical Research Letters, 37*(15).

Lefsky, M. A., Cohen, W., Acker, S., Parker, G. G., Spies, T., & Harding, D. (1999). Lidar remote sensing of the canopy structure and biophysical properties of Douglas-fir western hemlock forests. *Remote Sensing of Environment, 70*(3), 339–361.

Lefsky, M. A., Cohen, W. B., Parker, G. G., & Harding, D. J. (2002). Lidar remote sensing for ecosystem studies: Lidar, an emerging remote sensing technology that directly measures the three-dimensional distribution of plant canopies, can accurately estimate vegetation structural attributes and should be of particular interest to forest, landscape, and global ecologists. *Bioscience, 52*(1), 19−30.

Lefsky, M. A., Harding, D. J., Keller, M., Cohen, W. B., Carabajal, C. C., Espirito-Santo, F., Hunter, M. O., de Oliveira, R., Jr., & de Camargo, P. B. (2006). Correction to "Estimates of forest canopy height and aboveground biomass using ICESat". *Geophysical Research Letters, 32*(5), L05501.

Li, L., Guo, Q., Tao, S., Kelly, M., & Xu, G. (2015). Lidar with multi-temporal MODIS provide a means to upscale predictions of forest biomass. *ISPRS Journal of Photogrammetry and Remote Sensing, 102*, 198−208.

Lim, K., Treitz, P., Wulder, M., St-Onge, B., & Flood, M. (2003). LiDAR remote sensing of forest structure. *Progress in Physical Geography, 27*(1), 88−106.

Li, D., Wang, C., Hu, Y., & Liu, S. (2012). General review on remote sensing-based biomass estimation. *Geomatics and Information Science of Wuhan University, 37*(6), 631−635 (in Chinese).

Lu, D. (2006). The potential and challenge of remote sensing-based biomass estimation. *International Journal of Remote Sensing, 27*(7), 1297−1328.

Luckman, A., Baker, J., Kuplich, T. M., Yanasse, C.d. C. F., & Frery, A. C. (1997). A study of the relationship between radar backscatter and regenerating tropical forest biomass for spaceborne SAR instruments. *Remote Sensing of Environment, 60*(1), 1−13.

Malhi, Y., Phillips, O. L., Lloyd, J., Baker, T., Wright, J., Almeida, S., Arroyo, L., Frederiksen, T., Grace, J., & Higuchi, N. (2002). An international network to monitor the structure, composition and dynamics of Amazonian forests (RAINFOR). *Journal of Vegetation Science, 13*(3), 439−450.

Medvigy, D., Wofsy, S., Munger, J., Hollinger, D., & Moorcroft, P. (2009). Mechanistic scaling of ecosystem function and dynamics in space and time: Ecosystem demography model version 2. *Journal of Geophysical Research: Biogeosciences, 114*(G1).

Nelson, R., Krabill, W., & MacLean, G. (1984). Determining forest canopy characteristics using airborne laser data. *Remote Sensing of Environment, 15*(3), 201−212.

Pan, Y., Birdsey, R. A., Fang, J., Houghton, R., Kauppi, P. E., Kurz, W. A., Phillips, O. L., Shvidenko, A., Lewis, S. L., & Canadell, J. G. (2011). A large and persistent carbon sink in the world's forests. *Science, 333*(6045), 988−993.

Pang, Y., Li, Z., Chen, E., & Sun, G. (2005). Lidar remote sensing technology and its application in forestry. *Scientia Silvae Sinicae, 41*(3), 129 (in Chinese).

Purves, D., & Pacala, S. (2008). Predictive models of forest dynamics. *Science, 320*(5882), 1452−1453.

Saatchi, S. S., Harris, N. L., Brown, S., Lefsky, M., Mitchard, E. T., Salas, W., Zutta, B. R., Buermann, W., Lewis, S. L., & Hagen, S. (2011). Benchmark map of forest carbon stocks in tropical regions across three continents. *Proceedings of the National Academy of Sciences, 108*(24), 9899−9904.

Shugart, H., Saatchi, S., & Hall, F. (2010). Importance of structure and its measurement in quantifying function of forest ecosystems. *Journal of Geophysical Research: Biogeosciences, 115*(G2).

Sitch, S., Huntingford, C., Gedney, N., Levy, P., Lomas, M., Piao, S., Betts, R., Ciais, P., Cox, P., & Friedlingstein, P. (2008). Evaluation of the terrestrial carbon cycle, future plant geography and climate-carbon cycle feedbacks using five dynamic global vegetation models (DGVMs). *Global Change Biology, 14*(9), 2015−2039.

Sitch, S., Smith, B., Prentice, I. C., Arneth, A., Bondeau, A., Cramer, W., Kaplan, J. O., Levis, S., Lucht, W., & Sykes, M. T. (2003). Evaluation of ecosystem dynamics, plant geography and terrestrial carbon cycling in the LPJ dynamic global vegetation model. *Global Change Biology, 9*(2), 161−185.

Su, Y., Guo, Q., Xue, B., Hu, T., Alvarez, O., Tao, S., & Fang, J. (2016). Spatial distribution of forest aboveground biomass in China: Estimation through combination of spaceborne lidar, optical imagery, and forest inventory data. *Remote Sensing of Environment, 173*, 187−199.

Tao, H., Xu, S., Tian, Y., Li, Z., Ge, Y., Zhang, J., Wang, Y., Zhou, G., Deng, X., & Zhang, Z. (2022). Proximal and remote sensing in plant phenomics: Twenty years of progress, challenges and perspectives. *Plant Communications*, 100344.

Wang, X., Feng, Z., & Ouyang, Z. (2001). Vegetation carbon storage and density of forest ecosystems in China. *The Journal of Applied Ecology, 12*(1), 13−16 (in Chinese).

Wessman, C. A., Aber, J. D., Peterson, D. L., & Melillo, J. M. (1988). Remote sensing of canopy chemistry and nitrogen cycling in temperate forest ecosystems. *Nature, 335*(6186), 154−156.

White, J. D., Ryan, K. C., Key, C. C., & Running, S. W. (1996). Remote sensing of forest fire severity and vegetation recovery. *International Journal of Wildland Fire, 6*(3), 125−136.

Xue, B.-L., Guo, Q., Otto, A., Xiao, J., Tao, S., & Li, L. (2015). Global patterns, trends, and drivers of water use efficiency from 2000 to 2013. *Ecosphere, 6*(10), 1−18.

Yu, G., Wang, Q., & Zhu, X. (2011). Methods and uncertainties in evaluating the carbon budgets of regional terrestrial ecosystems. *Progress in Geography, 30*(1), 103−113.

CHAPTER 2

Working Principles of LiDAR

Contents

Light detection and ranging (LiDAR) systems comprise a set of instruments involving multiple disciplines, such as optoelectronics, mechanical engineering, navigation, and computer science. Understanding how each component of a LiDAR system works is fundamental to processing LiDAR data. This chapter introduces the physical principles of LiDAR. The hardware components, characteristics, and working principles of main LiDAR systems, including terrestrial, backpack, mobile, unmanned aerial vehicle, airborne, and spaceborne LiDAR systems, are described in detail.

LiDAR Principles, Processing and Applications in Forest Ecology
ISBN 978-0-12-823894-3
https://doi.org/10.1016/B978-0-12-823894-3.00002-5

2.1 Ranging Principle of LiDAR

The basic operational principle of LiDAR is the emission of laser pulses into the surrounding environment. As the speed of light in air (c) is a constant, the distance between a LiDAR sensor and an object (R) can be measured by recording the travel time of the emitted laser pulse (t). The pulse is then converted to R using the following equation:

$$R = \frac{1}{2} * c * t \tag{2.1}$$

At present, time–of-flight laser ranging is of two types: pulse laser ranging and phase laser ranging. The following two sections detail the differences between these two ranging methods.

2.1.1 Pulse Laser Ranging

Pulse laser ranging sensors measure distance by recording the time interval between a pulse emitted by the laser transmitter and received by the laser receiver (Fig. 2.1). Specifically, the distance measurement of the pulse laser ranging sensor consists of four steps, i.e., laser emission, laser detection, delay estimation, and time measurement.

Laser emission refers to the emission of a very narrow laser pulse from the laser pulse transmitter through the rotation and reflection of the scanning prism. Laser detection converts reflected laser pulses into electrical signals after a laser receiver has received them. Time delay estimation calculates the time delay of range measurement by processing the irregular laser echo signal, whereas time measurement uses a precision instrument

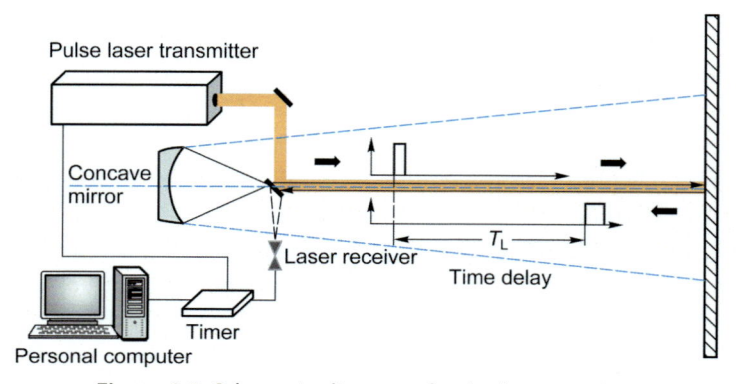

Figure 2.1 Schematic diagram of pulse laser ranging.

controlled by an atomic clock to measure the time interval between the laser echo and the main pulse through the distance counting method.

The range resolution ΔR of pulse ranging refers to the ability of LiDAR to distinguish different objects, which determines the size of the smallest object that can be detected. According to Eq. (2.1), the distance resolution is decided by the time resolution Δt:

$$\Delta R = \frac{1}{2}*c*\Delta t \tag{2.2}$$

During the operation of the LiDAR sensor, the accuracy of time recording is not constant because the working environment may influence the character and size of the laser pulse.

2.1.2 Phase Laser Ranging

Phase laser ranging modulates the laser using a continuous wave and calculates the distance by recording the phase difference between the reflected and transmitted waves (Fig. 2.2). The phase laser ranging principle is shown in the following equation:

$$R = \frac{1}{2}*c*t = \frac{1}{2}*c*\frac{\varphi}{2\pi}*T = \frac{c\varphi T}{4\pi} \tag{2.3}$$

where R represents the distance from the sensor to a target; c refers to the speed of light in the air; T represents the period of the continuous wave; and φ and t are the phase and time differences of the received and emitted beams, respectively. In practice, t is the time corresponding to the recorded phase difference plus the number of complete continuous waves:

$$t = \left(\frac{\varphi}{2\pi} + n\right)*T \tag{2.4}$$

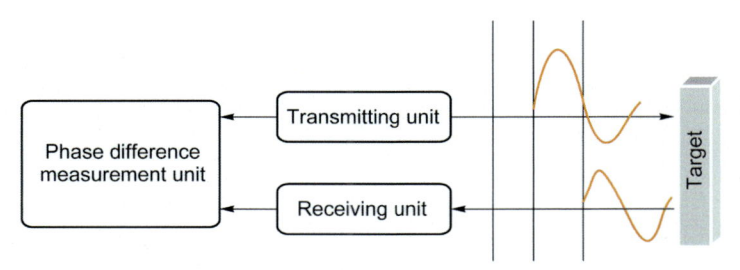

Figure 2.2 Schematic diagram of phase laser ranging.

where n represents the number of complete continuous waves in the process of laser emission to reception. Because the signal period T is inversely related to the signal frequency f, Eq. (2.4) can be converted to

$$R = \frac{c\varphi T}{4\pi} = \frac{c\varphi}{4\pi f} \tag{2.5}$$

The range resolution of a phase laser ranging sensor can be obtained by differentiating the distance R with regard to time t:

$$\Delta R = \frac{c}{4\pi f} * \Delta\varphi = \frac{1}{4\pi} * \lambda * \Delta\varphi \tag{2.6}$$

where λ represents the wavelength of a continuous wave. Compared with pulse laser ranging, phase laser ranging has the advantage of higher measurement accuracy, usually on the millimeter level. However, its detection distance is relatively short (less than 100 m) owing to the limitations of laser emission energy.

2.1.3 Ranging Accuracy

Ranging accuracy is an important indicator for evaluating a LiDAR system. Ranging accuracy depends on the ranging signals, which are mainly described by the pulse length or rise time and wavelength of the laser pulse. Generally, laser ranging accuracy is proportional to the square root of the signal-to-noise ratio (SNR) of the ranging signal. The higher the SNR, the higher the ranging accuracy. Various factors can affect SNR, including the input signal energy, noise bandwidth, radiation detector noise, and amplifier noise. Furthermore, other parameters can affect these factors (Fig. 2.3). When the received signal power is low and thermal noise is dominant, other noise may be ignored. In this case, the square root of the SNR is proportional to the power of the received signal. Because the noise bandwidth, phase shift bandwidth, and rise time of a pulse are inversely proportional to the measurement rate, the ranging accuracy can be expressed as follows:

$$\sigma_{R_{pulse}} \sim \frac{c}{2} t_{rise} \frac{\sqrt{B_{pulse}}}{P_{R_{peak}}} \tag{2.7}$$

$$\sigma_{R_{cw}} \sim \frac{\lambda_{short}}{4\pi} \frac{\sqrt{B_{cw}}}{P_{R_{av}}} \tag{2.8}$$

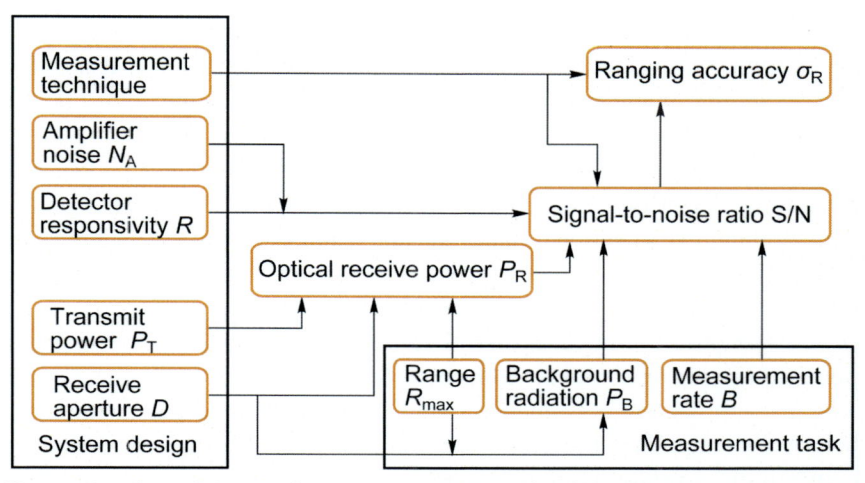

Figure 2.3 The influence of various parameters of a light detection and ranging (LiDAR) system on the signal-to-noise ratio and ranging accuracy. *(Revised from Wehr, A., & Lohr, U. (1999). Airborne laser scanning—an introduction and overview.* ISPRS Journal of Photogrammetry and Remote Sensing, 54(2–3), 68–82.)

where $\sigma_{R_{\text{pulse}}}$ refers to the ranging accuracy; $\sim$ represents a proportional relationship; and t_{rise}, B_{pulse} $P_{R_{\text{peak}}}$ $\sigma_{R_{\text{cw}}}$ λ_{short} B_{cw}, and $P_{R_{\text{av}}}$ represent the rise time, noise bandwidth of pulse ranging, peak power of the received pulse, ranging accuracy of the continuous wave, shortest wavelength of the continuous wave, noise bandwidth of phase ranging, and average power of the received continuous wave, respectively.

The mutual interference of pulse echoes during measurement is also an important factor affecting the ranging accuracy. To avoid this interference, a new echo is usually transmitted after the previous echoes are received. Therefore, the maximum interference-free distance R_{max} is calculated as follows:

$$R_{\text{max}} = \frac{1}{2} * c * PRF^{-1} \qquad (2.9)$$

where c is the propagation speed of light in the air, and *PRF* represents the pulse repetition frequency. To ensure that laser ranging is not affected by echo interference, the laser frequency cannot be too high, and thus, the point density of acquired data is limited. Practitioners have proposed solutions to the above issue from the standpoint of hardware design and post-processing of LiDAR data. The hardware-based solution combines multiple

LiDAR sensors with the same or various pulse repetition frequencies to acquire data. In such a system, all sensors maintain a certain distance from each other, with each responsible only for receiving the laser beam that it emits. However, equipment costs increase substantially as the number of LiDAR sensors increases.

Another way to solve the problem of echo interference is to use data processing algorithms. If the frequency of the LiDAR is increased rather than adding extra sensors, multiple-time-around (MTA) detections are caused (Fig. 2.4). MTA refers to the situation in which range ambiguity occurs when the return time of the echo exceeds one pulse transmission period (i.e., the period interval n between the transmitted and received pulse cannot be determined). Currently, MTA can be solved in three ways. The first approach is to estimate the initial value of n according to the flight height and then find the exact value of n by minimizing the difference between the continuous ranging value and the ranging value at the previous time node. The second approach is to estimate the flight height and the n value of each ranging point with an existing digital terrain model. The third approach is to solve n by adding random white noise, which minimizes the effect of added white noise on the data deformation estimated by the correct n value (Rieger & Ullrich, 2011).

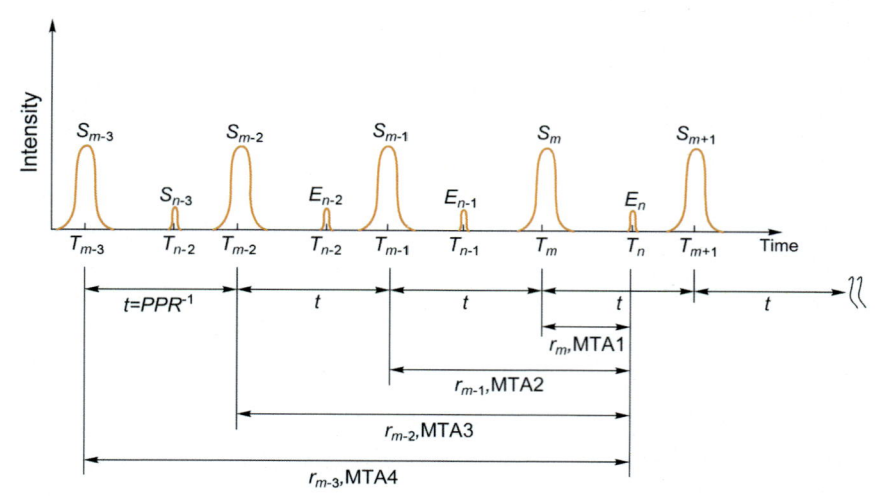

Figure 2.4 Schematic diagram of multiple-time-around (MTA). S_m represents the consecutively emitted laser pulse at time T_m, E_n represents the nth received target echo, r_m represents the interval of MTA of the mth transmitting wave, and t is the constant pulse repetition interval, which is the reciprocal of the pulse repetition rate (PPR).

2.2 Radiation Principle of LiDAR

2.2.1 Introduction to the LiDAR Equation

The LiDAR equation represents the change in laser energy during emission–reflection–reception according to the transmission principle of electromagnetic waves. Fig. 2.5 shows the transmission process of laser energy related to the receiving field of the laser emitter, the backscattered energy of the target, and the distance from the target to the laser sensor. Here, we introduce the derivation process of the LiDAR equation in detail.

Given that the energy emitted by the laser is P_e, the full angle of the receiving field of the laser emitter is α, and the distance from the target to the laser is S, the energy density of the scattered laser pulse can be defined as d:

$$d = \frac{P_e}{A_s} \tag{2.10}$$

where A_s represents the area of the target illuminated by the laser pulse, which can be expressed as

$$A_s = \pi \left(\frac{a^* S}{2}\right)^2 = \frac{\pi \alpha^2 S^2}{4} \tag{2.11}$$

Eq. (2.10) is then converted to

$$d = \frac{4P_e}{\pi \alpha^2 S^2} \tag{2.12}$$

Assuming that the laser pulse energy reaching the surface of the object with a certain reflectivity is scattered in all directions, and the projected area of the target object in the direction of the laser sensor is A, the scattering power P_s is defined as

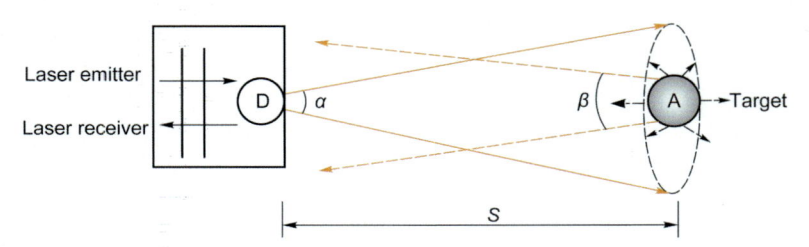

Figure 2.5 Schematic of laser pulse detection. α is the full angle of the receiving field of laser emitter D; β is the solid angle where the backscattered energy of target A is concentrated; and S is the distance from the target to the laser emitter.

$$P_s = d * A * \rho \tag{2.13}$$

where ρ is the reflectivity of the target. Assuming β is the solid angle in which the backscattered energy of the target is concentrated, and the diameter of the laser receiver is D, the energy density of the returned signal at the receiver (d_r) and the power received (P_r) are respectively calculated by

$$d_r = \frac{P_s}{\beta * S^2} \tag{2.14}$$

$$P_r = d_r * \pi * \left(\frac{D}{2}\right)^2 \tag{2.15}$$

Combining the above equations, P_r can be written as

$$P_r = \frac{d * A * \rho}{\beta * S^2} * \pi * \left(\frac{D}{2}\right)^2 = \frac{4P_e}{\pi \alpha^2 S^2} \frac{A * \rho}{\beta * S^2} \frac{\pi D^2}{4} = P_e * \frac{A \rho D^2}{\pi S^2} \tag{2.16}$$

2.2.2 Methods of Recording the Return Signal

All LiDAR systems are developed based on the LiDAR equation and can be further divided into two broad categories according to the method of recording the return signal, i.e., discrete-return system and full-waveform system (Fig. 2.6). Discrete-return systems identify returns from a pulse, and the criterion for collecting returns is based on the energy received by the laser receiver. According to the LiDAR equation, the received energy is positively related to the energy emitted by the laser, the effective area and reflectivity of the target, and the diameter of the laser receiver. It is negatively related to the emission angle, scattering angle, and ranging distance. Various echoes arrive at the receiver with different amounts of energy. If the received energy is greater than a given threshold, the echo signal is recorded as a discrete point. During laser transmission, because a laser pulse has a specific diameter, it is possible that only a part of the pulse encounters an object. Hence, one emitted laser pulse may return to the laser receiver as one or multiple returns. For example, flat ground typically corresponds to only one return, whereas complex forests may result in more than three returns. These returns provide the elevation and distribution information of different reflective surfaces within the laser spot. In general, the energy threshold prevents the discrete-return system from receiving unnecessary information, thereby reducing data storage and processing requirements. However, the system may miss subtle details of complex

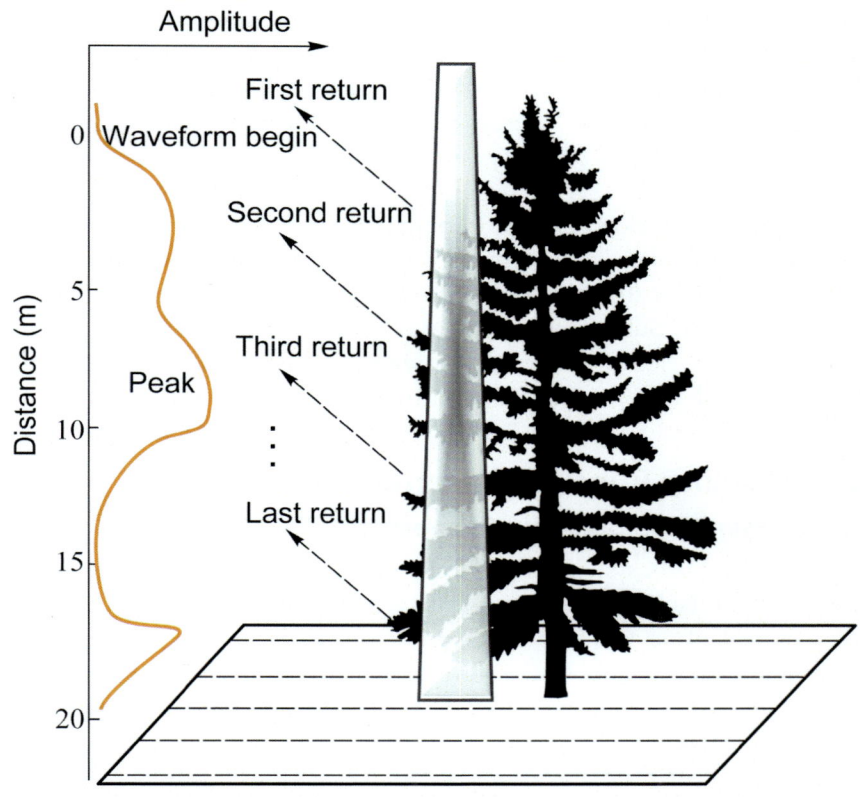

Figure 2.6 Comparison of the discrete-return and full-waveform LiDAR systems.

surfaces, which affects the accuracy of three-dimensional (3-D) reconstruction.

In contrast to discrete-return systems, full-waveform systems can record the complete waveform of backscattered signal echoes. Thus, full-waveform systems can obtain the detailed 3-D structural features of targets and provide additional physical attributes (e.g., intensity and amplitude) through the analysis of the wave profile (Duan & Xiao, 2011). However, the amount of waveform data is usually large, and the corresponding data processing is complex and difficult.

According to the size of the laser spot, full-waveform systems can be divided into small-footprint systems (0.2–3 m) and large-footprint systems (8–70 m). Small-footprint full-waveform systems can acquire a point cloud with high point density and precision. However, their equipment cost is high, and their low pulse energy results in relatively poor penetration ability of the emitted laser pulses. Hence, small-footprint full-waveform systems are more suitable for small-scale surveys. Compared with small-footprint full-waveform systems, large-footprint full-waveform systems have the advantage of high pulse energy, which enables wide coverage surveys. However, their data accuracy is slightly lower and usually spatially discontinuous.

Discrete point clouds are the current mainstream data for recording and storing LiDAR data. The first step of processing full-waveform data is usually to convert the waveform data into discrete point cloud data before further analysis. Subsequent chapters of this book focus on how to use LiDAR point clouds for forest ecosystem studies. The processing and analysis of full-waveform data are introduced in Chapters 4 and 9.

2.3 Working Principle of Terrestrial LiDAR

2.3.1 System Components of Terrestrial LiDAR

Terrestrial LiDAR systems are typically installed on stationary platforms on the Earth's surface. When collecting data, the laser scanner rotates on a two-dimensional plane while the laser prism rotates in a vertical direction to obtain 3-D information on the surrounding environment (Xi et al., 2012). As shown in Fig. 2.7, terrestrial LiDAR systems are typically installed on tripods and follow a "stop-and-go" work mode so that a complete point cloud of the scanned environment can be obtained through multiscan registration. Terrestrial LiDAR systems are commonly categorized into three types based on the field of view (FoV): panoramic, window, and hemispherical view scanners. Panoramic scanners have the largest FoV, providing a 360-degree horizontal view and a −30-degree to 90-degree vertical view.

In addition to the FoV, terrestrial LiDAR systems can be classified according to the scanning methods that differentiate the three types of scanners—line, photo, and panoramic. Regardless of the scanning method adopted, data scanning in terrestrial LiDAR is realized by controlling the scanner rotation and the rotation angle range. A typical panoramic scanning terrestrial LiDAR usually consists of a laser rangefinder, prism mirror, turning mirror, horizontal rotating platform, and vertical rotating platform (Fig. 2.8). The laser pulse emitted from the laser rangefinder is first deflected by the prism mirror. Its vertical emitter angle and horizontal emitter angle are then changed by, respectively, rotating the turning mirror on the vertical rotating platform and rotating the horizontal rotating platform on its own. In contrast to panoramic scanners, line scanners only rotate the

Figure 2.7 Schematic of the point cloud data obtained by a stationary terrestrial LiDAR system.

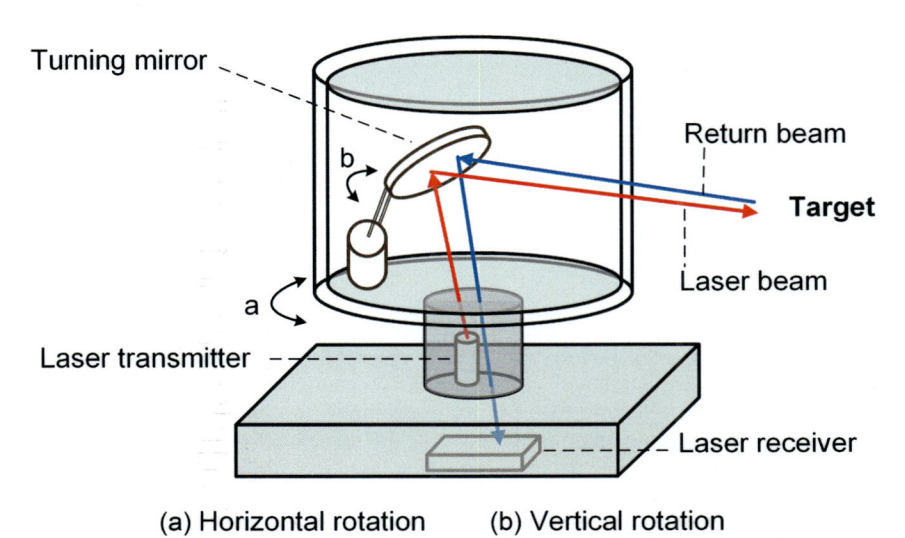

(a) Horizontal rotation   (b) Vertical rotation

Figure 2.8 Schematic diagram of horizontal and vertical scanning for a terrestrial LiDAR system.

prism mirror and do not rotate the scanner horizontally. Conversely, photo scanners only rotate the scanner horizontally by a specific angle.

In recent years, instead of adopting a vertical rotating platform to realize vertical scanning, the use of multiple laser sensors has become possible, which greatly reduces the weight and volume of the LiDAR scanner (Fig. 2.9). The most representative models are the VLP 16, HDL 32E, and HDL 64E LiDAR scanners produced by Velodyne (Velodyne Lidar, San Jose, CA, USA). These scanners contain multiple laser sensors in a vertical plane and use a unique set of calibration parameters to ensure that the origin of the range measurement for each laser is located at the scanner origin. Such devices require multiple laser sensors to ensure measuring accuracy. If the number of laser sensors decreases, the data resolution obtained by such devices in the vertical direction is greatly reduced.

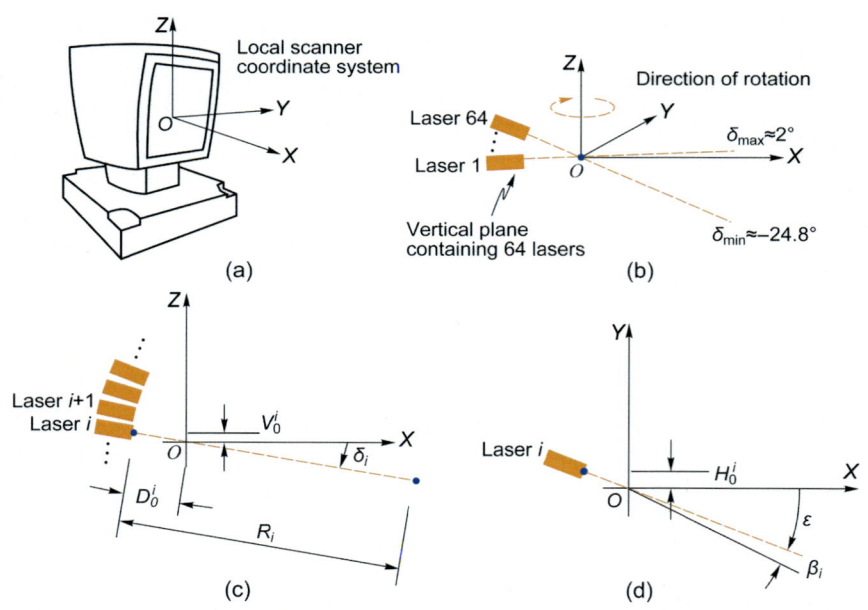

Figure 2.9 Schematic diagram of a terrestrial LiDAR scanner without a vertical rotating platform (taking the velodyne HDL 64E as an example). H_0^i represents the horizontal offset from scanner frame origin for laser i. V_0^i is the vertical offset from scanner frame origin for laser i. D_0^i is the distance offset for laser i. ε represents the horizontal orientation angle of laser i. β_i is the horizontal rotation correction for laser i. δ_{min} and δ_{max} represent the minimum and maximum vertical rotation correction, respectively. R_i is the raw distance measurement from laser i. *(Revised from Glennie, C., & Lichti, D. D. (2010). Static calibration and analysis of the Velodyne HDL-64E S2 for high accuracy mobile scanning.* Remote Sensing, *2(6), 1610–1624.)*

Since the early 2000s, terrestrial laser scanning (TLS) has matured into a robust technique for quantifying 3-D structures of forest ecosystems. However, much of the TLS technology has used high-end sensor units costing more than USD 40,000 putting this method out of reach for many potential users. In addition, many TLS systems are often too heavy to be mounted on infrastructure such as meteorological or flux towers. Fortunately, LiDAR has become the backbone of self-driving cars in recent years. Seizing on this great demand, manufacturers tend to develop quick, relatively inexpensive, and lightweight LiDAR sensors. Recent lightweight systems, e.g., Lipod (GreenValley Technology Co., Ltd., Beijing, China) (Fig. 2.10), make TLS easier and more affordable than ever before, and the estimated forest attributes such as tree height and diameter at breast height using these lightweight TLS systems are comparable with those estimated using high-end TLS systems. Moreover, these low-cost laser scanners have also been successfully mounted on some handhold platforms, which provide a new way for LiDAR data acquisition. A recent handhold instrument, the LiGrip (GreenValley Technology Co., Ltd., Beijing, China), integrates a Velodyne VLP-16 puck LiDAR sensor and a camera and has been successfully applied in many industry-specific projects, such as indoor modeling, forest inventory, and tunnel mapping (Fig. 2.11).

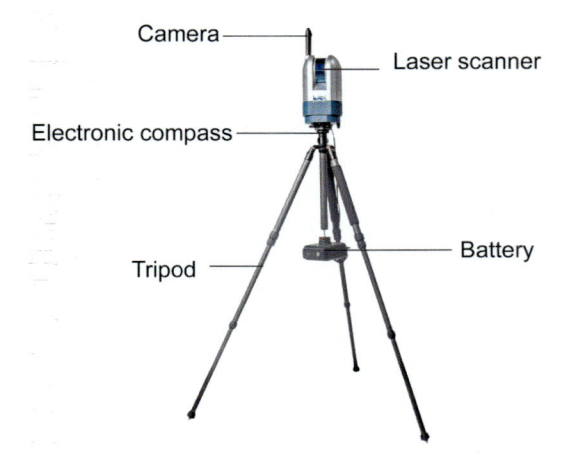

Figure 2.10 LiPod terrestrial LiDAR system developed by the authors research team.

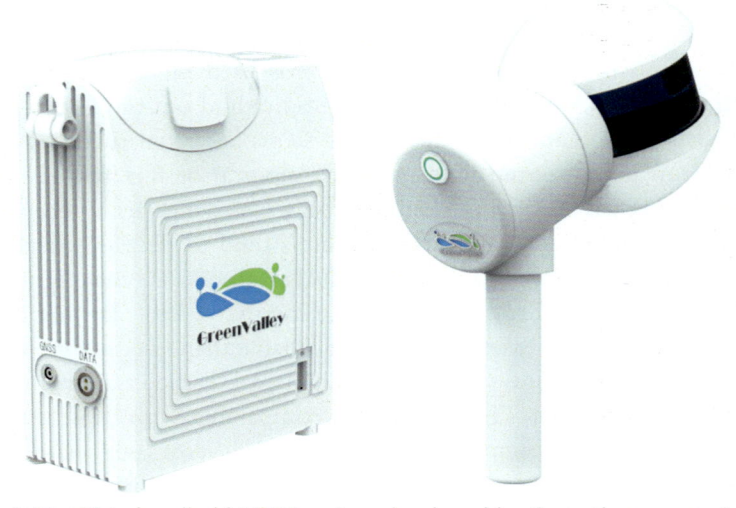

Figure 2.11 LiGrip handhold LiDAR system developed by the authors research team.

2.3.2 Laser Scanning Metrics for Terrestrial LiDAR Systems

The main scanning metrics of terrestrial LiDAR systems include the FoV, pulse frequency, scanning frequency, ranging capability, ranging accuracy, and the number of returns. The FoV of terrestrial LiDAR systems is the scanning angular range, which can be further divided into horizontal and vertical FoV. Pulse frequency refers to the number of laser beams emitted by the scanner per unit time, whereas scanning frequency represents the number of lines scanned per unit time. They jointly determine the number of laser points acquired in a single scan. Ranging capability is usually described by the nearest and farthest distances the LiDAR system can measure, which are mainly affected by pulse frequency. Ranging accuracy refers to the accuracy of LiDAR measurement, which is mainly affected by timing accuracy and the wavelength of the laser pulse. In general, compared with the pulse laser ranging system, the phase laser ranging system has higher ranging accuracy but with a much lower maximum ranging distance. The number of returns is a technical indicator only for discrete–return LiDAR systems. In forestry applications, multiple returns contain rich structural information on canopies and play an important role in extracting forest structural attributes.

2.4 Working Principle of Near-Surface LiDAR

Near-surface LiDAR has gained a great deal of interest in recent years. Compared with static TLS systems, near-surface LiDAR systems can greatly improve the efficiency of data acquisition because they are mounted on mobile platforms. This section introduces the system components and working principles of three main near-surface LiDAR systems: backpack, mobile, and unmanned aerial vehicle (UAV) mounted.

2.4.1 System Components of Backpack LiDAR

Backpack LiDAR systems are typically equipped with full inertial navigation system (INS) rather than a global navigation satellite system (GNSS)— only positioning system (Fig. 2.12). This means that these systems can use simultaneous localization and mapping (SLAM) technology to collect LiDAR data simultaneously, even in GNSS–denial environments, like indoor and forested areas (Figs. 2.13 and 2.14) (Qian et al., 2016). The application of SLAM technology enables backpack LiDAR systems to automatically register LiDAR scans without postprocessing, greatly improving data acquisition efficiency. Compared with static TLS systems, backpack LiDAR systems are estimated to be at least 10 times faster and are suitable for collecting data within large areas. A detailed introduction to SLAM technology is found in Section 5.1.2. This section introduces the work principles and system components of backpack LiDAR.

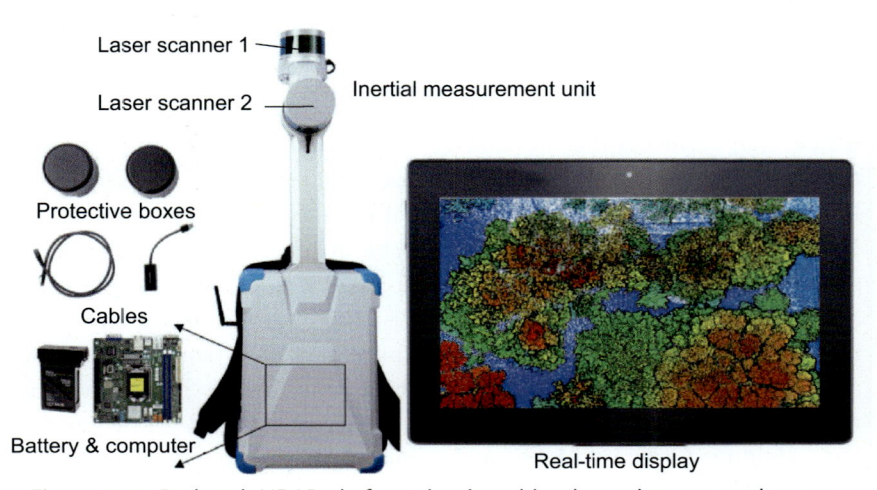

Figure 2.12 Backpack LiDAR platform developed by the authors research team.

Figure 2.13 Point cloud data of a lecture theater in Beijing collected using a backpack LiDAR system.

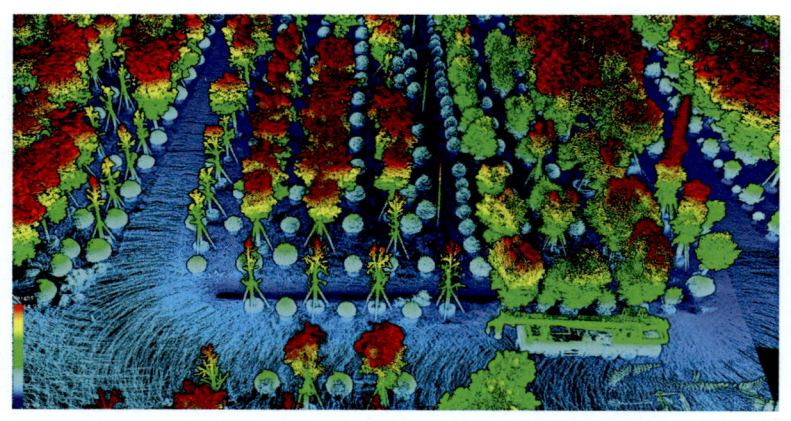

Figure 2.14 Point cloud data of a planation in Merced, California, collected with a backpack LiDAR system.

Backpack LiDAR systems are mainly composed of a laser scanner, an inertial measurement unit (IMU), a microcomputer, a tablet, and accessories (e.g., connection cables and a battery). Since the carrying weight of the operator is limited, current backpack LiDAR systems typically use lightweight laser scanners. As a representative of this category, the Velodyne Puck VLP-16 laser scanner is quite suitable for installation on a backpack system. This scanner has a vertical FoV of $\pm15°$ and a horizontal FoV of $360°$. It has a range measurement limit of 100 m and supports a measurement rate of 300,000 pulses per second, providing high–density point cloud acquisition capabilities. In recent years, manufacturers have developed mature laser scanners, such as R–Fans navigation LiDAR (SureStar Co., Ltd., Beijing, China) and Hesai Technology's 40-line hybrid solid-state LiDAR (Hesai Photonics Technology Co., Ltd., Shanghai, China). The emergence of these products has greatly enriched the types of laser scanners used in backpack systems. In addition, some recent studies have successfully placed these products on robotic dogs using the same techniques as backpack LiDAR systems (Fig. 2.15). Integrating LiDAR with robotics provides a new way to acquire LiDAR data in areas hard to access and conditions hazardous to people (e.g., isolated locations, collapsing buildings, and challenging terrains) as operators no longer need to be on site.

IMU is another core sensor in backpack LiDAR systems, which provides real-time attitude and orientation information for calculating walking trajectories. At present, IMU systems used in backpack platforms are primarily low-cost and lightweight. Microcomputers used in backpack LiDAR systems usually undertake tasks of real-time calculation of coordinates, registration, and storage of point clouds data. At the same time, these processing results can be displayed on a portable tablet for the user to view.

In addition to the hardware components mentioned above, GNSS receivers and cameras can be integrated into backpack LiDAR systems. When GNSS signals are available, GNSS receivers can provide the absolute coordinate information of the backpack LiDAR system. Moreover, by combining GNSS IMU information in the SLAM algorithm, the positioning accuracy of the point cloud can be further improved. For a backpack LiDAR system with a camera, the acquired point cloud can also contain the color information of ground targets, which is conducive to data postprocessing such as feature recognition and classification.

Figure 2.15 A backpack LiDAR system mounted on a robotic dog.

2.4.2 System Components of Mobile LiDAR

Mobile LiDAR is a method of collecting LiDAR data from a moving vehicle. Mobile LiDAR systems can acquire a high-density point cloud in a high-speed work mode along the road; therefore, they have been widely used to acquire the 3-D structures of roads, bridges, trees, and buildings in the city (Fig. 2.16). Such 3-D information strongly supports road mapping, building modeling, and urban ecosystem studies (Williams et al., 2013). In addition, the topographic information extracted from mobile LiDAR data can be used to estimate the risk of landslides and rockfalls along mountain roads (Michoud et al., 2015).

Many companies have developed commercial mobile LiDAR systems, such as the RIEGL VMX-450 mobile system (RIEGL Laser Measurement System GmbH, Horn, Austria) (Fig. 2.17). The authors research team also developed a mobile LiDAR system, as shown in Fig. 2.18. This system can automatically generate a high-precision point cloud by combining the information from the laser scanner, GNSS, and IMU sensors through the self-developed data processing software. Employing an advanced position and orientation system (POS) postprocessing algorithm, this software can guarantee data acquisition accuracy even when the GNSS signal is unavailable.

Figure 2.16 An example of the point cloud data acquired using a mobile LiDAR system.

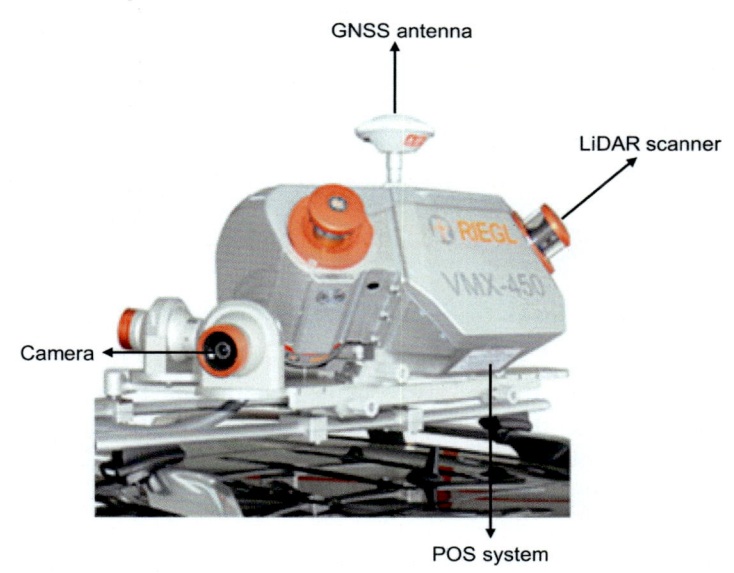

GNSS antenna

LiDAR scanner

Camera

POS system

Figure 2.17 RIEGL VMX-450 mobile system. GNSS and POS represent global navigation satellite system and position and orientation system, respectively. *(Revised from Mikrut, S., Kohut, P., Pyka, K., Tokarczyk, R., Barszcz, T., & Uhl, T. (2016). Mobile laser scanning systems for measuring the clearance gauge of railways: State of play, testing and outlook.* Sensors (Basel), 16*(5), 683.)*

Figure 2.18 The mobile LiDAR system developed by the authors research team.

Most mobile LiDAR systems consist of six essential parts: (1) a laser data acquisition system; (2) a navigation system; (3) a positioning system; (4) an image acquisition system; (5) a control system; and (6) a vehicle platform

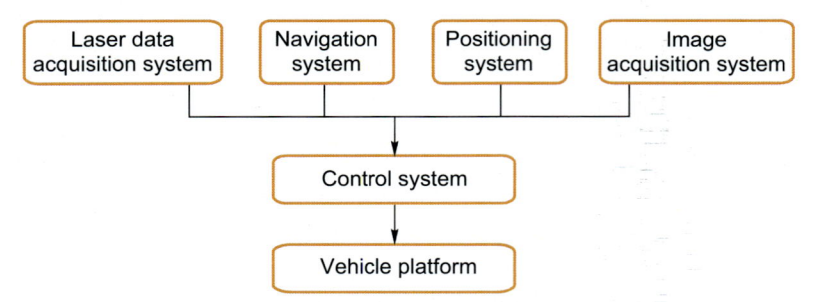

Figure 2.19 System components of typical mobile LiDAR systems.

(Fig. 2.19). The selection of laser scanners in mobile LiDAR systems is flexible, which can be either lightweight short/medium-range LiDAR sensors (e.g., Velodyne VLP16, HDL 32E, and HDL 64E scanner) or relatively bulky long-range LiDAR sensors (e.g., RIEGL VZ–400 and VZ–1000 scanner). The navigation system uses the dynamic differential GNSS technique. It uses two GNSS receivers, one on the base station and the other on the vehicle platform, to simultaneously receive the navigation and positioning signals of the same GNSS satellites to jointly determine the precise position of the vehicle platform. The positioning system uses the IMU to determine the vehicle's orientation, position, and speed. With the combination of the information from navigation and positioning systems, the 3-D position of each laser point can be calculated, as introduced in Chapter 5. The image acquisition system typically consists of a charge-coupled device (CCD) camera or a panoramic camera. Its main function is to assist the LiDAR sensor in collecting the texture information of the target objects.

2.4.3 System Components of Unmanned Aerial Vehicle LiDAR

UAVs, also known as drones, are a class of aircraft that can fly without a pilot. UAVs offer a relatively risk-free and low-cost way to quickly observe objects at a high spatial and temporal resolution that has opened a new distinct era of remote sensing. For these reasons, UAVs have attracted rapidly growing attention in academia and industry during the last decade. Many UAV companies have emerged, such as Parrot from France, AscTec and Microdrones from Germany, 3D Robotics from the United States, and DJI Innovation and EHang Intelligent Technology from China. In recent years, UAVs have become a popular platform for mounting LiDAR systems due to the significant reduction in the weight and size of laser scanners.

UAV LiDAR systems show obvious advantages in data collection efficiency since their movement is unobstructed by ground obstacles, and they fill the gap between terrestrial and airborne LiDAR systems.

In addition to UAVs, unmanned airships, motor gliders, and unmanned helicopters are also commonly used near-surface LiDAR platforms (Figs. 2.20 and 2.21). Compared with UAV platforms, these platforms have a larger payload, better endurance, and higher safety. In addition, these platforms do not require a specific landing area and can fly at low altitudes and speeds, thus supporting high–density and high–precision LiDAR data

Figure 2.20 An unmanned airship LiDAR system developed by the authors research team.

Figure 2.21 A motor glider LiDAR system developed by the authors research team.

acquisition. However, compared with UAV LiDAR systems, their cost is relatively high, with lower flexibility, mobility, and operability.

Compared with airborne LiDAR systems, point clouds acquired by UAV LiDAR systems typically have a higher point density (>100 pt/m^2). The data acquisition range of the UAV LiDAR depends on the selected UAV platform, which typically varies from 1 km^2 to 10 km^2. When the study area is too large, the efficiency and price—performance ratio of UAV LiDAR is lower than that of airborne LiDAR. Up to now, UAV LiDAR systems have been widely used to obtain the 3-D structure of different vegetation types and extract their structural attributes, such as canopy height, leaf area index, and biomass. As shown in Fig. 2.22, the 3-D structure of forests can be well detected using a self-developed UAV LiDAR system (Fig. 2.23). In addition to forest ecosystems, UAV LiDAR systems can also be used in agricultural and urban ecosystems. For example, UAV LiDAR systems can quickly obtain the crop height for growth analysis and obtain 3-D information on trees and buildings for urban modeling and urban ecological environment assessment (Fig. 2.24).

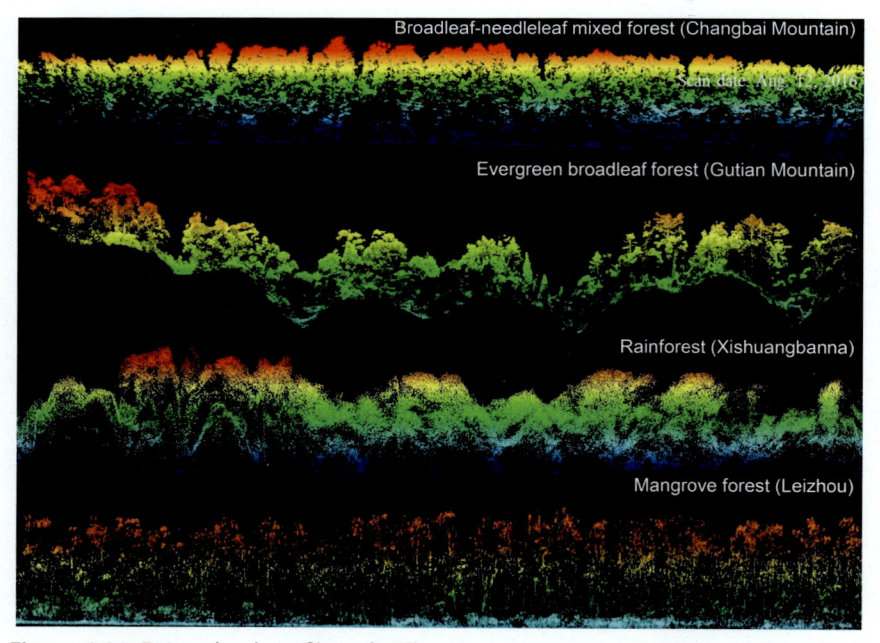

Figure 2.22 Point cloud profiles of different vegetation types obtained from (UAV LiDAR systems developed by the authors research team.

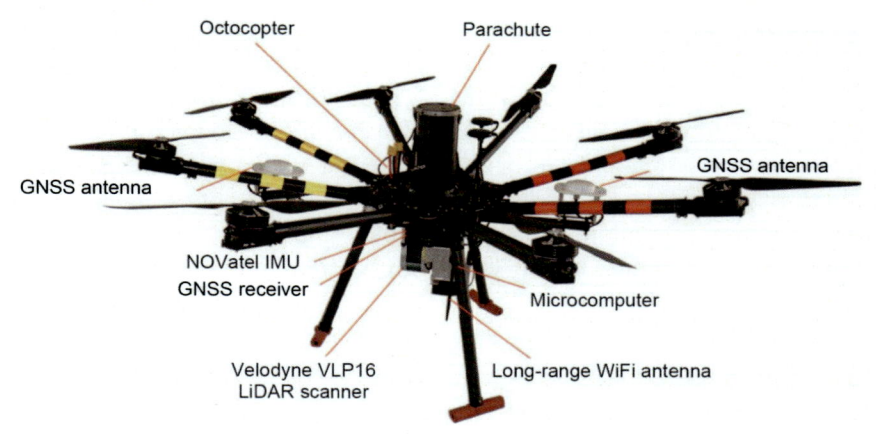

Figure 2.23 The UAV LiDAR system developed by the authors research team.

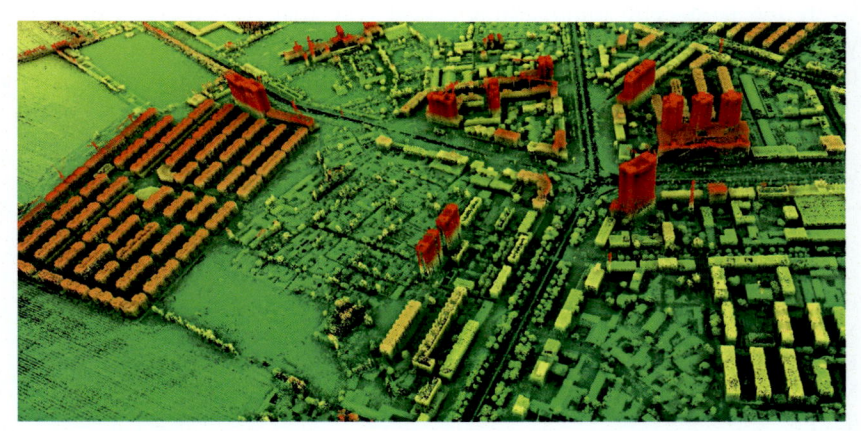

Figure 2.24 An example of urban point cloud data acquired in September 2016 using a motor glider LiDAR system in Hengyang City, Hunan Province, China.

UAV is the carrying platform for the UAV LiDAR system, and the selection of UAV has greatly influenced the stability and safety of the entire UAV LiDAR system, as well as final point cloud quality. Currently, UAVs can be classified in many ways, such as by weight, size, flight endurance, flying height, and wing type. For instance, Anderson and Gaston (2013) divided UAVs into micro, small, medium, and large UAVs based on flight distance, altitude, and endurance. Based on their classification criteria, most UAVs used in the literature for biodiversity monitoring and forest survey are micro or small UAVs. From a supervision perspective, the Civil Aviation Administration of China also classifies UAVs based on weight, payload, endurance, and flying height (Table 2.1). As a kind of aircraft, UAVs are also subject to rules and restrictions. Because most biodiversity research focuses on sparsely populated or uninhabited natural environments, taking China as an example, the operation of micro-UAVs (empty weight <7 kg) only requires the pilot to be responsible without specific license requirements. Therefore, micro-UAVs are currently the first choice for biodiversity-related surveys.

According to the wing type and flight principle, UAVs can be divided into fixed-wing and multirotor UAVs (Fig. 2.25). Compared with multirotor UAVs, fixed-wing UAVs typically fly faster and higher, with larger loads and longer endurance. In terms of takeoff and landing, although fixed-wing UAVs can use ejection equipment to solve the operational difficulty and site requirements of takeoff, their landing still requires relatively flat land. The multirotor UAV, a type that has emerged in recent years, has the advantages of hovering observation, simple operation, and low takeoff and landing requirements. Currently, multirotor UAVs mainly

Table 2.1 Specification comparison of different types of UAVs categorized by size.

	Micro UAV	Light UAV	Small UAV	Large UAV
Empty weight (kg)	<7	7−116	≤5700	>5700
Maximum payload (kg)	<5	5−30	≤50	200−900
Endurance (h)	<1	<2	<10	<48
Maximum flying height (km)	<0.25	<1	<4	3−20

(a)   (b)

Figure 2.25 Examples of (a) fixed-wing UAV and (b) multirotor UAV.

have 4, 6, or 8 rotors. The number of rotors determines the stability and payload of the UAV. According to existing research, fixed-wing UAVs are generally used for large-scale (coverage of a single fight >1 km^2) surveys, whereas multirotor UAVs are mainly used for small- or medium-scale surveys (coverage of a single fight <1 km^2). Since most forest-related research focuses on areas without suitable sites for the takeoff and landing of fixed-wing UAVs, multirotor UAVs have become the mainstream tool. A summary of the advantages and disadvantages of fixed-wing and multirotor UAVs is shown in Table 2.2.

The hardware components of the UAV LiDAR systems are like those of the mobile LiDAR systems. The main components of UAV LiDAR systems include laser scanners, GNSS systems, IMU systems, and microcomputers. Since the size and the weight of the laser scanner of a UAV system are limited by the payload of the UAV, practitioners often choose to mount lightweight sensors (e.g., Velodyne VLP16 and 32E and RIEGL VUX) on UAVs. Due to the high speed and large vibration of UAVs in flight, UAV LiDAR systems require high-precision IMU and GNSS devices. To obtain accurate navigation information, the GNSS system mounted on the UAV usually consists of a real-time differential GNSS system with two antennas. In the UAV LiDAR system, the microcomputer controls and deploys the above components uniformly, as well as carrying out data storage and real-time data processing.

In addition to the necessary hardware components above, a UAV LiDAR system is usually equipped with auxiliary devices such as a ground control station, a long-distance Wi-Fi unit (or 4G/5G module), a ground

Table 2.2 Advantages and disadvantages of fixed-wing and multirotor unmanned aerial vehicles (UAVs).

	Fixed-wing UAV	Multirotor UAV
Advantage	High flight speed Long range Long flight time Large payload Higher maximum flight altitude	Fewer requirements for takeoff sites Ability to hover for continuous observation of a specific target Easy operation
Disadvantage	Relatively difficult operation Restricted by take-off and landing sites	Small payload Limited endurance

GNSS base station, and high-resolution digital cameras. The ground control station is usually used to monitor and control the UAV platform and display the data collected by the UAV LiDAR system in real time. The long-distance Wi-Fi unit (or 4G/5G module) is usually used to control the UAV LiDAR system in real time and transfer the collected data obtained by the UAV LiDAR system to the ground control station, whereas ground GNSS base stations are designed to improve the accuracy of the real-time differential GNSS mounted on UAV LiDAR systems. In addition to obtaining color information of ground objects, UAV LiDAR systems carrying high-resolution digital cameras can also use vision-based obstacle avoidance methods to achieve autonomous cruising. For example, the recent LiAir X3 UAV LiDAR system (GreenValley Technology Co., Ltd., Beijing, China) has successfully realized fully autonomous powerline inspection through the integration of multiple sensors (Fig. 2.26).

2.4.4 System Metrics for Near-Surface LiDAR Systems

The system metrics for near-surface LiDAR systems vary greatly with platform types. For backpack and mobile LiDAR systems that utilize SLAM technology, relative position error (RPE) and absolute trajectory error (ATE) are the two most commonly used metrics for evaluating the positioning accuracy in SLAM. RPE determines the accuracy of loop closures of SLAM by measuring the difference between the estimated and true motion. Instead of evaluating relative pose differences, ATE requires

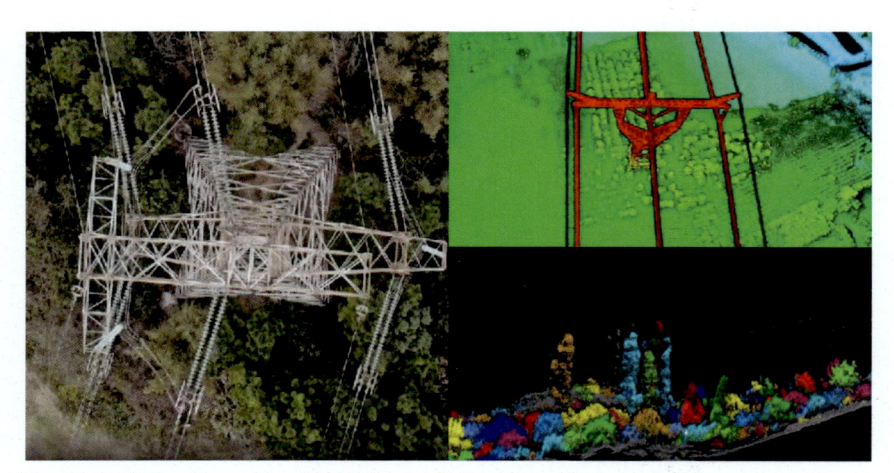

Figure 2.26 An example of UAV based fully autonomous powerline inspection by integrating multiple sensors.

absolute ground-truth poses and evaluates the absolute pose differences by linking the estimated poses with ground-truth poses using timestamps. As for UAV LiDAR systems, their system metrics are like those of airborne LiDAR systems, which are introduced in Section 2.5.2.

2.5 Working Principle of Airborne LiDAR

Airborne LiDAR systems are mainly carried on flying platforms such as UAVs, aerostats, and manned aircraft to obtain a wider range of 3-D information. Depending on the load ability of the aviation platforms, airborne LiDAR can also be integrated with other sensors (e.g., a high-resolution CCD, multispectral, and hyperspectral cameras) in the same system (Baltsavias, 1999; Li, 2012). Compared with terrestrial LiDAR, airborne LiDAR systems operate at higher altitudes. Hence all airborne LiDAR systems use pulse ranging scanners.

2.5.1 Systems Components of Airborne LiDAR

The composition of the airborne LiDAR system is much more complicated than that of terrestrial LiDAR. In addition to the operating system of the flying platform itself, the airborne LiDAR system includes a laser ranging system, a POS, and a data synchronization control system (Wehr & Lohr, 1999) (Fig. 2.27).

The laser ranging system consists of a ranging unit, an optical scanning unit, and a control processing system (Petrie, 2011). The ranging unit contains a laser transmitter and a laser receiver. When the position of the platform is fixed, the measurement achieved by the ranging unit is limited in a certain direction. The optical scanning unit addresses this issue through a prism and a rotating machine to control the direction of the laser emitted by the laser transmitter. Therefore, the angle of laser ranging is greatly expanded, and its actual scanning range depends on the configuration of the optical scanning unit (Lai, 2012). Currently, the scanning mechanisms of airborne LiDAR can be divided into four types, i.e., oscillating mirror, palmer scan, rotating polygon, and fiber scanner (Thiel & Wehr, 2004) (Fig. 2.28).

The scanning mechanism for airborne LiDAR determines the distribution of the laser footprints. The oscillating mirror method controls the laser angle by rotating the reflector mirror around the central axis with a fixed frequency, forming a periodically changing distribution of footprints on the ground. In general, oscillating mirrors produce a zigzag-line scan.

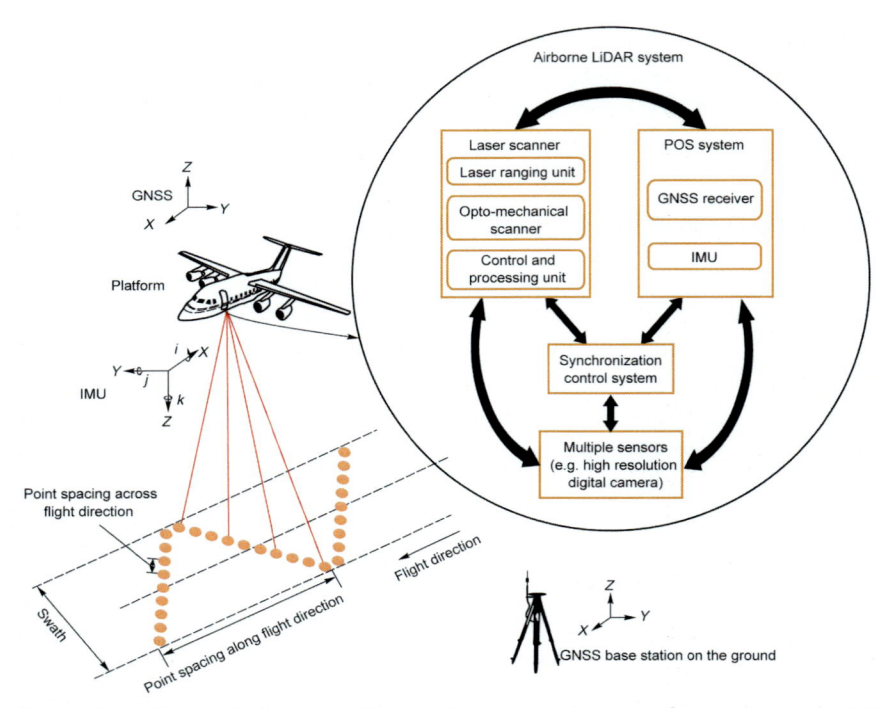

Figure 2.27 The typical composition and working principle of an airborne LiDAR system.

The resulting point cloud is unevenly distributed, with the central scanning strip having a greater point density than the edges (Fig. 2.29). The principle of the rotating polygon method is like that of the oscillating mirror method. However, because the step size of the rotation is controlled, the footprints are evenly distributed in a straight line. However, the scan angle of the rotating polygon method cannot be adjusted. The prism scan method uses a prism that can be rotated along an axis of rotation as a reflector. Because the prism is not perpendicular to the axis of rotation, the prism scan method produces an elliptical scan pattern. As the platform flies, the footprints form multiple ellipses along the flight direction. The mechanical structure of the prism scan is simple, making it suitable for high-speed operation. Therefore, the prism scan method has high scanning efficiency. Moreover, its footprints have a degree of overlap along the flight direction. Consequently, the prism scan method can increase the point density in the overlapping area and compensate for data in the occluded area (Shan & Toth, 2018).

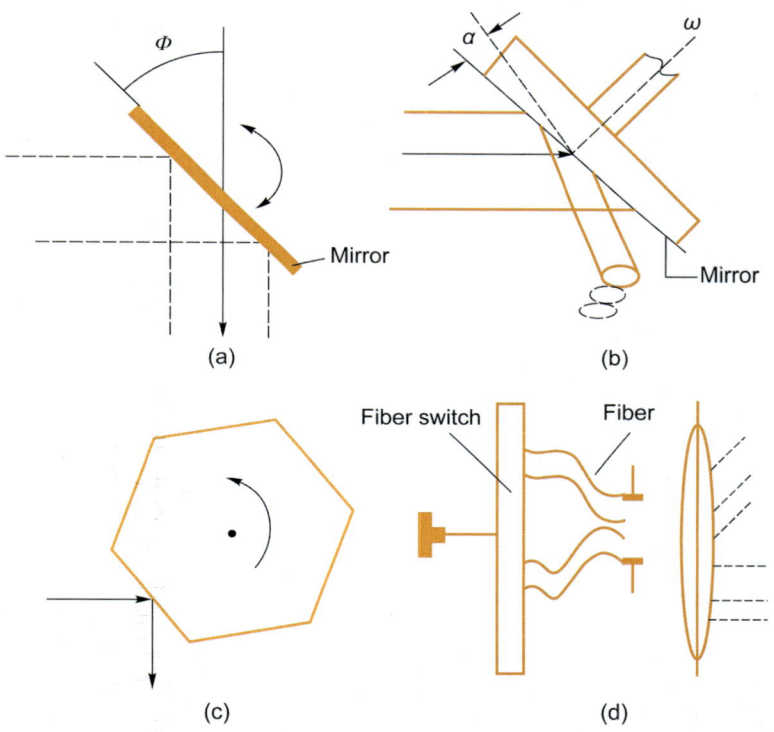

Figure 2.28 Four scanning modes of airborne LiDAR. (a) Oscillating mirror, (b) palmer scan, (c) rotating polygon, and (d) fiber scanner. Φ is the inclination angle of the mirror; α is the angle between the normal of the mirror and the rotation axis ω. *(Revised from Wehr, A., & Lohr, U. (1999). Airborne laser scanning—an introduction and overview.* ISPRS Journal of Photogrammetry and Remote Sensing, 54(2–3), 68–82.)

The fiber scanner is quite different from the other scanning methods. The fiber scanner consists of a transmitting array and a receiving array of fibers mounted in the focal plane of the transmitting and receiving lenses. With two rotating mirrors, each fiber in the transmitting and receiving paths is scanned synchronously in turn. These mirrors relay the light from the central fiber to a fiber in the fiber array mounted in a circle around the central fiber. This way, the optical signal from the transmitting fiber is linked to the corresponding fiber in the receiving path. This arrangement results in a series of scan lines on the ground parallel to the flight line. Owing to the small aperture of the fibers, fiber scanners can achieve high-speed scanning, resulting in high-density footprints. In addition, the footprints produced by fiber scanners are evenly distributed and can be acquired

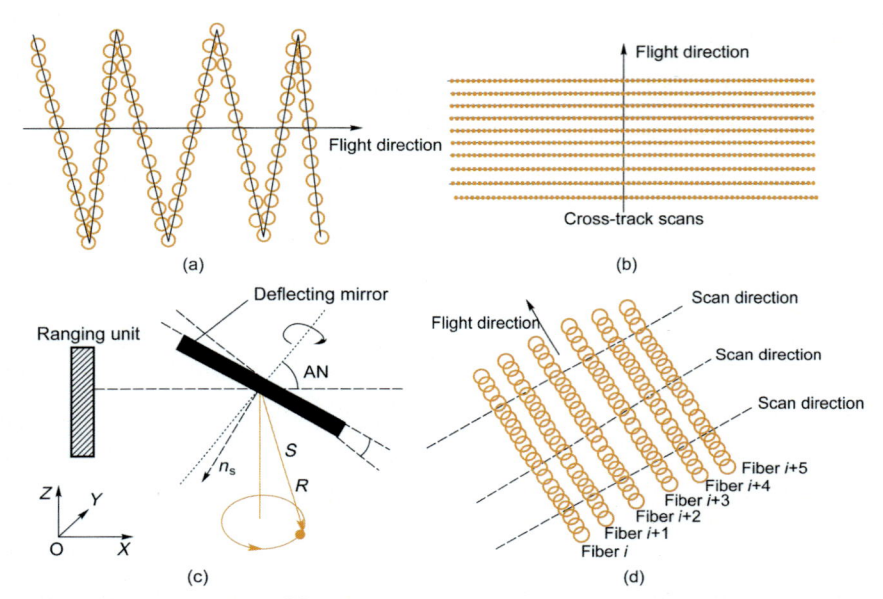

(a)

(b)

(c)

(d)

Figure 2.29 Distribution of footprints produced by airborne LiDAR with the mechanism of (a) oscillating mirror, (b) rotating polygon, (c) palmer scan, and (d) fiber scanner. AN is the angle between the laser pulse and the normal of the rotation axis; S is the distance from the center of the mirror to the ground; R is the radius of the circle formed by footprints on the ground; and n_s is the normal of the mirror. *(Revised from Wehr, A., & Lohr, U. (1999). Airborne laser scanning—an introduction and overview. ISPRS Journal of Photogrammetry and Remote Sensing, 54(2–3), 68–82.)*

in both front- and side-view modes. Therefore, all footprints collected sideways are available for use.

The POS of airborne LiDAR consists of an INS and a differential global navigation satellite system (DGNSS) (Fig. 2.30). The INS has seen rapid developments since its invention in the early 20th century. It is mainly used to measure the acceleration and angular velocity of moving objects. The core component of the INS is the IMU, typically comprising a triad of

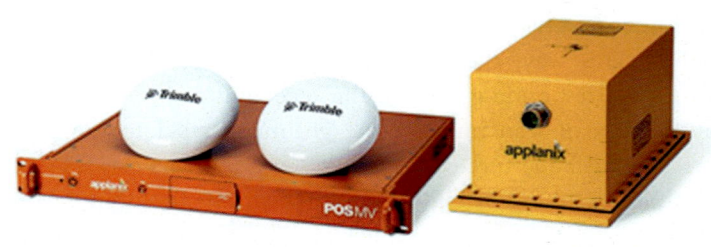

Figure 2.30 Example of a position and orientation system consisting of a differential global navigation satellite system and an inertial navigation system.

accelerometers, a triad of gyroscopes, and computing and processing components. The INS records the precise position, speed, altitude, and heading of the aircraft so that the flight path and the state of each moment of the aircraft can be estimated for subsequent data pretreatment.

The DGNSS is an advanced form of GNSS that provides greater positioning accuracy than standard GNSS. The position offered by GNSS is determined from the intersections of the range vectors measured between the receiver antenna and the satellites. At least four satellites are needed for position estimation. The more satellites in view of the receiver, the higher the position accuracy. Currently, GNSS positioning can be divided into two categories: single-point positioning and differential positioning. Single-point positioning provides 3-D coordinates using a single receiver, whereas differential positioning uses two or more receivers. The DGNSS in airborne LiDAR systems usually has 1–2 GNSS receivers on the plane, and one or more GNSS base stations will be set up synchronously on the ground according to the working area. By amalgamating signals from multiple GNSS receivers, DGNSS can provide very precise positioning accuracy.

Both GNSS and INS can be applied independently to the navigation and positioning of aircraft. However, when INS is used alone, substantial errors can accumulate owing to the increase in flight time, even with the most precise IMU. These errors will seriously affect the later airborne LiDAR data pretreatment. In addition, although the positioning accuracy of DGNSS is relatively good, GNSS signals may be lost during a high-speed flight, which results in a large positioning error within a short time. GNSS is also subject to factors in the external environment and electromagnetic interference. The POS combines both technologies to overcome their shortcomings. The POS can improve the stability of the INS and the dynamic performance and anti-jamming performance of DGNSS. Moreover, combining DGNSS and INS in an airborne LiDAR system can partially resolve the mismatch between the GNSS sampling and LiDAR scanning frequencies.

The data synchronization control system of a LiDAR system controls the data acquisition, synchronization, and recording of the above components (Fig. 2.31). In airborne LiDAR systems, although the sampling frequencies of GNSS receivers, IMUs, and laser scanners are not consistent, they can be fully synchronized using a clock control system (Williams et al., 2013). When other sensors (e.g., high-resolution CCD, thermal infrared, multispectral, and hyperspectral cameras) are included with the laser scanner

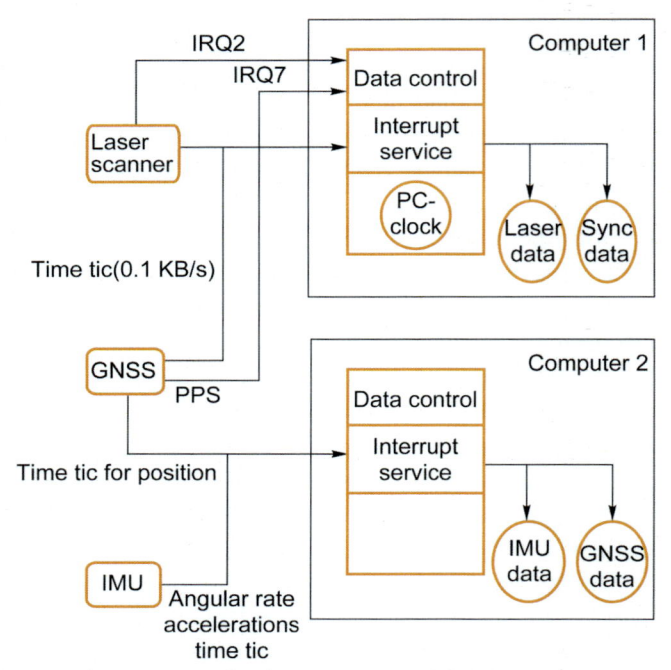

Figure 2.31 Synchronization of a laser scanner, global navigation satellite system (GNSS), and inertial measurement unit (IMU). *(Revised from Wehr, A., & Lohr, U. (1999). Airborne laser scanning—an introduction and overview. ISPRS Journal of Photogrammetry and Remote Sensing, 54(2–3), 68–82.)*

in the same airborne LiDAR system, the data synchronization control system also controls the functioning of these devices.

2.5.2 System Metrics for Airborne LiDAR Systems

The system metrics for airborne LiDAR systems include point density, flight height, FoV, scan width (SW), lateral overlap ratio, laser pulse frequency and power, the number of returns, and range resolution (Fig. 2.32). Data acquisition by an airborne laser system is a systematic project. The cost and quality of data acquisition depend on the parameter settings of the system.

Point density refers to the number of laser points per square meter, which can directly represent the quality of the acquired point cloud. Point density depends on many parameters, such as flight height, flight speed, and laser pulse frequency. In general, the point density, as well as the cost of data acquisition, is negatively correlated with flight height and flight speed and

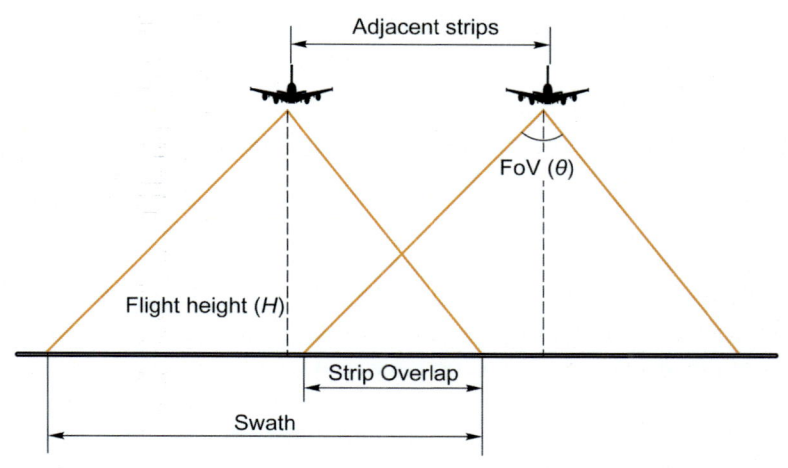

Figure 2.32 Schematic diagram of the parameters of airborne LiDAR systems. FoV represents the field of view.

positively correlated with the laser pulse frequency. The theoretical maximum and minimum values of flight height depend on laser pulse frequency, FoV, platform type, the terrain of the target area, and the safe distance between the laser and human eyes. In practice, the aircraft can choose an appropriate height between the maximum and minimum flight height according to the budget and the purpose of the flight campaign.

The SW and lateral overlap ratio are parameters that influence LiDAR data acquisition efficiency. SW is determined by the flight height (H) and the FoV (θ), whereas the lateral overlap ratio (ξ) depends on the SW and the distance between the two adjacent strips (e). The calculation of SW and lateral overlap ratio depends on the following equations:

$$SW = 2H\tan\frac{\theta}{2} \tag{2.17}$$

$$\xi = 1 - \frac{e}{SW} \tag{2.18}$$

The number of returns depends on laser scanner performance. In forested areas, a laser beam may touch leaves, branches, trunks, and the ground successively after emission. Therefore, the number of returns can better help us extract the internal structural parameters of the forest. The range resolution determines the shortest distance between objects that is detectable by LiDAR. Therefore, to identify the object to which the return belongs, the range resolution must be considered.

2.6 Working Principle of Spaceborne LiDAR

Spaceborne LiDAR systems have unique advantages in mapping the surface elevation of the Earth, Moon, and other planets, as they are aboard satellites in space and can usually provide global coverage. The study of spaceborne LiDAR systems dates back to the 1970s. The United States was the first country to conduct relevant research (Abshire et al., 2000), putting xenon lamp-pumped solid-state lasers on a series of Apollo spacecraft. In the early 1980s, flash-lamp pumped solid-state lasers became the main source of spaceborne laser ranging. In the 1990s, solid-state semiconductor lasers gradually replaced the flash-lamp pumped laser because of their advantages of long life, small size, and low energy consumption. The Geoscience Laser Altimeter System (GLAS), Lunar Orbiter Laser Altimeter, and China's Chang'e−1 laser altimeter all use solid-state semiconductor lasers (Li, 2004; Yu et al., 2013). In recent decades, spaceborne LiDAR systems have received broad attention in forest ecology studies, as they can measure the forest canopy height and other structural attributes at regional and even global scales. The GLAS aboard the Ice, Cloud, and land Elevation Satellite (ICESat), the Advanced Topographic Laser Altimetry System aboard ICESat-2, and the Global Ecosystem Dynamics Investigation instrument aboard the International Space Station are representative spaceborne LiDAR systems widely adopted for forest ecosystem studies. The main spaceborne LiDAR systems are shown in Table 2.3.

2.6.1 System Components of Spaceborne LiDAR

A spaceborne LiDAR system is composed of four subsystems, i.e., a transmitter subsystem, a receiver subsystem, a position orientation subsystem, and a control subsystem. The main function of the transmitter subsystem is to transmit laser beams; control their position, direction, and width in space; modulate their frequency, phase, pulse width, and amplitude to form continuous waves or pulses; and amplify the transmitted signal through the optical transmitting antenna to obtain the maximum energy from the detected target. The receiver subsystem receives the reflected and scattered signals and converts them into electrical signals. The main purpose of the position orientation subsystem is to record the location information of the satellite and the position information of the laser scanner for subsequent data

Table 2.3 Characteristics of spaceborne LiDAR systems.

	Launch time	Country	Observation target	Number of laser beams	Width of emitted laser (ns)	Interval between footprints (m)	Footprint size (m)	Precision of measured height (m)	Major mission
Clementine	1994	USA	Moon	1	10	100	250	40	Topography of moon
SLA-01/02	1996/1997	USA	Earth	1	10	750	100	1.5	Global elevation control point
MGS MOLA	1996	USA	Mars	1	5	330	160	10	Topography of mars
ICESat GLAS	2003	USA	Earth	1	6	170	70	0.15	Elevation of Earth's surface, especially for polar ice sheets
Messenger MLA	2006	USA	Mercury	1	6	—	—	<1.0	Topography of mercury
SELENE LALT	2007	Japan	Moon	1	17	—	—	5	Topography of moon
Chang'e−1	2007	China	Moon	1	5−7	1400	200	5	Topography of moon
Chandrayaan-1	2008	India	Moon	1	10	—	—	—	Topography of moon
LRO LOLA	2009	USA	Moon	5	5	25	5	0.1	Topography of moon
Chang'e−2	2010	China	Moon	1	10	—	40	5	Topography of moon, terrain of landing area

Continued

Table 2.3 Characteristics of spaceborne LiDAR systems.—cont'd

	Launch time	Country	Observation target	Number of laser beams	Width of emitted laser (ns)	Interval between footprints (m)	Footprint size (m)	Precision of measured height (m)	Major mission
BepiColombo MPO BELA	2018	Europe	Mercury	1	10	250	24	1.9	Topography of mercury
Gaofen-7	2018	China	Earth	2	7	2330	30	1	Generalized elevation control point
ICESat-2 ATLAS	2018	USA	Earth	6	1	0.7	10	0.1	Monitoring polar ice sheets and terrestrial ecosystem
GEDI	2018	USA	Earth	14	10	500	25	1	Measure forest biomass
CTECS	2022	China	Earth	5	7	200	25 ~ 30	1	Monitoring forest carbon stocks and generalization of elevation control point

Abbreviations: *ATLAS*, Altimetry System; *BELA*, BepiColombo Laser Altimeter; *CTECS*, Chinese Terrestrial Ecosystem Carbon Inventory Satellite; *GEDI*, Advanced Topographic Laser Global Ecosystem Dynamics Investigation; *GLAS*, Geoscience Laser Altimeter System; *ICESat*, Ice, Cloud, and land Elevation Satellite; *LALT*, Laser Altimeter; *LOLA*, Lunar Orbiter Laser Altimeter; *LRO*, Lunar Reconnaissance Orbiter; *MGS*, Mars Global Surveyor; *MLA*, Mercury Laser Altimeter; *MOLA*, Mars Orbiter Laser Altimeter; *MPO*, Mercury Planetary Orbiter; *SLA*, Shuttle Laser Altimeter.

pretreatment. The control system guides LiDAR scanner tracking and target scanning using angular velocity and angle information provided by the processor and transfers the output signal from the processor to other control centers through photoelectric communication (Abshire et al., 2000).

The basic principle of spaceborne LiDAR is similar to that of airborne LiDAR. The specific workflow is as follows. First, laser pulses with a power P_0 are continuously transmitted by the laser scanner installed on the satellite to the target. After the laser beams pass through the atmosphere or vacuum, they are reflected by the target, received by the receiving subsystem, and converted into echo pulses by photodetectors. According to the transit time ΔT of the laser pulse, the distance between the satellite and the detected target (R) can be measured (Fig. 2.33). Subsequently, the height of the target surface in the laser footprint can be calculated through the geometric radius of the satellite, the altitude of the satellite orbit, and the pointing angle of the laser scanner.

2.6.2 System Metrics for Spaceborne LiDAR Systems

The main parameters of spaceborne LiDAR systems include ranging accuracy, footprint size, pulse repetition frequency, along-track distance, across-track distance, and echo intensity. The ranging accuracy of spaceborne LiDAR is mainly determined by the timing accuracy and wavelength of the laser scanner. The footprint size represents the area on the Earth's surface irradiated by a laser pulse. It plays a crucial role in the accuracy of LiDAR data collected over sloped or flat terrain. If the laser pulse hits a flat surface, its footprint size can be defined by the following equation:

$$D = 2Z \cdot \tan\theta \tag{2.19}$$

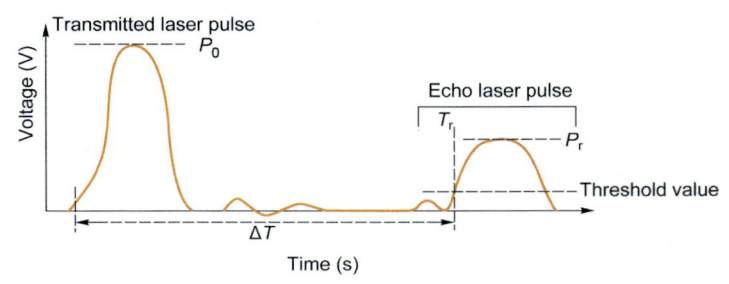

Figure 2.33 Schematic diagram of the measurement of transit time. T_r refers to the time during which the energy of the echo surpasses the threshold value. ΔT represents transit time. P_0, and P_r are the emitted pulse and received pulse with a power of P_0, respectively.

where D represents the footprint size, Z refers to the distance between the laser scanner and the irradiated surface, and θ represents beam divergence. Because satellites travel at high altitudes, the footprint size of the spaceborne LiDAR is typically large, even if the beam divergence is small. As mentioned before, the ranging accuracy is inversely proportional to footprint size and, therefore, also inversely proportional to beam divergence. Hence, a transmitting telescope is usually used as a beam expander in the spaceborne LiDAR system to reduce the beam divergence and improve the ranging accuracy. Pulse repetition frequency is the number of pulses in a repeating signal in a specific time unit. Adopting a high pulse repetition frequency helps increase the point density. Along-track and across-track distances refer to the spacing of footprints along and across the satellite track, respectively. These two parameters jointly determine the spatial resolution of spaceborne LiDAR systems. The echo intensity typically refers to the intensity and waveform characteristics of the targets detected by the spaceborne LiDAR. In practice, the surface reflectance of targets can be evaluated through the LiDAR equation and detected laser pulse energy. More details and comparisons of the system metrics and data products of different spaceborne LiDAR systems (such as waveform LiDAR and single-photon LiDAR) are introduced in Chapter 9.

2.7 Chapter Summary

The laser scanner is the core component of a LiDAR system. Understanding its ranging and radiation transfer principles is vital for knowing how LiDAR works. Various types of laser scanners have been developed. In general, pulse ranging LiDAR and phase ranging LiDAR have their advantages in terms of ranging distance and accuracy. With the development of LiDAR technology, different LiDAR systems, such as terrestrial, backpack, mobile, UAV, airborne, and spaceborne LiDAR systems, have been broadly applied in various fields. When selecting a LiDAR system, it is important to fully consider the characteristics of the research objects and purposes. Each LiDAR platform has different application fields. For example, terrestrial LiDAR and UAV LiDAR are suitable for forestry studies at the plot scale, whereas airborne and spaceborne LiDAR systems are more suited to acquiring 3-D structural information on forests at landscape and global scales.

References

Abshire, J. B., Sun, X., & Afzal, R. S. (2000). Mars orbiter laser altimeter: Receiver model and performance analysis. *Applied Optics, 39*(15), 2449–2460.

Anderson, K., & Gaston, K. J. (2013). Lightweight unmanned aerial vehicles will revolutionize spatial ecology. *Frontiers in Ecology and the Environment, 11*(3), 138–146.

Baltsavias, E. P. (1999). Airborne laser scanning: Existing systems and firms and other resources. *ISPRS Journal of Photogrammetry and Remote Sensing, 54*(2–3), 164–198.

Duan, Z., & Xiao, H. (2011). Review of forest parameter estimation methods for airborne LiDAR. *Forest Resources Management, 4*, 117–121 (in Chinese).

Glennie, C., & Lichti, D. D. (2010). Static calibration and analysis of the Velodyne HDL-64E S2 for high accuracy mobile scanning. *Remote Sensing, 2*(6), 1610–1624.

Lai, X. (2012). *Basic principle and application of airborne LiDAR* (in Chinese).

Li, S. (2004). Overview of the development of spaceborne laser altimeter. *Optical and Optoelectronic Technology, 2*(6), 4–6 (in Chinese).

Li, D. (2012). *Research on airborne LiDAR 3d imaging technology.* University of electronic science and technology (in Chinese).

Michoud, C., Carrea, D., Costa, S., Derron, M.-H., Jaboyedoff, M., Delacourt, C., Maquaire, O., Letortu, P., & Davidson, R. (2015). Landslide detection and monitoring capability of boat-based mobile laser scanning along Dieppe coastal cliffs, Normandy. *Landslides, 12*(2), 403–418.

Mikrut, S., Kohut, P., Pyka, K., Tokarczyk, R., Barszcz, T., & Uhl, T. (2016). Mobile laser scanning systems for measuring the clearance gauge of railways: State of play, testing and outlook. *Sensors (Basel), 16*(5), 683.

Petrie, G. (2011). Airborne topographic laser scanners. *GEOInformatics, 14*(1), 34.

Qian, C., Liu, H., Tang, J., Chen, Y., Kaartinen, H., Kukko, A., Zhu, L., Liang, X., Chen, L., & Hyyppä, J. (2016). An integrated GNSS/INS/LiDAR-SLAM positioning method for highly accurate forest stem mapping. *Remote Sensing, 9*(1), 3.

Rieger, P., & Ullrich, A. (2011). Resolving range ambiguities in high-repetition rate airborne lidar applications. *Electro-optical remote sensing, photonic technologies, and applications V.*

Shan, J., & Toth, C. K. (2018). *Topographic laser ranging and scanning: principles and processing.* CRC Press.

Thiel, K., & Wehr, A. (2004). Performance capabilities of laser scanners—an overview and measurement principle analysis. *International Archives of Photogrammetry, Remote Sensing and Spatial Information Sciences, 36*(8), 14–18.

Wehr, A., & Lohr, U. (1999). Airborne laser scanning—an introduction and overview. *ISPRS Journal of Photogrammetry and Remote Sensing, 54*(2–3), 68–82.

Williams, K., Olsen, M. J., Roe, G. V., & Glennie, C. (2013). Synthesis of transportation applications of mobile LiDAR. *Remote Sensing, 5*(9), 4652–4692.

Xi, X., Luo, D., Wang, F., & Wang, C. (2012). Review on the status and development of 3d laser scanning system on the ground. *Geospatial Information, 10*(6), 13–15 (in Chinese).

Yu, Z., Hou, X., & Zhou, Y. (2013). Development status of satellite-borne laser altimetry. *Advances in Laser and Optoelectronics, 50*(2), 52–61 (in Chinese).

CHAPTER 3

LiDAR Field Workflow and Systematic Error Sources

Contents

LiDAR Principles, Processing and Applications in Forest Ecology
ISBN 978-0-12-823894-3
https://doi.org/10.1016/B978-0-12-823894-3.00003-7

Although light detection and ranging (LiDAR) normally provides high-precision target measurement, systematic or random errors inevitably exist in the data acquisition process and may lead to invalid data acquisition. This chapter provides a detailed introduction to the workflow and error sources in terrestrial, backpack, mobile, unmanned aerial vehicle, and airborne LiDAR systems to help users avoid inappropriate operations and systematic errors during data acquisition.

3.1 Basic Operation of Terrestrial LiDAR

Terrestrial light detection and ranging (LiDAR) has been widely used in forest resource surveys by research institutes and universities because of its small size, high accuracy, safety and stability, strong operability, relatively simple scanning operations, and flexibility compared with airborne LiDAR systems. However, the operating range of terrestrial LiDAR is relatively narrow due to its static station scanning mode and the limitation of the laser emission energy, making it unsuitable for acquiring three-dimensional (3-D) data for large areas. This section introduces the recommended scanning operational process for terrestrial LiDAR, using the RIEGL VZ-400 (RIEGL Laser Measurement System GmbH, Horn, Austria) as an example.

3.1.1 Preparation for Terrestrial LiDAR Scanning

Preparation for terrestrial LiDAR scanning consists of three steps: collecting information about areas of interest, initial designing of the scanning route, and surveying the working area.

3.1.1.1 Information Collection for Areas of Interest

Information must be collected before data collection for a preliminary understanding of the areas of interest (e.g., size and shape). Information can be obtained from topographic maps and aerial or satellite images. Local climatic and meteorological information is needed to determine the best environmental and climatic conditions for equipment operation. Additionally, the distribution of roads and surface features within the working area should be investigated. Then the collected information is integrated to develop the scanning scheme and determine the scan locations.

3.1.1.2 Initial Design of Scanning Route

The scanning route should be designed according to the information collected about the areas of interest based on principles including complete coverage, uniform distribution of scanning stations, and convenient and nonrepeated routes. As shown in Fig. 3.1, the scanning route and stations can be marked on a topographic map or remote sensing imagery and inspected in the field to determine whether they are suitable for setting up terrestrial LiDAR scans.

3.1.1.3 Field Surveying of the Working Area

A field survey of the working area should be conducted based on the planned scanning route and distribution of the scan locations to ensure that the scanning work can be rationally and efficiently performed. During the field survey, the distribution of targets should be investigated—scan locations may be replanned according to the distribution of important targets. In addition, the scanning mode can be refined according to the importance of the targets. For targets with higher importance, additional scans can be collected to obtain more details, while for other targets, the scan density can be reduced to avoid redundancy.

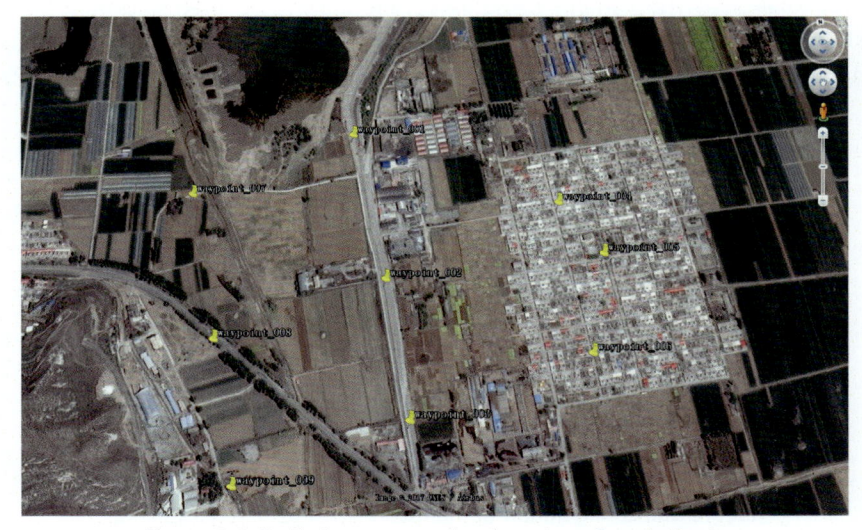

Figure 3.1 Preliminary route plan based on Google Earth.

3.1.2 Scanning Operational Planning

Based on the field survey, adjustments and modifications are made to the initial scanning scheme to achieve a reasonable distribution of scan locations covering the entire work area. In addition, the scanning route should connect all scans continuously without going back and forth. The heights of obstacles in the working area, such as powerlines and trees, must be considered to determine scan heights. Simultaneously, storage devices should be prepared based on the volume of data estimated by the pre-determined number of scans and the scanning mode for the areas of interest. After these preparations, data collection can be performed according to the scheme.

3.1.3 Terrestrial LiDAR Data Collection

3.1.3.1 Preparing and Checking the Scanning Equipment

First, the scanning system components should be connected according to the operating instructions. The components include a laser scanner, a camera (optional), and a computer. The scanner is used to collect 3-D LiDAR point clouds. Through continuous shooting, the camera acquires panoramic photos that can be matched with 3-D LiDAR point clouds to provide color (RGB) information. The computer is connected to the laser scanner to operate it and store and preprocess the collected data. Before data collection, the entire system must be run to determine how well it works in data acquisition, data storage, and registration between point clouds and two-dimensional (2-D) images. If 2-D images and point cloud data are not well matched, the camera must be recalibrated. The base station setting should then be scheduled to ensure no blind zone exists within the data acquisition area. Additionally, a certain overlap between stations (Fig. 3.2)

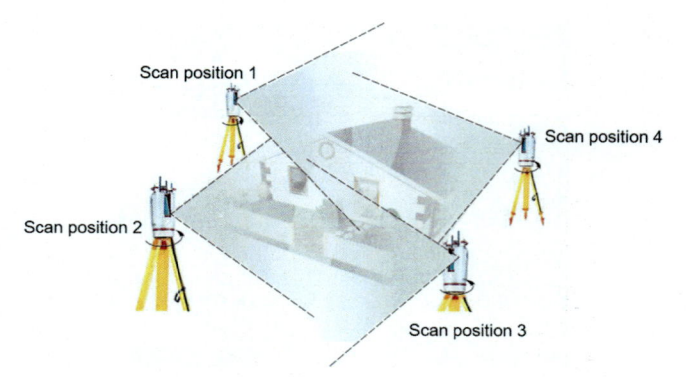

Figure 3.2 Configuration of multiscan terrestrial laser scanning.

should be guaranteed to ensure complete coverage of the working area and fulfill the requirements of subsequent data registration.

3.1.3.2 Detailed Scanning Process

High-reflectance targets should first be set up if the registration of multiple terrestrial laser scans is expected in the following data processing steps. The arrangement of high-reflectance targets is usually based on the following principles: (1) they should be evenly distributed throughout the area; (2) at least three targets are shared by two nearby terrestrial LiDAR scans; and (3) all targets should be set up with various heights in areas that can be easily found. After the targets are ready, the terrestrial LiDAR scanner should be set up. If a tripod is used as the platform, it should be ensured that the scanner is stable through the entire scanning process and is leveled. Then the scanner can be connected to the power supply and turned on (if the operating time is extended, spare batteries should be prepared in advance), and related scanning parameters should be set before starting a scan. During data collection, operators should stay away from the instantaneous field of view of the scanner to avoid data gaps caused by occlusion. The above-mentioned processes should be repeated in all planned scan locations to cover all areas of interest.

3.1.3.3 Rough Inspection of Data Quality

After finishing the data collection of each scan, data integrity and density should be checked using the computer. Data should be recollected if the required data quality is not achieved because of occlusion or low scanning density. If the data quality is satisfactory, operators can move the scanner to the next location and start a new scan.

3.1.4 Initial Data Inspection

The point cloud data of all individual scans are generally managed in a single project file after data collection over the entire working area. Because terrestrial LiDAR data acquisition is performed from scan to scan, the collected data of each scan are in an independent coordinate system (scanner's coordinate system). It is necessary to register the data from all scans into a unified coordinate system, that is, the projected coordinate system (PRCS). After registration, every station has an accurate relative position under the PRCS. Then the 3-D coordinates of all points can be converted to the geographic coordinate system through global registration, that is, to a global coordinate system.

After the initial data registration mentioned above, data quality inspection should be performed in two steps: global data inspection and local detail inspection. Data inspection at the global level refers to inspecting the integrity and overlap rate of the project data. Complete project data should fully cover the entire working area. Any gaps should be filled by performing supplementary scans. Inspecting the overlap rate ensures that adjacent scans have sufficient overlapping areas for the subsequent fine registration process. Otherwise, the accuracy of subsequent data processing might be directly affected. If significant data stratification or deflection appears after initial registration, additional scans are required. Local detail inspection consists of two aspects (Fig. 3.3). The first is checking the point density. If the point density is not sufficient to reflect critical details of target objects, additional scans should be collected at the original scan locations. The other aspect is checking the gaps in data. If there are gaps, supplementary scans are required. Most data gaps are caused by obstacles on the ground that cause blind zones in the scan. Therefore, one solution is to find a high location in the area or take advantage of a crane for supplementary measurements (Fig. 3.4).

3.2 Basic Operation of Mobile and Backpack LiDAR

3.2.1 Overview of Mobile and Backpack LiDAR Operation

In recent years, backpack and mobile LiDAR systems have received increasing attention. Compared with traditional terrestrial LiDAR systems, backpack and mobile LiDAR systems can greatly improve the efficiency of data acquisition. Compared with airborne LiDAR systems, backpack and mobile LiDAR platforms have higher flexibility and can greatly reduce the cost of data acquisition.

A mobile LiDAR system mounted on a vehicle usually consists of a long-range laser scanner, an inertial measurement unit (IMU), a global

Figure 3.3 Local detail inspection of terrestrial LiDAR point cloud data.

Figure 3.4 Point cloud data collection using terrestrial LiDAR on a crane to reduce occlusion.

navigation satellite system (GNSS), and a panoramic camera. It can simultaneously collect point cloud data and panoramic images for tasks such as terrain mapping, 3-D city modeling, and road asset censusing. The backpack LiDAR system uses a backpack as the carrier to achieve rapid and continuous acquisition of point cloud data. The essential components include one or more laser scanners, an IMU, a microcomputer, a GNSS receiver (optional), and a portable handheld tablet for data display and system control. Mobile LiDAR and backpack LiDAR share a similar measurement workflow, including (1) preliminary preparation that includes route surveying and route planning; (2) setting up a GNSS base station (optional for backpack or when using continuously operating reference stations (CORS)); (3) device initialization and data collection; and (4) data export and initial inspection. In the following sections, we use the LiMobile mobile LiDAR system (GreenValley Technology Co. Ltd., Beijing, China) and LiBackpack backpack LiDAR system (GreenValley Technology Co. Ltd., Beijing, China) as examples to introduce the detailed workflows of mobile and backpack LiDAR.

3.2.2 Preparation for Mobile and Backpack LiDAR Scanning

Before the scanning operation, a route survey should be conducted to examine the road conditions along the route and address the following issues: (1) Which routes are suitable for data collection? (2) Are there height restrictions (generally, the height from the LiDAR instrument aboard a vehicle to the ground is about 3 m, and for backpack LiDAR, it is around 2 m)? (3) How will areas without roads be accessed (especially important for operating backpack LiDAR in forests)? and (4) Which areas are of specific interest during data collection? During the route survey, areas with the above issues should be recorded to facilitate subsequent route planning.

3.2.3 Scanning Operational Planning
3.2.3.1 Mobile LiDAR System
3.2.3.1.1 Route Planning

Proper route planning can improve data collection efficiency and ensure IMU accuracy. Additionally, it can help to avoid missing or duplicated road sections. Route planning should address the following issues:

- In what order will the roads be scanned?
- Should the data collection be performed in one or two ways on the road, and should side roads be scanned as well?
- Where are the places for IMU static calibration before and after data acquisition?
- Where are the base stations set up during data collection? A flat open area (i.e., an area with good GPS signal reception) should be selected for the base station, which typically has an effective working range of 30 km.

3.2.3.1.2 Global Navigation Satellite System Base Station Setup

The GNSS base station should be set up in an open area free of vibration and away from interference from other signal sources. The base station should be set at known ground control points for projects with high accuracy requirements. The basic requirements for setting up base stations are as follows:

- Base stations should be placed at locations with sufficient space and easy access. The elevation angle of nearby obstacles in the field of view should be less than $10°-15°$ to ensure satellite signal reception.
- No large water bodies or objects that strongly reflect electromagnetic waves (e.g., metal billboards and oil cans) are nearby to reduce the impact of multipath effects on GPS signals.

- Base stations should be at least 50 m from high-voltage powerlines.

It is worth noting that base stations should be turned on 30 min before data collection to ensure data quality, and if a CORS network is used, the GNSS base station setup step can be ignored.

3.2.3.2 Backpack LiDAR System
3.2.3.2.1 Route Planning

The route planning of backpack LiDAR can be quite different for projects with dissimilar purposes and working areas. Here, we introduce several examples as guidelines for several issues to consider in backpack LiDAR route planning:

(1) Closing loops can improve data quality. As shown in Fig. 3.5, if buildings 1—3 are the targets, the recommended route is ① ② ③ ④ ⑤ ⑥ ⑦ or ⑤ ⑥ ⑦ ④ ① ② ③. Moreover, if the planned scanning route is ① ② ③ ④ ⑤ ⑥ ⑦, operators should continue to go from ⑦ to ③ or ④ for at least 5 m to ensure that the loop is closed.

(2) Maintain a straight path when possible. Curly routes may result in targets of interest on different sides of the route having significantly different point densities, which may influence the data processing that follows.

(3) Avoid repeated routes. Repeated routes may lead to certain targets being scanned multiple times, which may result in uneven point density distributions in the collected data. Moreover, repeated routes may lead to targets being scanned multiple times and have a "shadow" effect due to inevitable systematic errors.

(4) Large working sites should be broken into multiple small sections to reduce error accumulation. Operators can directly move from one divided section to another after closing the loops for the section.

(5) Walk on the same side of the road to close the loop in a round trip. Walking on two different sides back and forth may result in point cloud mismatch due to inevitable systematic errors. To scan both sides of the

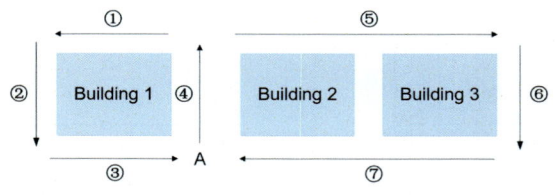

Figure 3.5 Examples of closing loops in backpack LiDAR route planning. Numbers represent the order of scanning.

road, two separate projects should be created and merged during data postprocessing.

(6) Plan the routes based on the purpose of the project. For example, if working in a 30 m × 30 m forest plot, the route designs in Fig. 3.6a are more appropriate for a forest plot with very dense vegetation, which can reduce the influence of occlusion effects. In contrast, the route designs in Fig. 3.6b are more appropriate for a forest with sparse vegetation, which can reduce redundant information and improve data collection efficiency.

(7) Once the routes are determined, it is recommended to strictly follow them during data collection to ensure data quality. Moreover, for data collection in forest plots, "Z" shape routes like in Fig. 3.6 are recommended.

3.2.3.2.2 Global Navigation Satellite System Base Station Setup

If the backpack LiDAR system has a GNSS receiver and the data collection project requires georeferenced point clouds, operators can set up a GNSS base station with a procedure similar to that described in the previous section for the mobile LiDAR system.

3.2.4 Mobile and Backpack LiDAR Data Collection

3.2.4.1 Mobile LiDAR System

(1) Equipment installation. Before performing mobile LiDAR data collection, operators should install the mobile LiDAR components, strictly following the hardware design.

(2) Mobile LiDAR data collection. After installing the mobile LiDAR system and turning it on, the system must usually be put in a static condition to perform self-inspection. For example, the LiMobile system

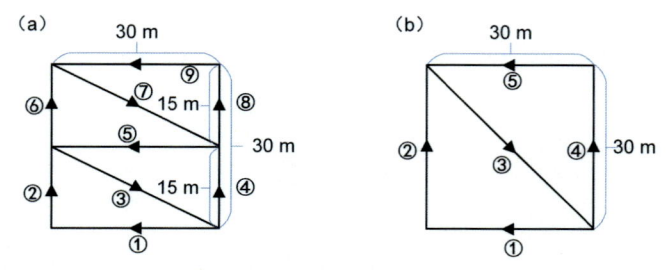

Figure 3.6 Examples of route planning for a backpack LiDAR system in (a) a dense forest plot and (b) a sparse forest plot, respectively. Numbers represent the order of scanning.

usually needs about 1 min for this step. After performing self-inspection, operators should check the device status indicators and wait for the various components (e.g., GNSS, IMU, and laser scanner) to be ready. Once the system is ready, the vehicle should be put into a static state for about 5—10 min to complete the IMU initialization process. The operators should then drive the vehicle in a figure "8" or around a building until the IMU alignment is completed. The heading angle error is usually less than 0.2° after IMU alignment. Finally, operators can specify the laser scanner parameters (e.g., scanning frequency, moving speed) based on the purpose of the project and begin to collect data along the predesigned routes.

(3) End data collection. When data collection is finished, the system should be put into a static condition for about 5 min to collect data before turning off the system. Some systems may require another figure "8" IMU alignment to be performed before ending data collection. Operators should follow the guidance of their specific system during data collection.

3.2.4.2 Backpack LiDAR System

The data collection procedure for backpack LiDAR is usually much simpler. Before data collection, operators should turn the system on and check whether the unit is working properly. For example, the LiBackpack system can be connected to mobile devices (e.g., phones and tablets) through Wi-Fi. Before collecting data, operators can examine the system status through mobile devices. When the system is ready, operators can collect data along the predesigned routes. If the system has a GNSS unit and the project requires georeferenced data, operators should perform a figure "8" IMU alignment in open areas before data collection, similar to the procedure for mobile LiDAR systems.

When collecting data, operators should restrict the movement of objects within the working area. If the working area is split into several small subareas, observable registration targets (e.g., specialized target spheres) should be set up as control points for the following data registration procedure. At least three targets should be observable within each subarea. Moreover, during data collection, operators should avoid direct sunlight, which may lead to noise points in the collected data. After data collection, operators should perform another figure "8" IMU alignment in open areas if the system has a GNSS unit and the project requires it. If not, operators can directly save the collected data and end the data collection procedure.

3.2.5 Initial Data Inspection

3.2.5.1 Mobile LiDAR System

After data acquisition, the quality of the collected data should be checked as soon as possible. In this process, the collected laser scanner, IMU, and GNSS measurements must first be resolved geometrically (i.e., deriving the 3-D coordinates of each point) because the originally collected data only have information about emission angle, distance, echo intensity, trajectory, vehicle attitude, etc. (Fig. 3.7). The principle and detailed procedure for resolving mobile LiDAR point clouds is introduced in Chapter 5.

After resolving the mobile LiDAR point clouds, operators should perform a quality inspection of the trajectory information. For sections with low positioning accuracy from the GNSS and IMU navigation information, operators should find the causes (e.g., close vegetation canopies or tall buildings), redesign the routes, and recollect the data in those sections. Once the trajectory accuracy is satisfactory, operators should visually examine the collected point clouds to determine whether there are areas not covered in the collected data and whether the collected point density meets the requirements of the project, similar to the initial data inspection of terrestrial LiDAR data. If any regions have the abovementioned issues, operators should recollect their data by adjusting the routes and system settings.

3.2.5.2 Backpack LiDAR System

Unlike mobile LiDAR systems, the real-time status and collected data of backpack LiDAR systems can be examined by their operators from connected mobile devices during data collection. Operators should carefully

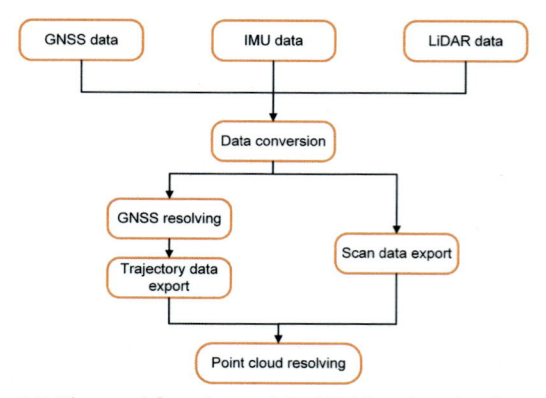

Figure 3.7 The workflow for mobile LiDAR point cloud resolving.

examine the real-time status to ensure the quality of collected data. Moreover, backpack LiDAR systems can usually output resolved LiDAR point clouds directly after data collection. Operators should export the data as soon as possible after data collection and examine the collected data visually. If the collected data have point cloud drifting issues, cannot fully cover the areas of interest, or do not meet the point density requirement of the project, the operators should redesign the route and recollect the data.

3.3 Basic Operation of Airborne and Unmanned Aerial Vehicle LiDAR

3.3.1 Overview of Airborne and Unmanned Aerial Vehicle LiDAR Operation

In this section, we primarily introduce the operational processes of airborne and UAV LiDAR systems. They share very similar data collection procedures except for the piloting of an aircraft and a UAV. For convenience, the operation of these two systems will be introduced with airborne LiDAR as the example unless specified. As described in Section 2.5 of Chapter 2, airborne LiDAR systems usually comprise an aircraft platform, a laser scanner, a camera, a position, an orientation system (POS) (including GNSS receivers and an IMU), and a synchronous control system. Because it integrates multiple subsystems, the data acquisition process of airborne LiDAR is a joint operation of various subsystems (Zhang, 2007). Therefore, the operating mode of the airborne LiDAR scanning system is determined by its composition and the operating principle of each subsystem.

The operational process of airborne LiDAR is similar to that of traditional aerial photogrammetry. However, its system integrates subsystems such as LiDAR sensors, high-resolution cameras, POS systems, and synchronous control systems. Hence, the fieldwork for airborne LiDAR is different from that for traditional aerial photogrammetry, from preliminary planning to final data processing.

The traditional aerial survey is only conducted using an aerial camera. The data quality requirements are represented by indicators that include overlapping ratios along and across flight lines and ground resolution. Therefore, the operational standard of aerial surveys is designed to meet the demands of these indicators. By contrast, an airborne LiDAR system usually contains both a LiDAR sensor and a high-resolution aerial camera, which requires a standard to meet quality requirements for both LiDAR and image data. The airborne LiDAR operating process mainly includes the preparation of aerial surveys, data acquisition, data resolving, and quality inspection. Among them, the preparation of aerial surveys occurs at the early stage of the process, data collection is the implementation stage, and data resolving and quality inspection are two essential steps to data quality control. Before the flight, a detailed flight plan arrangement should be conducted first, including setting the flight route, flight strip overlap, flight height, speed, and other parameters. Second, before flight, each sensor should be carefully examined to ensure system operability. Once the abovementioned steps are completed, data collection should be performed according to the scheduled routes. After data collection, the after-flight system parameters should be checked and calibrated. Although current mainstream airborne LiDAR equipment comes from various equipment manufacturers, the components and working principles are the same, and thus the operating process is similar, as shown in Fig. 3.8.

Compared with airborne LiDAR, the operational procedure of UAV LiDAR is simpler because of the smaller surveyed area and the easier operation of UAVs compared to aircraft. However, UAV LiDAR may require special attention to the working area size, terrain relief, and weather conditions. A too-large working area may lead to UAVs losing control during data collection. Moreover, the battery life of small UAVs is usually

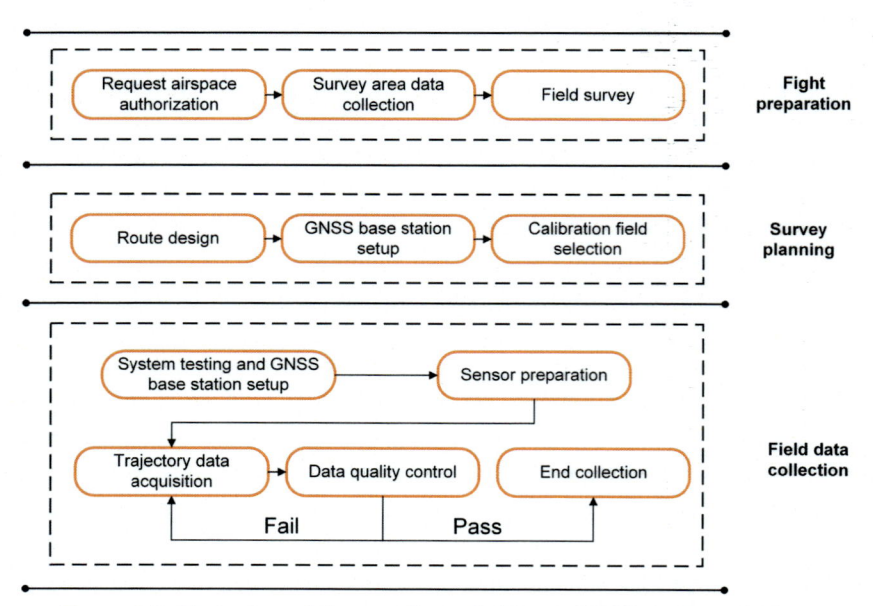

Figure 3.8 Illustration of the workflow of airborne LiDAR data acquisition.

shorter than 1 h, which may not fulfill the data collection mission of a large working area in one flight. In areas with steep relief, operators should pay special attention to the flight attitude to ensure the safety of UAVs (i.e., avoid crashing). Besides, small UAVs can hardly be operated under strong wind conditions.

3.3.2 Preparation for Aerial LiDAR Survey

The preliminary preparation for airborne LiDAR data acquisition is one of the most important steps of the entire LiDAR data acquisition process, which influences the data collection efficiency of the mission. This stage includes the application for airspace authorization, information collection of the areas of interest, and field surveys.

3.3.2.1 Application for Airspace Authorization

Before aerial survey execution, airspace authorization must be procured from relevant departments based on local regulations. This normally requires approval from the military, air force, civil aviation, etc. After permission is obtained, climatic conditions must be checked, and appropriate climatic conditions should be chosen for the aerial survey. This can ensure flight safety as well as improve the quality of collected data.

3.3.2.2 Information Collection for Areas of Interest

Before conducting the aerial survey, topographic maps, ground control points, vegetation information, remote sensing images, and other materials should be collected to analyze environmental and climatic complexity. These data are critical for conducting the following field surveys and designing flight strips (the distribution of observation targets may also be needed in designing flight strips). Climatic conditions are extremely important to determine the best flight period for data collection.

3.3.2.3 Field Survey

Field survey plays an essential role in the design of the aerial survey scheme. During the survey, ground control points should be identified because they are necessary for setting up GNSS base stations during the data collection. If there are no ground control points in the working areas, self-marked ground points should be set up and measured during the survey. With ground control points being identified or measured, operators should check whether GNSS receivers can receive GNSS signals normally at these locations, whether there are large water bodies or objects that may result in multipath effects, and whether there are strong electric signals that may influence GNSS signals. Additionally, vegetation coverage information and the reflectance of ground surface objects can be collected and recorded during the field survey.

3.3.3 Scanning Operational Planning

Flight planning is the basis of an aerial survey. It aims to make a detailed aerial survey scheme based on aerial photogrammetry standards and aerial survey tasks. The scheme mainly includes the coverage, time, and planned routes of the aerial survey, distribution of GNSS base stations, and selection of the ground calibration field. These factors directly affect the quality of the acquired LiDAR data and thus the benefits of the entire project. Moreover, an elaborate scheme is critical for the quality control of the entire scanning process.

3.3.3.1 Route Planning

Flight routes and their related parameters should be designed according to the required point density and working area size. The scanning angle should be set after analyzing the topography of the survey area. A large scanning angle suits areas with smooth terrain and low vegetation cover. In contrast, a small scanning angle is better for areas with complex terrain or high vegetation cover, as their impact on data quality cannot be ignored. Flight parameters such as flight height and speed are then determined based on the required point density, the specifications of the airborne LiDAR system, and a reasonable strip width and overlap ratio. Moreover, a maximum flight range of 30—40 km is recommended due to the accumulation of IMU drifting errors. Therefore, cross routes should be scheduled to ensure that each survey block contains one cross route during the flight. This is an important part of quality control and helps to verify the strip alignment accuracy during later data processing. Route planning aims to acquire data products that fulfill specific accuracy requirements and needs to follow general safety principles and consider efficiency and cost. Information about the working area should be collected and analyzed comprehensively to find the optimal route.

Route planning is usually achieved using the route planning system provided by manufacturers (Fig. 3.9). With the help of route planning software, route files can be generated using Google Earth or topographic maps on a scale of 1:10,000 or 1:50,000, including the route number and order of flight strips. Generally, parallel routes should be adopted to cover

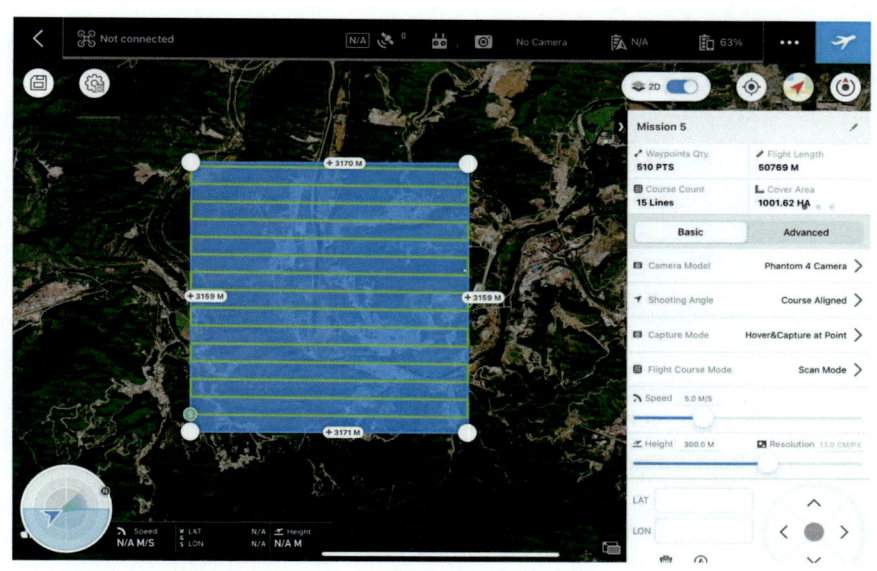

Figure 3.9 An example of flight route planning for unmanned aerial vehicle LiDAR scanning using DJI Pilot software.

the entire survey area, and an overlap rate of 10%—20% or more between adjacent strips should be ensured to avoid stratification or gaps in collected data. Current state-of-the-art route planning software can integrate digital elevation models (DEMs) to generate a flight route plan for the entire survey area, which is particularly useful for surveying mountainous regions. For example, the width of a flight strip changes with the terrain if the aircraft flies at a fixed altitude, which can be effectively avoided by adjusting flight routes based on DEM. After importing the route information into the airborne LiDAR system, it can collect data according to the predefined parameters during the aerial survey.

3.3.3.2 Setting Up Global Navigation Satellite System Base Stations

Differential GNSS (DGNSS) is usually required for aerial surveys. As shown in Fig. 3.10, GNSS base stations are set up on known ground control points and take continuous measurements to correct the measurements by the GNSS device aboard the aircraft (Zhang et al., 2014). In addition to errors caused by equipment installation, errors in DGNSS are the most significant error source in an airborne LiDAR system. Thus, the

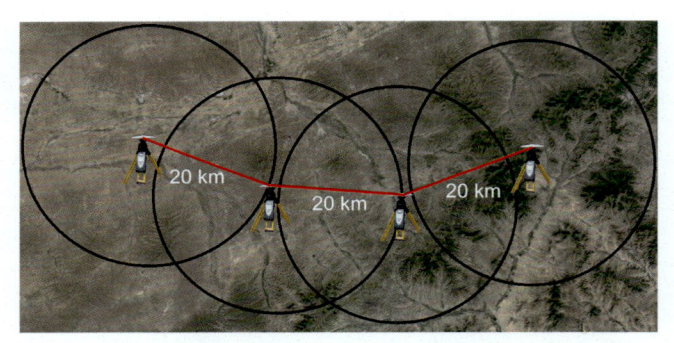

Figure 3.10 Example of setting the location of a base station.

data quality of GNSS can directly affect the accuracy of point cloud resolving.

For DGNSS error reduction, it is necessary to examine whether the number of ground control points can ensure GNSS base station coverage in the survey area. If the ground control points are insufficient to fulfill the survey requirements, more ground control points should be set up at reasonable positions where World Geodetic System 1984 (WGS-84) coordinates of grade C or higher are available. Moreover, the space between GNSS base stations should be such that the distance between the aircraft and a base station is always 30 km or less (Wei & Huang, 2007). GNSS base station distribution should be determined based on field surveys and adjusted if needed to make the whole scheme feasible and practical.

3.3.3.3 Selection of the Calibration Field

The calibration field is used to examine the equipment before and after data collection. Using the LiDAR data collected in the calibration field, specific eccentricity components and subtle divergences in installation angles between core system components can be calculated to correct errors throughout the system and improve data accuracy. The selected calibration field needs to be large enough so that the aircraft can acquire sufficient calibration data.

3.3.4 Airborne and Unmanned Aerial Vehicle LiDAR Data Collection

Based on the planned route, airborne and UAV LiDAR data collection is achieved through flight operations. This step directly affects the data quality, and thus, it is essential to coordinate its various components. During the entire flight, the system control and data storage units can provide real-time information about the system's status, through which operators can monitor the operating status of the entire system. As the accuracy of the final DEM is assessed against ground truth values, flat regions are usually adopted, such as playgrounds, flat roads, and car parks. Flat building roofs are also an ideal choice for position control. For each aerial survey, two or three targets are typically required for elevation control, with more required for horizontal position control.

3.3.4.1 System Test and Setting Up Global Navigation Satellite System Base Station on the Ground

Before the aerial survey, every unit of the LiDAR system should be tested and calibrated to ensure the stable operation of the entire system. According to the requirements of the aerial survey, all subsystems should be configured and undergo trial tests to ensure that data can be properly collected and stored. In addition, the GNSS base stations should be carefully set up with their heights precisely measured, which helps to guarantee the data quality of DGNSS measurements during the aerial survey.

3.3.4.2 Preparation and Sensor Adjustment

To ensure point cloud resolving accuracy, the GNSS base station should be turned on at least 30 min prior to aerial LiDAR data collection. After turning on the POS, the aircraft should perform a figure "8" maneuver and fly straight for a while to adjust the POS to its optimal working status. Then the aircraft can enter the working area and acquire LiDAR data. After data acquisition is completed, the POS should continue to run while the aircraft performs another inverse figure "8" maneuver and flies straight for a while. Then the POS can be turned off, with the GNSS base station working for another 30 min.

3.3.4.3 LiDAR Field Data Acquisition

The process of LiDAR data acquisition includes the simultaneous acquisition and storage of laser ranging, IMU, and GNSS data. In this process, each subsystem works independently according to predefined system parameters and is commonly controlled by the synchronization control system. During the flight, the laser scanner emits laser beams to the ground along the flight route, thereby forming a scan of the entire survey area. At the same time, the GNSS system and IMU system are used to collect position and attitude information for resolving LiDAR point clouds. It is worth noting that the difference between the planned and real flight height should not exceed 50 m during data acquisition, and flight speed should be limited within the design range. As for the steering, the aircraft should turn left or right alternately to avoid IMU error accumulation, and the heading angle of the aircraft should be smaller than 15° during turning. In addition, the laser pulse frequency should be adjusted according to the flight height and speed to ensure consistent point density.

3.3.5 Initial Data Inspection

After scanning, an immediate quality control process is required to check for missed areas. In this process, point cloud resolving should first be finished with the combination of the obtained laser ranging data, IMU data, and GNSS mobile station and reference station data. The workflow of the point cloud resolving can be divided into two steps:

(1) Resolve the GNSS mobile station, reference station, and IMU data to obtain high-accuracy trajectory data (Fig. 3.11);

(2) Calculate the point cloud coordinates by combining the postprocessed trajectory data and ranging data. It is worth noting that all the calibration processes (e.g., boresight calibration and atmospheric corrections) should be added during this step. Point cloud resolving also involves coordinate conversion. For example, the coordinate system of point clouds should be converted from the WGS-84 coordinate to the specified plane coordinate system (e.g., China Geodetic Coordinate System 2000).

After point cloud resolving, the following quality analysis should be performed:

(1) Check whether the obtained point cloud data completely cover the areas of interest.

(2) Check whether the point density meets the specified standard.

(3) Check whether the accuracy of the point cloud meets the specified standard.

If the data quality does not fulfill the above requirements, the data collector should analyze the reasons and design a supplementary data collection plan.

3.4 Error Sources in LiDAR Data Collection

Errors in LiDAR data can be divided into three types: coarse errors, random errors, and systematic errors. Coarse errors generally exist as outlier

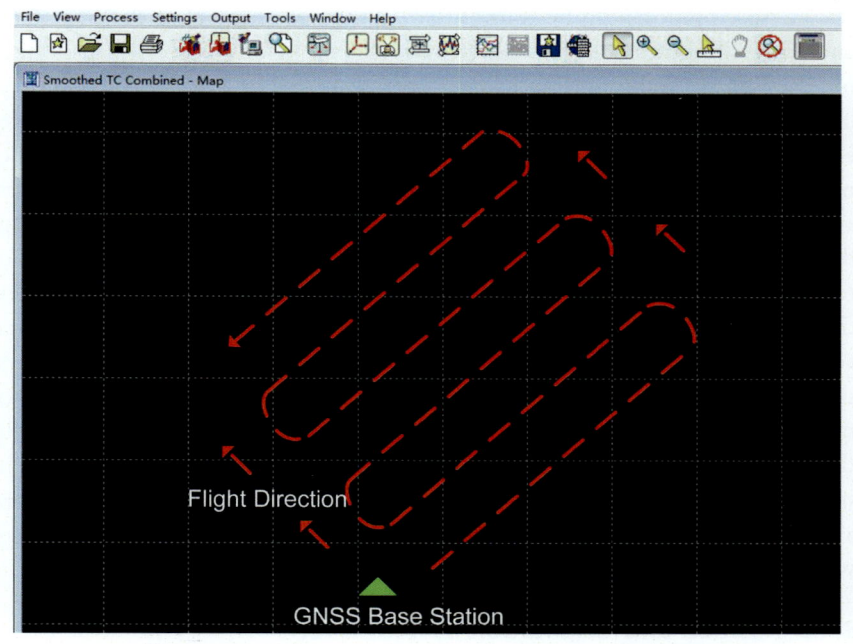

Figure 3.11 Schematic diagram of trajectory calculation.

points in point clouds that can be eliminated through a filtering procedure, while random errors can be eliminated using measurement adjustment methods. As for the systematic errors, they can be further classified into platform vibration error, POS error, scanner error, and system integration error (Shu & Xu, 2014).

3.4.1 Platform Vibration Error

Platform vibration error is a major error source for backpack, mobile, UAV, and airborne LiDAR platforms. Vibrations of backpack LiDAR are mainly caused by unsteady movements of the operator, especially when the operator is climbing up/down on ramps or stairs. Vibrations of mobile LiDAR can be caused by many factors, such as acceleration, braking, vehicle steering, undulating roads, and wind resistance. Moreover, opto-electronic components within mobile LiDAR systems (such as the focusing and zooming of optical sensors) can cause system vibrations. Similar to mobile LiDAR, vibrations in UAV and airborne LiDAR can be caused by many factors, such as high-frequency engine operation, turbulent aerodynamic flow, and aircraft takeoff and landing.

These vibrations can be transmitted to the core sensors of these systems and generally influence them in two ways. First, vibrations can cause their mechanical scanning module (e.g., laser scanner) to vibrate, which may interfere with the laser scanning operation and generate scanning angle errors, placement angle errors, etc. Second, vibrations can interfere with the optical axis of photoelectric measurement units (e.g., the GNSS receiver and IMU), affecting the stability of range measurements and generating flight trajectory errors. All these errors can lead to inaccurate point cloud resolving. Although some studies have successfully used a wavelet transform-based method to reduce the impact of vibration on LiDAR data acquisition (Cui, 2018), it may only work in limited conditions. Operators should still try to reduce platform vibrations during data collection to minimize their influence on LiDAR data resolving. Moreover, although the influence of platform vibration error on terrestrial LiDAR is much smaller than in the systems mentioned above, operators should still focus on avoiding areas under construction and weather conditions with strong winds.

3.4.2 Position and Orientation System Error

POS error is commonly seen in backpack, mobile, UAV, and airborne LiDAR systems. It usually includes GNSS and IMU errors, which are categorized as measurement errors. GNSS measurement data are used to calculate the position of LiDAR platforms, and its positioning accuracy directly affects the accuracy of the acquired LiDAR data (Zhang et al., 2009). GNSS positioning errors mainly include but are not limited to satellite clock difference, loss of the satellite lock, multipath effects, observational noise, insufficient satellite signals during data acquisition, and improper data postprocessing. It is difficult to eliminate GNSS errors in data postprocessing because GNSS errors may vary with the detection environment. The best solution for reducing GNSS errors is in the data acquisition phase. For example, the number of base stations can be increased and evenly distributed in the survey area.

IMU error is another factor affecting the positioning accuracy of the abovementioned LiDAR systems. The accuracy of IMU is also affected by its internal accelerometer error and gyro system offset, and these errors can be accumulated during the data collection process, which are the major sources of positioning errors of the abovementioned LiDAR systems. Moreover, the laser scanner, GNSS antenna, and IMU are usually rigidly connected to each other, and their relative positions usually stay stable during the data collection process. However, the offsets between them can hardly be measured, which may influence final positioning accuracy (Zhang et al., 2009).

Besides the abovementioned error sources, SLAM-induced POS error is common, especially in backpack and mobile LiDAR systems. A lack of registration features is one of its major causes, such as in grasslands or other scenarios with flat terrain and no other objects. Moreover, overlapped and unclosed routes can introduce errors into the SLAM algorithm, so these should be avoided when planning data collection routes.

Terrestrial LiDAR is rarely influenced by POS errors because it does not have a POS system. However, it may be easily influenced by data registration error, which refers to the registration process of multiscan terrestrial LiDAR data. Taking the RIEGL VZ-400 as an example, many users used the RiSCAN Pro software to register multiscan terrestrial LiDAR data using high-reflectance targets as tie points. During the registration process, the manual selection of high-reflectance targets, the number of selected tie points, the overlap ratio between adjacent scans, and the registration algorithm (e.g., the iterative closest point algorithm) can all influence the final registration accuracy.

3.4.3 Scanner Error

Scanner error exists in nearly all LiDAR systems because a laser scanner is a required component. As a core component, the measurement error of a laser sensor can be affected by many factors, including scanning angle, ranging, atmospheric refraction, and reflection errors.

Scanning angle error is caused by the deviation of the initial angle from the ideal state, which is usually caused by the hardware design and installation. The emitted laser pulse from a laser scanner is first deflected by the prism mirror, and then its vertical and horizontal emitter angles are changed by rotating the turning mirror on the vertical rotation platform and rotating the horizontal platform itself, respectively. However, because of the inertia of the rotating lens itself, the refracting angle differs from the theoretical value, that is, the torque error. Scanning angle errors can lead to pointing errors, instantaneous scanning angle errors, and scanning plane distortion.

Ranging error consists of the nonparallel error of transmitting and receiving laser signals introduced by signal processing within the laser scanner, the propagation time error of the laser pulse, and the vibration error of the scanner during operation. Moreover, laser scanners with long ranging distances and multiple-return recording capability tend to generate more noise when scanning objects close to them.

Atmospheric refraction error is caused by the atmospheric refraction effect in the laser propagation process. The influence of atmospheric refraction is related to the wavelength of the emitted laser pulse. When the wavelength is approximately 1 μm, the influence of atmospheric refraction on laser propagation is relatively small.

Reflection error refers to the error caused by the various reflectance characteristics of ground targets. After the emitted laser pulses reach the ground, laser pulses reflected by different targets may produce different reflection effects. Moreover, the diffuse reflection of laser pulses can generate many noises, while specular reflection by smooth surfaces can cause losses of laser pulse signals. In addition, terrain roughness and slope and vegetation types can affect the reflection of laser pulses.

3.4.4 System Integration Error

System integration error may have a bigger influence on backpack, mobile, UAV, and airborne LIDAR systems. Installation error and time synchronization error are the two primary system integration errors. For installation of the abovementioned LiDAR systems, the coordinate systems of the laser scanner, IMU, and GNSS antenna must be parallel. However, eccentricity and placement angle errors are inevitable during installation. These errors could significantly influence the accuracy of the point cloud resolving. Therefore, laboratory boresight calibration should be conducted after LiDAR system installation. Detailed information on boresight calibration is introduced in Chapter 5.

Synchronization error refers to time stamp differences in the data records of the GNSS, IMU, and laser scanner. To resolve the 3-D coordinates of point clouds, the navigation, positioning, and ranging information from the abovementioned core units should be unified as standard Coordinated Universal Time. However, the data sampling frequencies of these three units usually differ. For example, the sampling frequency of GNSS receivers is usually 1—20 Hz, the sampling frequency of IMUs is usually around 200 Hz, and the sampling frequency of laser scanners can reach more than 500 kHz. To match their records, time interpolation operations are usually needed. Interpolation errors induced during this process can reduce LiDAR point cloud resolving accuracy.

It should be noted that the abovementioned types of errors usually do not influence the LiDAR point cloud resolving accuracy alone. Instead, they may interact with each other to magnify the influence of a single error source. For example, the platform vibration error can increase the POS error and magnify the influence of systematic integration errors. Operators should not only focus on one type of error during the data collection and processing procedures.

3.5 Chapter Summary

Collecting LiDAR data in the field is an essential prerequisite for using LiDAR in ecological studies. It is the basis for extracting forest structural and functional attributes. Many factors can influence the data quality collected from LiDAR platforms, e.g., terrestrial, backpack, mobile, UAV, and aircraft. Although systematic errors related to core components of LiDAR systems are inevitable, operators should still try to avoid inducing random errors during the data collection and registration processes. Understanding the working principles and error sources of the different LiDAR platforms and formulating a standard data collection procedure will improve LiDAR data collection accuracy and thus improve the estimation accuracy of forest structural and functional attributes in the following analyses.

References

Cui, S. (2018). *Study on the algorithm of vibration response error processing for vehicle-borne LiDAR based on wavelet analysis*. Xi'an: Xi'an University of Technology (in Chinese).

Shu, R., & Xu, Z. (2014). *Principle of LiDAR imaging and method of motion error compensation*. Science Press (in Chinese).

Wei, E., & Huang, J. (2007). *GPS measurement operation and data processing*. Wuhan University Press (in Chinese).

Zhang, X. (2007). *Theory and method of airborne LiDAR measurement technology*. Wuhan University Press (in Chinese).

Zhang, D., Wu, W., & Wu, M. (2009). Calibration technology of airborne LiDAR. *Optics and Precision Engineering, 17*(11), 2806−2813 (in Chinese).

Zhang, G., Zhang, Z., & Yu, H. (2014). *Practical handbook for GPS RTK measurement technology*. China Communications Press (in Chinese).

CHAPTER 4

LiDAR Data Formats

Contents

Light detection and ranging (LiDAR) data are rendered in two primary formats, discrete point cloud and full-waveform. Discrete point cloud is the most commonly used format; it stores locations and other attributes of detected objects from discrete sampling of return laser echoes. Conversely, the full-waveform LiDAR data record is a continuous return signal with a time function that can potentially detect weak pulses and provide extra waveform parameters (e.g., echo amplitude and width). However, full-waveform LiDAR data can be more difficult to interpret visually and are usually decomposed into various components before use. This chapter introduces the storage, indexing, and reading methods of these two LiDAR data formats.

LiDAR Principles, Processing and Applications in Forest Ecology
ISBN 978-0-12-823894-3
https://doi.org/10.1016/B978-0-12-823894-3.00004-9

4.1 Format, Composition, and Characteristics of Point Cloud Data

4.1.1 Format and Composition of Point Cloud Data

The raw data obtained by a LiDAR system is often a massive set of points, also known as point cloud, that reflect the geometric information—e.g., three-dimensional (3-D) coordinates—and reflectance information (e.g., intensity of the reflected signal) of detected objects. Point cloud data are normally organized in the LAS or American Standard Code for Information Interchange (ASCII) format. In addition to these general formats, many companies have proprietary formats for storing LiDAR data. For example, RIEGL (RIEGL Laser Measurement System GmbH, Horn, Austria) uses its 3D data point cloud format, Trimble (Trimble Inc., Sunnyvale, CA, USA) uses its RWP format, Faro (Faro Technologies Inc., Lake Mary, FL, USA) uses its self-defined FLS format, and LiDAR360 (GreenValley Technology Co., Ltd., Beijing, China) uses its self-defined LiData format. However, using relevant software, all self-defined formats can be easily converted to general formats.

4.1.1.1 LAS Format

The LAS format was published by the American Society for Photogrammetry and Remote Sensing (ASPRS) in May 2003 and has become a standardized format for LiDAR data. The LAS format has been updated many times, and this section focuses on the newest version, LAS 1.4.

The LAS 1.4 file comprises a public header block, variable-length records, point data records, and extended variable-length records. The data types used in LAS 1.4 are char (1 byte), unsigned char (1 byte), short (2 bytes), unsigned short (2 bytes), long (4 bytes), unsigned long (4 bytes), long long (8 bytes), unsigned long long (8 bytes), float (4 bytes), double (8 bytes), and string.

(1) The fields included in the public header block are shown in Table 4.1. Any unnecessary or unused fields in the public header block must be set to zero. The specific meanings of the fields are as follows:

File Signature: This field must contain "LASF", as stipulated by the LAS specification. User software can make a preliminary determination of file types by checking these four characters.

File Source ID: The value range of this field is from 1 to 65,535. If LAS data are obtained from the original track file, the value is the route number. A zero value indicates that the data are not assigned an ID. In this case, the software will assign an ID.

Table 4.1 Fields in public header block.

Item	Format	Size (byte)	Required
File signature ("LASF")	Char[4]	4	*
File source ID	Unsigned short	2	*
Global encoding	Unsigned short	2	*
Project ID—GUID data 1	Unsigned long	4	
Project ID—GUID data 2	Unsigned short	2	
Project ID—GUID data 3	Unsigned short	2	
Project ID—GUID data 4	Unsigned char[8]	8	
Version major	Unsigned char	1	*
Version minor	Unsigned char	1	*
System identifier	char[32]	32	*
Generating software	char[32]	32	*
File creation day of year	Unsigned short	2	*
File creation year	Unsigned short	2	*
Header size	Unsigned short	2	*
Offset to point data	Unsigned long	4	*
Number of variable length records	Unsigned long	4	*
Point data record format	Unsigned char	1	*
Point data record length	Unsigned short	2	*
Legacy number of point records	Unsigned long	4	*
Legacy number of points by return	Unsigned long[5]	20	*
X scale factor	Double	8	*
Y scale factor	Double	8	*
Z scale factor	Double	8	*
X offset	Double	8	*
Y offset	Double	8	*
Z offset	Double	8	*
Max X	Double	8	*
Min X	Double	8	*
Max Y	Double	8	*
Min Y	Double	8	*
Max Z	Double	8	*
Min Z	Double	8	*
Start of waveform data packet record	Unsigned long long	8	*
Start of first extended variable length record	Unsigned long	8	*
Number of extended variable length record	Unsigned long	4	*

Continued

Table 4.1 Fields in public header block.—cont'd

Item	Format	Size (byte)	Required
Number of point records	Unsigned long	8	*
Number of points by return	Unsigned long [15]	120	*

Global Encoding: This field specifies some global properties of LAS files.

Project ID (GUID data): These four fields form a complete globally unique identifier (GUID) known as the project identifier (project ID), which is optional.

Version Number: This field consists of a major and a minor part jointly representing the version of the LAS format. For example, if the major part is 1 and the minor part is 4, the identified version is LAS 1.4.

System Identifier: System identification describes the source of LAS data.

Generating Software: This field describes the package and version used to create the LAS file (e.g., "TerraScan v-10.8", or "REALM v-4.2"). The remainder must be empty if the number of characters is fewer than 32.

File Creation Day of Year: This field is in unsigned short integer and indicate the creation date of the LAS file. January 1 is considered the first day in Greenwich Mean Time.

File Creation Year: This field uses four digits to represent the creation year of the LAS file.

Header Size: The size of the common file header in bytes. For LAS 1.4, the size of the common file header is 375 bytes.

Offset to point data: The number of bytes from the first field of the LAS file header to the point data record portion.

Number of Variable-Length Records: The size of the variable data area, which must be updated if the size of the variable data area changes.

Point Data Record Format: The format of the data recorded in the LAS file.

Point Data Record Length: The size of the point data record in bytes.

Legacy Number of Point Records: The total number of points recorded in the LAS file.

Legacy Number of Points by Return: The number of points recorded in each echo, including five unsigned long integer data that represent the number of points recorded in the first, second, third, fourth, and fifth echoes.

X, Y, Z scale factor: This field contains a double-precision floating-point value that describes the scaling factor for X, Y, and Z in the point data record.

X, Y, Z offset: The offset of X, Y, and Z, which can be used to calculate the actual coordinates of X, Y, and Z using Eq. (4.1):

$$X_{coordinate} = (X_{record} * X_{scale}) + X_{offset}$$
$$Y_{coordinate} = (Y_{record} * Y_{scale}) + Y_{offset} \quad (4.1)$$
$$Z_{coordinate} = (Z_{record} * Z_{scale}) + Z_{offset}$$

Max and Min X, Y, Z: The maximum and minimum values of X, Y, and Z coordinates in the LAS file.

Start of Waveform Data Packet Record: The offset from the LAS header to the first byte of the waveform data packet record in bytes.

Start of First Extended Variable-Length Record: The offset from the LAS header to the first extended variable-length record in bytes.

Number of Extended Variable-Length Records: The number of extended variable-length records stored in the LAS file after the point data record.

Number of point records: The total number of points recorded in the LAS file.

Number of points by return: The number of points recorded per echo. The first value is the number of points for the first echo, the second is the number of points for the second echo, etc. The maximum number is up to the 15th echo.

(2) Fields included in the variable-length records are shown in Table 4.2, and their specific meanings are as follows:

Reserved: This value must be set to 0.

Table 4.2 Fields in variable-length record.

Item	Format	Size (byte)	Required
Reserved	Unsigned short	2	
User ID	Char[16]	16	*
Record ID	Unsigned short	2	*
Record Length After Header	Unsigned short	2	*
Description	Char[32]	32	

User ID: This field is used to identify the user who generated the variable-length data record.

Record ID: The field depends on the User ID; there can be 0—65,535 record IDs for each User ID.

Record Length After Header: The number of bytes between the end of a header file and the beginning of a data record.

Description: The description information of the LAS file.

(3) Point data records.

The specific content of the point data record is determined by the Point Data Record Format ID. The format of Point Data Record Format 0 is the earliest version (Table 4.3), and their specific meanings are as follows:

X, Y, Z: The actual coordinate values calculated by Eq. (4.1) considering the scaling factor and offset.

Intensity: The echo intensity reflects the reflection characteristics of a detected object to the laser pulse. This field is optional.

Return Number: The echo sequence.

Number of Returns (given pulse): The total number of echoes.

Scan Direction Flag: Record of the direction of a laser scan, which is 1 for a forward scan and 0 for a reverse scan.

Edge of Flight Line: This field takes a value of 1 only when a point is the last point of a scan line.

Table 4.3 Point data record format (Format 0).

Item	Format	Size	Required
X	Long	4 bytes	*
Y	Long	4 bytes	*
Z	Long	4 bytes	*
Intensity	Unsigned short	2 bytes	
Return number	3 bits (bits 0—2)	3 bits	*
Number of returns (given pulse)	3 bits (bits 3—5)	3 bits	*
Scan direction flag	1 bit (bit 6)	1 bit	*
Edge of flight line	1 bit (bit 7)	1 bit	*
Classification	Unsigned char	1 byte	*
Scan angle rank (−90 to +90)—left side	Char	1 byte	*
User data	Unsigned char	1 byte	
Point source ID	Unsigned short	2 bytes	*

Classification: The category attributes of each point. If a point is not classified, the value must be set to 0. Table 4.4 shows the standard point cloud categories provided by ASPRS (for Point Data Record Formats 0–5).

Scan Angle Rank: This field records the sensor tilt angle, including roll, pitch, and yaw. The effective range is −90 degree to +90 degree with an accuracy of 1 degree. The value is 0 degree at the lowest point, −90 degrees at the left side of the flight direction, and +90 degrees at the right side of the flight direction.

User Data: The data in this field are determined by the user.

Point Source ID: This value records which file the point comes from and has a valid value range of 1–65,535.

With the development of LiDAR hardware and the expansion of the required parameters from data processing software, Point Data Record Formats 1–6 have gradually added the recorded content based on the Point Data Record Format 0 and developed into different upgraded versions of the point cloud record storage format, as follows:

Point Data Record Format 1: GPS Time (double, 8 bytes, required) is added based on Point Data Record Format 0.

Table 4.4 American Society for Photogrammetry and Remote Sensing (ASPRS) standard light detection and ranging (LiDAR) point classes (Formats 0–5).

Classification value	Meaning
0	Created, never classified
1	Unclassified
2	Ground
3	Low vegetation
4	Medium vegetation
5	High vegetation
6	Building
7	Low point (noise)
8	Model keypoint (mass point)
9	Water
10	Reserved for ASPRS definition
11	Reserved for ASPRS definition
12	Overlap points
13–31	Reserved for ASPRS definition

Point Data Record Format 2: Red (unsigned short, 2 bytes, required), green (unsigned short, 2 bytes, required), and blue (unsigned short, 2 bytes, required) are added based on Point Data Record Format 0, indicating the value of the red, green, and blue channels of each point. This can be used as auxiliary data for point cloud coloring.

Point Data Record Format 3: GPS time is added based on Point Data Record Format 2.

Point Data Record Format 4: Wave Packets are added based on Point Data Record Format 1, including the Wave Packet Descriptor Index (unsigned char, 1 byte, required), Byte Offset to Waveform Packet Data (unsigned long, 8 bytes, required), Waveform Packet Size in bytes (unsigned long, 4 bytes, required), Return Point Waveform Location (float, 4 bytes, required), X(t) (float, 4 bytes, Required), Y(t) (float, 4 bytes, required), and Z(t) (float, 4 bytes, required).

Point Data Record Format 5: Wave Packets are added based on Point Data Record Format 3.

Point Data Record Format 6: Classification Flags and Scanner Channel are added (Table 4.5). Classification Flags represent the special attribute of a point. Scanner Channel is used to indicate a multi-channel system. Channel

Table 4.5 Point data record (Format 6).

Item	Format	Size	Required
X	Long	4 bytes	*
Y	Long	4 bytes	*
Z	Long	4 bytes	*
Intensity	Unsigned short	2 bytes	
Return number	4 bits (bits 0–3)	4 bits	*
Number of returns (given pulse)	4 bits (bits 4–7)	4 bits	*
Classification flags	4 bits (bits 0–3)	4 bits	
Scanner channel	2 bits (bits 4–5)	2 bits	*
Scan direction flag	1 bit (bit 6)	1 bit	*
Edge of flight line	1 bit (bit 7)	1 bit	*
Classification	Unsigned char	1 byte	*
User data	Unsigned char	1 byte	
Scan angle	Short	2 bytes	*
Point source ID	Unsigned short	2 bytes	*
GPS time	Double	8 bytes	*

0 represents a single scanner and can support up to four channels (0–3). After Point Data Record Format 6, standard LiDAR point classes can also be updated to support more classes (Table 4.6).

Point Data Record Format 7: Add three color channels (*red*, *green*, and *blue*) (unsigned short, 2 bytes, required) based on Point Data Record Format 6.

Point Data Record Format 8: Add a near-infrared channel (unsigned short, 2 bytes, required) based on Point Data Record Format 7.

Point Data Record Format 9: Add Wave Packets based on Point Data Record Format 6.

Point Data Record Format 10: Add Wave Packets based on Point Data Record Format 7.

(4) Extended variable-length records are shown in Table 4.7.

Table 4.6 Standard LiDAR point classes (Formats 6–10).

Value	Significance
0	Default
1	Class
2	Ground
3	Low vegetation
4	Medium vegetation
5	High vegetation
6	Building
7	Low noise
8	Reserved definition
9	Water
10	Railway
11	Road surface
12	Reserved definition
13	Power line
14	Power line
15	Pylon
16	Power line
17	Bridge
18	High noise
19–63	Reserved definition
64–255	User-defined

Table 4.7 Extended variable-length record headers.

Name	Format	Size (byte)	Required
Reserved	Unsigned short	2	
User ID	Char[16]	16	*
Record ID	Unsigned short	2	*
Record length after header	Unsigned long long	8	*
Description	Char[32]	32	

4.1.1.2 ASCII Format

ASCII is another common format for point cloud data. ASCII has many formats, including ASC, XYZ, TXT, PTC, PTS, and PTX. An ASCII file generally has two parts. The first part is a header file to describe the data information, and the second part records the geometric coordinates, intensity, color, and other point information. Generally, each line corresponds to a point in the point cloud. The ASCII format is flexible and easy to read and write, and is widely used by hardware vendors. However, the read and write speeds of the ASCII format are slow, and it requires large storage space, making it difficult to store and process massive point cloud datasets.

4.1.2 Characteristics of Point Cloud Data

As a new 3-D data acquisition method, LiDAR is finding wide use in many applications, such as agriculture, forestry, and urban studies. An in-depth understanding of LiDAR data characteristics can help users better process and use LiDAR point cloud data. The characteristics of point clouds are as follows (Liang et al., 2005):

(1) Massive data: Laser scanners have a high sampling frequency and can acquire thousands of points in a few seconds.

(2) Three-dimensional data: Unlike traditional two-dimensional (2-D) images, laser scanners acquire 3-D coordinates of geo-objects.

(3) Uneven data distribution: Point density generally decreases as the distance from the scanner increases.

(4) Discrete data distribution: LiDAR point cloud data are discretely distributed.

(5) Intensity information: In addition to the 3-D coordinate information, laser scanners can also record the echo intensity information reflected by the object. The intensity information has been used in point cloud filtering and classification (Lu et al., 2014). However, the lack of necessary calibration means that the intensity information has not been widely used.

These characteristics of LiDAR point cloud data make it subject to some problems and challenges in subsequent data processing and applications (Liang et al., 2005):

(1) Lack of spectral and textural information: The ability of laser scanners to acquire 3-D information on geo-objects is a good complement to traditional remote sensing approaches that cannot directly measure height information. However, point cloud data lacks spectral and textural information on target objects. In practice, many researchers use LiDAR data with traditional optical images or hyperspectral data (Persson et al., 2004; Popescu & Wynne, 2004).

(2) Registration difficulties: For terrestrial laser scanners (TLSs), multi-angle and multi-site scanning is required to acquire data over large areas. Currently, the registration of multi-station TLS data is still based on the manual recognition of tie points from different TLS scans. However, in areas like forests, the wide variety of tree shapes makes it difficult to find precise tie points for registration.

(3) Incomplete point cloud data: The acquired point cloud data may be incomplete owing to the limited field of view and ranging distance of laser scanners, the complex characteristics of geo-objects, or occlusion effects among geo-objects. In forest areas, bottom–up TLS scanning and backpack LiDAR provide detailed information on tree stems and canopies, but it is difficult to obtain all upper canopy information; in contrast, the top–down scanning of drone and airborne laser scanners makes it easier to obtain upper canopy information, but understory information might be easily missed if canopy coverage is too high.

(4) Underestimation of tree height: An important application of LiDAR in forestry is to obtain tree structure attributes, among which tree height is fundamental. Many studies have shown that tree height estimations from LiDAR, especially drone or airborne LiDAR, are lower than ground truth measurements (Lefsky et al., 2002).

4.2 Indexing of Point Cloud Data

Discrete point clouds acquired by laser scanners are usually large in volume and unevenly distributed, which slows processing speed. Building a spatial index is an effective way to manage point clouds and improve data processing speed and efficiency. Spatial indexing uses a specific method to establish a mapping relationship between spatial locations and geo-objects and, therefore, to fulfill the goal of managing data with high efficiency in searching neighboring points. For example, the content of a book is equivalent to a spatial index. When we want to find the parts of interest in the book, we can first go through the content to find specific chapters and page numbers. Using this method, we do not need to start from the first page and look through the contents word by word.

The choice of the indexing structure of point cloud data depends on many factors, such as data type, data distribution, data query mode, and data update operations. Commonly used spatial indexing structures for point cloud data include regular grids, k-dimensional (k-d) trees, and octrees. This section describes the advantages and disadvantages of these three methods.

4.2.1 Regular Grid

Regular grid divides the space occupied by a point cloud into equal-sized grids and records the points in each grid. The procedure for establishing a regular grid index is to calculate the bounding box of a point cloud data $(X_{min}, Y_{min}, Z_{min}, X_{max}, Y_{max}, Z_{max})$ and set an appropriate grid size L. Then, the index number (i, j, k) of a point (x_i, y_i, z_i) can be calculated as

$$i = \frac{x_i - X_{min}}{L}, j = \frac{y_i - Y_{min}}{L}, k = \frac{z_i - Z_{min}}{L} \qquad (4.2)$$

A unique index value can be established for each point in the point cloud data using Eq. (4.2), and each index value corresponds to some points in the space. Thus, the indexing relationship between the point cloud and the regular grid is established (Fig. 4.1).

The principles and operation of the regular grid algorithm are relatively simple and easy to implement. The method is suitable for cases where points are evenly distributed. A finer grid division can result in greater search accuracy but requires more storage space and search time. Additionally, it has the disadvantages of having only one indexing resolution, large data redundancy, and high maintenance difficulty.

4.2.2 *k*-d Tree

The k-d tree, proposed by Bentley (1975), is a data structure for splitting data in k-d space and can be regarded as a binary tree in k-d space. The

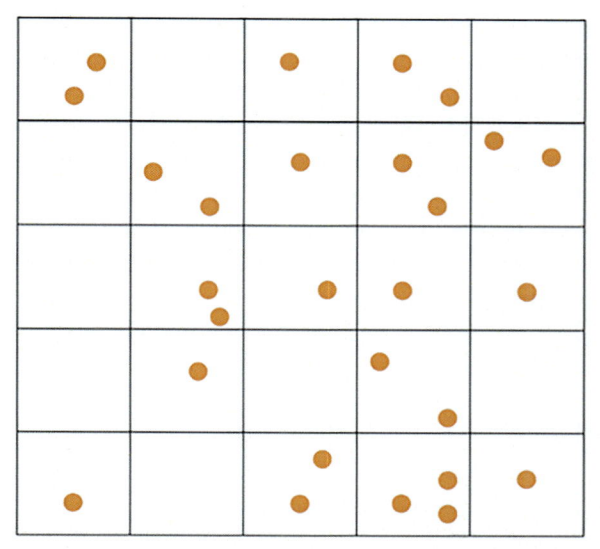

Figure 4.1 Illustration of regular grid indexing method; orange points are point cloud data.

internal nodes of a k-d tree contain an attribute P and a value V, with the attribute P corresponding to a spatial dimension. A space is first divided into two parts, i.e., $P \leq V$ and $P > V$, and then a new P and V are selected in each divided part to further divide the space. For 3-D point cloud data, three attributes, X, Y, and Z, are selected for division. As shown in Fig. 4.2, the X attribute is first selected for division, and the part where X is less than or equal to the specified value is divided into the left subtree; the part where

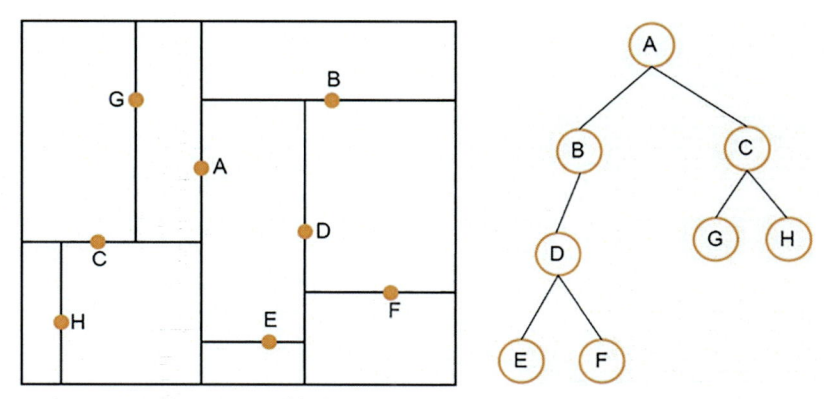

Figure 4.2 Illustration of k-d tree indexing method. $A{-}H$ represent nodes.

X is greater than the specified value is divided into the right subtree. Then, the Y attribute in each subspace is selected to be divided, and a new child node is obtained. Next, the Z attribute in each subspace is selected to be divided, and a new child node is obtained. After all the attributes are used, each subspace continues to be divided based on the first attribute until the subspace contains only one node.

The k-d tree method establishes the neighboring relationship between points, which provides a great advantage in point searches and neighborhood searches (Otepka et al., 2013). However, when dealing with a large amount of point cloud data, establishing such neighboring relationships is time-consuming and inefficient.

4.2.3 Octree

An octree uses the bounding box occupied by a point cloud as the root node and divides the bounding box into eight smaller cubes of equal size. If a small cube is empty, it is used as a leaf node and is not further divided. Otherwise, it is a new parent node and then divided again into eight cubes. All cubes are divided iteratively according to this rule until they do not contain more than one point or the user-defined number of points (Fig. 4.3).

Octree indexing has several advantages. For example, an octree is a hierarchical tree structure whose nodes do not intersect, meaning that the different subregions corresponding to each node do not overlap. Another advantage is that it preserves semantic information about the point cloud. Finally, an octree can be used to optimize the visualization of point cloud data. The rendering of 3-D point cloud data is very time-consuming, and an octree can be used to filter points to render a specific scene instead of

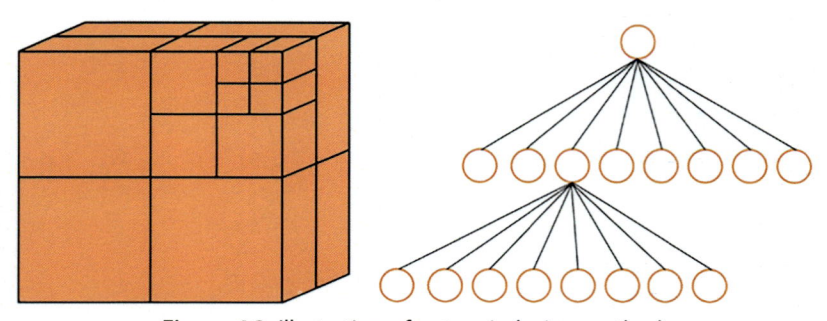

Figure 4.3 Illustration of octree indexing method.

rendering all points (Mosa et al., 2012). The octree algorithm, which continuously divides the 3-D space into eight parts, is simple to operate. However, when point cloud data are numerous and unevenly distributed, the depth of the octree is increased, which may result in inefficient operations, such as queries.

4.3 Reading Point Cloud Data

Because LiDAR data processing is still being developed, programming is usually needed to meet user requirements. This book uses Python as an example programming language to explain basic point cloud data processing algorithms. Reading data is a prerequisite for point cloud data processing, but LAS is a binary file that cannot be directly opened by a text editor like other ASCII files. LAS files can only be read by professional software or programming methods. This section focuses on how to read LiDAR point cloud data and establish a point cloud data index using Python.

4.3.1 Why Python?

Since being created in the early 1990s by Guido Van Rossum, Python has been widely applied in system management and web programming. Python was developed from the ABC programming language, which was designed for nonprofessional programmers. Python also includes the advantages of Modula-3 and the programming habits of the Unix shell and C. Python is one of the most popular programming languages in use today.

Python is completely free and available to all operating systems. Many open-source scientific computing libraries also provide a Python interface. Users can install Python and several extension libraries on any computer. Owing to the simplicity, legibility, and extensibility of Python, more research institutions are using it for scientific computing, and more universities are adopting Python to teach computer programming. Moreover, many Python-specific scientific computing libraries have been developed, such as NumPy, SciPy, and Matplotlib, which provide Python with the capabilities of fast array processing, numerical operations, and scientific drawing. Python can also call on the geospatial data processing library from the Open Source Geospatial Foundation, which allows it to process raster data and vector data and reproject spatial data. LiDAR data are important geospatial data that require support from scientific computing libraries. Therefore, the development environment of Python and its numerous extension libraries are ideal for analyzing and processing LiDAR data.

4.3.2 Basic Syntax of Python

Like many other programming languages, Python has a fixed syntax that must be followed when writing programs. Here, we focus on the application of LiDAR in forest ecology. Some basic Python syntax is introduced below for readers with little experience in Python programming. Readers can learn more comprehensive Python syntax and examples from professional Python textbooks.

According to the Python syntax, identifiers (such as the names of variables, constants, functions, and statement blocks) can be composed of letters, numbers, and underscores but cannot begin with numbers. Identifiers in Python are case-sensitive, and identifiers that begin with an underscore have special meanings. For example, an identifier beginning with a single underscore (e.g., _foo) represents a class attribute that cannot be directly accessed but can be accessed only through the interface provided by the class; an identifier beginning with a double-underscore (e.g., __foo) represents a private member of the class; an identifier both beginning and ending with a double-underscore (e.g., __foo__) represents a special identifier for a particular method in Python, such as __init__(), representing the constructor function of a class. The biggest difference between Python and other programming languages is that the hierarchy of Python statements is controlled by indentation. Incorrect indentation can cause errors and thus requires particular attention from Python novices.

4.3.2.1 Python Variable Types

The basic variable types include Number, String, List, Tuple, and Dictionary, of which Number includes int, long, float, and complex variations. Different variable types can be converted through conversion functions.

4.3.2.2 Basic Operations and Flow Control of Python

Python supports basic arithmetic, comparison, assignment, logical, and bit operators like other programming languages. As an object-oriented language, Python also supports member operators, identity operators, and matrix operations from other libraries. Python primarily uses conditional statements and loop statements to realize process flows, such as "if/else", "while", and "for" statements.

4.3.2.3 Important Libraries Used in this Book

In addition to using built-in Python functions, third-party extension libraries, e.g., NumPy, SciPy, Matplotlib, and OSGeo, might also need to be imported to process LiDAR data. Here, we introduced some common libraries that are used in LiDAR data processing.

(1) NumPy

Although List, a basic Python data type, can be used as a multidimensional array, it is not efficient for numerical calculations. Moreover, the built-in array module supports only one-dimensional rather than multidimensional arrays and does not include efficient arithmetic functions. Therefore, List is not suitable for numerical operations. NumPy is a popular library for scientific computing in Python that can be used to store and process large matrices. NumPy contains powerful N-dimensional array objects, mature function libraries, toolkits for integrating C, C++, and Fortran code, and practical functions for linear algebra, Fourier transforms, and random number generation. Using NumPy is equivalent to converting Python into a free and powerful MATLAB system.

(2) SciPy

SciPy is a numerical calculation library that includes mathematical functions common to science and engineering applications (e.g., linear algebra, numerical solutions of ordinary differential equations, signal processing, image processing, and sparse matrices). Because SciPy is based on the NumPy library, NumPy must be installed before SciPy. This book uses some mathematical functions in SciPy as examples to show how to extract vegetation attributes from LiDAR point cloud data.

(3) Matplotlib

Matplotlib is the most widely used 2-D plotting library in Python. Users can easily use it to visualize data, and the library supports various image output formats. This book uses examples from Matplotlib to show how to visualize LiDAR point cloud data and the corresponding extracted vegetation attributes.

(4) OSGeo

OSGeo is an open-source geographic information processing library developed by the Open Source Geospatial Foundation that supports multiple programming languages. Several OSGeo libraries are available in Python, including GDAL, OGR, and OSR. Among them, GDAL is designed to read, write, and process spatial raster data, OGR is designed to read, write, and process vector data, and OSR is designed to read, write, and process the projection information of spatial data. LiDAR data record 3-D spatial information. The data can also be used to generate various 2-D spatial products that can then be processed using OSGeo libraries.

4.3.3 Reading LiDAR Point Cloud Data Python

LAS, a standard data format for LiDAR data, stores data in a binary format according to specific coding rules, facilitating data exchange and storage. The advantage of the binary format is that it takes up less space and can be read faster. The disadvantage is that it cannot be viewed directly. The encoding rules of LAS were introduced in Section 4.1, and it can be read using the Python binary function library, struct. Because the struct library involves binary encoding, which is unfamiliar to many readers, here we introduce a third-party library named laspy as an alternative for reading LAS data.

Laspy is a Python library for reading, modifying, and writing LAS files, as well as LAZ files, in versions from 1.0 to 1.4. Laspy includes a set of command-line tools that can perform basic file operations, such as format conversion, format verification, and file comparison. Laspy supports Python 2.6+ and 3.5+ and can be installed using pip tools. Laspy only uses NumPy; therefore, it can run on Linux, OS X, and Windows platforms as long as NumPy is installed.

The following code shows an example of reading a LAS file and building a k-d tree index. Note that this example code used laspy 1.0, which is slightly different from reading LAS files using laspy 2.0.

```
#The following code is to build a k-d tree index.
#Coding: utf-8
#Import laspy package
#Notice this code used the laspy 1.0 version
import laspy
import numpy as np
#Las file location; ". /" represents the current working directory
lasfile = ". /read. las"
#Open the las file;"r" means in read-only mode
inFile = laspy. file. File(lasfile, mode = "r")
#Read the coordinates of the data
x. y. z = inFile. x, inFile. y, inFile. z
#Read the classification information
classfication = inFile. raw_classification
#Read the echo information
return_num = inFile. return_num
#Read the scan angle information
scan_angle_rank = inFile. scan_angle_rank
#Build indexing
#Kd-tree indexing, which uses the spatial library in SciPy library
#The principle of kd-tree indexing can be referred to Section 4.2
#This example creates an index based on horizontal coordinates
from scipy import spatial
#Encapsulate x, y. If you're building a three-dimensional index, encapsulate x, y, z
lasdata = zip(x, y)
#Build kd-tree indexing
tree = spatial. KDTree(lasdata)
#Find out how many points are in a circle with a radius of 1m from the center point (323000,4102251)
aa = tree. query_ball_point(np. array([323000, 4102251]) , 1)
#Displays the searched x and y cloud
print x[aa], y[aa]
```

After reading, LiDAR point cloud data can be displayed in 3-D space based on point attributes, such as elevation, height, and intensity (Fig. 4.4). Fig. 4.4a shows a point cloud displayed by intensity (reflecting the reflectivity of ground objects to some extent) using the LiDAR360 software. Fig. 4.4b shows the same point cloud displayed by elevation, in which the colors from blue to red represent that the elevation changes from low to high. Fig. 4.4c shows the same point cloud displayed by elevation using the Eye-Dome Lighting (EDL) enhancing mode. EDL is an image-based data rendering technique. It uses shadows to render data interactively, which can

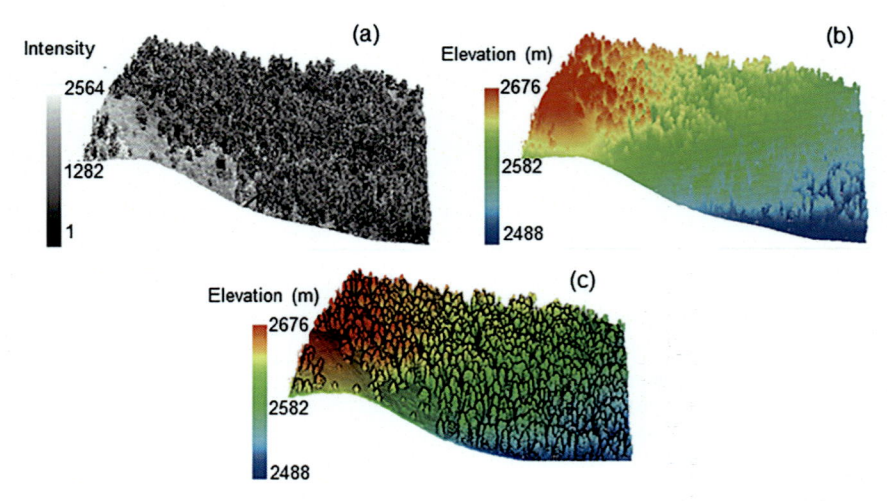

Figure 4.4 LiDAR point cloud displayed by (a) intensity, (b) elevation, and (c) elevation using eye-dome lighting rendering model. (Acquisition date: September 2007; location: Sierra Nevada mountains, California, USA.)

enhance the information depth of the point cloud data and make visualization more realistic.

4.4 Reading Full-Waveform Data

4.4.1 Introduction to Full-Waveform LiDAR Data

In contrast to discrete LiDAR systems that only sample a fixed number of returns, a full–waveform LiDAR system records the entire returned beam with its energy distribution, called a waveform. From the perspective of continuous energy distribution, full-waveform LiDAR data are the origin of discrete LiDAR point cloud data. Therefore, full-waveform LiDAR data can be converted to discrete LiDAR point cloud data (Fig. 4.5) (Mallet & Bretar, 2009). This is usually used as the strategy for full-waveform LiDAR data processing. Although full-waveform LiDAR data can provide more detailed information than discrete point cloud data (Fig. 4.6) (Mallet & Bretar, 2009), understanding how to extract and use this information is still being explored.

Full-waveform LiDAR data can be divided into two categories, i.e., large-footprint data and small-footprint data. Large-footprint LiDAR data are usually acquired from spaceborne LiDAR systems with a footprint size

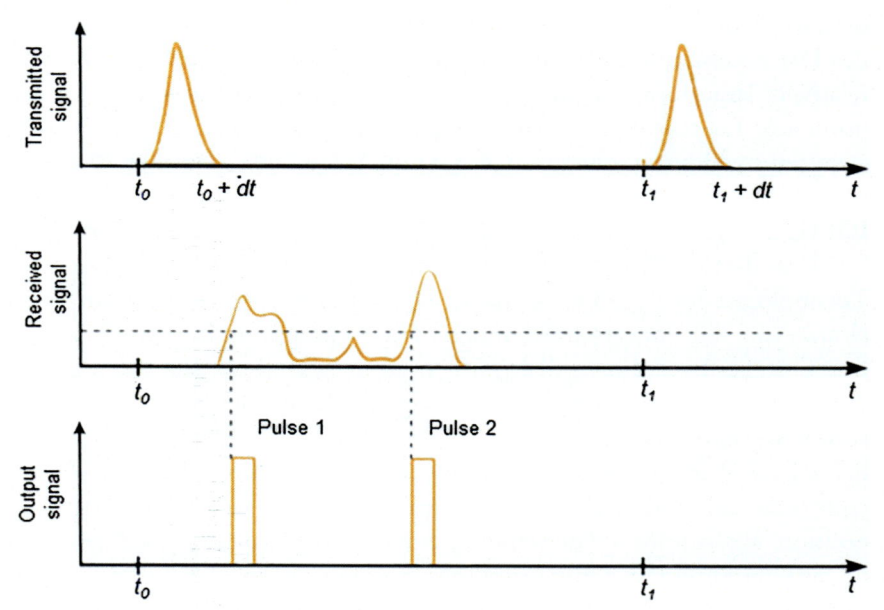

Figure 4.5 Simplified pulse emission (*above*) and corresponding received signal (*middle*). Two significant peaks are detected with the threshold method (*middle* and *bottom*). Two echoes are generated for this pulse instead of four. *(From Mallet, C., & Bretar, F. (2009). Full-waveform topographic lidar: State-of-the-art. ISPRS Journal of Photogrammetry and Remote Sensing, 64(1), 1—16.)*

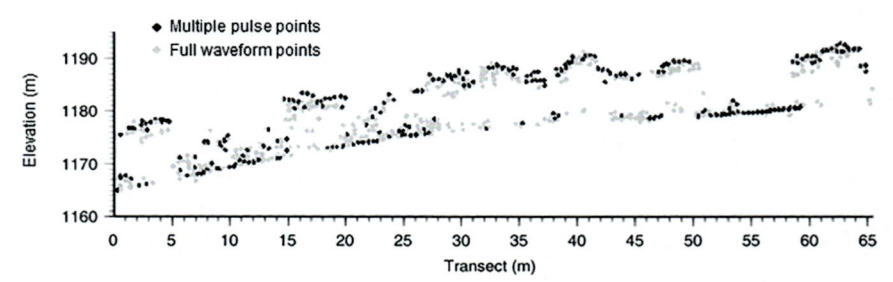

Figure 4.6 Comparison of data over a vegetated area acquired using full-waveform and multiple-pulse LiDAR. *(From Mallet, C., & Bretar, F. (2009). Full-waveform topographic lidar: State-of-the-art. ISPRS Journal of Photogrammetry and Remote Sensing, 64(1), 1—16.)*

of tens of meters, such as the Global Ecosystem Dynamics Investigation (GEDI) system and the Geoscience Laser Altimeter System (GLAS). The relatively large footprint size and sparse footprint distribution make them unsuitable for conversion to point clouds. Small-footprint data are usually acquired by airborne or terrestrial systems. Since 2004, commercial small-footprint waveform LiDAR hardware is emerging from companies like RIEGL, TopoSys (TopoSys GmbH, Biberach an der Riß, Germany), TopEye/Blom (TopEye, Gothenburg, Sweden), Optech (Teledyne Technologies Inc, Quebec, Canada), etc. These systems record both point clouds and the energy distributions of the emitted and received beams. Many studies have shown that small-footprint waveform data can improve the density and positioning accuracy of point cloud data, as well as provide additional useful parameters such as echo amplitude and width. Accurately decomposing the waveform data is essential for further use. Many hardware companies only provide raw waveform data and do not provide decomposition algorithms. Understanding how to decompose waveform data remains an important topic in LiDAR data processing.

4.4.2 Common Full-Waveform LiDAR Data Formats

Compared with point cloud data, there are fewer common data formats for full-waveform LiDAR data, especially large-footprint LiDAR data. Most large-footprint LiDAR data are stored with a specific binary file defined by the hardware manufacturer or another common file exchange format. For example, the LiDAR data obtained by Scanning LiDAR Imager of Canopies by Echo Recovery is saved as a ".DAT" file, and GLAS and GEDI products use a hierarchical data format (HDF) file to store footprint location, return energy, and other information.

Two data formats are commonly used for small-footprint waveform data, LAS and PulseWaves. As mentioned in Section 4.1.1, LAS supports waveform data storage after version 1.3. The waveform data can be embedded in the LAS file or stored as an auxiliary file, ".wdp". PulseWaves is an open-source data format for storing georeferenced full-waveform LiDAR data released by rapidlasso GmbH in December 2011. Pulse-Waves is now integrated into RIEGL's RiPROCESS as the new standard for delivering full-waveform data. Moreover, NEON (National Ecological Observatory Network) also uses this format to share the Slant Range

Waveform LiDAR data product. PulseWaves consists of two types of files—".plz" pulse files and ".wvz" wave files. The pulse files store the information of emitted laser pulses, such as a georeferenced origin and target point. Wave files contain the actual samples of the outgoing and returning waveform shapes for the digitized sections of the emitted and received waveforms.

To be consistent with Section 4.1.1, we focus on introducing the waveform data of LAS files in this section. The common information of waveform data is stored in the public header area and defined variable-length records. In the public header area, the "Start of Waveform Data Packet Record" records the number of bytes from the beginning of the LAS file to the first byte of the waveform data (Table 4.1). The description of waveform data is stored in defined variable-length records; the LAS file can store a maximum of 225 waveform packet descriptors (Table 4.1). These descriptors include basic information about the waveform packets used to read and convert raw waveform records, such as bits per sample, number of samples, temporal sample spacing, and digitizer gain and offset (Table 4.8).

As mentioned in Section 4.1.1, point formats 4, 5, 9, and 10 can store waveform packet information. Compared with other point formats, these formats add four additional fields to read the waveform's description, location, and size, as described below:

(1) Wave Packet Descriptor Index: This index relates to the specific user-defined records that describe the waveform information at that point. An index value of 0 indicates that the laser point does not have the corresponding waveform data.

Table 4.8 Waveform packet descriptor field format.

Item	Format	Size
Bits per sample	Unsigned char	1 byte
Waveform compression type	Unsigned char	1 byte
Number of samples	Unsigned long	4 bytes
Temporal sample spacing	Unsigned long	4 bytes
Digitizer gain	Double	8 bits
Digitizer offset	Double	8 bits

(2) Byte Offset to Waveform Packet Data: This value represents the starting position of the LiDAR point waveform packet. If the waveform is stored in the LAS file, the position of the waveform packet in the file should add the "Start of Waveform Data Packet Record" from the common header area. Otherwise, this value is the position of the waveform data offset when the waveform is stored in the auxiliary file.

(3) Waveform Packet Size in Bytes: This value represents the size of the waveform packet associated with this return.

(4) Return Point Location: This field represents the picosecond value from the time the laser is emitted (the starting digitization value) to the time the reflected pulse is received.

4.4.3 Reading Full-Waveform Data

Currently, Python's laspy library does not support reading full-waveform data. Still, it can be used to read basic information in the LAS file and then use Python to directly read the raw waveform records from the binary file. There are four steps to read a specific point's waveform data:

(1) Read the variable-length records to obtain the descriptor of the waveform packet.

(2) Read the common header block to get the "Wave Data Packet Recording Start Position" if waveform data are stored in the LAS file. This step can be omitted if waveform data are stored in the auxiliary file.

(3) Read the point record to obtain the information on the waveform data for each point. If we only focus on the waveform amplitude, only "Waveform Data Offset", "Wave Packet Descriptor Index", and "Wave Data Packet Size" are mandatory. With this information, we can locate the position of the waveform data packet in the file and obtain an accurate data length.

(4) Use the digitizer gain and offset stored in the waveform packet descriptor to convert the raw digitized value to a voltage with this formula: VOLTS = OFFSET + GAIN * Raw_Waveform_Amplitude.

The following Python code is an example of reading a specific point's waveform data from a LAS file. Note that this example code used laspy 1.0—this is slightly different from reading LAS files with laspy 2.0.

```python
# The following code is for reading waveform data from a LAS file.

#Coding: utf-8

#Import laspy and matplotlib packages

#Notice this code used laspy version 1.0

import numpy as np

import laspy

import matplotlib.pyplot as plt

# example files

pfix = '100429_152240_2535pt_UTM'

fn_las = pfix + '.las'

fn_wdp = pfix + '.wdp'

# read entire files

fp = laspy.file.File(fn_las)

# read waveform packet descriptors (WPD) from

# variable length records (VLR)

wpds = []

vlrs = fp.header.vlrs

for vlr in vlrs:

    if 99 < vlr.record_id and vlr.record_id < 355:

        wpds.append(vlr.parsed_body)

wpds = np.array(wpds)

with open(fn_wdp, 'rb') as bf:

    #in this example, we only load the waveform data of Point 15

    i = 14

    print(i)

    #if wave_packet_desc_index equel 0, there is no waveform data with this point.

    if fp.wave_packet_desc_index[i] == 0:

        print("this point do not contain waveform data")
```

```python
else:
    #read the descriptor from the VARIABLE LENGTH RECORDS
    sample_bits,wf_type,n_samples,sample_freq, \
        gain, offset = wpds[fp.wave_packet_desc_index[i]-1]
    #read the offset of the waveform data
    #Notice that if waveform data is stored in the LAS file, the offset should add the waveform data
starting location from the header area
    off = fp.byte_offset_to_waveform_data[i]
    bf.seek(off)
    # read the size of point 15's waveform packet
    wf_packet_size = fp.waveform_packet_size[i]
    b = bf.read(wf_packet_size)
    #check the bytes of sample and read the waveform packet
    nbits = int(sample_bits/8)
    if nbits == 1:
        a = np.frombuffer(b, dtype = np.uint8, count = int(n_samples))
    elif nbits == 2:
        a = np.frombuffer(b, dtype = np.uint16, count = int(n_samples))
    elif nbits == 4:
        a = np.frombuffer(b, dtype = np.uint32, count = int(n_samples))
    elif nbits == 8:
        a = np.frombuffer(b, dtype = np.uint64, count = int(n_samples))
    else:
        print("error in calculating nbits the nbits")
        exit(-1)
    #Convert the digital value into voltage
    wave_form_volitage = a * gain + offset
    # plot the waveform data
    fig, ax = plt.subplots()
    ax.plot(np.arange(n_samples) * sample_freq, wave_form_volitage)
    plt.show()
```

4.5 Chapter Summary

Discrete point cloud and full-waveform are the two major LiDAR data formats. Point cloud is the most commonly used data format for current LiDAR systems and is usually stored in LAS or ASCII formats. The ASCII format is flexible and easy to read and write, making it useful and intuitive for novices who are learning and exploring point cloud data. The LAS format is more powerful than ASCII by virtue of its faster read speeds and richer information content. However, a certain foundation in programming is required to process LAS data conveniently. The Python language, which is open-source and incorporates many extension libraries, is quite suitable for the primary analysis and processing of LiDAR data. Full-waveform LiDAR data can be divided into two categories, large footprint and small footprint. No common formats are available for storing large-footprint data, whereas small-footprint full-waveform data are often stored as LAS or PulseWaves files. Both formats have good rigor and openness, although reading full-waveform data is more complex than reading point cloud data.

References

Bentley, J. L. (1975). Multidimensional binary search trees used for associative searching. *Communications of the ACM, 18*(9), 509–517.

Lefsky, M. A., Cohen, W. B., Parker, G. G., & Harding, D. J. (2002). Lidar remote sensing for ecosystem studies: Lidar, an emerging remote sensing technology that directly measures the three-dimensional distribution of plant canopies, can accurately estimate vegetation structural attributes and should be of particular interest to forest, landscape, and global ecologists. *BioScience, 52*(1), 19–30.

Liang, X., Zhang, J., Li, H., & Yan, P. (2005). Representations of LIDAR data in different applications. *Remote Sensing Information, 6*, 60–64 (in Chinese).

Lu, X., Guo, Q., Li, W., & Flanagan, J. (2014). A bottom-up approach to segment individual deciduous trees using leaf-off lidar point cloud data. *Isprs Journal of Photogrammetry and Remote Sensing, 94*, 1–12.

Mallet, C., & Bretar, F. (2009). Full-waveform topographic lidar: State-of-the-art. *ISPRS Journal of Photogrammetry and Remote Sensing, 64*(1), 1–16.

Mosa, A. S. M., Schön, B., Bertolotto, M., & Laefer, D. F. (2012). Evaluating the benefits of octree-based indexing for LiDAR data. *Photogrammetric Engineering & Remote Sensing, 78*(9), 927–934.

Otepka, J., Ghuffar, S., Waldhauser, C., Hochreiter, R., & Pfeifer, N. (2013). Georeferenced point clouds: A survey of features and point cloud management. *ISPRS International Journal of Geo-Information, 2*(4), 1038–1065.

Persson, Å., Holmgren, J., Söderman, U., & Olsson, H. (2004). Tree species classification of individual trees in Sweden by combining high resolution laser data with high resolution near-infrared digital images. *International Archives of Photogrammetry, Remote Sensing and Spatial Information Sciences, 36*(8), 204–207.

Popescu, S. C., & Wynne, R. H. (2004). Seeing the trees in the forest. *Photogrammetric Engineering & Remote Sensing, 70*(5), 589–604.

CHAPTER 5

Data Preprocessing and Feature Extraction

Contents

Various light detection and ranging (LiDAR) systems (e.g., terrestrial, mobile, drone, airborne, and spaceborne LiDAR) are now available, and their data, usually point clouds or waveforms, have distinct characteristics. LiDAR data processing may involve a complex workflow, particularly when datasets acquired by different systems must be used together. This

LiDAR Principles, Processing and Applications in Forest Ecology
ISBN 978-0-12-823894-3
https://doi.org/10.1016/B978-0-12-823894-3.00005-0

chapter introduces the preprocessing and feature extraction methods of point clouds. Because point clouds are the more commonly used data format, here we will use point clouds rather than waveforms as an illustration unless otherwise specified. Preprocessing aims to generate ready-for-use LiDAR data, usually including resolving, registration, and denoising steps. Specifically, resolving aims to convert raw data into points with coordinates in a specific geographic coordinate system; registration aims to transform point cloud coordinates from multiple scans (or flight lines) or multiple LiDAR systems into the same coordinate system; and denoising aims to remove abnormal or noisy points caused by the scanner or the environment. Feature extraction aims to identify global or local features, such as color, surface normal, and curvature, that can reflect point cloud characteristics. These features can then be used in the classification step, which aims to segment and divide point cloud data into different classes to obtain the required information. The objective and focus of feature extraction and classification may vary by specific application. In forest applications, for instance, we may focus on distinguishing tree points from ground points and further separating tree points into branch and leaf points to extract forest structural attributes; in urban applications, planners may be more interested in finding features to separate buildings, trees, and roads. We use forest applications as examples to illustrate the abovementioned preprocessing and feature extraction steps in the sections that follow.

5.1 Point Cloud Resolving

Point cloud resolving is an essential step of light detection and ranging (LiDAR) data preprocessing, which involves large numbers of complex calculations, such as resolving LiDAR range information, deriving position and orientation system data, and performing coordinate transformation. This section introduces four resolving methods: geometric, simultaneous localization and mapping (SLAM)-based, global navigation satellite system-inertial measurement unit (GNSS-IMU) navigation, and integrated.

5.1.1 Geometric Approach

Resolving point clouds using the geometric approach involves calculating the relative spatial coordinates of a target point P by using its distance S to the scanner, the horizontal scanning angle α, and the vertical scanning angle β of a laser pulse measured synchronously by the control encoder (Fig. 5.1). Its spherical coordinates (S, α, β) can be transformed into the Cartesian coordinate system (X_P, Y_P, Z_P), where the X-axis and Y-axis are perpendicular in the horizontal plane, and the Z-axis is perpendicular to the XOY

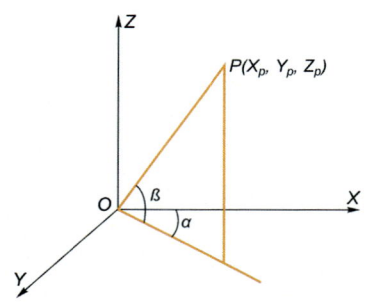

Figure 5.1 Principle of the geometric approach for resolving the coordinate of a target point P. α and β are, respectively, the horizontal and vertical scanning angles of the laser pulse.

plane. The Cartesian coordinate system is a left–handed coordinate system, and the transformation can be expressed by Eq. (5.1):

$$\begin{cases} X_P = S\cos\beta\cos\alpha \\ Y_P = S\cos\beta\sin\alpha \\ Z_P = S\sin\beta \end{cases} \tag{5.1}$$

If the GNSS information of the scanner is recorded, the absolute co-ordinates of each target point can then be obtained through the above equation. The geometric approach is a straightforward method for resolving point clouds and is commonly used in terrestrial LiDAR systems.

5.1.2 Simultaneous Localization and Mapping Approach

Moving LiDAR systems (e.g., backpack, mobile, unmanned aerial vehicle (UAV), and airborne LiDAR) usually require resolving the point clouds while the platform is moving. Therefore, obtaining real-time trajectory and attitude information on the platform is critical to the data resolving process. This step can usually be achieved through the SLAM algorithm and GNSS-IMU navigation approach. SLAM is often used for positioning and navigating a robot or autonomous vehicle and has the advantage of not requiring GNSS positioning information (Durrant-Whyte & Bailey, 2006). SLAM can deal with the problem of constructing or updating a map of an unknown environment while simultaneously keeping track of an agent's location within it. The positioning accuracy depends on unbiased maps, which require high-precision sensors during the survey. As the sensor operation continues, errors in distance and direction caused by sensor movement will accumulate, resulting in large positioning errors. LiDAR is one of the most reliable operational sensors owing to its accurate distance and direction measurements. Therefore, the combination of SLAM and LiDAR provides a unique way to obtain three-dimensional (3-D) information about a GNSS-denial environment (e.g., in forests with dense vegetation canopies).

SLAM uses a series of LiDAR observations O_t (including distance and direction) in a discrete-time domain (with a time interval t) to calibrate the sensor trajectory x_t and update the map surveying m_t. Thus, it can be represented as a problem of maximizing the probability $P(m_t, x_t | O_{(1:t)})$. SLAM includes four steps: feature extraction, data association, state estimation, and map updating (Fig. 5.2). In feature extraction, outstanding landmarks (also called feature points) in an observed environment are identified to retrieve the sensor location. Many algorithms can be used to identify these landmarks in LiDAR point clouds. One of the most commonly used algorithms is the random sample consensus (RANSAC), which extracts feature points by searching for extreme points in the point cloud. Therefore, it is unsuitable for feature extraction of smooth surfaces because it requires an obvious difference between the distance measurements of two adjacent laser beams. RANSAC randomly selects a subset of a LiDAR point cloud, fits a line using the least squares approach, and finds points next to the line. Doing this iteratively, the algorithm can identify linear features in the point cloud and extract feature points.

Data association indicates matching landmarks extracted from different times and is usually accomplished by finding the feature with the shortest distance to the landmark. After feature extraction, a newly extracted landmark is tried to be associated with those already existing in the feature set. If the distance between a new landmark and its closest existing landmark already in the feature set satisfies a specific distance criterion in the subsequent verification, these two landmarks are associated; otherwise, the new landmark is added to the feature set. The subsequent verification is usually completed using the extended Kalman filter (EKF), and the Euclidean distance to the closest landmark can be adopted as a simple

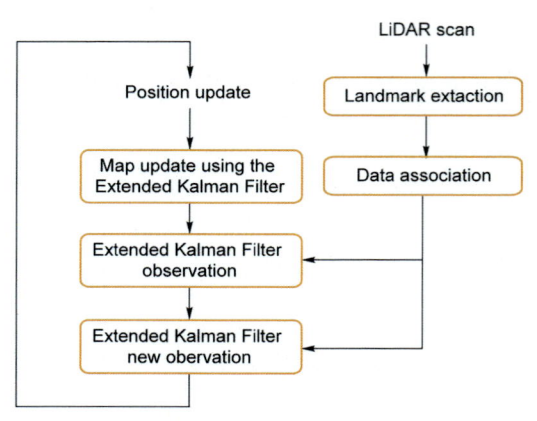

Figure 5.2 The workflow of the simultaneous localization and mapping algorithm.

criterion. Other distance measures, such as the Mahalanobis distance, may perform better, but the computational complexity may also greatly increase.

State estimation is used to obtain the optimal trajectory information of the LiDAR sensor based on the matched landmarks at different times and the trajectory information recorded by the IMU. Using the space resection principle, the spatial position of the sensor at time t can be inferred from the landmarks retained from the feature extraction and data association steps. Time-series state estimation is applied to optimize the trajectory information of the sensor and can be written as

$$P(x_t|o_{1:t}, m_t) = \sum_{m_{t-1}} P(o_t|x_t, m_t) \sum_{x_{t-1}} \frac{P(x_{t-1}|m_t, o_{1:t-1})}{Z} \tag{5.2}$$

At present, the most commonly used optimization method for state estimation is the EKF method, which can predict the optimal position and velocity of an object by observing its position from a limited set of noise-contaminated data. The basic assumption of EKF is that the position and velocity errors of an object are random and follow a Gaussian distribution. The state information of the sensor in the EKF is expressed as a vector named system state, which includes its position, orientation, and the positions of all observed landmarks in the environment. The optimization of this discrete-time system can be solved recursively through linear minimum-variance unbiased estimation. Compared with the original Kalman filter algorithm, EKF performs linearization estimation based on the first-order Taylor expansion of the nonlinear Gaussian distribution and usually performs better for nonlinear data.

Map updating is used to recalculate the positioning information of LiDAR points according to the optimized trajectory and velocity information derived from the state estimation step. To achieve an optimized overall estimate for all measurements, this process is repeated each time when a new observation is added, and *a posterior* verification is applied to the matched landmarks through data association. If the updated associated landmark satisfies the following criteria (Eq. 5.3), the association is considered successful; otherwise, the updated landmark is identified as a new landmark and added to the landmark dataset,

$$m_i^{\mathrm{T}} S_i^{-1} m_i \le \lambda \tag{5.3}$$

where m_i is the observation after updating the landmark, and S_i is the covariance matrix of the landmark.

5.1.3 Global Navigation Satellite System-Inertial Measurement Unit Navigation Approach

As mentioned, the GNSS-IMU navigation approach is the other frequently used method for resolving point clouds obtained by moving LiDAR systems, especially for UAV LiDAR and airborne LiDAR. It uses GNSS to obtain the precise 3-D spatial positions of a laser scanner. It uses IMU to measure the posture of the scanner (expressed as three angles φ, ω and κ, representing roll, pitch, and yaw, respectively). Thus, the 3-D position of each laser sampling point on the ground can be calculated according to geometric principles.

Taking airborne LiDAR as an example, the scanning principles of the airborne LiDAR system are shown in Section 2.5. Given the spatial co-ordinates of a point O_s (X_s, Y_s, Z_s) and the vector $\boldsymbol{S}$ from O_s to a point P on the ground, the coordinates (X, Y, Z) of a point P can be derived by adding vector $\boldsymbol{S}$ to point O_s:

$$
\begin{bmatrix} X \\ Y \\ Z \end{bmatrix} = \begin{bmatrix} X_s \\ Y_s \\ Z_s \end{bmatrix} + \boldsymbol{A} \begin{bmatrix} x \\ y \\ z \end{bmatrix} \tag{5.4}
$$

where $\boldsymbol{A} = \begin{bmatrix} a_1 & a_2 & a_3 \\ b_1 & b_2 & b_3 \\ c_1 & c_2 & c_3 \end{bmatrix}$ represents the transformation matrix from the

sensor coordinate system to the ground coordinate system, which is a function of the sensor posture $(\varphi, \omega, \kappa)$. This formula (Eq. 5.4) is also known as the general imaging equation, which establishes the relationship between the coordinates (x, y, z) of a point in the sensor coordinate system and its coordinates (X, Y, Z) in the ground coordinate system. According to the basic principles of photogrammetry, the relationship between the two corresponding points and the projection center of the sensor shows a collinear relationship, which can be described by the collinear equation. The imaging equation can be used to analyze errors in the estimated parameters.

For an airborne LiDAR system, point O_s is the center of the sensor, and its coordinates in the ground coordinate system are (X_s, Y_s, Z_s), which can be measured by GNSS. The coordinates of point P in the ground coordinate system are (X, Y, Z), and its coordinates in the sensor coordinate system are (x, y, z). The sensor posture $(\varphi, \omega, \kappa)$ can be measured by the high-precision IMU. Some other flight parameters need to be estimated by calibration, such as the deviation between the optical projection center of the laser rangefinder and the phase center of the GNSS antenna, the position deviation of the IMU relative to the GNSS antenna, and the divergence in the postures of the IMU and the platform.

Based on the imaging equation, seven reference coordinate systems are involved in 3-D measurements using airborne LiDAR, namely the coordinate system of laser beams, the laser scanning coordinate system, the airborne platform coordinate system, the inertial platform coordinate system, the local horizontal coordinate system (reference coordinate system of the Earth tangent plane), the local vertical coordinate system, and the geodetic coordinate system. Three-dimensional measurements using airborne LiDAR aim to complete the transformation between these seven coordinate systems and eventually obtain the 3-D coordinates of a target in the geodetic coordinate system.

If the distance between the laser emitter and the target reflecting a laser pulse is R, the coordinates of the laser footprint are $(x_{LV}, y_{LV}, z_{LV})^{T}$ in the laser beam coordinate system, which is expressed as

$$
\begin{bmatrix} x_{LV} \\ y_{LV} \\ z_{LV} \end{bmatrix} = \begin{bmatrix} 0 \\ 0 \\ R \end{bmatrix} \tag{5.5}
$$

The relationship between the laser beam coordinate system and the laser scanning coordinate system is shown in Fig. 2.32, where θ is the instantaneous scanning angle between the z_{LV} and z_{L} axes. According to Fig. 2.32, the coordinates of the laser footprint in the laser scanning coordinate system $(x_{L}, y_{L}, z_{L})^{T}$ can be expressed as

$$
\begin{bmatrix} x_{L} \\ y_{L} \\ z_{L} \end{bmatrix} = \begin{bmatrix} 1 & 0 & 0 \\ 0 & \cos\theta_i & -\sin\theta_i \\ 0 & \sin\theta_i & \cos\theta_i \end{bmatrix} \cdot \begin{bmatrix} 0 \\ 0 \\ R \end{bmatrix} \tag{5.6}
$$

where θ_i is the instantaneous scanning angle of the ith laser beam, which is generally obtained by dividing the system scanning angle θ into N segments with N for the number of laser beams in a continuous scanning line using Eq. (5.7):

$$
\theta_i = \frac{\theta}{2} - i \cdot \frac{\theta}{N-1} \tag{5.7}
$$

The laser scanner and IMU are installed on an airborne platform. Ideally, the corresponding axes of the laser scanning, internal platform, and airborne platform coordinate systems should be parallel. However, unavoidable divergences caused by installation errors exist among the three coordinate systems. In the coordinate transformation, the coordinates of a laser footprint in the laser scanning coordinate system can be directly converted to its coordinates in the inertial platform coordinate system. Assuming that the angles between the x, y, z axes of the laser scanning and inertial platform coordinate systems are α, β, γ, and the shift between their origins is $(\Delta x_L, \Delta y_L, \Delta z_L)^T$, the coordinates $(x_I, y_I, z_I)^T$ of the laser footprint in the inertial platform coordinate system can be expressed as

$$\begin{bmatrix} x_I \\ y_I \\ z_I \end{bmatrix} = A_I \cdot \begin{bmatrix} x_L \\ y_L \\ z_L \end{bmatrix} + \begin{bmatrix} \Delta x_L \\ \Delta y_L \\ \Delta z_L \end{bmatrix} \qquad (5.8)$$

where A_I is a transformation matrix with $A_I = A(\alpha) \cdot A(\beta) \cdot A(\gamma)$:

$$A(\alpha) = \begin{bmatrix} 1 & 0 & 0 \\ 0 & \cos\alpha & -\sin\alpha \\ 0 & \sin\alpha & \cos\alpha \end{bmatrix}$$

$$A(\beta) = \begin{bmatrix} \cos\beta & 0 & \sin\beta \\ 0 & 1 & 0 \\ -\sin\beta & 0 & \cos\beta \end{bmatrix}$$

$$A(\gamma) = \begin{bmatrix} \cos\gamma & -\sin\gamma & 0 \\ \sin\gamma & \cos\gamma & 0 \\ 0 & 0 & 1 \end{bmatrix}$$

IMU records the flight attitude and posture $(\varphi, \omega, \kappa)$ of an aircraft when a laser beam is emitted. Assuming that the shift of origins between the inertial platform coordinate system and the local horizontal coordinate system is $(\Delta x_I, \Delta y_I, \Delta z_I)^T$, the coordinates of the laser footprint in the local horizontal coordinate system $(x_{GL}, y_{GL}, z_{GL})^T$ can be expressed as

$$\begin{bmatrix} x_{GL} \\ y_{GL} \\ z_{GL} \end{bmatrix} = A_{GL} \cdot \begin{bmatrix} x_I \\ y_I \\ z_I \end{bmatrix} - A_{GL} \cdot \begin{bmatrix} \Delta x_I \\ \Delta y_I \\ \Delta z_I \end{bmatrix} \qquad (5.9)$$

where A_{GL} is a transformation matrix, and $A_{GL} = A(\varphi) \cdot A(\omega) \cdot A(\kappa)$,

$$A(\varphi) = \begin{bmatrix} 1 & 0 & 0 \\ 0 & \cos\varphi & -\sin\varphi \\ 0 & \sin\varphi & \cos\varphi \end{bmatrix}$$

$$A(\omega) = \begin{bmatrix} \cos\omega & 0 & \sin\omega \\ 0 & 1 & 0 \\ -\sin\omega & 0 & \cos\omega \end{bmatrix}$$

$$A(\kappa) = \begin{bmatrix} \cos\kappa & -\sin\kappa & 0 \\ \sin\kappa & \cos\kappa & 0 \\ 0 & 0 & 1 \end{bmatrix}$$

After obtaining the coordinates of the laser footprint in the local horizontal coordinate system, its coordinates in the geodetic coordinate system $(x_{\mathrm{WGS}}, y_{\mathrm{WGS}}, z_{\mathrm{WGS}})^{\mathrm{T}}$ can be obtained by applying rotation and shifting using Eq. (5.10).

$$\begin{bmatrix} x_{\mathrm{WGS}} \\ y_{\mathrm{WGS}} \\ z_{\mathrm{WGS}} \end{bmatrix} = A_{\mathrm{WGS}} \cdot A_{\mathrm{GH}} \cdot \begin{bmatrix} x_{\mathrm{GL}} \\ y_{\mathrm{GL}} \\ z_{\mathrm{GL}} \end{bmatrix} + \begin{bmatrix} x_{\mathrm{GPS}} \\ y_{\mathrm{GPS}} \\ z_{\mathrm{GPS}} \end{bmatrix} \tag{5.10}$$

where A_{WGS} is the rotation matrix about the longitude and latitude of the footprint, and A_{GH} is a transformation matrix to address the vertical deviation between the local horizontal and vertical coordinate systems.

In summary, the 3-D geodetic coordinates of the target point can be expressed using Eq. (5.11):

$$\begin{bmatrix} x_{\mathrm{WGS}} \\ y_{\mathrm{WGS}} \\ z_{\mathrm{WGS}} \end{bmatrix} = A_{\mathrm{WGS}} \cdot A_{\mathrm{GH}} \cdot A_{\mathrm{GL}} \cdot \left(A_{\mathrm{I}} \cdot A_{\mathrm{L}} \cdot \begin{bmatrix} 0 \\ 0 \\ R \end{bmatrix} + \begin{bmatrix} \Delta x_{\mathrm{L}} \\ \Delta y_{\mathrm{L}} \\ \Delta z_{\mathrm{L}} \end{bmatrix} - \begin{bmatrix} \Delta x_{\mathrm{I}} \\ \Delta y_{\mathrm{I}} \\ \Delta z_{\mathrm{I}} \end{bmatrix} \right)$$

$$+ \begin{bmatrix} x_{\mathrm{GPS}} \\ y_{\mathrm{GPS}} \\ z_{\mathrm{GPS}} \end{bmatrix} \tag{5.11}$$

where A_L is a transformation matrix related to the instantaneous scanning angle:

$$A_L = \begin{bmatrix} 1 & 0 & 0 \\ 0 & \cos\theta_i & -\sin\theta_i \\ 0 & \sin\theta_i & \cos\theta_i \end{bmatrix}$$

5.1.4 Integrated Approach

In complex urban and forest environments, the GNSS signal can be intermittently lost owing to skyscrapers or tree canopies. Thus, the positioning accuracy of GNSS is severely degraded, and the posture of the LiDAR sensor must be determined by integrating the GNSS, inertial navigation system (INS), and SLAM. This integrated approach is often used in mobile and backpack LiDAR scanning systems. The laser scanner mounted on a mobile or backpack platform continuously records the traveled distance of a laser pulse and its index in the trajectory. The direction of a laser pulse can be determined from its index value, and the 3-D spatial coordinates of the corresponding point are calculated by combining the calibration parameters of the mobile system and the integrated position and posture information.

Using mobile LiDAR as an example, given the vehicle position (X_G, Y_G, Z_G) obtained by integrating GNSS/INS/SLAM, the posture of the laser scanner obtained by the IMU ($\left(\theta_y, \theta_p, \theta_r\right)$ in Fig. 5.3), and the relative coordinates (X_l, Y_l, Z_l) of the target point P with regard to the laser scanner, the coordinates of P can be calculated using Eq. (5.12).

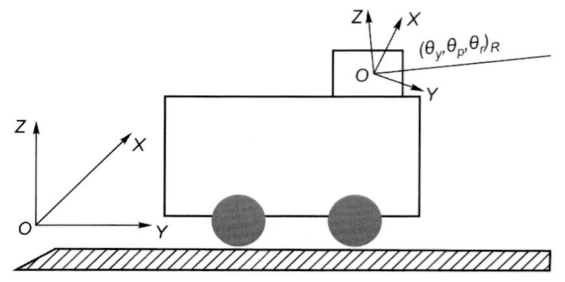

Figure 5.3 Schematic diagram of coordinate transformation of the mobile light detection and ranging (LiDAR) system. The symbols $\left(\theta_y, \theta_p, \theta_r\right)$ are the angles indicating the posture of the laser scanner with regard to the axes of the geodetic coordinate system.

$$\begin{bmatrix} X \\ Y \\ Z \end{bmatrix} = \begin{bmatrix} a_1 & a_2 & a_3 \\ a_4 & a_5 & a_6 \\ a_7 & a_8 & a_9 \end{bmatrix} \begin{bmatrix} X_1 \\ Y_1 \\ Z_1 \end{bmatrix} + \begin{bmatrix} X_G \\ Y_G \\ Z_G \end{bmatrix} \tag{5.12}$$

$$\begin{cases} a_1 = \cos\theta_y \cos\theta_r + \sin\theta_p \sin\theta_r \\ a_2 = \sin\theta_y \cos\theta_r - \cos\theta_y \sin\theta_p \sin\theta_r \\ a_3 = \cos\theta_p \sin\theta_r \\ a_4 = -\cos\theta_p \sin\theta_y \\ a_5 = \cos\theta_y \cos\theta_p \\ a_6 = \sin\theta_p \\ a_7 = \sin\theta_y \sin\theta_p \cos\theta_r - \cos\theta_y \sin\theta_r \\ a_8 = -\cos\theta_y \sin\theta_p \cos\theta_r - \sin\theta_y \sin\theta_r \\ a_9 = \cos\theta_p \cos\theta_r \end{cases}$$

From this equation, we can see that one of the key steps in calculating the absolute coordinates of point cloud data is to obtain the real-time posture of the LiDAR sensor and the real-time travel trajectory of the vehicle. The real-time posture of the sensor can be obtained from measurements using the IMU. The real-time trajectory of the vehicle can be derived by integrating GNSS, INS, and SLAM. The EKF and graph optimization are often used to integrate GNSS, INS, and SLAM measurements. Qian et al. (2016) utilized information such as the heading angle and velocity extracted from a GNSS/INS to improve the positioning accuracy of SLAM in a forest environment. Li et al. (2020) incorporated the feedback from mapping into positioning in SLAM and significantly improved the smoothness of the estimated trajectory with six degrees of freedom in a complex urban environment.

In the above calculation, we assume that the centers of different sensors, such as the laser scanner, IMU, or GNSS, are in the same spatial position and the same plane. However, there is a certain spatial shift between different sensors owing to their physical sizes (Fig. 5.4). If this is not systematically corrected, a substantial mismatch and bias may occur in the obtained point cloud data (Fig. 5.5). To address this problem, divergences in the positions and postures between the LiDAR scanner and IMU, the LiDAR scanner and GNSS, and the GNSS and IMU are estimated during data processing. They are further calibrated and verified using the static plane-based functional model (SPFM) after completing the assembly. The

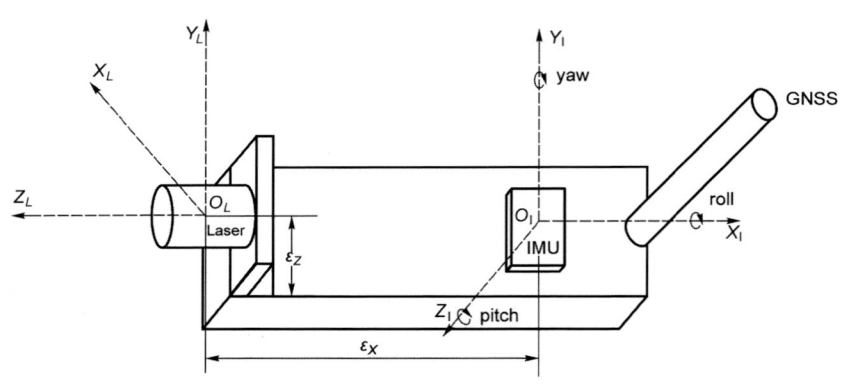

Figure 5.4 Schematic diagram of position deviation ε between different components of a mobile LiDAR system. ε_Z is the vertical position deviation, and ε_X is the horizontal position deviation. *GNSS*, global navigation satellite system; *IMU*, inertial measurement unit.

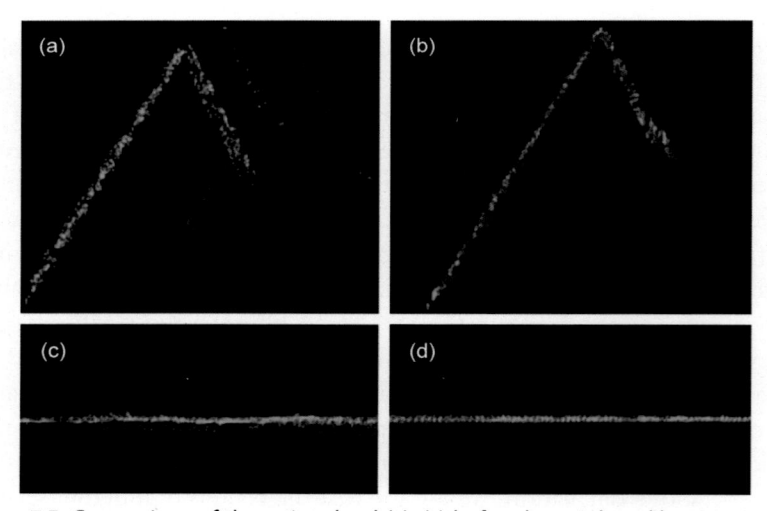

Figure 5.5 Comparison of the point cloud (a), (c) before boresight calibration and (b), (d) after boresight calibration.

SPFM assumes that the LiDAR points on the same static plane should also be on the same plane mathematically:

$$\left\langle \overrightarrow{g_k}, \begin{bmatrix} \overrightarrow{r} \\ l \end{bmatrix} \right\rangle = 0 \tag{5.13}$$

where $\overrightarrow{g}_k = [g_1, g_2, g_3, g_4]^{\mathrm{T}}$ is the static plane; $\overrightarrow{r}$ are the geographic coordinates of a point that should be on the plane; l are its coordinates in the local coordinate system of that scan. The link between l and $\vec{r}$ can be expressed as

$$\vec{r} = \boldsymbol{R}(w, p, k)\, \vec{l} + \boldsymbol{t} \tag{5.14}$$

where $\boldsymbol{R}(w, p, k)$ and $\boldsymbol{t}$ are the rotation matrix and shift vector between the local coordinate system of a scan and the geographic coordinate system. By conducting multiple scans at different locations, the same static plane is scanned multiple times, and the required parameters can be calibrated and verified using the above two equations. If the error is too large, we can apply the least squares method to further optimize the parameters.

At present, some commercial software can resolve point cloud data acquired from various platforms. For example, the LiGeoreference software provides data–resolving functions for various LiDAR systems and can also resolve colored point clouds in combination with image data (Fig. 5.6). In addition, such commercial software generally supports user–defined data acquisition and resolving. It provides an interface to third–party software, which offers great convenience to users and reduces the difficulty of point cloud resolving.

5.2 Point Cloud Registration

Point cloud registration is the process of aligning two or more 3-D point clouds covering areas with overlaps. The point cloud registration process

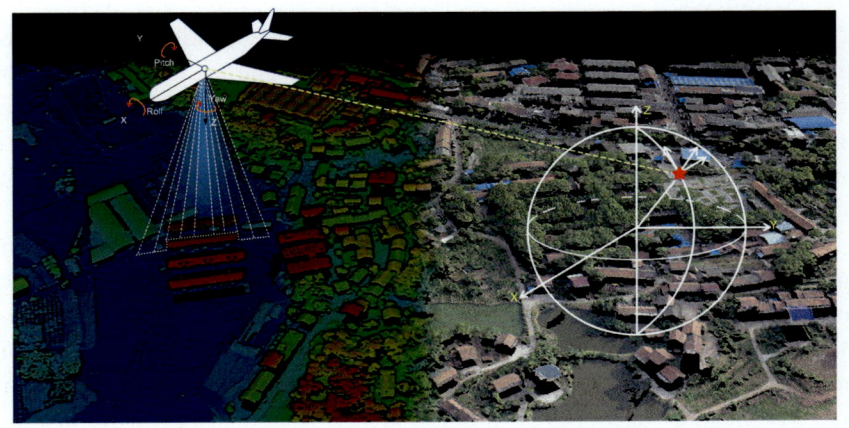

Figure 5.6 An example of commercial software (LiGeoreference) for resolving point cloud data acquired using different LiDAR systems.

includes three steps, which are preprocessing, registration, and alignment and stitching. Specific registration methods for point clouds acquired by different LiDAR systems may vary due to the distinct characteristics of those systems. In this section, we focus on the point cloud registration methods for multiscan terrestrial LiDAR, mobile LiDAR, airborne LiDAR, and multiplatform LiDAR data.

5.2.1 Multiscan Terrestrial LiDAR Data Registration

Multiple scans are often preferred in terrestrial laser scanning to acquire complete data. Point cloud data from different scanning locations have different coordinate systems and need to be transformed into the same coordinate system using the following transformation matrix:

$$M = \begin{pmatrix} r_{11} & r_{12} & r_{13} & t_1 \\ r_{21} & r_{22} & r_{23} & t_2 \\ r_{31} & r_{32} & r_{33} & t_3 \\ 0 & 0 & 0 & 1 \end{pmatrix} \tag{5.15}$$

where $r_{11}-r_{33}$ are parameters related to rotation and t_1-t_3 are parameters related to translation. Point cloud registration determines the transformation matrix between different coordinate systems. Coarse registration and fine registration are two of the steps in point cloud registration. The former makes the point clouds from different scans as close as possible using rough methods such as manual registration, geometric feature-based registration, and RANSAC.

Manual registration is often the simplest coarse registration method. It is accomplished by manually selecting tie points. The basic principles for selecting tie points are that (1) they should be evenly distributed over the whole scanning area; (2) they should be from obvious and stable features such as road intersections and building vertices; and (3) at least three pairs of tie points should be selected.

Registration using geometric features calculates the transformation matrix between point clouds from different scans according to local geometric features, such as curvature, surface normal, and curvature change. Bae (2006) proposed an algorithm based on geometric features and a neighborhood search. First, the normal vector and curvature change for two point cloud datasets, C_1 and C_2, are calculated using the K-nearest neighborhood. Second, the points in C_1 with a curvature change greater than a given threshold are selected as the initial points, and two thresholds for normal vector and curvature are then defined to find the corresponding points in C_2. Third, the initial transformation matrix is calculated for these corresponding points. This process is repeated until the registration error is lower than a specific threshold. Guan et al. (2020) proposed a marker–free registration method that identified shaded areas from the raw point cloud of a single terrestrial laser scanner (TLS) scan and then adopted them as key features to automatically register TLS data from multiple scans (Fig. 5.7).

The RANSAC algorithm calculates the initial transformation matrix for two point cloud datasets, P and Q, to be registered using three randomly selected pairs of tie points. Then points with a distance divergence between P and Q lower than a given threshold are identified and counted. If there are enough points, the transformation is accepted. Otherwise, the process is repeated until satisfactory accuracy is achieved. Aiger et al. (2008) improved this method by randomly selecting four coplanar points from P and then

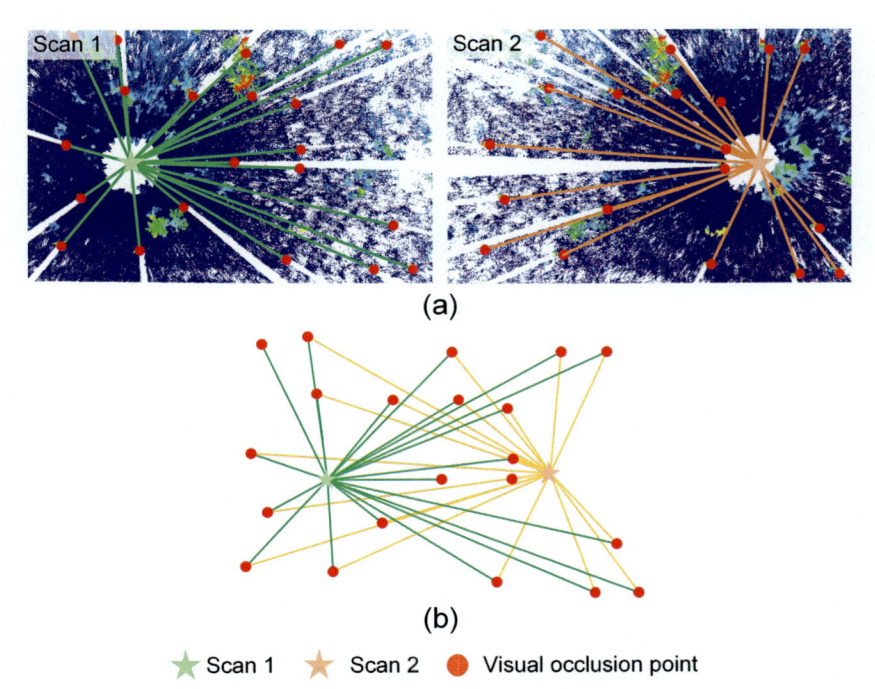

Figure 5.7 Schematic diagram of the marker-free registration method for automatically registering multiple scans of forests: (a) shaded areas and identified tree trunks in a single terrestrial laser scanner (TLS) scan using the occlusion effect of trunks; (b) the spatial relationship between trunk points from two neighboring TLS scans.

identifying the corresponding points in Q, which reduced the time for random selection of the initial points.

After the rough registration of point clouds, fine registration is needed to further improve the registration accuracy. The most representative fine registration algorithm is the iterative closest point (ICP) (Besl & McKay, 1992). The ICP iteratively calculates the transformation matrix by searching for pairs of the nearest points from two point cloud datasets, P and Q ($P \in Q$), to minimize the registration error. The basic concept is as follows. If there are N_P points in P, first for each point P_i in P, its corresponding point Q_i is found in Q. Second, rotation matrix $\boldsymbol{R}$ and translation vector $\boldsymbol{T}$ are obtained by solving the transformation matrix using the identified point pairs. Third, $\boldsymbol{R}$ and $\boldsymbol{T}$ are applied to points in P to check whether $f(\boldsymbol{R}, \boldsymbol{T})$ satisfies the following criteria. If so, the algorithm terminates; otherwise, the above steps are repeated:

$$f(\boldsymbol{R}, \boldsymbol{T}) = \min\left(\sum_{i=1}^{N_P} \left\| (\boldsymbol{R}P_i + \boldsymbol{t}) - Q_i \right\|^2 \right) \qquad (5.16)$$

The key step of the ICP is to find the corresponding points in two TLS scans and calculate the rotation matrix and translation vector. Its advantages include good registration accuracy and broad applicability. However, limitations also exist: (1) the ICP requires that the two TLS datasets partially overlap, which may not be the case in practical applications; (2) searching for all tie points in two datasets is computationally intensive; and (3) the original ICP uses the Euclidean distance between two points as the objective function in optimization, which may result in a local minimum. More studies have been conducted on improving the ICP algorithm to address these limitations (Greenspan & Yurick, 2003; Levoy & Rusinkiewicz, 2001). The main improvements include (1) removing the background and edge points in preprocessing; (2) improving the method for finding tie points and removing incorrectly identified pairs according to specific constraints; (3) sampling the original point cloud to reduce the number of points involved in the calculation and improving the efficiency of data indexing; and (4) solving the rotation matrix $\boldsymbol{R}$ and translation vector T using various methods such as unit quaternion, singular value decomposition, an orthogonal matrix, and dual quaternions.

The point cloud data of a magnolia tree are shown as an example in Fig. 5.8a and b. These data were collected from a forest farm in the Fangshan District, Beijing, China, using a RIEGL VZ-400 TLS (RIEGL

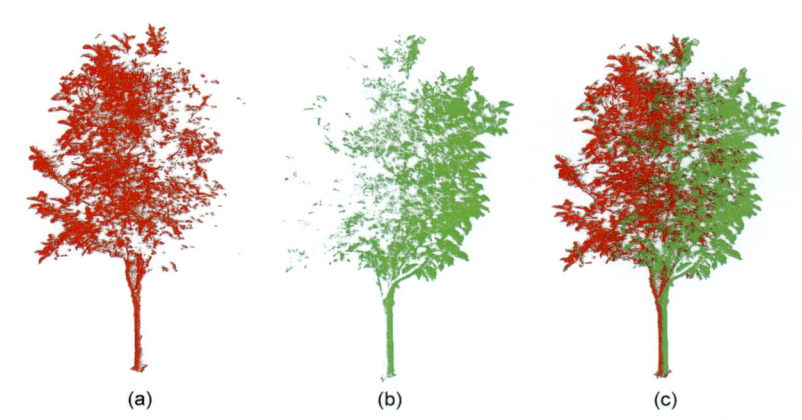

(a)   (b)   (c)

Figure 5.8 The workflow of point cloud registration. (a) point cloud from the first scan; (b) point cloud from the second scan; and (c) point cloud after registration.

Laser Measurement System GmbH, Horn, Austria) at two scanning positions and registered (including rough registration and fine registration) using the manual method, with a registration error of within 1 cm (Fig. 5.8c). Using the marker-free automatic registration method proposed by Guan et al. (2020), TLS point clouds collected in various forest types with different structures were successfully registered with an error smaller than 1 cm (Fig. 5.9). This method has high efficiency and registration accuracy while maintaining robustness under complex forest conditions.

5.2.2 Mobile LiDAR Data Registration

Mobile LiDAR systems equipped with GNSS receivers and IMU are often used to acquire point cloud data in urban and forest environments. Unfortunately, the integrated GNSS-IMU solution does not perform consistently well in complex urban or forest areas owing to the influence of multipath effects in GNSS signals caused by tall buildings or tree canopies, the loss of GNSS signals in GNSS-denial environments (e.g., tunnels), and the accumulation of positioning errors. Hence, SLAM is often integrated to improve positioning accuracy. These issues can be addressed through registration.

Although the deformation of the entire mobile LiDAR point cloud is not linear, we can assume that there is no deformation within a segment of the point cloud and then correct the nonrigid transformation between segments. To do this, mobile LiDAR scanning (MLS) data are first segmented according to the error distribution and overlap rate. For an MLS

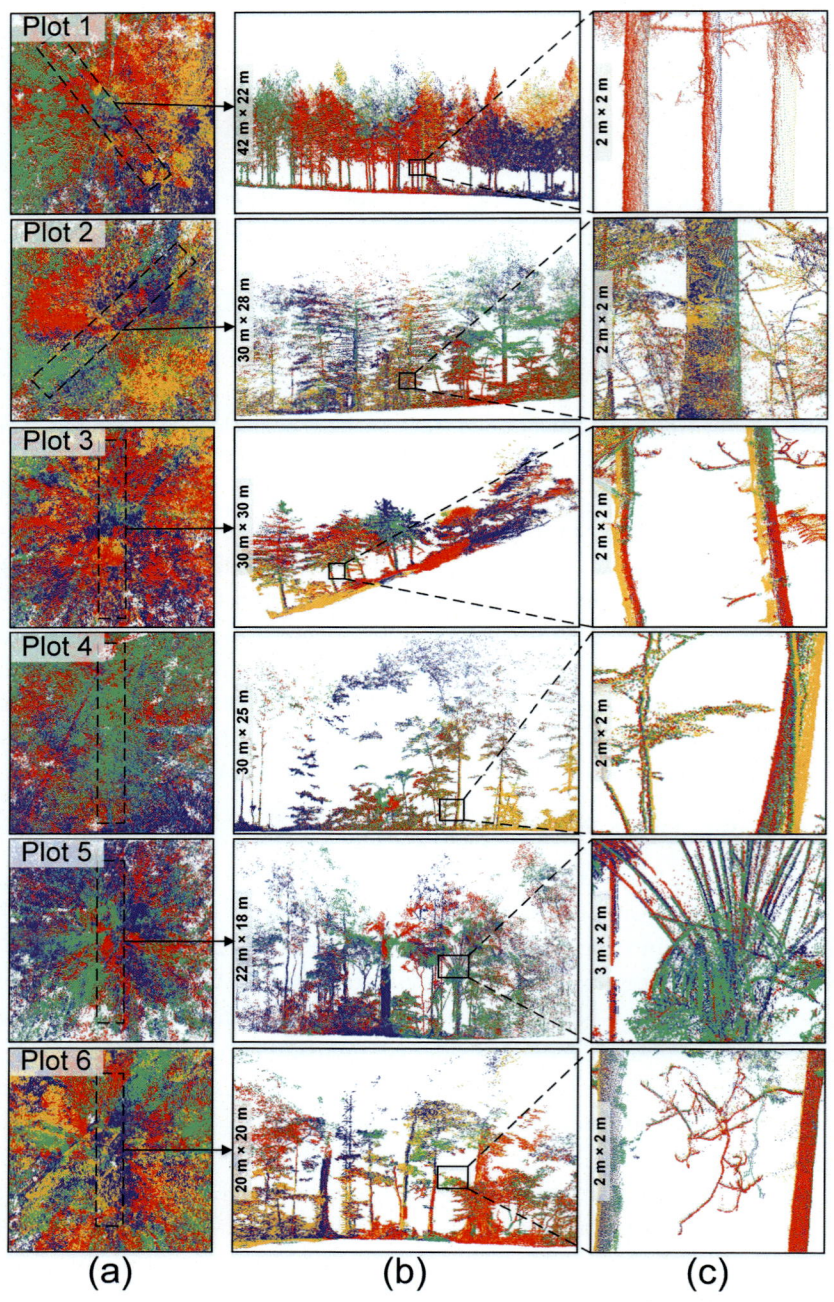

Figure 5.9 (a) Registered terrestrial laser scanner data at one planted temperate coniferous forest plot (plot 1), three temperate mixed conifer and broadleaf forest plots (plots 2—4), and two rainforest plots (plots 5—6) using the marker-free automatic registration method; (b) point cloud profile examples in each plot; and (c) enlarged point cloud segments in the profiles. The numbers presented on the left short side of (b) and (c) are the sizes of the profiles or segments.

trajectory, the bounding box of each segment can be calculated, and whether the bounding box of a segment intersects with the previous one along the trajectory is checked. If they intersect and their overlap rate exceeds a user-defined threshold, they are treated as an initial matched pair of segments.

For these initially matched pairs, the poles and high-reflectance planar targets evenly distributed in the environment are detected based on local geometric features, and then feature-based point cloud registration using ICP is performed. The initial transformation matrix is calculated by adopting the position shift in the centers of paired segments as the initial translation vector and the angle between the axes of their bounding boxes in the XY plane as the initial yaw angle. Then tie points in the paired segments are obtained by searching for the nearest point of the same object, namely the pole or planar target, within a neighborhood of 1 m. Finally, the algorithm is applied to each pair of overlapped segments to obtain the rigid transformation matrix and the root–mean–square error (RMSE) e_{rmse} of the position of features after registration:

$$e_{rmse} = \frac{1}{N} \sum_{i=1}^{N} \sqrt{\left(d_{Line2Line}\right)_i^2 + \left(d_{Plane2Plane}\right)_i^2} \tag{5.17}$$

where N is the number of geometric features; $d_{Line2Line}$ and $d_{Plane2Plane}$ are the line-to-line and plane-to-plane errors, respectively, which are calculated using Eqs. (5.18)–(5.20).

The point-to-point distance error metric $d_{P2P}\left(P_i, P_j\right)$ is

$$d_{P2P}\left(P_i, P_j\right) = \sqrt{\left(P_i - P_j\right)^2} \tag{5.18}$$

The point-to-line distance error metric $d_{P2Line}\left(P_i, L_j\right)$ is

$$d_{P2Line}\left(P_i, L_j\right) = \frac{\left(P_i - L_{j,1}\right) \times \left(P_i - L_{j,2}\right)}{\left\|L_{j,2} - L_{j,1}\right\|} \tag{5.19}$$

The point-to-plane distance error metric $d_{P2Plane}\left(P_i, P_j\right)$ is

$$d_{P2Plane}\left(P_i, P_j\right) = \left(P_i - P_j\right) \cdot P_j^n \tag{5.20}$$

where P_j^n is the normal vector of P_j. The angle distance $d_\theta\left(L_i, L_j\right)$ of two line segments is defined as

$$d_\theta\left(L_i, L_j\right) = \min\left(\|L_i\|, \|L_j\|\right) \cdot \sin\left(\theta\left(L_i, L_j\right)\right) \tag{5.21}$$

where $\sin\left(\theta\left(L_i, L_j\right)\right)$ is the sine of the angle between two lines. Because the extracted pairs of tie lines are not exactly parallel, it is necessary to quantify this angle and its impact on registration.

The line-to-line distance error metric $d_{\text{Line2Line}}\left(L_i, L_j\right)$ is defined as

$$d_{\text{Line2Line}}\left(L_i, L_j\right) = \sqrt{\left(d_\theta\left(L_i, L_j\right)\right)^2 + \left(d_{\text{P2Line}}\left(P_i, L_j\right)\right)^2} \qquad (5.22)$$

Similarly, the plane-to-plane distance error metric $d_{\text{Plane2Plane}}\left(P_i, P_j\right)$ is defined as

$$d_{\text{Plane2Plane}}\left(P_i, P_j\right) = \sqrt{\left(d_\theta\left(P_i^n, P_j^n\right)\right)^2 + \left(d_{\text{P2Plane}}\left(P_i, P_j\right)\right)^2} \qquad (5.23)$$

After feature-based point cloud registration, the relative constraint between each pair of overlapping areas is obtained. Then global trajectory optimization is used to improve the overall accuracy of the whole MLS point cloud registration. The global posture graph $G(x, c)$ is composed of nodes to be solved (x, corresponding to the segments) and constraints c between nodes, where $x = \left(x_1^{\text{T}}, \cdots, x_n^{\text{T}}\right)^{\text{T}}$ is a series of posture nodes in the trajectory. Ω_{ij} is the weight matrix of the constraint between the ith and jth nodes. If the ith and jth are continuous nodes, c_{ij} is the relative posture obtained from the initial trajectory. If they are paired posture nodes in the overlapping area, c_{ij} is the relative posture obtained from local registration. $e\left(x_i, x_j, c_{ij}\right)$ is an error function to evaluate the difference between the optimized and observed postures. Finally, the global posture graph optimization can be regarded as a nonlinear least squares problem:

$$x^* = \arg\min_x G(x, c)$$
$$G(x, c) = \sum_{i,j} e\left(x_i, x_j, c_{ij}\right)^{\text{T}} \Omega_{ij} e\left(x_i, x_j, c_{ij}\right) \qquad (5.24)$$

The weight matrix $\Omega_{i,j}$ is determined by the variance of the observed node postures. Position and posture errors are usually small in open areas but can be large under dense canopies or next to tall buildings. These errors can be evaluated using trajectory post-processing software. Thus, $\Omega_{i,j}$ is defined as follows (Eq. 5.25):

$$\Omega_{ij} = \frac{I}{\left(\sigma_{\text{pose}}^2 \cdot e_{ij}\right)} \qquad (5.25)$$

where I is an identity matrix; e_{ij} is 1.0 or e_{rmse}; σ^2_{pose} contains the variance of positional and rotational errors. This large-scale nonlinear least squares problem is solved using Ceres Solver to produce the registered point cloud data (Fig. 5.10).

5.2.3 Airborne LiDAR Data Registration

As described in Chapter 3, each flight path of an airborne LiDAR system only covers a certain strip of the ground owing to the limitation of flight height and the effective field-of-view of the scanner. Thus, multiple route flights are often required to obtain full coverage of the study area. A certain overlap rate (usually 10%—20% at least) should be guaranteed between adjacent flight strips. After point cloud resolving, a complete dataset covering the entire study area can be obtained. However, errors in system calibration, posture measurement, and GNSS positioning during point cloud resolving can result in a position shift of the same object in point clouds from different flight strips (Fig. 5.11). Strip alignment aims to correct this shift and improve the alignment precision.

Generally, strip alignment uses the same features, such as elevation in the overlapping area between strips, to evaluate and eliminate the deformation of each strip. Early strip alignment methods were mainly one-dimensional adjustments and assumed that the deviation between strips was mainly caused by errors in elevation. On the basis of this assumption, an adjustment algorithm using three parameters, including divergences in the elevation,

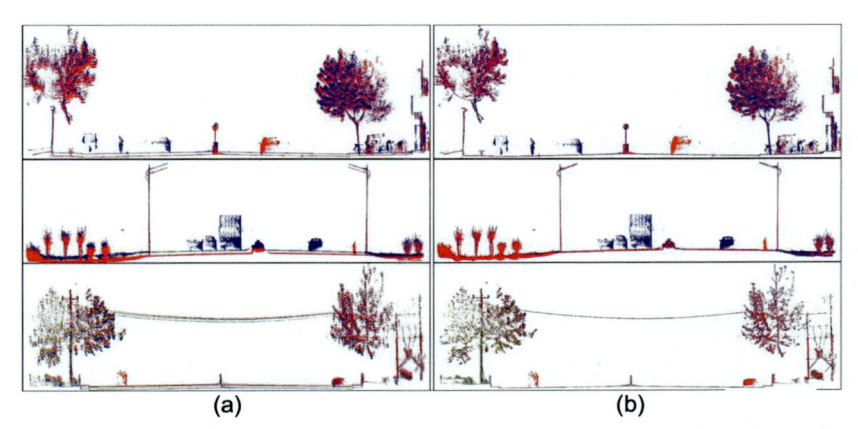

Figure 5.10 An example of registration of mobile LiDAR scanning data in an urban environment. (a) The raw point clouds from multiple trajectories, and (b) the point cloud after registration.

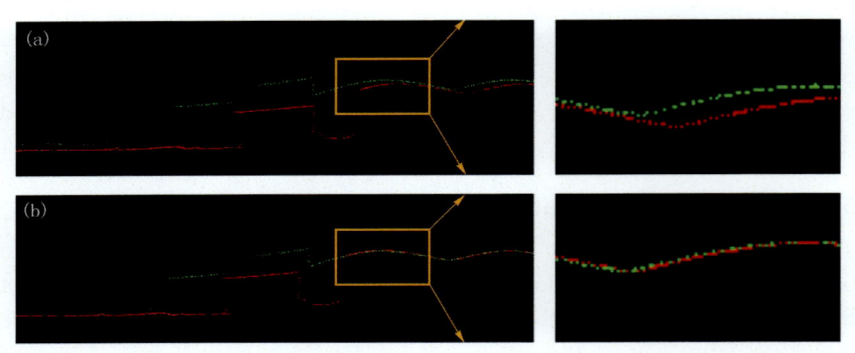

Figure 5.11 Comparison of airborne LiDAR point clouds (a) before and (b) after strip alignment.

azimuth angle, and inclination angle of different flight trajectories, was proposed (Crombaghs et al., 2000; Tung, 2005). However, position errors in surfaces are also unavoidable in point cloud resolving, making it challenging for one-dimensional adjustment methods to meet the precision requirement of strip alignment (Vosselman & Haralick, 1996). Thus, the development of 3-D strip alignment methods has become a research focus. In the following example, we use airborne LiDAR data to introduce the principles and steps of 3-D strip alignment.

The alignment accuracy of a strip should be validated first before applying strip alignment. If the alignment accuracy does not meet the precision requirement, strip alignment is required. The alignment accuracy is usually evaluated using the relative elevation accuracy in the overlap region of adjacent strips. One of the commonly used metrics is the elevation discrepancy between digital elevation models of the overlapping region derived from different strips. A lower elevation discrepancy indicates higher alignment accuracy. Another commonly used metric is the difference in the mean and variance of elevation values from randomly sampled points in the overlapping region. If the difference in the elevation values is small, we can assume that only systematic errors exist in the aligned point cloud data of a strip (Latypov, 2002).

Strip alignment includes four steps: selecting the overlapping region of adjacent strips, matching tie points, selecting and solving the model of strip alignment, and finally applying the adjustment model (Fig. 5.12).

The main purpose of selecting overlapping regions is to avoid unnecessary computation and improve the efficiency of strip alignment. Specific criteria, such as avoiding vegetation, including continuous and smooth

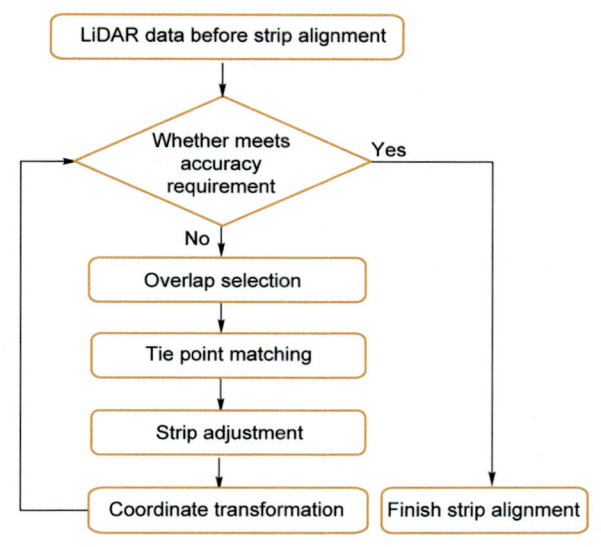

Figure 5.12 The workflow for strip alignment.

ground, and selecting evenly distributed patches, are commonly adopted to select suitable regions from overlapping strips.

Matching tie points in the selected overlapping regions provides input data for the adjustment model in the next step. A variety of methods are available, such as the regular grid method (Kraus et al., 2006), the least squares method based on irregular triangulation (Maas, 2000), the least squares 3-D surface matching method (Akca & Gruen, 2006), and ICP. ICP is widely used to match tie points because of its high precision in pattern recognition (see Section 5.2.1).

Taking the spatial coordinates of matched tie points as input data, the optimal strip alignment model is determined by calculating and optimizing the strip alignment parameters to achieve the minimum position deviation for point clouds in the overlapping region:

$$\min_{T_k} \frac{1}{n} \left\| \sum_{i=1}^{n} P_j - T_k(P_k) \right\| \tag{5.26}$$

where P_j and P_k are the selected tie points in the overlapping region of the jth and kth strips, respectively; T_k is the transformation matrix for depicting strip deformation; n is the number of selected points. In a 3-D strip alignment method, the transformation matrix T_k consists of nine parameters, namely three translation parameters, three posture parameters, and three drift parameters. At present, the least squares adjustment is commonly used, which aims to calculate the optimal parameters in T_k by minimizing the error equation constructed from input tie points.

After determining the optimal transformation parameters, the transformation matrix is applied to correct all the points of each strip and the ensemble of all strips. After the correction is completed, it is necessary to further check the accuracy of strip alignment. If the accuracy requirement is fulfilled, strip alignment is completed. Otherwise, the above steps must be repeated to achieve the required accuracy.

5.2.4 Multiplatform LiDAR Data Registration

LiDAR data acquired by scanners mounted on different platforms have specific characteristics in terms of their spatial and temporal scale. Spaceborne LiDAR data usually provide the widest coverage and can also provide waveform information. Its worldwide coverage makes spaceborne LiDAR most suitable for estimating forest structural properties at a global scale. Airborne LiDAR data can cover a wide area with sufficient accuracy for precisely extracting the topographic information on ground surfaces at a large scale. Hence, it is widely used to produce digital elevation products in forests and mountain areas. Mobile LiDAR is suitable for acquiring high-precision point cloud data from roads in urban environments owing to its high flexibility and mobility. Similarly, backpack LiDAR is a portable system with a scanner and accessories assembled in a backpack and has been widely used for surveys in urban and forest environments. Terrestrial LiDAR usually provides the highest positioning accuracy and point density, which is particularly suitable for fine-resolution scanning in small areas, such as extracting 3-D structural attributes of individual plants with complex and diverse forms or even constructing their 3-D models (Guo et al., 2014).

At present, airborne, terrestrial, and backpack LiDAR are the most widely used systems for extracting the 3-D structure parameters of forests.

Airborne LiDAR systems can acquire point clouds for forests at a large scale to extract a wide range of vertical structure parameters. Owing to the low point density and occlusion from canopies, however, it is not usually possible to obtain accurate and complete information about the subcanopy vegetation (Liu et al., 2011). Terrestrial and backpack LiDAR systems can acquire point clouds of forest stands with high precision and high point density, although it is difficult for them to capture the upper canopy in dense forests with complex structures owing to occlusion and their limited scanning range. The LiDAR data acquired by these different systems are highly complementary as they correspond to specific scanning angles and have different coverage, data integrity, and precision. Therefore, many studies have been reported on the fusion of airborne and terrestrial (or backpack) LiDAR data to derive forest structure attributes at a larger scale with higher data integrity and accuracy. For instance, Guan et al. (2019) proposed an automatic registration method for backpack and UAV LiDAR data based on the assumption that each forest stand had a unique tree distribution pattern. The method consists of five steps, including individual tree segmentation, generation of a triangulated irregular network (TIN) based on the locations of individual trees, TIN matching, coarse registration, and fine registration (Fig. 5.13). TIN matching uses a voting strategy according to the similarity of tree locations extracted from backpack and UAV LiDAR data.

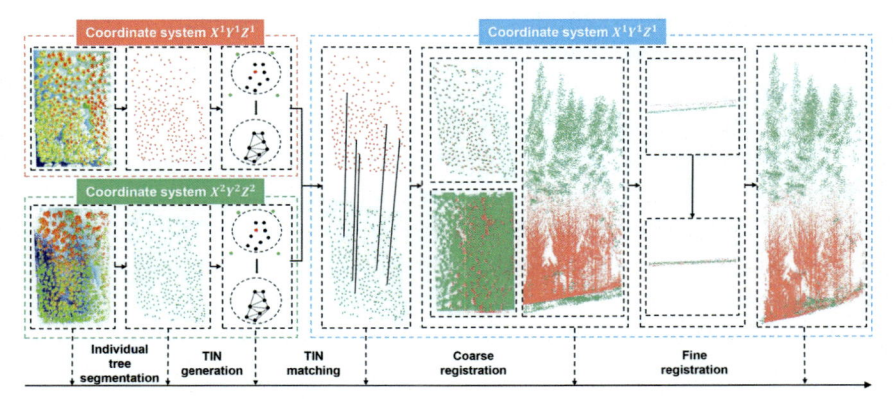

Figure 5.13 Five-step process of the multiplatform LiDAR registration framework. Red points represent the LiDAR data acquired from the side-view backpack LiDAR, and green points represent the unmanned aerial vehicle LiDAR data. *TIN*, triangulated irregular network.

5.3 LiDAR Point Cloud Denoising

The existence of noise points is unavoidable in point cloud data owing to the precision of scanners, the characteristics of targets, and environmental factors such as wind. Removing noise points during preprocessing can better restore the actual shape of objects and improve the accuracy of subsequent point cloud processing. Existing denoising algorithms can be divided into the following categories: spatial distribution–based, depth-based, cluster-based, distance-based, and density-based algorithms (Papa-dimitriou et al., 2003). In this section, we mainly introduce three commonly used methods, including spatial distribution–based, cluster-based, and density-based denoising algorithms.

5.3.1 Denoising Algorithms Based on the Spatial Distribution of Points

These algorithms assume that the spatial distribution of all points follows a standard spatial distribution model such as normal or Poisson distribution, and hence, they treat points far from the standard distribution as noise points. The statistical outlier removal (SOR) filter provided in the Point Cloud Library adopts a typical denoising algorithm based on the spatial distribution of points (Zhu et al., 2012). The basic steps are as follows. For each point, first, its K-nearest neighboring points are found, and the mean μ and standard deviation σ of the distances between these points are calcu-lated. Then, the spatial distribution of these points is assumed to be a Gaussian distribution with μ and σ^2 as its mean and variance. Finally, the average distance between this point and its K neighboring points is calculated and compared with a predefined threshold based on μ and σ, usually being $\mu + \sigma \times$ multiplier. If the average distance is beyond the threshold, this point is regarded as an outlier and removed.

The following is a case study. The point cloud data for magnolia trees at a forest farm in Fangshan, Beijing, were collected using a RIEGL VZ-400 TLS and denoised using the SOR filter (Fig. 5.14b and d). The figure shows that noise points between the leaves and branches are well identified and removed.

A Python script for SOR is shown in the following text box.

```python
#The following SOR algorithm implemented in Python is used to remove noise points from the
sample data.
# -*- coding: UTF-8 -*-
import laspy
import numpy as np
from scipy import spatial
#las file location Sample data is airborne data, but the algorithm is general
lasfile = "./sor.las"
#Open las file
inFile = laspy.file.File(lasfile, mode = "r")
#Reading the XYZ coordinates of the Lidar point cloud
x,y,z = inFile.x,inFile.y,inFile.z
# Encapsulating x,y
lasdata = zip(x,y,z)
tree = spatial.KDTree(lasdata)
#Set the selection of several times SD as threshold and neighborhood points
sigma=1    #SD
K=51        # Neighborhood point number
#Create an array of storage distances for each point neighborhood
k_dist=np.zeros_like(x)
#Find points with distances less than
for i in range(len(x)):
        dist.index =tree.query(np.array([x[i],y[i],z[i]]), K)
k_dist[i] = np.sum(dist)
#Determine the maximum threshold of noise points
max_distance = np.mean(k_dist) + sigma*np.std(k_dist)
#Point cloud XYZ value after noise points removal
x=x[k_dist<max_distance]
y=y[k_dist<max_distance]
z=z[k_dist<max_distance]
```

5.3.2 Cluster-Based Denoising Algorithm

Schall et al. (2005) proposed a point cloud denoising algorithm based on kernel density estimation, which uses the following steps. For a given point cloud dataset P (P_i, $i = 1,2,3 \ldots n$), first, a least squares planar fitting is performed for each point P_i and its neighboring points. Then, a local likelihood function L_i based on the squared distances from these points to the fitted plane is constructed for each point, and the overall likelihood

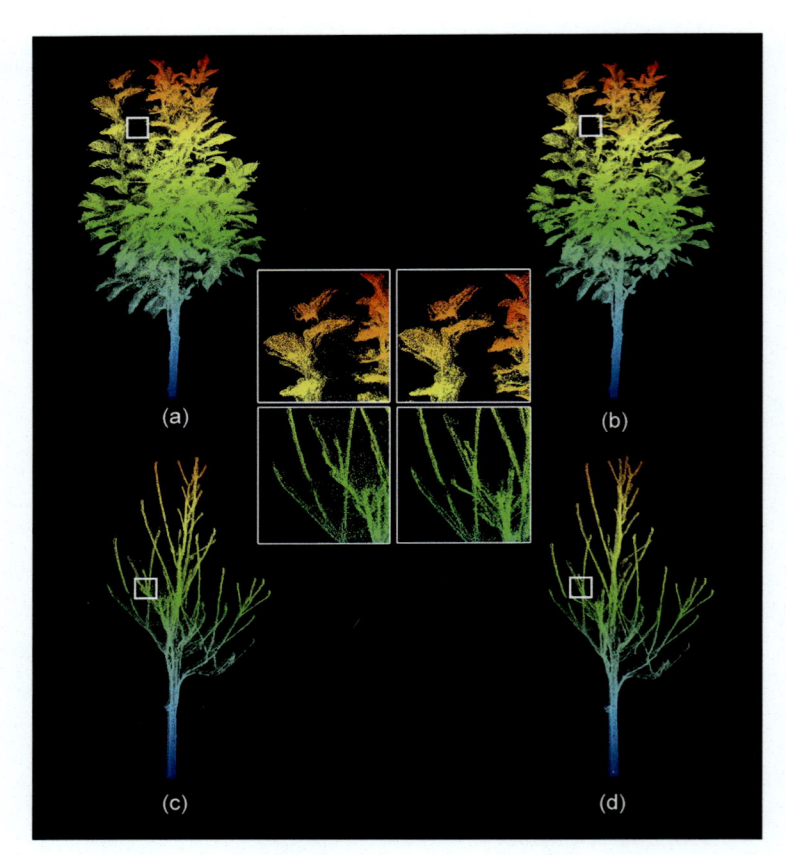

Figure 5.14 An example of point cloud denoising. (a) and (c) are the raw leaf-on and leaf-off point clouds of a magnolia tree, respectively, and (b) and (d) are denoised results of (a) and (c), respectively.

function L is calculated by weighting L_i. After obtaining L, each sample point is moved onto the most likely sampling plane through an iterative approach, and the iteration is repeated until a specific condition is met. Because a fixed kernel size is not suitable for data with different sampling densities, a principal component analysis is performed for each point and its neighboring points to obtain an adaptive kernel size. This algorithm usually works well and can detect outliers quickly.

5.3.3 Density-Based Denoising Algorithm

A density–based denoising algorithm was proposed by Breunig et al. (2000). The following concepts and steps are involved in this algorithm. (1) The K-distance of a point P, denoted as $K\_dist(P)$, is the maximum distance between P and its K neighboring points. (2) The K-distance neighborhood of a point P is the set of points whose distance to P is not greater than $K\_\text{dist}(P)$:

$$N_{K\_\text{dist}(P)}(P) = \{Q \in D \mid d(P, Q) \leq K\_\text{dist}(P)\} \tag{5.27}$$

where D is the point cloud data to be processed, and $d(P, Q)$ is the distance between point P and Q. These eligible points are called the nearest neighboring points of P. (3) The reachability distance of a point P with respect to point O is calculated:

$$\text{Reach\_dist}(P, O) = \max\{K_{\text{dist}(O)}, d(P, O)\} \tag{5.28}$$

If P is far from O, the reachability distance is the actual distance between them; otherwise, it is $K\_\text{dist}(O)$. (4) Then, the local reachability density of a point P is calculated:

$$\text{Lrd}_{\text{Minpts}}(P) = 1 / \left(\frac{\sum_{O \in N_{\text{Minpts}}(P)} \text{Reach\_dist}_{\text{Minpts}}(P, O)}{\left| N_{\text{Minpts}}(P) \right|} \right) \tag{5.29}$$

where Minpts is the number of nearest neighbors used in defining the local neighborhood of P. (5) Finally, the local outlier factor (LOF) is obtained:

$$\text{LOF}_{\text{Minpts}}(P) = \frac{\sum_{O \in N_{\text{Minpts}}(P)} \dfrac{\text{Lrd}_{\text{Minpts}}(O)}{\text{Lrd}_{\text{Minpts}}(P)}}{\left| N_{\text{Minpts}}(P) \right|} \tag{5.30}$$

Points with a LOF value greater than one are regarded as noise points. This algorithm does not require *a priori* knowledge of the scanning area and can be applied to point cloud data with different point densities.

5.4 Point Cloud Feature Extraction

The features in point cloud data indicate the surface and edge characteristics of objects and can be classified into geometric features, statistical features, and topological features. Geometric features include the surface normal and curvature. Statistical features include vertical point distribution, density-based features, and curvature-based features. Topological features include the critical point and skeleton.

5.4.1 Geometric Features

Geometric features are usually calculated based on a point and its neighboring points (Fig. 5.15), which can be defined using different shapes. For a given point P (x, y, z), its neighborhood can be defined as (1) a cylindrical neighborhood containing points that are within a circle of radius r centered at point P after projection onto a two-dimensional plane, i.e., $\left\{Q|\left(x_q - x\right)^2 + \left(y_q - y\right)^2 \leq r^2\right\}$ (where $\left(x_q, y_q, z_q\right)$ are coordinates of the points in the neighborhood Q); (2) a spherical neighborhood, consisting of points within a sphere of radius r centered at P, i.e., $\left\{Q|\left(x_q - x\right)^2 + \left(y_q - y\right)^2 + \left(z_q - z\right)^2 \leq r^2\right\}$; and (3) a K-nearest neighborhood, containing the nearest K points to P (Guo et al., 2015; Weinmann et al., 2015).

The basic local geometric features of a point are its linearity, planarity, and scattering. These features are derived from the neighborhood matrix:

$$S = \begin{pmatrix} x_1 & y_1 & z_1 \\ x_2 & y_2 & z_2 \\ \vdots & \vdots & \vdots \\ x_n & y_n & z_n \end{pmatrix} \tag{5.31}$$

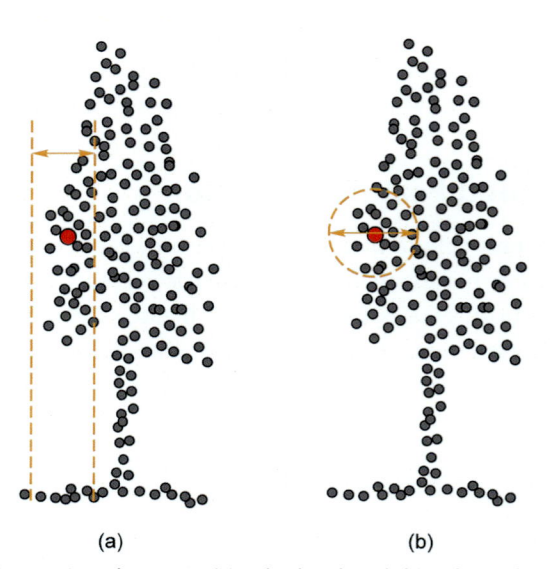

(a)   (b)

Figure 5.15 Comparison between (a) cylindrical and (b) spherical neighborhoods.

where n is the number of points in the neighborhood. After normalization, we can obtain

$$
\overline{S} = \begin{pmatrix} x_1 & y_1 & z_1 \\ x_2 & y_2 & z_2 \\ \vdots & \vdots & \vdots \\ x_n & y_n & z_n \end{pmatrix} - \begin{pmatrix} \overline{x} & \overline{y} & \overline{z} \\ \overline{x} & \overline{y} & \overline{z} \\ \vdots & \vdots & \vdots \\ \overline{x} & \overline{y} & \overline{z} \end{pmatrix} \tag{5.32}
$$

where $(\overline{x}\ \ \overline{y}\ \ \overline{z})$ is the center of S. The covariance matrix can be obtained using

$$
C = \frac{1}{n}\overline{S}^{\mathrm{T}}\overline{S} = \begin{pmatrix} \sigma_{xx} & \sigma_{xy} & \sigma_{xz} \\ \sigma_{xy} & \sigma_{yy} & \sigma_{yz} \\ \sigma_{xz} & \sigma_{yz} & \sigma_{zz} \end{pmatrix} \tag{5.33}
$$

This covariance matrix can also be decomposed as:

$$
C = \begin{pmatrix} \sigma_{xx} & \sigma_{xy} & \sigma_{xz} \\ \sigma_{xy} & \sigma_{yy} & \sigma_{yz} \\ \sigma_{xz} & \sigma_{yz} & \sigma_{zz} \end{pmatrix} = M^{\mathrm{T}}\begin{pmatrix} \lambda_1 & 0 & 0 \\ 0 & \lambda_2 & 0 \\ 0 & 0 & \lambda_3 \end{pmatrix}M \tag{5.34}
$$

where $(\lambda_1, \lambda_2, \lambda_3)$ are three eigenvalues with $\lambda_1 \geq \lambda_2 \geq \lambda_3 > 0$; M consists of the corresponding eigenvectors (v_1, v_2, v_3). Then the 3-D geometric features of a point (linearity (D^{L}), planarity (D^{P}), and scattering (D^{S}) are defined as

$$
D^{\mathrm{L}} = \frac{\sqrt{\lambda_1} - \sqrt{\lambda_2}}{\sqrt{\lambda_1}}, \quad D^{\mathrm{P}} = \frac{\sqrt{\lambda_2} - \sqrt{\lambda_3}}{\sqrt{\lambda_1}}, \quad D^{\mathrm{S}} = \frac{\sqrt{\lambda_3}}{\sqrt{\lambda_1}}, \tag{5.35}
$$

In addition, the eigenvector corresponding to the minimum eigenvalue λ_3 is adopted as the normal vector of this point. The curvature of a point can also be calculated from the eigenvalues of the covariance matrix. Hoppe et al. (1992) proposed a method based on covariance analysis to estimate the curvature of a point. Each eigenvalue of the covariance matrix represents the spatial variation of its related eigenvector along the corresponding direction, and curvature estimation is used to quantify the deviation of the tangent plane formed by ν_1 and ν_2. Therefore, the ratio of λ_3 and the sum of all eigenvalues is adopted as the curvature:

$$\text{Cur}(p_i) = \frac{\lambda_3}{\sum_{j=1}^{3} \lambda_j} \tag{5.36}$$

The curvature of a point p_i can also be estimated using the normal vectors of p_i and its neighboring points.

$$\text{Cur}(p_i) = \frac{1}{k} \sum_{j=1}^{k} \left\| \overrightarrow{n_{p_i}} - \overrightarrow{n_{\text{neighbor}(j,p_i)}} \right\| \tag{5.37}$$

where $\boldsymbol{n}_{p_i}$ is the normal vector of p_i, and $\overrightarrow{n_{\text{neighbor}(j,p_i)}}$ is the normal vector of the jth neighborhood point of p_i.

5.4.2 Statistical Features

Statistical features are based on the specific statistics of point attributes within a defined neighborhood, such as point density, vertical point distribution, and the number of returns. The fast point feature histogram (FPFH) is a comprehensive statistical feature that integrates information regarding the spatial interconnections between a point and its neighboring points into a histogram (Liu et al., 2019; Rusu et al., 2009). The normal vectors $\boldsymbol{n}$ and $\boldsymbol{n}'$ for each pair of points q and q' within a spherical neighborhood are calculated (with q being the point with a smaller angle between its associated normal and the line connecting the point, i.e., $\boldsymbol{n} \cdot (q' - q) \geq \boldsymbol{n}' \cdot (q - q')$), and a Darboux $\boldsymbol{uvw}$ frame is defined:

$$\begin{cases} \boldsymbol{u} = \boldsymbol{n} \\ \boldsymbol{v} = (q' - q) \times \boldsymbol{u} \\ \boldsymbol{w} = \boldsymbol{u} \times \boldsymbol{v} \end{cases} \tag{5.38}$$

Subsequently, the angular variations of $\boldsymbol{n}$ and $\boldsymbol{n}'$ can be represented as $(\alpha, \varphi, \theta)$ (as shown in Fig. 5.16); their relation is defined as

$$\begin{cases} \alpha = \boldsymbol{v} \cdot \boldsymbol{n}' \\ \varphi = \boldsymbol{u} \cdot \dfrac{(q' - q)}{\|q' - q\|_2} \\ \theta = \arctan(\boldsymbol{w} \cdot \boldsymbol{n}', \ \boldsymbol{u} \cdot \boldsymbol{n}') \end{cases} \tag{5.39}$$

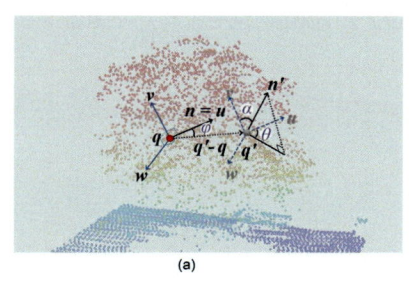 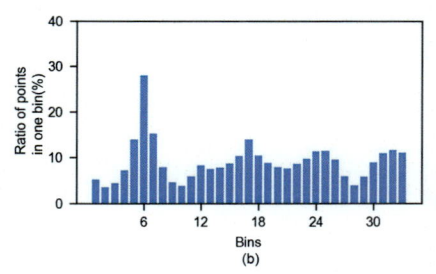

(a)   (b)

Figure 5.16 Schematic diagram for (a) calculating the fast point feature histogram (PFH) and (b) an example of fast PFH. *(From Liu, X., Chen, Y., Li, S., Cheng, L., & Li, M. (2019). Hierarchical classification of urban ALS data by using geometry and intensity information.* Sensors (Basel), *19(20), 4583.)*

To simplify the computation of histogram features, a set of the angle variations $(\alpha, \varphi, \theta)$ for q and each point in its spherical neighborhood is calculated first and aggregated into three histograms, which are further concatenated to form a simplified point feature histogram (SPFH). The final FPFH is the weighted sum of SPFH in the K-neighborhood of q, as shown in Eq. (5.40):

$$\mathrm{FPFH}(q) = \mathrm{SPFH}(q) + \frac{1}{|N(q)|} \sum_{q' \in N(q)} \frac{\mathrm{SPFH}(q')}{q' - q_2} \tag{5.40}$$

where $N(q)$ represents the K-neighborhood of q and $|N(q)|$ is the number of points in $N(q)$.

5.4.3 Topological Features

Edges and boundaries are basic topological features. Bae et al. (2007) used normal vector, curvature, and variance of curvature to detect edges and extract boundaries in point cloud data. Edges are detected on the basis that edges and corners usually have greater curvature than other objects. Thus, if the curvature of a point $\mathrm{Cur}(p_i)$ is greater than a given threshold, it is marked as an edge point. Otherwise, it is not an edge point:

$$\mathrm{edge\_or\_not}(p_i) = \begin{cases} \mathrm{true}, & \text{if } \mathrm{Cur}(p_i) > \text{threshold} \\ \mathrm{false}, & \text{if } \mathrm{Cur}(p_i) \leq \text{threshold} \end{cases}$$

The threshold $= \mathrm{Cur}(p_i)_{\mathrm{mean}} + \mathrm{Cur}(p_i)_{\mathrm{std}}$ usually works well when the data are relatively simple and smooth. For complex data, Bae et al. (2007) adopted the variance of point curvature to identify whether the point belonged to a plane. The variance of point curvature is calculated using Eq. (5.41).

$$\mathrm{Var}_{\mathrm{Cur}(p_i)} = \frac{1}{n}\sum_{i=1}^{n}\left(\mathrm{Cur}(p_i) - \mathrm{Cur}(p_i)_{\mathrm{mean}}\right)^2 \qquad (5.41)$$

and the identification rule is

$$\mathrm{surface\_no\_not}(p_i) = \begin{cases} \mathrm{true}, & \text{if } \mathrm{Var}_{\mathrm{Cur}(p_i)} < \mathrm{threshold} \\ \mathrm{false}, & \text{if } \mathrm{Var}_{\mathrm{Cur}(p_i)} \geq \mathrm{threshold} \end{cases}$$

where the threshold is also a tolerance that can be set to 10^{-4} for a terrestrial LiDAR system. This edge detection method can recognize rough surfaces or surfaces with a sharp change of surface normal, while curvature-based methods are only suitable for smooth surfaces and can easily misclassify rough surfaces as curved surfaces.

Some studies have tried to extract boundaries by detecting the deviation between a point and the geometric center of its neighboring points. The geometric center of the neighboring points will be close to a nonboundary point, while the geometric center of the neighboring points for a boundary point will deviate. By using the larger two eigenvalues (λ_1 and λ_2) and their corresponding eigenvectors, Bae et al. (2007) defined a two-dimensional ellipse enclosing the region. The boundary point can be detected using the relative distance from a point to the boundary region.

$$c(p_i) = \sum_{i=1}^{2} \frac{e_i(p_i - p_{\mathrm{icentroid}})}{\lambda_i} \qquad (5.42)$$

where p_i are the coordinates of the ith point, and $p_{\mathrm{icentroid}}$ are the coordinates of the geometric center of the neighborhood of the ith point. Again, a threshold is used to determine whether a point is a boundary point:

$$\mathrm{boundary\_or\_not}(p_i) = \begin{cases} \mathrm{true}, & \text{if } c(p_i) > c(\mathrm{elli}) \\ \mathrm{false}, & \text{if } c(p_i) \leq c(\mathrm{elli}) \end{cases}$$

where the threshold $c(\mathrm{elli})$ can be a percentage of the total distance or determined by a chi-square test to find an appropriate confidence interval.

5.5 Point Cloud Classification

Point cloud data usually contain a large number of objects with distinct geometric features, such as the ground, trees, and buildings. Although the categories may vary in different applications, point cloud classification methods are often developed from similar general principles such as model fitting, region growing, clustering, hierarchical strategy, machine learning, and deep learning.

5.5.1 Point Cloud Classification Based on Model Fitting

Many artificial objects can be described using specific geometric shapes. For example, the facade of a building can be regarded as a plane, while poles and tree trunks can be represented using cylinders. All these shapes can be depicted using a few parameters. Hence, classification based on model fitting is also known as parameter-based classification. Two of the most used methods are the Hough transform and RANSAC.

(1) Hough transform

The Hough transform (Ballard, 1981; Hough, 1962) is one of the most popular feature extraction methods in image analysis and computer vision, and it is broadly used to detect features such as lines, circles, and ellipses from two-dimensional images. The Hough transform significantly reduces computational complexity by describing simple geometric objects using parameters. For instance, in a Cartesian coordinate system, the equation of a straight line is expressed as

$$Y = kX + b \tag{5.43}$$

where k is the slope and b is the intercept. Therefore, a line can be represented by a point (k, b) in the k-b parameter space. However, vertical lines pose a challenge and result in unbounded values of k. Thus, the line is converted to another parameter space:

$$d = X\cos\alpha + Y\sin\alpha \tag{5.44}$$

where d is the distance from the origin to the nearest point on the line, and α is the angle between the perpendicular of the line (through the origin) and the X-axis (Fig. 5.17).

The purpose of detecting a line in an image is to find all pixels on the line. The coordinates (X, Y) of each pixel are known, and d and α are variables. If (d, α) are derived according to each (X, Y), we can transform the image from the Cartesian coordinate system to the Hough space polar coordinate system. This transformation is called the Hough transform.

Simonse et al. (2003) use a two-dimensional Hough transform to extract the diameter at breast height (DBH) of trees from point cloud data. First, the points at 1.25−1.35 m above ground are extracted and interpolated into

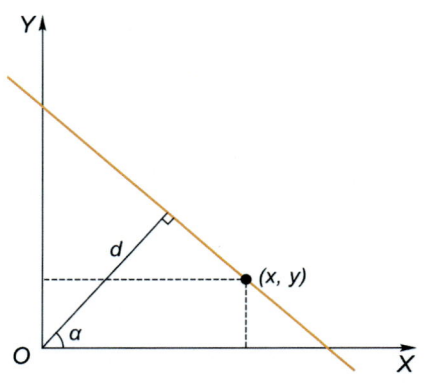

Figure 5.17 Line detection based on hough transformation. α is the angle between the vertical line from the origin to the straight line and the X-axis. d is the distance from the origin to the nearest point on the line.

a raster image with a 1 cm grid. As the diameter needs to be provided when detecting a circle using the Hough transform, the DBH is initially set to 100 cm and decreases with intervals of 10 cm. If a circle is correctly detected, the optimal diameter value is adopted as the DBH.

Vosselman and Maas (2010) reviewed several classical methods for extracting geometric shapes or smooth surfaces from point cloud data, including the 3-D Hough transform. This is an extension of the two-dimensional Hough transform and can be used to detect planes, cylinders, and spheres in 3-D space. For instance, a nonvertical plane can be expressed as

$$Z = a_x X + a_y Y + d \qquad (5.45)$$

where a_x and a_y are the slope of the plane regarding the X and Y axes; d is the intercept of the plane with the Z-axis. The above three parameters can define the parameter space, i.e., each point (a_x, a_y, d) in the parameter space corresponds to a plane in the object space, and each point (X, Y, Z) in the object space also corresponds to a plane in the parameter space. Thus, the task of detecting planes in point cloud data can be converted to the following task. After mapping all points to planes in the parameter space, the number of planes intersecting at a point in the parameter space is equal to the number of points on the plane in the object space. Therefore, the parameters of a plane in the object space can be determined by the intersection points of planes in the parameter space. Moreover, if the normal vectors can be calculated accurately, the computation in the 3-D

Hough transform can be accelerated, and its credibility can also be increased. To reduce the dimension of the parameter space and the memory requirement, this plane detection process can further be divided into two steps: detecting the plane normal vector and calculating the distance between a plane and the origin.

Borrmann et al. (2011) reviewed different Hough transform methods, including the standard, adaptive probability, asymptotic probability, and random Hough transforms, and compared their pros and cons in plane detection for point cloud data.

(2) RANSAC

RANSAC is a method for estimating the optimal parameter model from the observed data (Fischler & Bolles, 1981). Its basic steps are as follows. First, a subset of points is randomly sampled from the entire dataset to derive an initial candidate parameter model, with points in the subset as the initial candidate points. Then the deviation between other points and the candidate model is calculated. If the deviation of a point is smaller than a given threshold, it is also regarded as a candidate point and added to the candidate model. If the number of candidate points satisfies a specific threshold, the candidate model is reasonable, and then all the current candidate points are used to fit the candidate model again. If the new model is better than the original model, the original model is replaced; otherwise, the original model is retained. These steps are repeated until the candidate point is set and the performance of the candidate model becomes stable.

For example, there is a potential linear feature in the point cloud data, but many noise points also exist (Fig. 5.18a). RANSAC randomly selects two points (or a few points) as the initial seed points (red points in Fig. 5.18b) to fit a line and calculates the distance h from each remaining point to the fitted line (Fig. 5.18c). A remaining point with an h smaller than a given threshold is treated as a neighboring point to the current seed points (Fig. 5.18d) and a new seed point for the next iteration. The linear model is fitted again using the new seed points (Fig. 5.18e). The above process is iterated until the change in the fitted line is negligible or within tolerance, and the final model is adopted.

Boulaassal et al. (2007) used RANSAC to extract the building facade from a terrestrial LiDAR point cloud. First, a few points are randomly selected from the input data for calculating model parameters. Then the performance of the model is evaluated using a series of criteria, such as tolerance, cost function, or the standard deviation of distances of extracted points to the fitted plane. Finally, a model with more points than the others

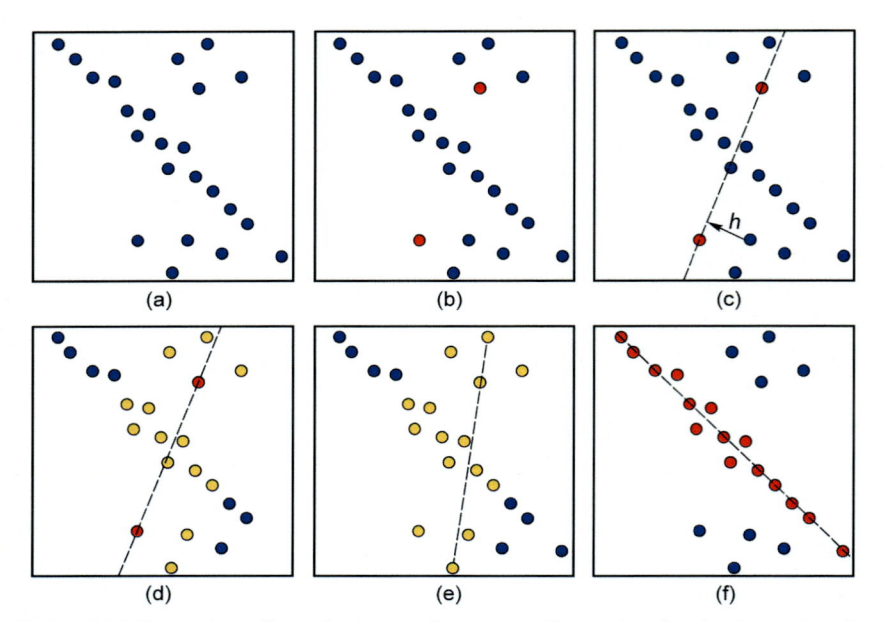

Figure 5.18 Extracting a linear feature and corresponding point cloud points using the random sampling consistency algorithm. (a) A set of unclassified points, (b) randomly selected initial two seed points (red points), (c) the distance h from rest points to the fitted line, (d) the selected seed points that meet the condition, (e) the updated fitted line based on the new seed points, and (f) the final fitted line.

is selected as the best plane, and the calculation is terminated when the possibility of finding a better model decreases. In this process, the minimum number of selected points N_{min} to fit the initial model depends on the type of the model to be fitted (such as lines, planes, cylinders, or spheres) and should generally be greater than the number of model parameters. N_{min} can be determined according to the following Eq. (5.46) (Fischler & Bolles, 1981):

$$N_{min} = \frac{\log(1 - p)}{\log(1 - (1 - e)^m)} \tag{5.46}$$

where p is the probability that at least a subset of good observations can be found from the initial points, given that a certain percentage (e) of the initial points are erroneous observations; m is the number of model parameters.

Schnabel et al. (2007) extract planes, spheres, cones, cylinders, and rings from point clouds using RANSAC. In their study, the number of initial points is reduced by incorporating the normal vector of each point and the compatible points of each candidate model using a standard score that contains two parameters: one is the maximum distance from compatible points to the plane, and the other is the deviation between the point normal vectors and the plane normal vector.

5.5.2 Point Cloud Classification Based on Region Growing

Vosselman et al. (2004) proposed a widely used a region-growing algorithm to extract surfaces, which consists of two major steps: selecting the seed surface and growing the surface following specific criteria. This algorithm determines the seed surface by fitting a plane and calculating the residual errors for each point and its neighboring points within a specific range. If the squared sum of residuals is lower than a given threshold, points with the lowest squared sum of residuals compose the seed surface. However, this method is sensitive to outliers and abnormal points. For these cases, the seed surface can be determined using plane fitting based on robust least squares or the Hough transform. Surface growth usually obeys the following criteria. (1) Proximity of points: only points close to one of the planes can be added to the plane. For 2.5-dimensional data, this proximity can be determined by checking whether a candidate point can be connected to points in a plane through a Delaunay triangle network; (2) local plane: a plane is fitted for each point and its neighboring points on the seed surface. Only a candidate point whose distance to the fitted plane is smaller than a given threshold d is accepted (Fig. 5.19); and (3) normal vector: a candidate point is added to the current plane only if the angle between the normal vectors of the candidate point and its nearest neighboring point is smaller than a given threshold. As shown in Fig. 5.19, if the yellow surface is determined, the distance d of a candidate point (red) to the yellow plane is calculated. In addition, the angle θ between the normal vectors of the candidate point and its adjacent point (blue) on the yellow plane is calculated. If both d and θ satisfy specific requirements, the candidate point is added to the yellow plane; otherwise, not.

Pu and Vosselman (2006) adopted the region-growing algorithm from Vosselman et al. (2004) to perform initial point cloud segmentation, extracted important features of the segments, such as size, position, orientation, and topological structure, and finally used these features as constraints to precisely extract different building sectors such as walls, windows, and doors. Xu et al. (2009) proposed a point cloud segmentation method for terrestrial LiDAR data based on graph theory and region growing,

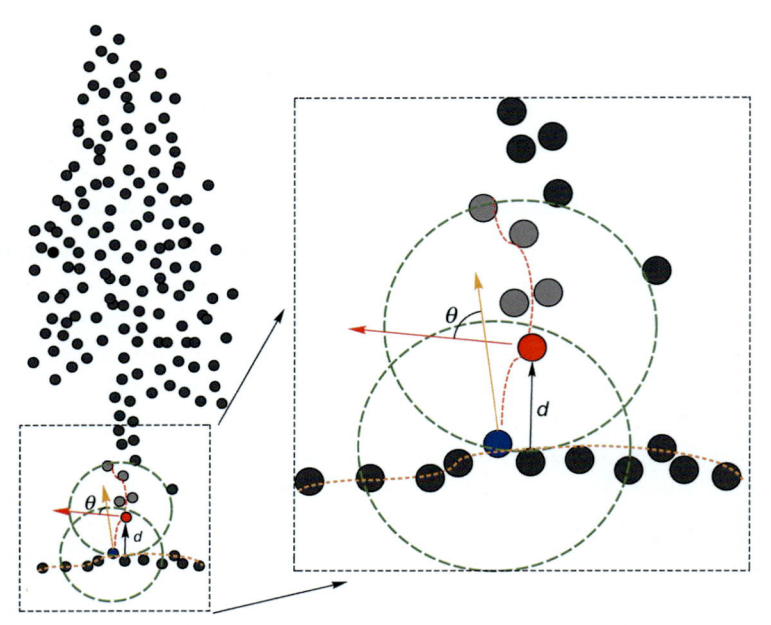

Figure 5.19 Schematic diagram of the region-growing algorithm. d is the distance from a candidate point (red) to the seed plane. θ is the angle between normal vectors of the candidate point and its adjacent point (blue) on the plane.

where the reflectance image created from point cloud data is segmented based on graph theory to obtain seed points. Then, the point cloud data are segmented based on region growing and the obtained seed points.

Because traditional region–growing algorithms are based on a single point, it is computationally intensive and time-consuming to apply them to 3-D point cloud datasets that usually have a large volume of data. Vo et al. (2015) proposed an octree-based region-growing algorithm for point cloud segmentation, which performs segmentation from coarse to fine scales. Its workflow is as follows. First, a minimum cube enclosing the point cloud data is determined and acts as the root node of the octree. Second, the root node is recursively divided into eight child nodes until they meet specific termination criteria. The salient features, including residuals and normal vectors, are calculated for each node, and coarse segmentation is performed using region growing. Finally, the nodes located at the boundary are identified from the coarse segmentation results, and fine segmentation is performed on nodes within a given buffer zone from the boundary.

Although segmentation based on region growing is simple and usually works well for uniformly connected regions, its accuracy and efficiency largely depend on the choice of initial points or regions. Universal and effective guidelines for the selection of initial points or regions are still not available (Teboul et al., 2010).

5.5.3 Clustering-Based Point Cloud Classification

Clustering is an unsupervised classification approach that automatically divides objects—in this case, points—into different groups according to specific rules and object features. Clustering methods can be roughly divided into two categories: hierarchical clustering and partitional clustering (Jain et al., 1999). Their main difference is that partitional clustering produces individual independent clusters, while hierarchical clustering creates clusters in a hierarchical treelike structure, in which the tree nodes are split into clusters at different levels according to specific rules. For partitional clustering, the biggest challenge lies in determining the number of clusters in advance. Currently, there are several different methods for estimating the optimal number of clusters, such as the elbow, average silhouette, and gap statistic methods. In general, partitional clustering is more suitable for large point cloud datasets. In contrast, hierarchical clustering is not suitable for large datasets because of its heavy computational burden. Classical clustering methods include K-means clustering, density-based spatial clustering of applications with noise, and fuzzy clustering. We will introduce the first two in the following sections.

(1) K-means clustering

MacQueen (1967) proposed the K-means clustering method by dividing N–dimensional data into K clusters. First, K points are randomly selected from the point cloud data as the initial clustering centers, with the number of clusters K specified by the user. Then each point is categorized into the nearest cluster, and the cluster centers are recalculated. This step is repeated until the cluster centers in a new iteration do not change or the change is within a given threshold. This algorithm is easy to implement and often achieves high efficiency and accuracy, particularly for data with contrasting objects. There are a few limitations that should be noted. For example, the number of clusters K needs to be determined in advance, and it may be challenging to provide an appropriate value without prior knowledge. In addition, the selection of initial clustering centers has a significant impact on clustering results, and the iterations could be time-consuming for a large dataset.

An example of K-means clustering is provided in Fig. 5.20. The original point cloud data are unclassified (Fig. 5.20a), and two initial cluster centers are randomly selected (Fig. 5.20b). Each point is grouped into one of the two clusters according to its distance to the cluster center (Fig. 5.20c). Then the center of each cluster is recalculated based on the current points belonging to it (Fig. 5.20d), and every point is reclustered based on the updated cluster centers (Fig. 5.20e). The above process is iterated until the two cluster centers are stable and the final clusters are obtained (Fig. 5.20f).

Kashani and Graettinger (2015) tested the K-means clustering algorithm to detect buildings damaged by strong winds. Three criteria, including the within-cluster sum of squares (SS_w), Calinski–Harabasz index (CH), and Davies–Bouldin index (DBI), were used to obtain the optimal K value. The values of SS_w will vary considerably if K is lower than its optimal value, while the values of SS_w do not change significantly when K exceeds its optimal value. Hence, there is an elbow point in the plot of SS_w values versus the number of clusters, which corresponds to the optimal K. The

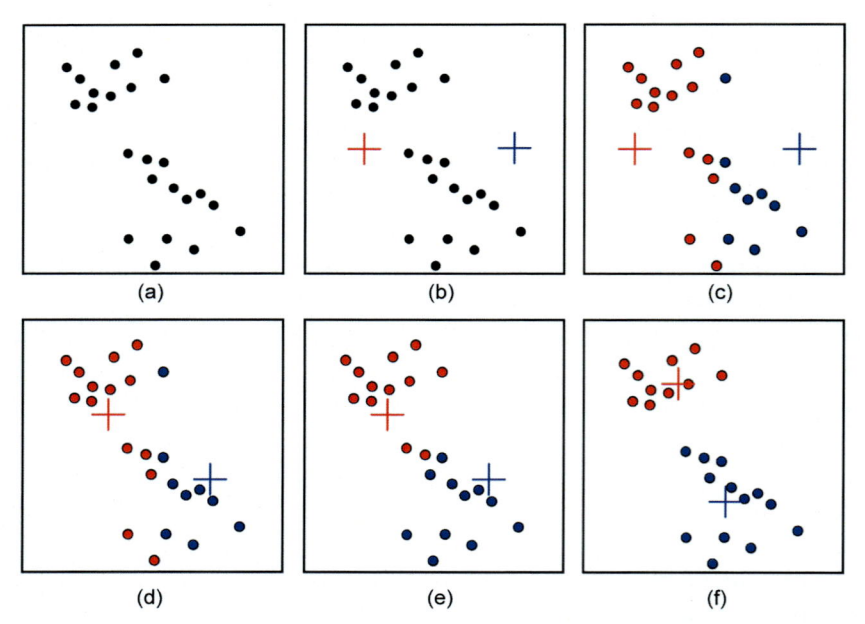

Figure 5.20 The processing pipeline of the K-means clustering algorithm. (a) A set of unclassified points, (b) randomly selected initial centers of the two clusters, (c) the clusters obtained after the first iteration, (d) the updated centers of the two clusters, (e) the updated clusters based on the new cluster centers, and (f) the final clusters and their centers.

elbow point is defined as the point with the largest distance to the connection line between the first and last points in the plot of SS_w versus K. Then the CH index can be calculated using Eq. (5.47).

$$CH = \frac{SS_b}{SS_w} \cdot \frac{n - K}{K - 1} \tag{5.47}$$

where SS_b is the between–cluster sum of squares; n is the number of points in the dataset. An ideal clustering result is reflected by a high SS_b value, a low SS_w value, and a high CH value. Therefore, the optimal K can also be determined by finding the highest CH value. The DBI index is defined as:

$$DBI = \frac{1}{K} \sum_{i=1}^{K} D_i \tag{5.48}$$

where D_i is the ratio of the intra–class and inter–class distances of the i th cluster. A satisfactory clustering result should produce a large inter–class distance and a small intra–class distance. Therefore, the K value corresponding to the lowest DBI value can be selected as the optimal number of clusters.

(2) Density-based spatial clustering of applications with noise

Density-based clustering of applications with noise (DBSCAN) is a density-based clustering algorithm that relies on the following concepts and definitions (Ester et al., 1996). (1) The Eps-neighborhood of a point: the Eps-neighborhood of a point P refers to the neighboring points of P where the distance to P is not greater than Eps. (2) Core points: if the number of points within the Eps-neighborhood of a point is greater than a given threshold (MinPts, in this case, MinPts = 3), the point is a core point, e.g., P in Fig. 5.21a; otherwise, it is a border point, such as Q in Fig. 5.21b. (3) Direct density-reachable: for a sample set D, if point Q is located in the Eps-neighborhood of a core point P, Q is directly density-reachable from P (Fig. 5.21c). (4) Density-reachable: if there is a chain of points P_1, P_2, ..., P_n, $P_1 = P$, $P_n = Q$, for $P_i \in D$ ($1 \leq i \leq n$) such that P_{i+1} is directly density-reachable from P_i, then Q is density-reachable from P. Density-reachability is a canonical extension of direct density-reachability. This relationship is transitive but not symmetrical. Only core points are density-reachable from each other. As shown in Fig. 5.21c, P, P_2, and P_3 are core points, and Q is a border point. Q is density-reachable from P, but not vice versa. (5) Density-connected: if both P and Q are density-reachable from point O, then P is density-connected to Q and vice versa. This is a symmetrical relationship, as shown in Fig. 5.21d. (6) Cluster: assuming that D is a dataset of points,

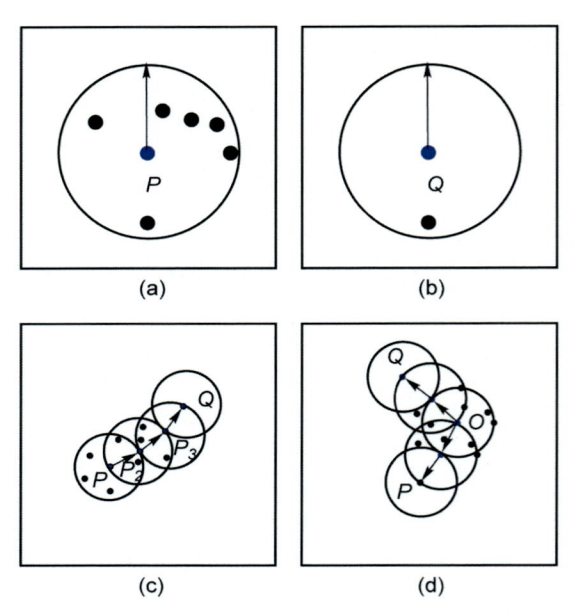

Figure 5.21 The processing pipeline of the density-based spatial clustering of applications with noise algorithm. (a) A core point P of a dataset; (b) a border point Q of a dataset; (c) Q is directly density-reachable from P, with P_2 and P_3 also being core points; (d) both P and Q are density-reachable from point O, and P is density-connected to Q.

cluster C is a nonempty subset of D and satisfies the following conditions: (i) for any point P and point Q, if $P \in C$ and Q is density-reachable from P, then $Q \in C$; and (ii) for any P, if $Q \in C$, P is density-connected to Q. (7) Noise: if C_1, C_2, ..., C_k are the clusters in D with parameters Eps_i and $MinPts_i$, $i = 1, 2, ..., k$, the set of points in D not belonging to any cluster C_i is the noise.

The objective of DBSCAN is to find the largest set of density-connected points. To find a cluster, DBSCAN first selects a core point P from the dataset and retrieves all points that are density-reachable from P. Then all core points within the Eps–neighborhood of P are examined, and new points that are density-reachable from them are found and added to the cluster. This is repeated until no new points can be merged into the cluster. In the next step, a new core point that has not been clustered is selected, and the above process is repeated. Finally, all clusters are identified, and the remaining points not belonging to any cluster are taken to be noise points. DBSCAN does not require a predefined number of clusters and can identify noise points and clusters of an arbitrary shape.

5.5.4 Hierarchical Point Cloud Classification

Hierarchical point cloud classification is usually a rule-based classification, which relies on a series of expert rules to recognize different classes. An example of the hierarchical classification of point cloud data in an urban ecosystem is provided by Carlberg et al. (2009). Hierarchical classification starts by extracting objects that can be easily recognized. For example, water strongly absorbs near-infrared radiation (1064 nm or 1550 nm), which results in gaps or sparse points in the point cloud. This can be used to classify the water bodies first. Ground points are distinguished from nonground points in the next step, which is also referred to as point cloud filtering. More details can be found in Chapter 6. Then, vegetation and artificial objects can be classified using properly designed rules based on their distinct size and shape features. For instance, artificial objects, such as poles or building roofs, usually have noticeable linear or planar features, while the points of vegetation are more scattered.

After this coarse classification, the vegetation points are further classified into grass, shrub, or trees using structural characteristics such as height. It is worth noting that the accuracy of this step relies heavily on the accuracy of the preceding coarse segmentation. After the ground and vegetation points are identified, the remaining points stem mainly from artificial objects and are often clustered in the scene. Thus, quick and optimal segmentation can be achieved using the connected component label method. The unlabeled clusters are then recognized based on their features, such as the size, shape, and height distribution of segments. Shape recognition rules can also be based on the linearity, planarity, and scattering of a cluster (Carlberg et al., 2009).

5.5.5 Machine Learning-Based Point Cloud Classification

Supervised machine learning methods are commonly used in point cloud classification to learn the characteristics of different classes within a given feature space from labeled training data. The trained classifier can be used to predict the label for each point in a new dataset. Three typical supervised classification methods—decision tree (DT), support vector classification (SVC), and random forest (RF)—are introduced in this section.

(1) DT

A DT predicts the label of a target point after learning a series of simple decision rules that are in a treelike structure with different levels. The dataset is split into subsets at each node of the DT according to a tested attribute. The Gini index or entropy is often used to sort the attributes and estimate threshold values for attributes at different levels. The terminal leaf nodes correspond to different classes. The predicted class probability of an unlabeled point is the fraction of samples of the same class in a leaf $\widehat{\boldsymbol{L}}_{DT}$

$$\widehat{\boldsymbol{L}}_{DT} = \frac{\boldsymbol{M}}{\text{sum}(\boldsymbol{M})} \tag{5.49}$$

where $\boldsymbol{M}$ is a vector containing the number of samples per class. The predicted class of the unlabeled point $\widehat{l}_{DT}$ is the class with the highest probability.

$$\widehat{l}_{DT} = \arg\max_{c \in \mathscr{L}} \widehat{\boldsymbol{L}}_{DT}(c) \tag{5.50}$$

where $\mathscr{L}$ is the label space.

(2) SVC

For a binary classification problem, SVC tries to find a hyperplane that can correctly separate the data.

$$f(x) = \boldsymbol{\omega}^{T}\phi(\boldsymbol{x}) + b \tag{5.51}$$

where $\boldsymbol{\omega}$ and b are parameters of the hyperplane, and $\phi(\boldsymbol{x})$ maps the input data $\boldsymbol{x}$ into a higher-dimensional space to deal with data that are linearly inseparable. Many hyperplanes may correctly classify the data, and SVC aims to find the best hyperplane that maximizes its distance to the nearest point on both sides, which is expressed as:

$$\min_{\omega,b,\zeta} \frac{1}{2} \boldsymbol{\omega}^{T}\boldsymbol{\omega} + C\sum_{i=1}^{n} \zeta_i$$
$$\text{subject to } y_i(\boldsymbol{\omega}^{T}\phi(\boldsymbol{x}_i) + b) \geq 1 - \zeta_i, \tag{5.52}$$
$$\zeta_i \geq 0, i = 1, ..., n$$

where $C > 0$ is a regularization parameter; and ζ_i is a slack variable. This objective function can be solved using Lagrange multipliers and sequential minimal optimization (Chang, 2011) and $\boldsymbol{\omega} = \sum_{i=1}^{n} y_i\alpha_i\phi(\boldsymbol{x}_i)^{T}$. For a given point, its binary label can be determined using Eq. (5.53).

$$\widehat{l}_{\text{SVC}} = \text{sgn}\left(\sum_{i=1}^{n} y_i \alpha_i \phi(\mathbf{x}_i)^{\text{T}} \phi(\mathbf{x}) + b \right)$$

$$= \text{sgn}\left(\sum_{i=1}^{n} y_i \alpha_i \kappa(\mathbf{x}_i, \mathbf{x}) + b \right) \tag{5.53}$$

where $\kappa(\cdot, \cdot)$ is a kernel function that is introduced to reduce the computational burden of $\phi(\mathbf{x}_i)^{\text{T}} \phi(\mathbf{x})$. A variety of kernels are available, and the radial basis function is one of the most popular.

SVC can be easily extended to handle a multiclass task by using the one-vs.-one strategy. Traditional SVC does not provide a probability estimate for the predicted class. Platt scaling can be used in SVC to estimate the probability that a given input belongs to a specific class for both binary and multiclass tasks (Swami & Jain, 2013; Wu et al., 2003).

(3) RF

RF is an ensemble of unpruned DTs (Breiman, 2001; Chehata et al., 2009). Each DT is constructed from a subset of the training data, which is generated using bootstrap sampling. When splitting the data at a node in a DT, it is better to base the partition on a random subset of the attributes rather than all attributes, such that different DTs can have various structures. The predicted probability vector $\widehat{\mathbf{L}}_{\text{RF}}$ of an input sample is a vote by all DTs $\widehat{\mathbf{L}}_{\text{DT}}$ in the RF, weighted by their probability estimates.

$$\widehat{\mathbf{L}}_{\text{RF}} = \frac{1}{n} \sum_{i=1}^{n} \widehat{\mathbf{L}}_{\text{DT}i} \tag{5.54}$$

where $\widehat{\mathbf{L}}_{\text{DT}i}$ is the predicted probability vector from a DT, and n is the number of DTs in the RF. The predicted class $\widehat{l}_{\text{RF}}$ is the one with the highest mean probability estimate across the DTs.

$$\widehat{l}_{\text{RF}} = \underset{c \in \mathscr{L}}{\text{argmax}} \widehat{\mathbf{L}}_{\text{RF}}(c) \tag{5.55}$$

where $\mathscr{L}$ is the label space.

5.5.6 Deep Learning-Based Point Cloud Classification

Deep neural networks (DNNs) are a branch of neural networks that have developed rapidly in the past few years and have achieved the best performance in many computer vision and remote sensing classification tasks due to their ability to extract spatial-temporal features from massive data. DNNs have been used for processing LiDAR data in various aspects, such as classification and segmentation. In particular, 3-D DNNs such as PointNet (Qi et al., 2017), PointCNN (Li et al., 2018), and submanifold sparse convolutional networks (Graham et al., 2018) have broken through the bottlenecks of point cloud classification caused by the sparseness and randomness of point cloud data and provide promising "end-to-end" approaches to extract and learn the features in point cloud data.

PointNet is a new type of neural network that directly processes point clouds. It directly takes unordered point clouds as input and outputs class labels for the input points. PointCNN extends the typical convolutional neural networks for regular data to deal with irregular and disordered point cloud data. Experiments show that PointCNN can achieve equivalent or better performance compared with other advanced methods on a variety of challenging benchmark tasks. Hence, 3-D DNNs have been broadly applied in LiDAR point cloud processing and classification and are still evolving rapidly. Yousefhussien et al. (2018) used a revised PointNet model and proposed a fully convolutional architecture for semantically classifying 3-D point clouds with spectral information. This model could learn pointwise and blockwise features from point clouds with varying densities and improve the accuracy of point cloud classification in an urban environment. Zhang et al. (2018) fused 2-D images and 3-D point clouds to create a revised DNN approach to classify 3-D urban scenes. Using this approach, the fine features of buildings could be extracted efficiently, which was suitable for large-scale scenes. By projecting 3-D point clouds of individual trees to raster images from different perspectives, the classification of tree species was also achieved using DNNs (Zou et al., 2017). However, such projection or other types of data conversion may result in feature loss. Chen et al. (2021) proposed a new deep-learning framework for classifying tree species from point clouds, which used different sampling methods to extract species-specific local features and retain global features. The extracted spatial features and attribute features were then mapped to a high-dimensional space to complete tree species classification. This framework did not convert the input point cloud data into other types and could improve the accuracy of tree species classification by 2%−6% compared with other 3-D deep learning models. Guo et al. (2019) provided a comprehensive review of deep learning methods for 3-D point cloud classification (Fig. 5.22), and Guo et al. (2020) and Jin et al. (2020) discussed the challenges and opportunities for applying deep learning in ecological studies.

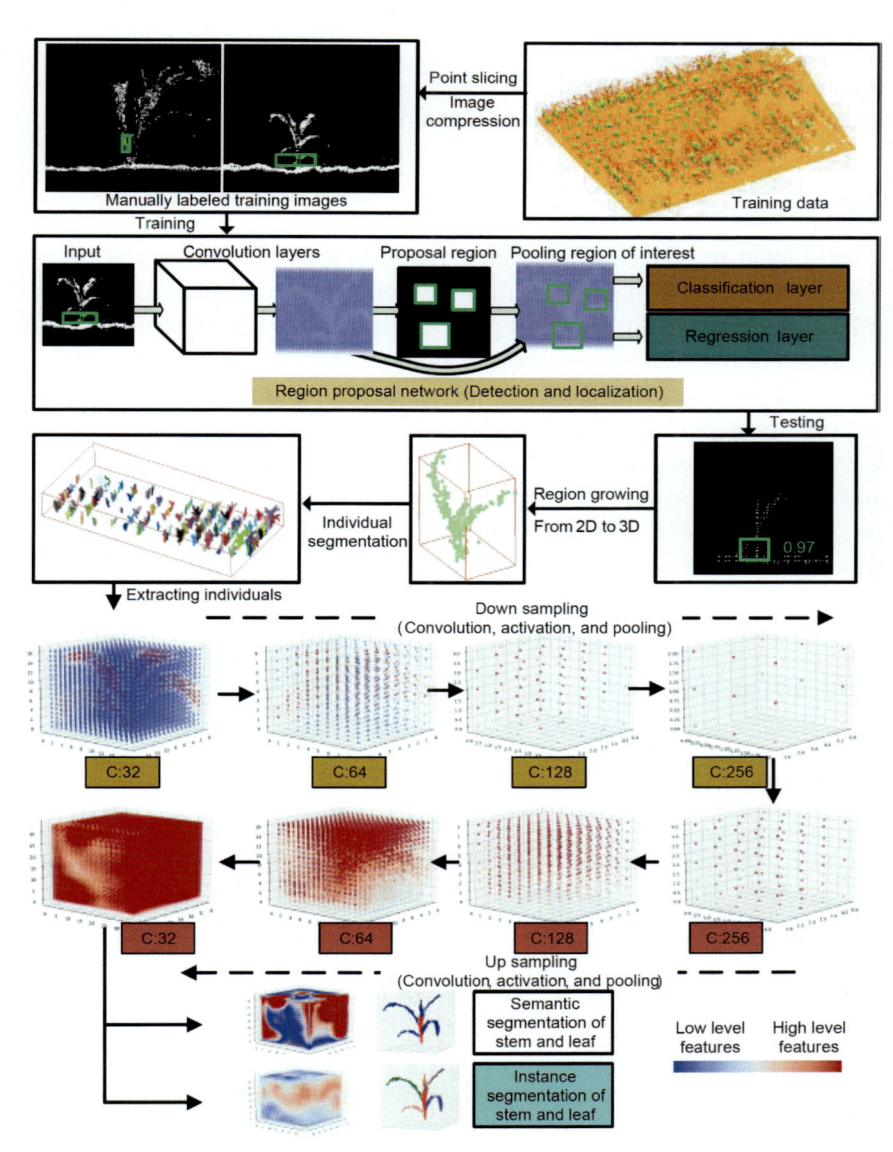

Figure 5.22 Applications of deep learning in 3-D point cloud classification of individual plants and organs.

Along with the development of DNNs, future research could endeavor to better fuse LiDAR data and other kinds of big data, such as satellite radar data, to improve classification accuracy. Efforts should also be given to clarify why deep learning can achieve satisfying results and develop hybrid models by coupling data- and physical-process-driven models for the ultimate purpose of improving ecosystem applications dominated by long-range spatial connections across multiple timescales.

5.6 Chapter Summary

This chapter provides a detailed description of LiDAR data preprocessing, including point cloud resolving, registration, feature extraction, and classification. These preprocessing steps are essential for extracting the structural attributes of forests from multisource point cloud data. Many state-of-the-art techniques and methods have been applied in point cloud preprocessing, such as integrated SLAM and GNSS-IMU for resolving and deep learning for point cloud classification. These different approaches have advantages and limitations in specific applications. Therefore, there is a growing research focus to take their advantages to develop a more comprehensive approach that can more effectively mine the substantial amount of information in point cloud datasets.

References

Aiger, D., Mitra, N. J., & Cohen-Or, D. (2008). 4-points congruent sets for robust pairwise surface registration. In *ACM SIGGRAPH 2008 papers* (pp. 1—10).

Akca, D., & Gruen, A. (2006). *Recent advances in least squares 3D surface matching.*

Bae, K.-H. (2006). *Automated registration of unorganised point clouds from terrestrial laser scanners.* Curtin University.

Bae, K. H., Belton, D., & Lichti, D. D. (2007). Pre-processing procedures for raw point clouds from terrestrial laser scanners. *Journal of Spatial Science, 52*(2), 65—74.

Ballard, D. H. (1981). Generalizing the Hough transform to detect arbitrary shapes. *Pattern Recognition, 13*(2), 111—122.

Besl, P., & McKay, N. (1992). *Method for registration of 3-D shapes* (Vol 1611). SPIE.

Borrmann, D., Elseberg, J., Lingemann, K., & Nüchter, A. (2011). The 3D Hough transform for plane detection in point clouds: A review and a new accumulator design. *3D Research, 2*(2), 3.

Boulaassal, H., Landes, T., Grussenmeyer, P., & Tarsha-Kurdi, F. (2007, 2007-09). Automatic segmentation of building facades using Terrestrial Laser Data. [International archives of photogrammetry, remote sensing and spatial information systems]. ISPRS workshop on laser scanning 2007 and SilviLaser 2007, Espoo, Finland.

Breiman, L. (2001). Random forests. *Machine Learning, 45*(1), 5—32.

Breunig, M. M., Kriegel, H.-P., Ng, R. T., & Sander, J. (2000). *Lof: Identifying density-based local outliers.* Proceedings of the 2000 ACM SIGMOD international conference on management of data, Dallas, Texas, USA.

Carlberg, M., Gao, P. R., Chen, G., Zakhor, A., & Ieee. (2009, Nov 07-10). Classifying urban landscape in aerial lidar using 3D shape analysis. *IEEE international conference on image processing ICIP* [2009 16th IEEE international conference on image processing, vols. 1—6]. 16th IEEE International Conference on Image Processing, Cairo, EGYPT.

Chang, C.-C., & Chih-Jen, L. (2011). LIBSVM: A library for support vector machines. *ACM Transactions on Intelligent Systems and Technology, 2*(3), 27:21–27:27.

Chehata, N., Guo, L., & Mallet, C. (2009). *Airborne lidar feature selection for urban classification using random forests*. Paris, France: Laserscanning.

Chen, J., Chen, Y., & Liu, Z. (2021). Classification of typical tree species in laser point cloud based on deep learning. *Remote Sensing, 13*(23), 4750.

Crombaghs, M., Brügelmann, R., & de Min, E. J. (2000). On the adjustment of overlapping strips of laser altimeter height data. *International Archives of Photogrammetry and Remote Sensing, 33*(B3/1), 230–237.

Durrant-Whyte, H., & Bailey, T. (2006). Simultaneous localization and mapping: Part I. *IEEE Robotics & Automation Magazine, 13*(2), 99–110.

Ester, M., Kriegel, H.-P., Sander, J., & Xu, X. (1996). A density-based algorithm for discovering clusters in large spatial databases with noise. In *kdd*.

Fischler, M. A., & Bolles, R. C. (1981). Random sample consensus: A paradigm for model fitting with applications to image analysis and automated cartography. *Communications of the ACM, 24*(6), 381–395.

Graham, B., Engelcke, M., & Van Der Maaten, L. (2018). 3d semantic segmentation with submanifold sparse convolutional networks. In *Proceedings of the IEEE conference on computer vision and pattern recognition*.

Greenspan, M., & Yurick, M. (2003). Approximate kd tree search for efficient ICP. In *Fourth international conference on 3-D digital imaging and modeling, 2003. 3DIM 2003. Proceedings*.

Guan, H., Su, Y., Hu, T., Wang, R., Ma, Q., Yang, Q., Sun, X., Li, Y., Jin, S., & Zhang, J. (2019). A novel framework to automatically fuse multiplatform LiDAR data in forest environments based on tree locations. *IEEE Transactions on Geoscience and Remote Sensing, 58*(3), 2165–2177.

Guan, H., Su, Y., Sun, X., Xu, G., Li, W., Ma, Q., Wu, X., Wu, J., Liu, L., & Guo, Q. (2020). A marker-free method for registering multi-scan terrestrial laser scanning data in forest environments. *ISPRS Journal of Photogrammetry and Remote Sensing, 166*, 82–94.

Guo, B., Huang, X., Zhang, F., & Sohn, G. (2015). Classification of airborne laser scanning data using JointBoost. *ISPRS Journal of Photogrammetry and Remote Sensing, 100*, 71–83.

Guo, Q., Jin, S., Li, M., Yang, Q., Xu, K., Ju, Y., Zhang, J., Xuan, J., Liu, J., & Su, Y. (2020). Application of deep learning in ecological resource research: Theories, methods, and challenges. *Science China Earth Sciences, 63*(10), 1457–1474.

Guo, Q., Liu, J., Tao, S., Xue, B., Li, L., Xu, G., Li, W., Wu, F., Li, Y., Chen, L., & Pang, S. (2014). Perspectives and prospects of LiDAR in forest ecosystem monitoring and modeling. *Chinese Science Bulletin, 59*(6), 459–479.

Guo, Y., Wang, H., Hu, Q., Liu, H., Liu, L., & Bennamoun, M. (2019). Deep learning for 3D point clouds: A survey. *IEEE Transactions on Pattern Analysis and Machine Intelligence, 43*(12), 4338–4364.

Hoppe, H., DeRose, T., Duchamp, T., McDonald, J., & Stuetzle, W. (1992). Surface reconstruction from unorganized points. In *Proceedings of the 19th annual conference on computer graphics and interactive techniques*.

Hough, P. V. (1962). Method and means for recognizing complex patterns. *US Patent, 3*(6).

Jain, A. K., Murty, M. N., & Flynn, P. J. (1999). Data clustering: A review. *ACM Computing Surveys (CSUR), 31*(3), 264–323.

Jin, S., Guo, Q., Li, M., Yang, Q., Xu, K., Ju, Y., Zhang, J., Xuan, J., Su, Y., Xu, Q., & Liu, Y. (2020). Application of deep learning in ecological resource research: Theories, methods, and challenges. *Science China Earth Science, 63*, 1457–1474.

Kashani, A. G., & Graettinger, A. J. (2015). Cluster-based roof covering damage detection in ground-based lidar data. *Automation in Construction, 58*, 19–27.

Kraus, K., Ressl, C., & Roncat, A. (2006). *Least squares matching for airborne laser scanner data.* Fifth International Symposium Turkish-German Joint Geodetic Days "Geodesy and Geoinformation in the Service of our Daily Life", (pp. 1–7).

Latypov, D. (2002). Estimating relative lidar accuracy information from overlapping flight lines. *Isprs Journal of Photogrammetry and Remote Sensing, 56*(4), 236–245.

Levoy, M., & Rusinkiewicz, S. (2001). Efficient variants of the ICP algorithm. In *Proceedings of the Third International Conference on 3-D Digital Imaging and Modeling,* (pp. 145–152).

Li, Y., Bu, R., Sun, M., Wu, W., Di, X., & Chen, B. (2018). Pointcnn: Convolution on x-transformed points. *Advances in Neural Information Processing Systems, 31.*

Li, S., Li, G., Wang, L., & Qin, Y. (2020). SLAM integrated mobile mapping system in complex urban environments. *Isprs Journal of Photogrammetry and Remote Sensing, 166,* 316–332.

Liu, X., Chen, Y., Li, S., Cheng, L., & Li, M. (2019). Hierarchical classification of urban ALS data by using geometry and intensity information. *Sensors (Basel), 19*(20), 4583.

Liu, S., Xiong, J., & Wu, B. (2011). ETWatch: A method of multi-resolution ET data fusion. *Journal of Remote Sensing, 15*(2), 255–269.

Maas, H.-G. (2000). Least-squares matching with airborne laserscanning data in a TIN structure. *International Archives of Photogrammetry and Remote Sensing, 33*(B3/1; PART 3), 548–555.

MacQueen, J. (1967). *Classification and analysis of multivariate observations.*

Papadimitriou, S., Kitagawa, H., Gibbons, P. B., & Faloutsos, C. (March 5–8, 2003). Loci: Fast outlier detection using the local correlation integral. *Proceedings 19th international conference on data engineering (cat. No.03CH37405).*

Pu, S., & Vosselman, G. (2006). Automatic extraction of building features from terrestrial laser scanning. *International Archives of Photogrammetry, Remote Sensing and Spatial Information Sciences, 36*(5), 25–27.

Qian, C., Liu, H., Tang, J., Chen, Y., Kaartinen, H., Kukko, A., Zhu, L., Liang, X., Chen, L., & Hyyppä, J. (2016). An integrated GNSS/INS/LiDAR-SLAM positioning method for highly accurate forest stem mapping. *Remote Sensing, 9*(1), 3.

Qi, C. R., Su, H., Mo, K., & Guibas, L. J. (2017). Pointnet: Deep learning on point sets for 3d classification and segmentation. In *Proceedings of the IEEE conference on computer vision and pattern recognition.*

Rusu, R. B., Blodow, N., & Beetz, M. (2009). Fast point feature histograms (FPFH) for 3D registration. In *2009 IEEE international conference on robotics and automation.*

Schall, O., Belyaev, A., & Seidel, H. P. (June 21–22, 2005). Robust filtering of noisy scattered point data. *Proceedings Eurographics/IEEE VGTC Symposium Point-Based Graphics.*

Schnabel, R., Wahl, R., & Klein, R. (2007). *Efficient RANSAC for point-cloud shape detection.* Computer graphics forum.

Simonse, M., Aschoff, T., Spiecker, H., & Thies, M. (2003). Automatic determination of forest inventory parameters using terrestrial laser scanning. In *Proceedings of the scandlaser scientific workshop on airborne laser scanning of forests.*

Swami, A., & Jain, R. (2013). Scikit-learn: Machine learning in Python. *Journal of Machine Learning Research, 12,* 2825–2830.

Teboul, O., Simon, L., Koutsourakis, P., & Paragios, N. (2010). Segmentation of building facades using procedural shape priors. In *2010 IEEE computer society conference on computer vision and pattern recognition.*

Tung, C.-H. (2005). *Airborne LIDAR systematic error analysis and strip adjustment master thesis for the department of geomatics.* National Cheng Kung.

Vo, A. V., Linh, T. H., Laefer, D. F., & Bertolotto, M. (2015). Octree-based region growing for point cloud segmentation. *Isprs Journal of Photogrammetry and Remote Sensing, 104,* 88–100.

Vosselman, G., Gorte, B. G., Sithole, G., & Rabbani, T. (2004). Recognising structure in laser scanner point clouds. *International Archives of Photogrammetry, Remote Sensing and Spatial Information Sciences, 46*(8), 33–38.

Vosselman, G., & Haralick, R. M. (1996). Performance analysis of line and circle fitting in digital images. In *Proc. Workshop Performance Characteristics of Vision Algorithms*.

Vosselman, G., & Maas, H. G. (2010). *Airborne and terrestrial laser scanning*. CRC Press.

Weinmann, M., Jutzi, B., Hinz, S., & Mallet, C. (2015). Semantic point cloud interpretation based on optimal neighborhoods, relevant features and efficient classifiers. *ISPRS Journal of Photogrammetry and Remote Sensing, 105*, 286–304.

Wu, T.-F., Lin, C.-J., & Weng, R. (2003). Probability estimates for multi-class classification by pairwise coupling. *Advances in Neural Information Processing Systems, 16*.

Xu, W.-x., Kang, Z.-z., & Jiang, T. (2009). Segmentation approach for terrestrial point clouds based on the integration of graph theory and region growing. In *2009 joint urban remote sensing event*.

Yousefhussien, M., Kelbe, D. J., Ientilucci, E. J., & Salvaggio, C. (2018). A multi-scale fully convolutional network for semantic labeling of 3D point clouds. *ISPRS Journal of Photogrammetry and Remote Sensing, 143*, 191–204.

Zhang, R., Li, G., Li, M., & Wang, L. (2018). Fusion of images and point clouds for the semantic segmentation of large-scale 3D scenes based on deep learning. *ISPRS Journal of Photogrammetry and Remote Sensing, 143*, 85–96.

Zhu, D., Guo, H., & Su, W. (2012). *Tutorials on point cloud library*. Beihang University Press.

Zou, X., Cheng, M., Wang, C., Xia, Y., & Li, J. (2017). Tree classification in complex forest point clouds based on deep learning. *IEEE Geoscience and Remote Sensing Letters, 14*(12), 2360–2364.

CHAPTER 6

LiDAR Data Filtering and Digital Elevation Model Generation

Contents

A digital elevation model (DEM) is a 3-D digital cartographic dataset representing terrain. Elevation has an important influence on ecosystem

LiDAR Principles, Processing and Applications in Forest Ecology
ISBN 978-0-12-823894-3
https://doi.org/10.1016/B978-0-12-823894-3.00006-2

processes. Hence, acquiring an accurate DEM is essential for forest ecology studies. Because laser pulses can penetrate vegetation and obtain information about terrain, light detection and ranging (LiDAR) has revolutionized the generation of high-resolution and high-precision DEMs. The prerequisite of LiDAR-based DEM generation is to separate the ground points from the nonground points, a process known as LiDAR data filtering. This chapter provides an overview of mainstream filtering algorithms and analyzes their effectiveness in various environmental settings. We also introduce the methods for LiDAR-based DEM generation and discuss the factors that affect the accuracy of DEM generation.

6.1 Introduction to LiDAR Data Filtering

6.1.1 Basic Concepts

Laser pulses emitted from light detection and ranging (LiDAR) systems are intercepted and reflected by objects both on and above the ground, including natural vegetation, artificial objects, and moving objects. As shown in Fig. 6.1, the information obtained by a LiDAR system typically includes the following categories:

- terrain surfaces, such as natural abiotic surfaces (e.g., bare soil, sand, and rock outcrops) and artificial surfaces (e.g., roads)—point clouds of

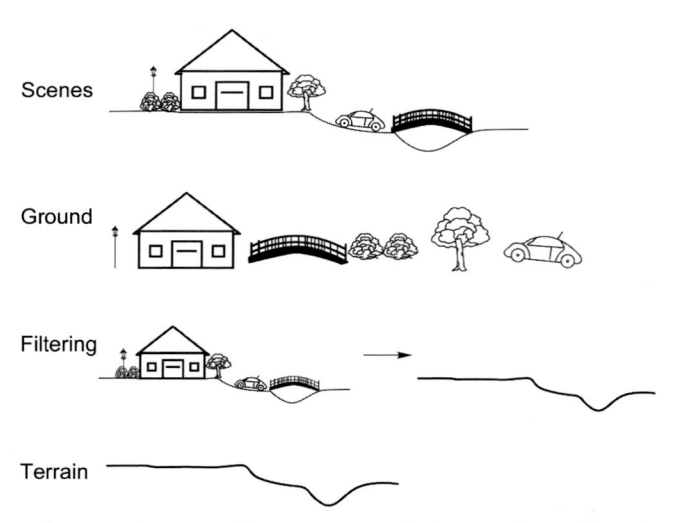

Figure 6.1 Schematic diagram of the categories of objects collected by light detection and ranging (LiDAR).

terrain surfaces are preserved after filtering and form the basis for generating a digital elevation model (DEM);

- vegetation, such as forests, shrubs, grasslands, and crops;
- artificial ground objects, such as buildings, electricity poles and pylons, and bridges—artificial ground objects can be further divided into separable and associated ground objects; separable ground objects, such as power lines and buildings, have an obvious boundary on the terrain surface; conversely, the surfaces of associated ground objects (e.g., bridges) are merged with the terrain surface;
- moving objects, such as vehicles, pedestrians, and animals;
- noise, such as high-altitude noise and low-altitude noise—high-altitude noise is mainly caused by suspended particulate matter or flying objects (e.g., birds and airplanes), whereas low-altitude noise is mainly caused by the multipath effect of laser echoes.

LiDAR data filtering aims to identify ground points from LiDAR data. After filtering, the raw point cloud data are labeled as ground and non-ground points. The main difference between LiDAR data filtering and classification is that LiDAR data classification further separates the non-ground points into categories, including vegetation, buildings, and roads. LiDAR data filtering and classification differ in their purpose and methods (Sánchez-Lopera & Lerma, 2014). After filtering, the identified ground points can be used to generate a DEM using interpolation algorithms. A DEM is a three-dimensional (3-D) digital expression of terrain elevation. It is widely used in topographical mapping, urban modeling, hydrological and ecological research, forest management, environmental monitoring, disaster prediction, and other studies (Griffin, 1990). By interpolating the first echoes of LiDAR data, a digital surface model (DSM) can also be generated. In contrast to a DEM, a DSM is a digital expression of the surface morphology of all objects, including ground objects (Fig. 6.2).

6.1.2 Challenges in LiDAR Data Filtering

Many LiDAR data filtering algorithms have been proposed, and most perform well in simple scenes. In complex areas, however, LiDAR data filtering accuracy is reduced. The robustness of current LiDAR filtering methods requires further improvement. The main challenges of LiDAR data filtering are as follows:

- *Low-altitude noise.* Low-altitude noise typically exists in point cloud data in an independent or clustered form. Owing to the closeness of the

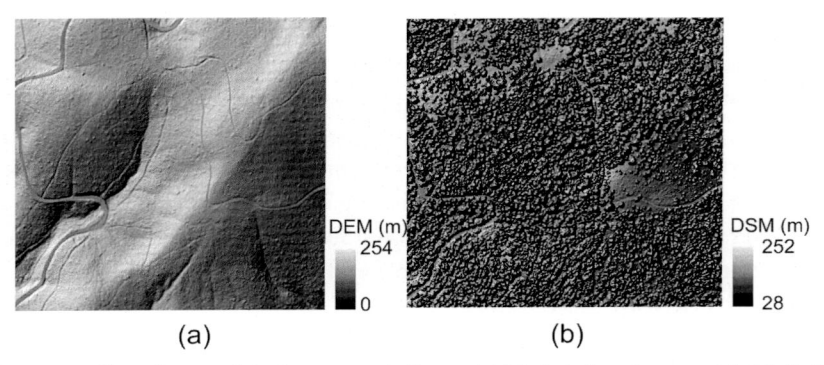

DEM (m)
254

0

(a)

DSM (m)
252

28

(b)

Figure 6.2 Elevation model of a mountain forest: (a) Digital elevation model; (b) digital surface model. *DEM*, digital elevation model; *DSM*, digital surface model. *(Data acquisition platform: Airborne; acquisition time: September 2007; location: Sierra Nevada mountains, USA.)*

terrain surface and low-altitude noise, low-altitude noise can be easily misclassified as ground points, leading to erroneous filtering.

- *Dense vegetation.* In areas of dense vegetation, it is difficult for laser pulses to penetrate the canopy to acquire ground information. Thus, the paucity of ground points may cause filtering errors.
- *Low objects.* Low objects, especially shrubs and grasslands, are often misclassified as terrain surface as they have a low height and are close to the ground.
- *Steep terrain.* In steeply sloping areas, ground points may be misclassified as nonground points, resulting in a flattened DEM. If the elevation of the area of interest changes abruptly, such as along cliffs and fracture lines, a DEM generated with low-accuracy filtering may not reflect the actual topographic changes.
- *Missing data.* During LiDAR data acquisition, missing data issues may occur when LiDAR pulses encounter water areas or other objects that do not reflect echoes. An insufficient overlap of adjacent strips and certain types of terrain (e.g., cliffs) may also cause missing data issues. Missing data issues may lead to discontinuous terrain in the point cloud and increase filtering errors.
- *Complex scenarios.* Most filtering algorithms have poor adaptability to complex scenarios. Complex scenarios, such as buildings with varying sizes and shapes, dense vegetation, rivers, hills, mountains, and narrow streets, may greatly decrease filtering accuracy.

- *Attachments.* Attachments (e.g., bridges) are difficult to be distinguished from terrain because they are usually closely connected to the ground.

6.2 Introduction to Filtering Methods

Filtering is a vital procedure in the LiDAR data processing workflow. Researchers have used various principles to develop filtering algorithms over the past two or three decades. These filtering algorithms can be divided into four categories: slope-, region-, surface-, and clustering-based (Sithole & Vosselman, 2004). In recent years, deep learning techniques have been developed, bringing new opportunities for LiDAR data filtering. This section introduces some of the most representative and widely used filtering methods.

6.2.1 Slope-Based Filtering Methods

The assumption of the slope-based filtering algorithms is that large height differences among adjacent points are caused by ground objects rather than steep terrain and that the higher point is a nonground point. The closer the two points are within a certain height difference, the greater the probability that the higher point belongs to an object rather than the ground (Fig. 6.3).

Vosselman (2000) proposed a slope-based filtering method in which the height difference between two ground points is used to determine the optimal filter function $\Delta h_{\max}(d)$. To solve the filter function, Vosselman

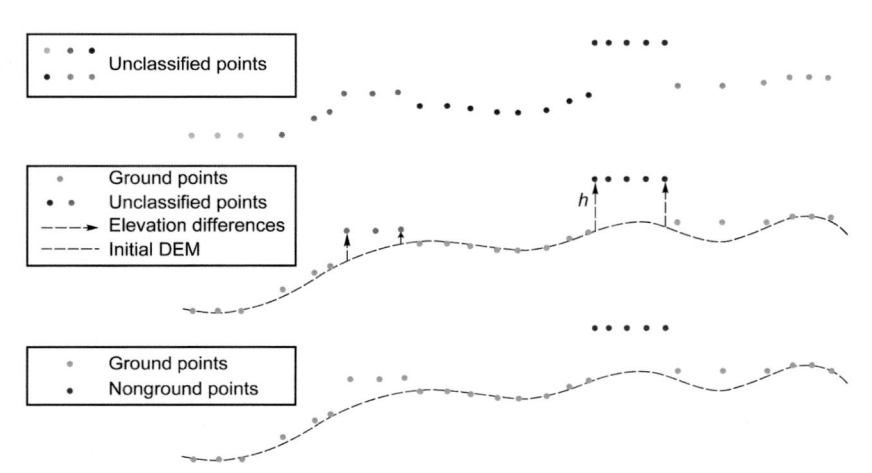

Figure 6.3 Schematic diagram of the slope-based filtering algorithm. *DEM*, digital elevation model; *h*, elevation difference.

(2000) introduced two methods. The first approach is based on the observation that most slopes are not steeper than 30%. With a 95% confidence interval, the filter function $\Delta h_{\max}(d)$ can be defined as follows:

$$\Delta h_{\max}(d) = 0.3d + 1.65\sqrt{2}\sigma \tag{6.1}$$

where d is the distance between a point pair (p_i, p_j). Then, the set of ground points for generating DEM could be defined as

$$\text{DEM} = \left\{ p_i \in A \middle| \forall\, p_i \in A : h_{p_i} - h_{p_j} \leq \Delta h_{\max}\big(d\big(p_i, p_j\big)\big) \right\} \tag{6.2}$$

where A refers to the set of all points. According to Eq. (6.3), a point p_i is classified as a terrain point if the height differences between p_i and all possible p_j are all smaller than $\Delta h_{\max}\big(d\big(p_i, p_j\big)\big)$. However, it is difficult to specify a threshold when the terrain changes abruptly.

The second approach is to use a training method to obtain the shape characteristics of the terrain. To ensure training accuracy, sample points should be selected carefully so they represent true ground points. The filter function is trained using the maximum height difference of the sample point pairs. Let $f(\Delta h)$ be the cumulative probability distribution for the height differences between sample points at distance d, then $F_{\max}(\Delta h) = F(\Delta h)^N$ represents the maximum cumulative probability distribution of N independent point pairs in the training samples at a height difference Δh. Therefore, the probability density function of the maximum height difference can be acquired by

$$f_{\max}(\Delta h) = \frac{\partial F_{\max}(\Delta h)}{\partial \Delta h} = NF(\Delta h)^{N-1} f(\Delta h) \tag{6.3}$$

Although this training method can preserve the terrain well, it tends to misclassify the nonground points as ground points.

Wang and Tseng (2010a) proposed an adaptive bilateral filtering algorithm based on slope. The filter works in multiple directions and thus can avoid the excessive filtering of steep areas. The results show that this method can preserve the topographic features of regions with substantial slope changes. Susaki (2012) generated an initial DEM using a set of slope parameters and extracted the initial ground points by calculating the local maximum slope and performing a plane estimation. Then, they used an iterative process to update the slope parameters until an optimal set of ground points was found. This method performed well in urban areas with many streets and buildings. Overall, the slope-based filtering methods are suitable for areas with flat slopes but have relatively poor accuracy in steeply sloping areas.

6.2.2 Morphological Filtering Methods

The theoretical basis of morphological filtering methods is mathematical morphology, which uses structuring elements to extract the topological structure information from images or point clouds (Haralick & Shapiro, 1992). Four basic morphological operations—namely dilation, erosion, opening, and closing—form the foundation of mathematical morphology. Let f be the target image and b be a structuring element. The erosion of f by leads to a subset of f in which b can be fully included in f. (Eq. 6.4). The dilation of f gives a set ensuring that the intersection between f and b is not empty (Eq. 6.5). Opening is erosion followed by dilation (Eq. 6.6), and closing is dilation followed by erosion (Eq. 6.7). Both opening and closing have the effect of smoothing f. As illustrated in Fig. 6.4, opening removes small objects from the foreground of an image (usually seen as large objects) and places them in the background, whereas closing removes small holes in the foreground, changing small "islands" of the background into the foreground:

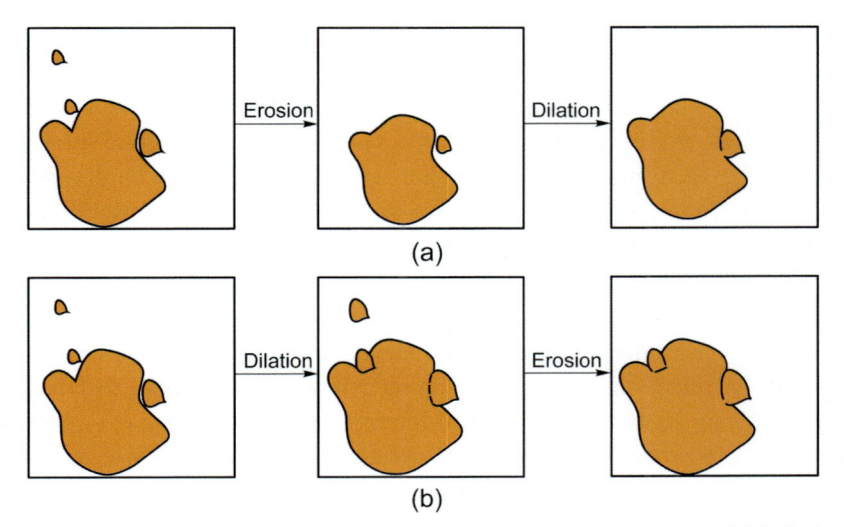

(a)

(b)

Figure 6.4 Illustration of basic morphological operations: (a) opening and (b) closing.

$$f \ominus b = \{x | b(x) \subset f\} \tag{6.4}$$

$$f \oplus b = \{x | b(x) \cap f \neq \phi\} \tag{6.5}$$

$$f \circ b = (f \ominus b) \oplus b \tag{6.6}$$

$$f \cdot b = (f \oplus b) \ominus b \tag{6.7}$$

The size and shape of the chosen structuring element depend on the topological structure to be extracted from the dataset. Generally, structuring elements can be divided into symmetrical elements and asymmetrical elements. Because asymmetrical elements can cause shifts in results, current morphological operations are mainly based on symmetrical shapes such as disks, squares, diamonds, hexagons, or line segments.

The workflow of morphological filtering methods typically contains the following steps. First, the raw LiDAR data is rasterized into grids, and the lowest point in each grid is extracted. All grid cells are initially marked as ground cells. During rasterization, interpolation is usually used to fill empty cells. Second, the slope of each ground cell is calculated according to the neighboring ground cells. The elevation difference threshold of each ground cell is subsequently defined according to the slope. Third, the rasterized image is fed into the opening operation with a specified sliding window. If the elevation difference of a ground cell before and after the opening operation is larger than the elevation difference threshold, the corresponding cell is marked as a nonground cell; otherwise, it remains a ground cell. Fourth, the second and third steps are repeated with an increasing sliding window size until the predefined maximum size is reached. Last, points from raw LiDAR data in the same horizontal space are selected for each ground cell, and their elevation difference to the cell is calculated. If the elevation difference is smaller than a specific threshold, the corresponding point will be marked as a ground point.

The first step of morphological filtering is point cloud rasterization, which aims to convert a point cloud to a raster with a certain grid size, in which the cell value is defined as the elevation of the lowest point within the cell. Kilian et al. (1996) used sliding windows from small to large to implement successive morphological filtering of points. After each opening operation, the weight of each cell is defined by the window size, and the final filtering result is determined by the weights of cells from the final iteration. Zhang et al. (2003) rasterized a point cloud by extracting the minimum value of each grid. They then used an opening operation to process the generated grid and calculated the elevation difference of the grid before and after the opening operation. If the elevation difference was smaller than a given threshold, the corresponding cell was deemed a ground cell. After that, they increased the size of the sliding windows iteratively according to the elevation difference and repeated the opening operation to obtain the final filtering results. Their method assumed that the terrain slope is continuous within a certain range. However, this assumption may not be valid in some environments, especially in complex scenarios. Hence, this method is unsuitable for urban areas where the topography varies greatly.

Chen et al. (2013) improved the algorithm proposed by Zhang et al. (2003) using the following process:

(1) *Rasterizing point clouds and filling empty cells.* The point cloud data are rasterized into a binary raster with a grid size of 1 m × 1 m. If a cell is not associated with any points, its value is defined as 0; otherwise, its value is defined as 1. Then, an opening operation using a disk structuring element is applied to the binary raster to locate the holes in the raster. The equation of the structuring element is shown in Eq. (6.8), in which r refers to the radius, d refers to the average point density, and c represents the grid size. After the opening operation, the values of small holes are replaced by their nearest values, whereas the values of large holes are the lowest value on the boundary:

$$r = \left(\frac{1}{d}\right)^{0.5} \times \frac{1}{c} \tag{6.8}$$

(2) *Removal of tree points.* A smaller sliding window is applied to remove discrete data, such as trees, to obtain a relatively smooth terrain surface.

(3) *Removal of low-altitude noise.* High-altitude noises can be removed by a morphological opening operation. However, the removal of low-altitude noise is much more complex. First, an extended local minimum method is used to select a set of points, the heights of which are lower than a certain height threshold h. These points may either be terrain points or low-altitude noise points. To remove terrain points from this point set, the lowest points area with the minimum values is calculated. If the area is smaller than an area threshold a, the corresponding points are referred to as low-altitude noise points; otherwise, they are referred to as terrain points. After removing low-altitude noise, the empty cells are filled using the method in step (1).

(4) *Removal of building points.* The size of the sliding window for the opening operation is enlarged iteratively. If the height difference of a region
between two consecutive opening operations is less than 1 m and the
area of this region is less than a certain area threshold, this region is
regarded as terrain; otherwise, it is allocated to a subsequent judgment
procedure to decide whether it is a building. This judgment procedure
assumes that the abnormal elevation change caused by a building usually has a large height difference along the boundary, in contrast to the
abnormal elevation change caused by topographic relief. If the boundary of a remaining region conforms to the abovementioned pattern, it is
marked as a building, and the results of the opening operation are
updated. The sliding window size is gradually increased, and the opening operation is repeated until all buildings have been identified.

(5) *Labeling ground points.* By subtracting the initial raster from the final
raster, a cell is marked as a ground cell if the height difference of this
cell is less than 0.5 m. For each ground cell, if the elevation difference
from its associated point to the cell is small than 0.5 m, the corresponding point is marked as a ground point.

Pingel et al. (2013) improved the morphological filtering algorithm
from the following three aspects: (1) filling holes in the raster data using a
resampling technique; (2) using a linear growth method instead of an
exponential method to increase the size of the sliding window; and (3)
simplifying the process of determining a slope threshold.

One advantage of morphological filtering methods is that they can be
easily implemented. In morphological filtering, multiple opening operations that iteratively increase the sliding window size perform better than a
single opening operation. However, the threshold of elevation difference
and the maximum size of the sliding window in morphological filtering
need to be defined manually. Another factor that hinders the use of
morphological filtering methods is the difficulty of distinguishing regions
with steep terrain and regions with ground objects. The elevation differences of these regions before and after the opening operation are similar,
resulting in low-accuracy filtering if using a simple elevation difference
threshold. Moreover, the rasterization of the point cloud may cause additional errors. A larger grid size may cause more ground points to be misclassified as nonground points, while a smaller grid size may cause the
incomplete filtering of nonground points.

The following code illustrates a morphological filtering algorithm
implemented in Python language.

```python
# The following code is a morphological filtering algorithm implemented using Python language
# -*- coding: UTF-8 -*-
# Import necessary libraries for morphological filtering
import laspy
import numpy as np
import pandas as pd
from scipy import ndimage
#las file location
lasfile = "./filter.las"
# Open the las file
inFile = laspy.file.File(lasfile, mode = "r")
# Get the coordinates of point cloud
x,y,z = inFile.x,inFile.y,inFile.z
# Get the data range of point cloud
xmin,xmax=min(x),max(x)
ymin,ymax=min(y),max(y)
#rasterize
# Set resolution 0.5m * 0.5m
pixelWidth,pixelHeight=0.5,-0.5
nrow=int(np.ceil(abs((ymax-ymin)/pixelHeight)))
ncol=int(np.ceil(abs((xmax-xmin)/pixelWidth)))
# Set the initial grid
int_dem_mask = np.zeros((nrow,ncol),dtype='float')
int_dem_mask[:,:] = -9999.0
# Set the grid start point
xOrigin = xmin
yOrigin = ymax
# Calculate the position of the point cloud in the grid
xOffset = (x - xOrigin) / pixelWidth
xOffset = xOffset.astype(int)
yOffset = (y - yOrigin) / pixelHeight
yOffset = yOffset.astype(int)
# Find the lowest elevation value in each grid
points = np.column_stack([xOffset,yOffset,z])
pts = pd.DataFrame(points,columns=['Off_x','Off_y','z'])
min_raster=np.asarray(pts.pivot_table(columns='Off_x',index='Off_y',values='z',aggfunc='min'))
```

```python
# Assign none value grid with its adjacent point value
index = np.where(np.isnan(min_raster))
for i in range(len(index[0])):
    ix,iy=index[0][i],index[1][i]
    # Search for the domain data. If no data is found, increase the stride size by 2 and continue to search
    Fill_Flag = True
    win_size = 1
    while Fill_Flag:
        win_size = 1 + 2
        x_index = range(ix-(win_size-1)/2, ix+(win_size-1)/2+1)*win_size
        y_index = np.repeat(range(iy-(win_size-1)/2, iy+(win_size-1)/2+1),win_size)
        # Remove the outer points of the grid to avoid program errors
        mask_x =  ((np.asarray(x_index)<0) | (np.asarray(x_index)>ncol -1))
        mask_y =  ((np.asarray(y_index)<0) | (np.asarray(y_index)>nrow -1))
        win_mask = mask_x + mask_y
        win_index = ([np.asarray(x_index)[~win_mask],np.asarray(y_index)[~win_mask]])
        win_data = min_raster[win_index]
        #Remove null data NaN
        win_data = win_data[~np.isnan(win_data)]
        if len(win_data) !=0:
            min_raster[ix,iy]= win_data[0]
            min_raster[ix,iy]= np.mean(win_data)
            Fill_Flag = False
# Set the height difference parameter
dh_max,dh0 = 1.0.1
# Set slope boosting
s=0.7
# Set the number of window iterations for morphological filtering
for k in range(1.11):
    # Sets the initial window size
    b=1 # Linear growth
    wk = 2*k*b+1
    # Open operation
    tmp_raster = ndimage.grey_opening(min_raster,size=(wk,wk))
    # Calculates the height difference based on the window size
    if wk <= 3:
```

```
        dh = dh0
    else:
        wk_l = 2*(k-1)*b+1
        dh = s*(wk-wk_l)*pixelWidth + dh0
    if dh > dh_max:
        dh=dh_max
    # Mark ground points
    for i_row in range(nrow):
        for i_col in range(ncol):
            # Obtain index of points in grid
            cls_index = np.where((points[:,0]==i_col) & (points[:,1]==i_row))[0]
            # Judge the distance threshold
            dh_index = np.where((points[cls_index,2]-tmp_raster[i_col,i_row]) > dh)
            if len(dh_index[0])>0:
                pts_cls[cls_index[dh_index]]=1
    # Display the number of ground points after each iteration
    print 'Ground points:' + str(len(np.where(pts_cls[:]==0)))
# Finally, the pts_cls variable stores information about which points are ground and nonground.
```

6.2.3 Interpolation-Based Filtering Algorithms

The workflow of interpolation-based filtering algorithms is as follows. First, an initial terrain surface is generated. Then, the unclassified points are determined according to specific criteria, and new ground points are added iteratively until an accurate terrain surface is generated after satisfying the requirements. Kraus and Pfeifer (1998) proposed an iterative least-squares interpolation-based filtering algorithm based on linear prediction (Kraus & Mikhail, 1972), which includes the following steps:

(1) A curved surface is computed using least-squares interpolation with equal weights for all points. The curved surface can be treated as an initial terrain surface between the actual terrain surface and the land surface.

(2) The weight of each point is computed using the residual v_i from the point to the curved surface. The terrain points tend to have negative residuals, whereas the vegetation points are more likely to have slightly negative or positive residuals. The value of weight W_i is calculated as follows:

$$W_i = \begin{cases} 1 & v_i \leq g \\ \dfrac{1}{1 + \left(a + (v_i - g)^b\right)} & g < v_i \leq g + w \\ 0 & g + w < v_i \end{cases} \qquad (6.9)$$

where parameters a and b determine the slope of the weight, and g is a shift value. In practice, a and b can be set as 1 and 4, respectively. The shift value g is usually negative and determines the weight of a point. Points with negative residuals are more likely to be terrain points and, therefore, have a weight of one. Points with large positive residuals greater than $g + w$ are given a weight of zero. Kraus and Pfeifer (1998) used an adaptive method to change the value of g for each iteration. The value of parameter w is relatively small, indicating that only points close to the curved surface have significant impact on the final result.

(3) After calculating the weights, points with a weight of 0 are eliminated in the interpolation to update the curved surface. The remaining points are used to calculate a new curved surface.

(4) The above process is repeated until the final curved surface reflects the true terrain. The workflow of this interpolation-based filtering algorithm is shown in Fig. 6.5.

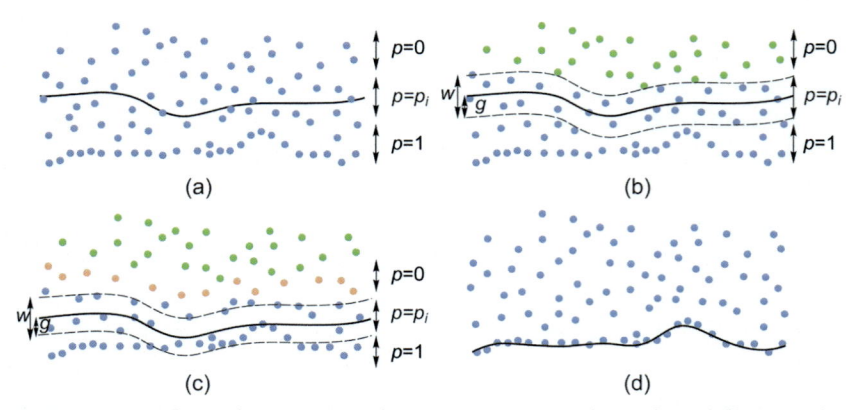

Figure 6.5 Workflow of the iterative least-squares interpolation-based filtering algorithm. (a) Linear least-squares interpolation of all points with equal weights to obtain initial terrain surface. (b) Calculation of residuals of all points to the interpolated surface and calculation of weight W_i according to the residuals and values of g and w. (c) Removal of the points with a weight of zero followed by least-squares interpolation of the remaining points to obtain a new terrain surface. (d) Final terrain surface obtained from multiple iterations of (b) and (c).

Although the least-squares interpolation-based algorithm can create an accurate terrain surface, a prerequisite is homogeneous distribution of terrain points and vegetation points. Therefore, this method is unsuitable for urban areas and steeply sloping terrain in which points are usually distributed in clusters with few ground points.

Lee and Younan (2003) replaced the traditional least-squares method with a normalized least-squares method for filtering. Chen et al. (2013) used a multiresolution hierarchical classification algorithm for filtering and designed three levels for classifying ground and nonground points based on the elevation differences between sampling points and the interpolated terrain surface. At each level, a thin plate spline is used to iteratively generate a terrain surface until no further ground points can be classified. The classified ground points are then used to update the terrain surface in the subsequent iterations. From the low to the high level of the hierarchy, they used the results of the first layer to determine the data type in the second layer and repeated this process for the second and third layers. This method does not require prior knowledge of the complexity of the terrain surface, and it works well in flat and steeply sloping areas.

Kobler et al. (2007) proposed a repetitive interpolation method for point cloud filtering in forested environments. The workflow of this method is summarized as follows:

(1) The proposed method is applied after initial filtering, removing noise and most nonground points.
(2) Multiple sets of points are sampled independently and randomly from the filtered point cloud during each iteration. Based on the samples, ground elevation at these locations is estimated using spatial interpolation methods. A global mean offset can be estimated by averaging all ground elevation estimates at the selected sample locations. Subsequently, the final elevation at each sample location can be computed by adding the global mean offset to the lowest elevation at each sample location. After repeating the above processes several times, the real ground elevation can be obtained.

Mongus and Žalik (2012) used a sliding window with an iteratively decreasing size to implement thin plate spline interpolation for filtering. Their method uses a top-hat transformation to solve the issue of discontinuous point cloud heights caused by ground objects such as buildings. During the interpolation, a multiresolution hierarchical structure is used to organize the control points of the thin plate splines, and the nonground points are filtered according to the residuals of the points after each iteration. The elevations of these identified nonground points are replaced with the interpolated values from the thin plate spline interpolation. The above processes are then repeated until the required resolution of the terrain raster is obtained. This method performs well in areas where point cloud filtering is challenging (e.g., areas with distinct topographic features or areas consisting of objects with various sizes and shapes).

6.2.4 Progressive Densification Filtering Methods

The principle of progressive densification filtering is to select the initial ground points to construct the initial terrain according to a predetermined strategy and then iteratively increase the ground points according to specific criteria until all points are classified. Axelsson (2000) proposed a progressive triangulated irregular network (TIN) densification algorithm. First, a grid is used to extract the lowest point of each grid cell as the ground point P_G. An initial terrain surface is obtained from the ground points extracted by TIN generation. Subsequently, unclassified points $P_{unlabelled}$ are projected to the TIN and examined in turn. For each P_i, if its distance to the nearest TIN facet is less than the predefined distance threshold, and if the angle between the TIN facet and the line connecting the point with the facet's closest vertex is less than a threshold angle, P_i is determined as the ground point (Fig. 6.6). Then, the ground point P_i is used to rebuild the TIN. The above processes are iterated multiple times until no new ground points are added. The workflow of this method is shown in Fig. 6.6.

The classic progressive TIN densification method is sensitive to steep terrains and broken line features. Zhang and Lin (2013) improved the method by adding several initial seed points based on a region-growing method. To obtain a more accurate terrain surface, we proposed an improved progressive TIN densification method in forested environments by combining progressive TIN densification with a morphological method (Zhao et al., 2016). The workflow of this method is shown in Fig. 6.7. First, an opening operation eliminates most nonground points in the initial seed points. Second, a translation plane fitting method removes points with large residuals. Afterward, the remaining points are treated as seed points for a downward progressive TIN densification procedure to improve the ability to deal with slope variations. Lastly, an upward densification procedure is performed to obtain the final filtered results.

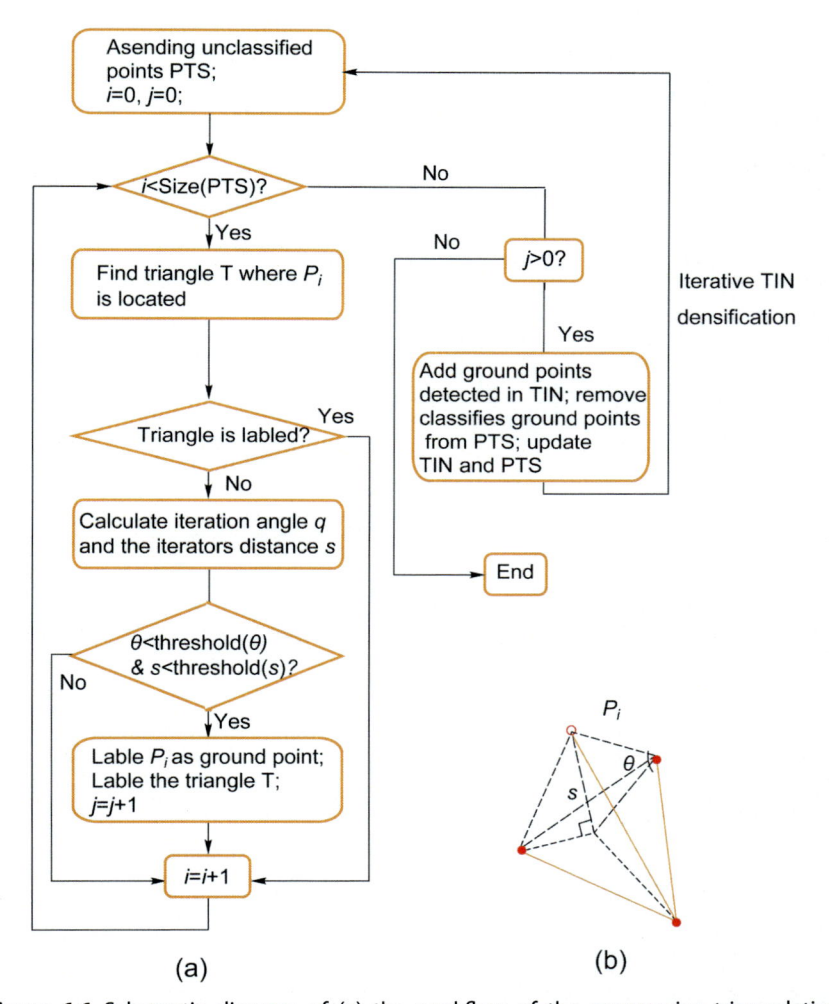

Figure 6.6 Schematic diagram of (a) the workflow of the progressive triangulation densification and (b) the updating of parameters. i, index of points in PTS; j, number of points added to TIN in each iteration; *PTS*, ascending unclassified points; *size(PTS)*, size of PTS; *threshold* (θ), angle threshold; *threshold*(s), distance threshold; *TIN*, triangulated irregular network.

The improved progressive TIN densification method was validated in 15 500 m × 500 m sites. These sites were randomly selected in the southern Sierra Nevada mountains, California, USA, in which the dominant tree species are white fir (*Abies concolor*), ponderosa pine (*Pinus ponderosa*), incense cedar (*Calocedrus decurrens*), sugar pine (*Pinus lambertiana*),

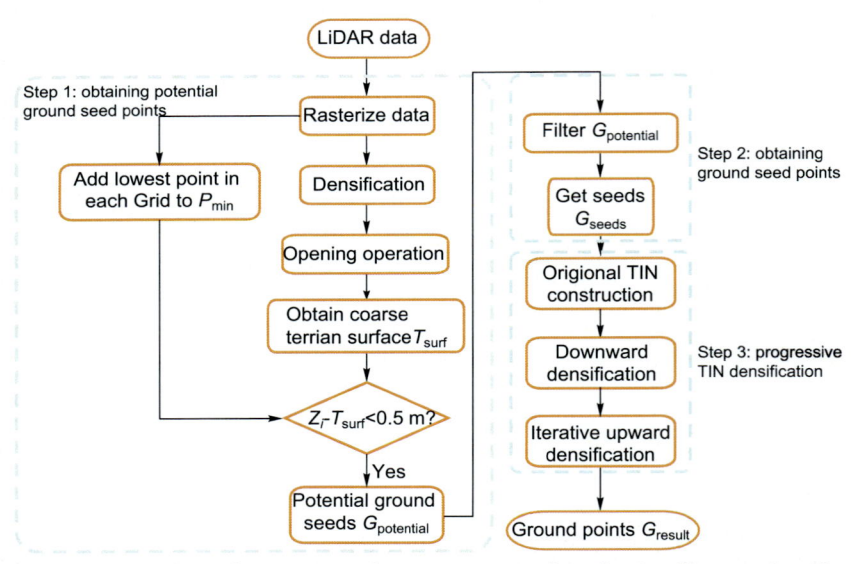

Figure 6.7 Workflow of the improved progressive TIN densification filtering algorithm. $G_{potential}$, potential ground seed points; G_{seeds} represents ground seed points; G_{result}, ground points; P_{min}, set of the lowest point in each grid; P_i, point in P_{min}; TIN, triangulated irregular network.

giant sequoia (*Sequoiadendron giganteum*), black oak (*Quercus kelloggii*), and canyon live oak (*Quercus chrysolepis*). These study sites contain a variety of topographic conditions, with the mean elevation ranging from below 450 m to over 3000 m and the mean slope from flat (below 10°) to precipitous (over 40°). Moreover, vegetation conditions differ greatly. The mean canopy cover varies from less than 20% to 90%, and the mean tree height varies from approximately 5 m to 20 m. This method has been integrated into the commercial software, LiDAR360 (GreenValley Technology Co., Ltd., Beijing, China). Compared with the traditional progressive TIN densification method, the improvements of this method include the following aspects:

(1) A morphological opening operation and residuals from a translation plane fitting are used to obtain the seed points. For the classic progressive TIN densification, the grid size is set at a relatively large value to ensure each cell contains at least one ground point, resulting in a small number of unevenly distributed initial seed points. With such a set of seed points, the subsequent seed point densification completely depends on the iterative angle and iteration distance. Hence, the misclassification of ground points into nonground points can occur more frequently. The improved method uses a 3 m morphological sliding window to obtain a sufficient number of evenly distributed seed points, which overcomes the issues of seed point generation in traditional progressive TIN densification methods and increases the filtering speed.

(2) A downward densification is used after constructing the triangulation. After the construction of the initial TIN-based digital terrain model (DTM), most of the unclassified points are above the TIN, while a few are below the TIN. The ground points below the TIN are misclassified as nonground points. The downward densification improves the initial TIN-based DTM as much as possible to approximate the real terrain, enabling topographical coverage and retaining finer details, especially for hilltops and steep slopes.

6.2.5 Filtering Methods Based on Segmentation

The first step of filtering methods based on segmentation is to use a region-growing method to cluster the point cloud and construct surfaces according to the common geometric and statistical features (e.g., elevation differences, slope, normal vectors, and residuals of the neighborhood fitting surface) of the input data (e.g., grid, TIN, and point cloud). Next, a regional classification is performed according to a threshold of elevation difference, smoothness, structural relationship, geometric features, or attribute information. The process does not rely on a single point but considers the similarity of ground objects. Then, the filtered results can be identified with various image segmentation methods such as edge detection, region growing, and clustering based on image similarity or discontinuity. As shown in Fig. 6.8, the roof patches are relatively regular after segmentation, whereas the vegetation patches are relatively irregular and contain fewer points, and ground patches have extensibility in space.

In region growing, the type of neighbor search is determined by the input data. If the input data is a grid, it is best to search for neighboring cells. If the input data is a TIN, the neighboring relationship is determined by the common edges between the triangles in the TIN. If the input data are a point cloud, the search is for neighboring points within a certain spatial distance.

Sithole (2005) organized point cloud data according to vertical segments, then divided the points in the segment into straight lines according

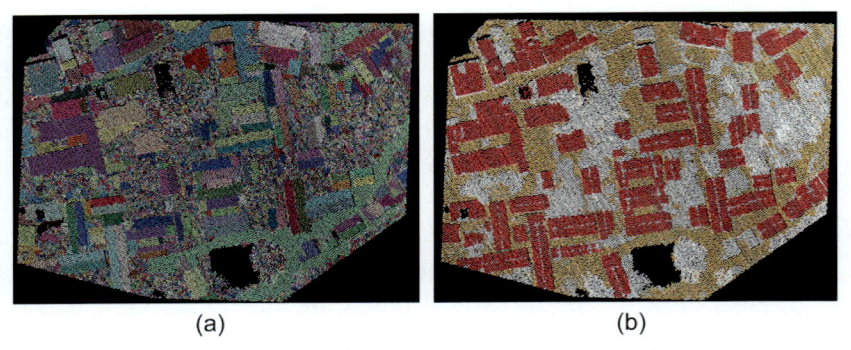

(a)   (b)

Figure 6.8 Results from segmentation-based filtering methods: (a) The raw airborne point cloud, (b) the filtered point cloud.

to their proximity in different directions and generated patches according to the common points between straight lines. Afterward, the generated patches were classified into different types. Tóvári and Pfeifer (2005) proposed a region–growing method based on normal vectors and spatial similarities. The main parameters in the region-growing method are the similarity of normal vectors and the distance between a candidate point and the adjusting plane. Wang and Tseng (2010b) proposed a split–and–merge segmentation based on an octree structure. They estimated the optimal parameters through a plane fitting method and extracted the geomatic properties (e.g., intersections of planes) of segmented planes to achieve point cloud filtering. This method performed well for filtering airborne and terrestrial LiDAR data.

Sánchez-Lopera and Lerma (2014) added an angular classification method to region growing to distinguish buildings, vegetation, ground, and other objects in urban areas. Their method is suitable for areas dominated by objects with obvious geometric and morphological features, such as towns and cities. In contrast, the complex terrain of forested areas does not have clear uniform characteristics and, therefore, poses a challenge for the clustering and classification of point clouds.

6.2.6 Filtering Methods Combining Other Information

Currently, most filtering algorithms use elevation differences between adjacent points but rarely consider the response characteristics of ground objects to laser pulses. Hence, it is hard to distinguish objects with similar elevations in the same region. With the development of new hardware, more LiDAR systems can provide intensity and waveform information of laser echo signals. These technologies complement traditional LiDAR scanning, which provides only discrete point cloud data (Goepfert et al., 2008). Moreover, image data acquired simultaneously with LiDAR data can aid in point cloud filtering. Fusing auxiliary information such as intensity, waveform, and imagery is promising for point cloud filtering and improving the filtering accuracy.

6.2.6.1 Fusion with Intensity Information

In addition to the 3-D coordinates of objects, laser scanners can also record the intensity information representing the energy of laser pulses reflected by the scanned objects. Lai and Wan (2007) proposed a method that combines intensity and distance information from LiDAR data to extract a DEM in rural areas. First, they denoised the intensity image using a flatness-based mean filtering method (Lai et al., 2005; Lai & Wan, 2007). Next, they selected points with an intensity between I_1 and I_2 as initial ground points and used a least-squares method to fit the initial terrain surface based on the selected points. Then, they updated the terrain surface by iteratively fitting it based on the corrected values of each coefficient of the fitted surface. This iteration was repeated until the correction values were less than the given threshold or the number of iteration times exceeded the maximum number of iterations.

Goepfert et al. (2008) found that DEMs produced in densely vegetated areas may be overestimated because it is difficult for the laser pulses to penetrate the vegetation. They proposed a filtering algorithm that combined the intensity information and the distribution characteristics of multiple echoes. Their method improved the interpolation filtering algorithm proposed by Kraus and Pfeifer (1998). First, they fitted an initial surface model based on all points with equal weights and calculated the residuals of each point. Then, they iteratively updated the fitted model based on the residuals until the maximum number of iterations was met.

Wang and Glenn (2009) proposed a filtering algorithm for forested environments that combined the intensity and elevation information of the point cloud. First, they normalized the intensity of the point cloud and generated a binary intensity image using a threshold of intensity. Points with an intensity value greater than the threshold were classified as vegetation points, whereas points with an intensity value smaller than the threshold were classified as ground points. Next, they assumed that the frequency distribution of the point elevations was bimodal Gaussian and fitted two Gaussian curves to represent the distribution of ground and vegetation points, respectively. Then, a grid with 4 m × 4 m cells was used to calculate the average and standard deviation of the Gaussian distribution and analyze the type of points in each cell. Lastly, a 4 m resolution DEM is obtained using an inverse distance weight interpolation method.

Many factors affect the intensity of a point cloud from laser emission to reception, including the transmitting pulse frequency, divergence angle, pulse width, scanning angle, atmospheric conditions, distance, and topographical variations. Even for the same objects, their point cloud intensities may vary with the data acquisition time. Meanwhile, different objects may have similar intensities owing to various factors, which affects the recognition of ground objects. Hence, the echo intensity must be corrected before data analysis. Theoretically, the correction of echo intensity requires the measured values of various influencing factors, which is difficult to achieve in practice. Currently, no standard is available for correcting and unifying the intensity of LiDAR data. This is the main factor that hinders the use of intensity information in LiDAR data.

6.2.6.2 Fusion with Full-Waveform Information

Compared with a discrete LiDAR system, waveform LiDAR provides richer information, including the wave width and amplitude. Doneus et al. (2008) proved that wave width and amplitude could be adopted to distinguish ground and nonground points in areas with low vegetation. The points with a large wave width are treated as nonground points, whereas the points with a small wave width are treated as ground points. Doneus and Briese (2006) obtained 3-D coordinates, wave amplitudes, and wave widths from waveform data through waveform decomposition (Wagner et al., 2006) and used an interpolation-based method to filter the LiDAR data in forests. Hu et al. (2011) proposed a method for generating a large-scale DEM based on full-waveform data. (1) First, they used a Gaussian decomposition algorithm to obtain a discrete point cloud with echo locations, wave widths, and other information. (2) Next, they extracted ground seed points through a series of morphological opening operations and used these points to generate a TIN. The TIN model was used to detect weak echoes from ground points that could not be detected by Gaussian decomposition. (3) The detected echoes were added to the last echoes detected by Gaussian decomposition. Then, they repeated (2) and (3) until no new ground points were detected. This method extracted up to 30% more ground points under shrubs and trees than the traditional Gaussian decomposition method, and the DEM generated from the filtered ground points contained more topographic details than the DEM generated by the TerraModeler software manufactured by the Terrasolid (Terrasolid, Helsinki, Southern Finland, Finland).

6.2.6.3 Fusion With Optical Images

Traditional optical remote sensing methods can provide object location, size, shape, and texture but lack information about 3-D features. Compared with traditional optical remote sensing methods, LiDAR has an obvious advantage in obtaining the 3-D coordinates of ground objects. However, it lacks spectral and texture information on those objects. The fusion of LiDAR data and optical images can provide 3-D structural, color, and texture features, which have great potential in the classification of tree species or extraction of forest parameters in forest studies (Persson et al., 2004; Popescu & Wynne, 2004).

Bretar and Chehata (2007) proposed a method for generating the DEM of forested areas by combining LiDAR intensity and optical images. They used the LiDAR intensity of the near-infrared band. They extracted the hybrid normalized difference vegetation index (HNDVI) based on the LiDAR intensity and red bands of the optical images. The calculation of HNDVI is as follows:

$$HNDVI = \frac{intensity - R}{intensity + R} \tag{6.10}$$

where *intensity* refers to the intensity value of LiDAR data and R refers to the red band of the optical image. A vegetation mask was obtained by thresholding HNDVI. Then, an adaptive morphological filtering algorithm was used to separate the vegetation points and ground points in the masked vegetation areas. The results show that combined LiDAR intensity and optimal images can better identify the ground points in vegetated areas.

Brattberg and Tolt (2008) proposed a classification algorithm based on combining LiDAR data and optical images. First, the point cloud was interpolated to generate a 0.25 m resolution raster image, and an initial DEM model was obtained using the region-growing algorithm. Then three segmentation algorithms were used for classification: (1) segmentation using multiple echo information, shape, and elevation variation features for classification; (2) segmentation to detect large areas that do not include ground points and mark flat areas as potential buildings; and (3) segmentation using principal component analysis to mark the connected regions. The authors then used the majority voting approach to fuse the results of the three segmentation algorithms. Lastly, they used optical images to further correct the classification results. The final results show that combining LiDAR data and optical images significantly improves classification accuracy.

At present, most LiDAR systems are equipped with digital cameras that can simultaneously acquire high-resolution optical images and LiDAR point data. Combining the advantages of LiDAR data and optical images can effectively improve the accuracy of filtering.

6.2.7 Deep Learning–Based Filtering Methods

Currently, one issue common to existing filtering methods that must be addressed is their reliance on handcrafted rules or statistically derived thresholds. Thus, current filtering methods may be not robust in complex areas. In recent years, the development of deep learning techniques has provided an opportunity to solve this issue. Pioneering studies have demonstrated the feasibility of deep learning methods for filtering LiDAR data. Schmohl and Sörgel (2019) proposed an encoder–decoder architecture with sparse submanifold convolutional networks for filtering airborne LiDAR data. However, owing to the challenges of large-scale voxelized representation, this method still faces issues relating to limited resolution and sampling range. Yousefhussien et al. (2018) used a multiscale fully convolutional neural network modified from PointNet to carry out semantic point segmentation in urban and rural areas. In their segmentation, nine classes were defined: power lines, low vegetation, impervious surfaces, cars, fences/hedges, roofs, facades, shrubs, and trees. They obtained promising results when classifying unnormalized points.

Overall, the application of deep learning technologies in LiDAR data filtering is still in the early stage, and relevant studies are rare. The author's research team proposed a point-based fully convolutional neural network (PFCN) for filtering ground points from airborne LiDAR data in forested environments (Jin et al., 2020). The PFCN consists of three parts: an input layer, hidden layers, and output layers (Fig. 6.9). The input layer data are an $N \times M$ matrix, where N is the number of points and M is the number of attributes of each point. The input data are normalized using the following equation:

$$[X, Y, Z] = \frac{[X, Y, Z] - \text{Min}([X, Y, Z])}{\text{Max}([X, Y, Z]) - \text{Min}([X, Y, Z])} \tag{6.11}$$

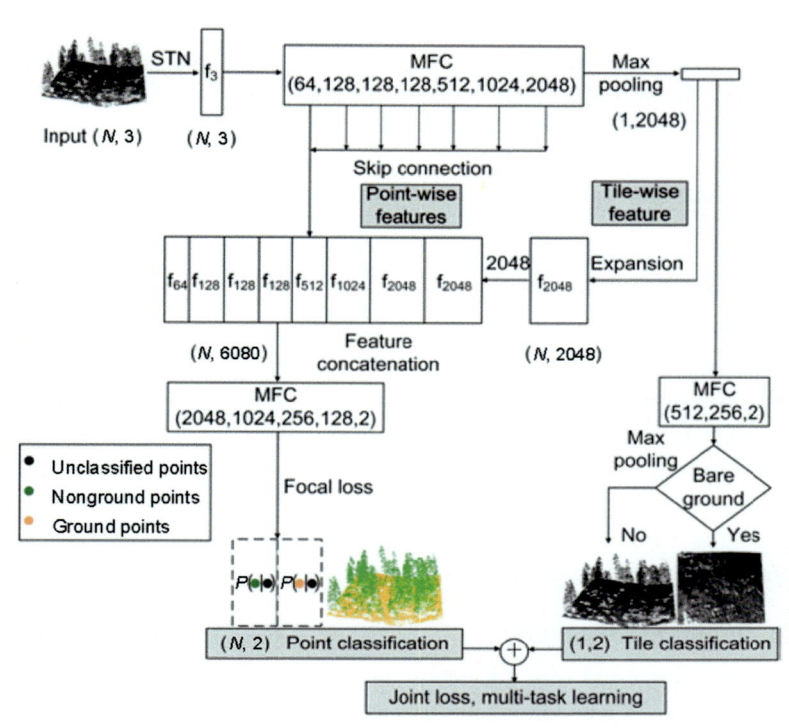

Figure 6.9 Architecture of point-based fully convolutional neural network. The input layer is the normalized and unclassified point cloud in black. Hidden layers contain a spatial transformer network and a set of fully convolutional multilayers (MFCs). MFCs extract pointwise and tilewise features for multitask learning, including a point classification task and a tile classification task.

where X, Y, and Z represent the 3-D coordinates of the point cloud. After normalization, the normalized data are fed into the hidden layers, which consist of a spatial transformer network and three multilayer fully convolutional (MFC) layers. The spatial transformer network is a data–dependent network that can align points in a canonical space before feature extraction. Each MFC layer consists of numerous filters, and each filter performs a 1-D convolution in the feature dimension with 1-D batch normalization and activation. The first MFC has seven layers of size 64, 128, 128, 128, 512, 1024, and 2048 for extracting pointwise and tilewise features. The second MFC uses these features for point cloud classification, and the third MFC outputs an $N \times 2$ matrix representing the probability of each point being classified as ground or nonground. The results show that the proposed PFCN method performs better than most existing filtering methods and is robust for different vegetation densities and topographical conditions.

6.3 Accuracy Evaluation Methods

Accuracy evaluation of LiDAR data filtering consists of quantitative and qualitative evaluation. Visual inspection and random sampling are the most common methods for evaluating accuracy. Comparing the filtered results with existing land-cover maps is also an effective approach (Meng et al., 2010).

6.3.1 Quantitative Evaluation

The main statistics for quantitatively evaluating the filtering accuracy are Type I errors (*TI*, ground points misclassified as nonground points), Type II errors (*TII*, nonground points misclassified as ground points), total error (*TE*), and kappa coefficient (*kp*). The mathematic definitions of the abovementioned statistics are shown in Table 6.1. A common strategy to generate reference data for evaluating filtering is to manually inspect and modify the initial filtered points generated automatically or semi-automatically by a software program.

6.3.2 Qualitative Evaluation

It has been suggested that where the error occurs is more important than the error itself. For example, misclassification in particular locations (e.g., peaks and depressions) and the misclassification of high objects as ground points can cause substantial errors in the generated terrain. Thus, accurately evaluating filtering results based solely on quantitative analysis is difficult. Some studies have addressed this issue by interpolating the filtered ground points into a DEM and using the interpolated DEM to evaluate the filtering results through visual comparison with a reference DEM (Caruso & Quarta,

Table 6.1 Mathematic definitions of Type I errors, Type II errors, total error, and kappa coefficient.

		Filtering result		Mathematic definition	
		Ground points	Nonground points	$TI = b/(a+b)$	$P_o = (a+d)/e$
Reference data	Ground points	a	b	$TII = c/(c+d)$	$P_c = ((a+b) \times (a+c) + (c+d) \times (b+d)/e^2$
	Nonground points	c	d	$TE = (b+c)/e$	$kp = (P_o - P_c)/(1-P_c)$

Note that: a is the number of ground points that are correctly classified; b is the number of ground points that are misclassified as nonground points; c is the number of nonground points that are misclassified as ground points; d is the number of nonground points that are correctly classified; *TI*, Type I errors; *TII*, Type II errors; *TE*, total error; *kp*, kappa coefficient; $e = a+b+c+d$.

1998). In addition to this visual inspection method, the root–mean–square error (RMSE) between the interpolated DEM and reference DEM can be used to evaluate the filtering results (Eq. 6.12):

$$\text{RMSE} = \sqrt{\frac{\sum_{i=1}^{n}\left(Z_i^{\text{filtered}} - Z_i^{\text{reference}}\right)^2}{n}} \tag{6.12}$$

where Z_i^{filtered} is the elevation of the filtering results, and $Z_i^{\text{reference}}$ is the elevation of the corresponding reference data.

6.4 Digital Elevation Model Generation

A DEM can be expressed by a regular grid or a TIN. Regular grids are equally spaced in the horizontal direction, and the elevations of the cells are represented by pixel values, which is suitable for flat terrain. A TIN is much more complex than a regular grid. The elevations are represented by the attribute values of triangle nodes, which is more suitable for areas with large terrain fluctuations.

6.4.1 Digital Elevation Model Interpolation Methods

Because ground points are unevenly distributed discrete 3-D points, a spatial interpolation procedure is needed to convert these discrete points into a continuous DEM layer. Spatial interpolation is the process of using points with known values to estimate values for other unknown points (Chen & Chang, 2003). Commonly used interpolation methods include inverse distance weighting (IDW), natural neighbor (NN), Kriging, and spline interpolation. Users should choose the appropriate interpolation method according to their data features.

6.4.1.1 Inverse Distance Weighting Interpolation

This method is based on the principle that points closer to the prediction location have more influence on prediction. When assigning the weights of known points in the interpolation, the weight of a known point is the inverse of the distance between the prediction location (unknown point) and the known point. The general equation of IDW can be expressed as:

$$Z_p = \sum_{i=1}^{n}\left(d_i^{-e} \times Z_i\right) / \sum_{i=1}^{n} d_i^{-e} \tag{6.13}$$

where Z_p represents the unknown point, Z_i represents the value of the i th known point, n represents the number of known points, d_i represents the Euclidean distance from the unknown point to point i, and e represents the power of distance.

The weight calculation is controlled by the power of distance. When the power is one, it means that the change rate of values among points is constant, which belongs to the category of linear interpolation. By defining a higher power, greater weight can be assigned to the nearest point. The resulting surface will have more detail but be less smooth. In contrast, a smaller power gives greater emphasis to more distant points, resulting in an interpolated surface that is smoother but with fewer details (Guo et al., 2010).

Many studies have shown that interpolation has the best performance when the power of IDW is defined as two (Chaplot et al., 2006). In addition to the power, the weight calculation is also controlled by the number of known points n involved in the prediction. The known points involved in the interpolation can be selected according to a fixed or variable radius. The value of a variable radius can be determined by setting the expected number of known points (usually more than 12).

6.4.1.2 Natural Neighbor Interpolation

NN interpolation is a simple local interpolation method that predicts the values of unknown points based on NNs. This method is based on Voronoi tessellation. Initially, a Voronoi diagram (also known as Thiessen polygons) is constructed using all the points. Next, a new Thiessen polygon is created in the interpolation. Then, the overlapping points between this new polygon and the initial polygons are used for prediction. The weights of these overlapping points in the interpolation are proportional to the overlapping area of the Thiessen polygons (Eq. 6.14):

$$Z_p = \sum_{i=1}^{n} \frac{a_i}{a} Z_i \qquad (6.14)$$

where Z_p represents the predicted value, Z_i represents the value of known point i, a_i represents the area of the overlapping area between Thiessen polygons of the i th sample point and the unknown points, and a represents the area of the Thiessen polygon of the unknown points. Because this method only considers the known points around the predicted locations and the predicted value is within the range of the values of known points,

NN interpolation performs well for points with an uneven spatial distribution.

6.4.1.3 Kriging Interpolation

Kriging is an advanced geostatistical procedure that follows the theory of regional change. It considers spatial variation to be statistically consistent across the surface of the region (Guo et al., 2010). The basic principle of Kriging interpolation is to assume that the spatial variation of an attribute is neither completely random nor completely determined but is code-termined by spatial autocorrelation factors, offsets, and random error.

Kriging is considered to be the best unbiased linear interpolation method. "Best" means the smallest variance between the true value and the predicted value. "Unbiased" means that the expectation of errors equals zero. "Linear" means the predicted value is a linear combination of weighted known values. According to different explanations of spatial autocorrelation and whether there is an unbiased estimate, many Kriging interpolation methods have been developed, including ordinary Kriging, block Kriging, global Kriging, and co-Kriging interpolation. Ordinary Kriging interpolation assumes no deviation and only considers spatial-related factors. The steps of ordinary Kriging interpolation are summarized as follows:

(1) A semivariogram function is created based on known points to determine the spatial dependence of the autocorrelation model. A semivariogram is mathematically expressed by Eq. (6.15):

$$\gamma(d) = \frac{1}{2n} \sum_{i=1}^{n} \{Z(x_i) - Z(x_i + d)\}^2 \tag{6.15}$$

where d is the distance between control points, n is the number of pairs of points whose distance is d, x_i is the location of the ith point, and Z is the control point elevation value.

(2) A semivariogram figure is drawn according to different spacing d.

(3) The best theoretical variation model is selected according to the semivariogram figure. Commonly used theoretical models are the spherical model, the exponential model, and the Gaussian model.

(4) The fitting model is used to predict the value of an unknown point as follows:

$$Z_p = \sum_{i=1}^{n} \omega_i Z_i \tag{6.16}$$

where Z_p represents the predicted value, Z_i represents the value of point i, and ω_i represents the weight of point i.

6.4.1.4 Radial Basis Function Interpolation

The radial basis function (RBF) is a series of exact interpolation techniques where the basic function is composed of the mathematical functions of a single variable. The single variable in the RBF refers to the distance (d) between the known point and the predicted point. The RBF includes the complex logarithmic function, complex quadratic function, inverted complex quadratic function, natural cubic spline function, thin plate spline function, and tension spline function. The RBF is suitable for large datasets and areas of flat terrain.

The thin plate spline function is a commonly used interpolation method that uses mathematical functions to minimize the curvature of the entire surface so that a smooth surface passes through the control points accurately. The mathematical expression of the thin plate spline function is as follows:

$$Z(x, y) = \sum R_i d_i^2 \log d_i + ax + by + c \tag{6.17}$$

where (x, y) is the coordinates of the predicted point; $d_i^2 = (x - x_i)^2 + (y - y_i)^2$; and (x_i, y_i) is the coordinates of the control point i.

The thin plate spline function consists of two parts: $ax + by + c$, which represents the local trend function; and $\sum R_i d_i^2 \log d_i$, which represents the basic function of the surface with minimum curvature. The correlation coefficients R_i, a, b, and c are determined by the following linear equation:

$$\begin{cases} \sum R_i d_i^2 \log d_i + ax + by + c = f_i \\ \sum_{i=1}^{n} R_i = 0 \\ \sum_{i=1}^{n} R_i x_i = 0 \\ \sum_{i=1}^{n} R_i y_i = 0 \end{cases} \tag{6.18}$$

where f_i represents the elevation of the control point i. Various plate spline functions have been developed, including thin plate tension spline, regular spline, and regular tension spline.

In summary, different interpolation methods are suitable for different scenarios. For example, the IDW interpolation is suitable for scenarios with a high density of evenly distributed and known points. However, its predicted values are limited to those between the known maximum and minimum points. Therefore, IDW is not suitable for interpolating areas with large elevation changes (e.g., valleys). Moreover, IDW interpolation tends to overestimate the elevation in forested areas and underestimate it in low–meadow areas. Therefore, the IDW method should be used with caution according to the characteristics of the target area.

Spline interpolation can obtain a smooth surface that passes through all known points and avoids information loss in the interpolation of elevation models from discrete point cloud data (Aguilar et al., 2010). In contrast to the IDW method, the predicted values from spline interpolation can be larger/smaller than the maximum/minimum value of the known points. Therefore, spline interpolation is suitable for areas with large terrain fluctuations. The Kriging interpolation method accounts for the distance influence and considers spatial autocorrelation between the known points. Therefore, the accuracy of the Kriging interpolation is better than the IDW interpolation, despite the different point densities.

In addition to interpolation methods, data acquisition methods (e.g., the setting of scan parameters) and data preprocessing (e.g., denoising and resampling) can affect interpolation accuracy. Abramov and McEwen (2004) analyzed the relationship between DEM resolution and interpolation accuracy. They found that the optimal interpolation methods for generating a small (e.g., 82 pixels/degree), medium (e.g., 250 pixels/degree), and high-resolution (e.g., 1000 pixels/degree) DEM were the spline function method, linear interpolation method, and NN method, respectively. Although the optimal interpolation method varies with the DEM resolution, the interpolation accuracy of most methods decreases with a gradual increase in DEM resolution (e.g., 1.5 m, 1.0 m, 0.5 m). Therefore, users should consider various factors to choose the optimal interpolation method for practical applications.

6.4.2 Digital Elevation Model Error Sources

The errors in the DEM products generated from LiDAR data can be caused by many factors, including the LiDAR system, point cloud processing, geographical factors, and DEM resolution (Hodgson et al., 2005).

6.4.2.1 *Errors from the LiDAR System*

The measurement errors of a LiDAR system consist of three components (i.e., the sensor, the aircraft, and the navigation system). They can be divided into vertical errors (also called height errors) and horizontal errors (also called position errors). Both vertical and horizontal errors are mainly caused by the misidentification of the position of the returned laser pulse and the positioning errors of the global navigation satellite system (GNSS)/inertial navigation system (Hodgson & Bresnahan, 2004). Factors influencing the accuracy of DEM generation from the LiDAR scanner include pulse frequency, wavelength, and scanning angle. A larger pulse frequency can result in a higher point density. In the same area, the greater the scanning angle, the higher the point density. Moreover, the influence of the scanning angle also depends on the fluctuation of the terrain. For example, when the scanning angle is greater than $10°$ in a steeply sloping area, the expansion of the blind scanning area may result in a decrease in the point cloud coverage, thereby affecting DEM interpolation accuracy. In contrast, the influence of scanning angle can be ignored in flat terrains with a scanning angle of less than $20°$. Flying altitude and speed affect the data quality as well. Given a low flying altitude (e.g., 100 m) and a high point density (e.g., 6 pt/m^2), the effect of the scanning angle on DEM accuracy can be neglected (Ahokas et al., 2005). As the flight altitude increases, the point cloud density decreases, and the size of the footprint that reaches the ground increases (a smaller footprint can more easily penetrate the vegetation canopy to reach the ground). This may result in a decrease in the accuracy of the generated DEM. An increase in the flying speed leads to an increase in pulse interval and a decrease in point density, ultimately affecting the quality of the DEM products. Moreover, navigation errors from the GNSS and IMU system installed on the airborne platform influence the calculation of point cloud coordinates and, thus, the accuracy of the generated DEM.

If the LiDAR data have higher resolution (smaller point spacing), a finer terrain description can be obtained. Thus, by increasing the pulse frequency, narrowing the scanning angle, reducing the flying height, or increasing the number of scans in the same area, the quality of data collection and the generated DEM can be improved (Aguilar et al., 2010).

6.4.2.2 Errors from Point Cloud Processing

Elements of point cloud processing that affect DEM accuracy include denoising, resampling, filtering, and interpolation. In filtering, omission and commission errors may affect the accuracy of the generated DEM products, while the point density after resampling is one of the main factors affecting the DEM accuracy (Guo et al., 2010). Generally speaking, the higher the point density, the higher the DEM accuracy. LiDAR data can provide high-density ground points for generating high-resolution and high-precision DEMs. However, owing to the irregular distribution of point cloud data, there are redundant points from repeated data acquisitions in the same area (Liu et al., 2007), substantially increasing the calculation burden. In addition, redundant points contribute less to improving DEM accuracy. Therefore, removing redundant points is necessary to ensure DEM accuracy (Guo et al., 2010).

Liu et al. (2007) resampled raw point cloud data into a series of datasets with different point densities—100%, 75%, 50%, 25%, 10%, 5%, and 1%—of the raw point cloud. They then used the IDW method to interpolate a series of 5 m resolution DEMs from the resampled datasets and used the DEM generated from the raw point cloud as a reference to evaluate the influence of point density on the accuracy of DEM generation. Their results showed no significant difference in DEM accuracy with a point density of 50%, but the required DEM generation time could be reduced by 50%. Therefore, an appropriate reduction of point density does not affect DEM accuracy. However, the choice of the point density for resampling should consider factors such as the point density of the raw point cloud, terrain features, interpolation methods, and DEM resolution. For example, DEMs with different resolutions have different sensitivities to point density. The higher the DEM resolution, the greater the effect of point density on the accuracy of DEM generation.

6.4.2.3 Errors from Geographical Factors

Geographical factors include seasons, land-cover types, and terrain characteristics.

- Season types

    The timing of LiDAR data acquisition can have a substantial impact on the point clouds obtained in tall, dense deciduous forests (Hyyppä et al., 2005). In summer, dense branches and leaves reduce ground visibility, which affects the ratio of pulse penetration through the canopy to the ground. The acquired point cloud, therefore, contains many vegetation points but few ground points. Seasonal differences are not as evident in coniferous forest canopies. Laser pulses can penetrate vegetation in early spring, late autumn, or winter, obtaining a high density of ground points (Su & Bork, 2006). Thus, the acquired topographic data quality during these seasons is higher than in summer. Therefore, if possible, LiDAR data should be collected in early spring, late autumn, or winter.

- Land–cover types

    Hodgson and Bresnahan (2004) studied the influence of five different land-cover types on DEM generation. The five land-cover types were sidewalks, low grasses (height less than 8 cm), high grasses (height less than 90 cm), shrubs, evergreens, and deciduous forests. The results showed that tall and large trees in evergreen and deciduous forests reduced the ratio of pulses penetrating the canopies, thereby decreasing the accuracy of the generated DEM. Moreover, they found that denser understory vegetation could also reduce the quality of the DEM. Cowen et al. (2000) reported a clear linear relationship between vegetation coverage and ground point density in coniferous forests and that the orientation and shape of leaves in deciduous forests could affect the quality of LiDAR data. When the laser beam passes through horizontally oriented leaves, the light intensity is usually weakened, making it difficult to obtain ground points and causing an overestimate of ground elevation. Because coniferous trees have a small leaf area, their leaf orientation has less influence on the accuracy of DEM generation (Su & Bork, 2006). In grasslands, because the leaf orientation of grasses is usually vertical to the ground, laser pulses may be repeatedly reflected by leaves, which prolongs the echo time and results in an underestimation of elevation.

- Terrain characteristics

Terrain characteristics can affect DEM accuracy significantly. DEM accuracy varies linearly with terrain complexity. The maximum error of a DEM often occurs in areas of horizontal strata or steep terrain (Hodgson et al., 2005).

6.4.2.4 Errors from the Digital Elevation Model Resolution

DEM resolution refers to the smallest unit used to record the elevation value. A higher resolution can enrich the details of the terrain, but it can substantially increase the data size of the DEM. Therefore, decisions on the DEM resolution should consider factors such as point density, terrain complexity, computational efficiency, and the usage of the DEM product. For example, a low-resolution DEM generated from high-density ground points requires less disk space for storage, but it may reduce the value of the raw point cloud. Therefore, the choice of resolution is critical in DEM generation. Typically, the number of cells in the DEM grid should approximate the number of ground points in the entire area (Eq. 6.19):

$$R = \sqrt{\frac{A}{n}} \tag{6.19}$$

where A represents the area, n represents the number of points, and R represents the resolution.

When the DEM resolution is relatively high, the precision of DEM-derived products (e.g., slope and curvature) can be improved correspondingly. However, with different DEM resolutions, the contribution of different factors to DEM errors varies greatly. For example, when the DEM resolution is low, the influence of point density on DEM accuracy is greatly reduced (Guo et al., 2010).

6.4.3 Digital Elevation Model Accuracy Analysis

Constructing a mathematical model to evaluate the accuracy of a DEM is more complicated than establishing a terrain surface model. The mathematical model needs to consider factors such as topographic variability, interpolation methods, and the quality of raw LiDAR data (e.g., accuracy, point density, and distribution mode). Using the factors affecting DEM error (Section 6.4.2) as an example, the mathematical model of DEM accuracy can be expressed as follows:

$$A_{(DEM)} = f(S, M, R, P, D_s, D_n, O) \tag{6.20}$$

where $A_{(DEM)}$ is the accuracy of the DEM; S refers to the surface features of DEM; M refers to the interpolation method; R represents the surface features of terrain; P, D_s, and D_n are the accuracy, distribution, and density of the raw point cloud, respectively; and O refers to other factors.

The elevation error of a DEM is a synthesis of various propagation errors during the establishment of the model. Because of the uneven distribution of ground points in the target area, it is necessary to interpolate them to obtain a DEM for the whole area. The selection of the interpolation method and rasterization process can add errors to the final elevation. The surface features of a terrain determine the difficulty of representing that terrain. The interpolation of ground points may be more accurate in areas with small variations. In contrast, it is more difficult to invert all the variation trends in areas with steep terrain by interpolating the ground points. Therefore, errors in the raw point cloud may be propagated into the generated DEM during the modeling process.

The evaluation indicators, propagation rules, and expression forms of errors and accuracy are important issues in spatial data uncertainty theory research. The mathematical model of DEM error should reflect the amount, spatial distribution, and spatial correlation of the DEM error. However, the complexity of spatial data makes it difficult to express the DEM error model in a simple formula. A common solution is to establish an accuracy model according to the research purpose and application (e.g., using an indicator to describe the errors, establishing an autocorrelation model according to the spatial correlation among local units, and using a visualization method to represent the spatial structure and pattern of errors).

Many indicators have been proposed, including the RMSE, mean error, and standard deviation (Eqs. 6.21—6.23). These indicators mainly describe the DEM errors statistically, so the actual elevation must be obtained for the calculation:

$$\text{RMSE} = \sqrt{\frac{\sum_1^n \left(Z_i^{\text{pre}} - Z_i^{\text{real}} \right)^2}{n}} \tag{6.21}$$

$$\text{ME} = \frac{\sum_1^n \left(Z_i^{\text{pre}} - Z_i^{\text{real}} \right)}{n} \tag{6.22}$$

$$\text{SD} = \sqrt{\frac{\sum_1^n \left(Z_i^{\text{pre}} - Z_i^{\text{real}} - \text{ME} \right)^2}{n}} \tag{6.23}$$

where ME represents mean error, SD represents standard deviation, Z^{pre} represents interpolated elevation, Z^{real} represents actual elevation, and n represents the number of points used in the interpolation.

DEM errors are spatially autocorrelated, which systematically influences the analysis of DEM accuracy in different terrain. Autocorrelation may cause a biased evaluation of DEM accuracy if a single and nonspatial statistical model is adopted. Because understanding the spatial distribution of DEM errors is critical for accurately predicting the errors in the DEM, a description of the autocorrelation of errors is important. Typically, the autocorrelation of errors is expressed by Moran's I (Eq. 6.24):

$$I = \frac{\sum_{i=1}^{n}\sum_{j=1}^{n} w_{ij}(z_i - \text{ME})(z_j - \text{ME})}{\text{SD}^2 \sum_{i=1}^{n}\sum_{j=1}^{n} w_{ij}} \tag{6.24}$$

where I refers to Moran's I, and w_{ij} is the weight of adjacent cells. Moran's I reflects the variance of the error and the change of the mean error value. The Moran's I value ranges from -1 to 1. If the Moran's I value is 1, it means that the error is strongly positively correlated; if the value of Moran's I is -1, it means that the error is strongly negatively correlated; and if Moran's I is 0, the error is randomly distributed.

The numerical and autocorrelation models can quantitatively describe the overall DEM error; however, they cannot express the spatial distribution of DEM errors. Through visualization, the abstract data can be transformed into concrete and vivid graphs, which help analyze the uncertainty distribution of DEM errors and provide a reference to improve and use DEM data. Visualization tools, such as shading, grayscale, and 3-D stereograms, are important for studying the uncertainty of DEM errors. The error model of terrain parameters can be obtained by stacking the errors in different terrain data (e.g., slope, aspect, and elevation) into overlapping layers. The spatial distribution of errors not exhibited by the statistical model can be represented by different gray levels or colors.

Below, we use a study from the Sierra Nevada mountains, California, USA, as an example to investigate the effects of topographic variability, point density, and DEM resolution on the accuracy of different interpolation methods (Guo et al., 2010). The five interpolation methods evaluated are NN, IDW, TIN, spline function, ordinary Kriging, and universal

Kriging. The RMSE obtained by 10-fold cross-validation is used for accuracy verification. Our study consisted of the following aspects:

(1) The influence of different topographic variability on DEM accuracy with various DEM resolutions (i.e., 0.5 m, 1 m, 1.5 m, and 10 m). The variability of the terrain is expressed by the coefficient of variation (CV) (Eq. 6.25):

$$CV = \frac{\sqrt{\sum (Z_j - \overline{Z})^2 / n}}{\overline{Z}} \qquad (6.25)$$

where Z_j represents the elevation of the jth point, $\overline{Z}$ represents the average elevation, and n represents the number of points. It is worth noting that there is a linear, positive correlation between terrain fluctuation and slope—that is, the larger the slope, the greater the terrain fluctuation. To some extent, terrain variation may include the influence of the slope.

(2) The influence of different interpolation methods on the DEM accuracy at various resolutions (i.e., 0.5 m, 1 m, 1.5 m, and 10 m) and point densities (i.e., 90%, 80%, 70%, 60%, 50%, 40%, 30%, 20%, and 10%). Different point densities are achieved by subsampling points.

(3) The effect of different interpolation methods on DEM accuracy at various resolutions (i.e., 0.5 m, 1 m, 1.5 m, and 10 m).

Our results show that Kriging interpolation is the optimal interpolation method at different resolutions and terrain variabilities (Fig. 6.10), and CV has a positive effect on the error in DEMs. The IDW, NN, and TIN methods are more susceptible to terrain fluctuations than the Kriging interpolation method for DEMs at high spatial resolution (0.5 and 1 m). For DEMs at a low spatial resolution (5 and 10 m), the choice of method has less impact on accuracy (except for the spline interpolation method).

Kriging is also the optimal interpolation method with different resolutions and point densities (Fig. 6.11). When generating high spatial resolution DEMs (0.5 m and 1 m), the higher the point density, the higher the DEM accuracy. In other words, RMSE was exponentially correlated with point density, or RMSE was negatively linearly correlated with the $\log_e$ density. When the point density accounted for over 70% of the raw data, the differences in the DEMs generated from different interpolation methods could be ignored. We recommend using the IDW, NN, and TIN methods with high point density data because of their high efficiency. When generating a DEM of low resolution (5 and 10 m), the interpolation

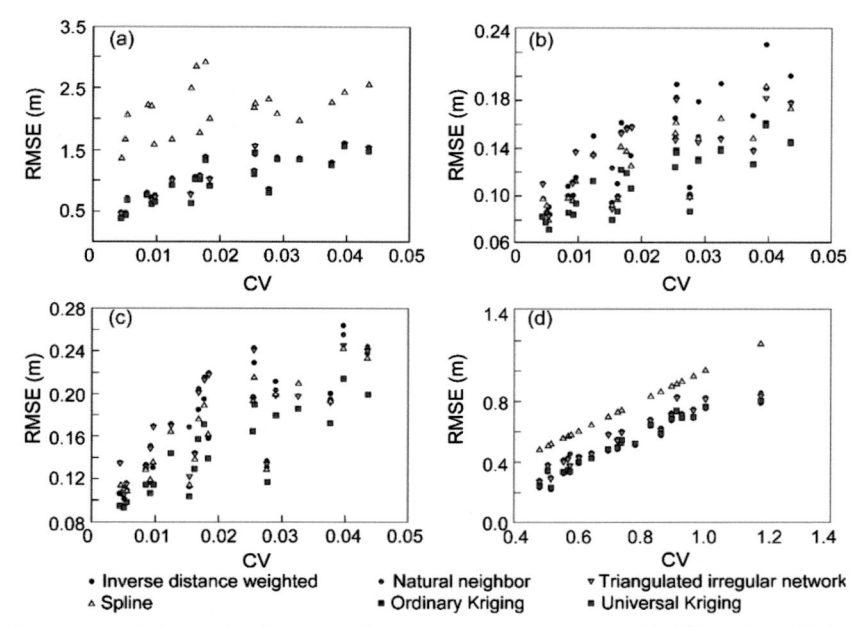

Figure 6.10 Relationship between the root-mean-square error (RMSE) and coefficient of variation (CV) at various digital elevation model resolutions: (a) 0.5 m, (b) 1 m, (c) 5 m, and (d) 10 m.

method (except for the spline method) and point density have less effect on DEM accuracy.

By comparing the relationship between the RMSE and DEM resolution of different interpolation methods, we also found that Kriging interpolation was the best in generating high spatial resolution DEMs (0.5 and 1 m; Fig. 6.12). For the generation of low-resolution DEMs, any of the interpolation methods can be chosen (except the spline method).

According to the results of the above three experiments, we found that RMSE has a positive linear correlation with CV and a negative linear correlation with $\log_e$density. Moreover, we found that a multiple variable regression model can be used to explain the variability of RMSE (Eq. 6.26):

$$RMSE = a \times CV + b \times \log_e\text{density} + C \qquad (6.26)$$

where a and b represent the correlation coefficients of CV and $\log_e$density, respectively, and C is a constant.

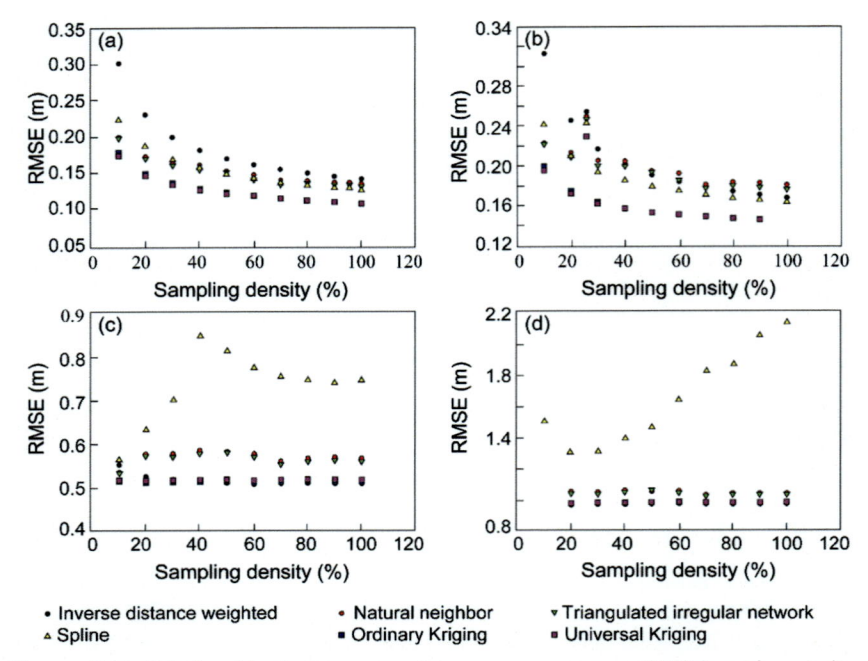

Figure 6.11 Relationship between root-mean-square error (RMSE) and sampling density at various digital elevation model resolutions: (a) 0.5 m, (B) 1 m, (c) 5 m, and (d) 10 m.

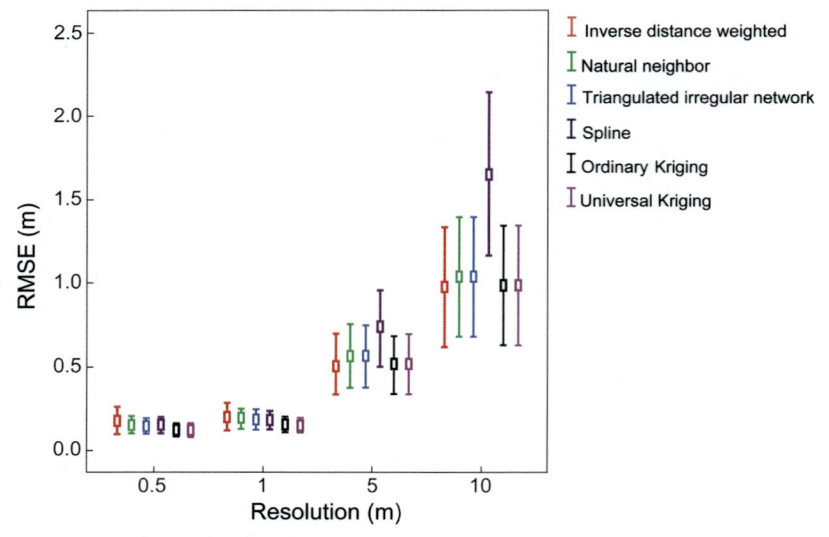

Figure 6.12 Relationship between root-mean-square error (RMSE) and interpolation methods at various digital elevation model resolutions.

6.5 Chapter Summary

Point cloud filtering is one of the prerequisites for LiDAR data processing. The quality of filtering directly influences the accuracy of subsequent terrain and forest parameter extractions. Currently, there are many point cloud filtering algorithms, each with its advantages and disadvantages. Readers are encouraged to select filtering algorithms appropriate for their LiDAR data collection quality and the topographic variability of their study areas. For example, in mountainous areas with high forest coverage, the improved progressive TIN densification algorithm is more robust and accurate than most filtering algorithms. Accurate ground points obtained by filtering ensure the generation of high-precision DEMs. Interpolation algorithms also have a great impact on the accuracy of generated DEMs. The Kriging method is an effective and accurate approach to interpolate a DEM from ground points in different environmental settings.

References

Abramov, O., & McEwen, A. (2004). An evaluation of interpolation methods for Mars Orbiter Laser Altimeter (MOLA) data. *International Journal of Remote Sensing, 25*(3), 669—676.

Aguilar, F. J., Mills, J. P., Delgado, J., Aguilar, M. A., Negreiros, J., & Pérez, J. L. (2010). Modelling vertical error in LiDAR-derived digital elevation models. *ISPRS Journal of Photogrammetry and Remote Sensing, 65*(1), 103—110.

Ahokas, E., Yu, X., Oksanen, J., Hyyppä, J., Kaartinen, H., & Hyyppä, H. (2005). Optimization of the scanning angle for countrywide laser scanning. *International Archives of the Photogrammetry, Remote Sensing and Spatial Information Sciences, 36*.

Axelsson, P. (2000). DEM generation from laser scanner data using adaptive TIN models. *International Archives of Photogrammetry and Remote Sensing, 33*, 110—117.

Brattberg, O., & Tolt, G. (2008). Terrain classification using airborne lidar data and aerial imagery. *International Archives of the Photogrammetry, Remote Sensing and Spatial Information Sciences, 37*(B3b), 261—266.

Bretar, F., & Chehata, N. (2007). Digital terrain model on vegetated areas: Joint use of airborne lidar data and optical images. In *Proc. of photogrammetric image analysis (PIA)*. Munchen, Germany: ISPRS.

Caruso, C., & Quarta, F. (1998). Interpolation methods comparison. *Computers & Mathematics with Applications, 35*(12), 109—126.

Chaplot, V., Darboux, F., Bourennane, H., Leguédois, S., Silvera, N., & Phachomphon, K. (2006). Accuracy of interpolation techniques for the derivation of digital elevation models in relation to landform types and data density. *Geomorphology, 77*(1—2), 126—141.

Chen, J., & Chang, K.t. (2003). *Introduction to geographic information systems*. China Science Publishing & Media Ltd.

Chen, C., Li, Y., Li, W., & Dai, H. (2013). A multiresolution hierarchical classification algorithm for filtering airborne LiDAR data. *ISPRS Journal of Photogrammetry and Remote Sensing, 82*, 1—9.

Cowen, D. J., Jensen, J. R., Hendrix, C., Hodgson, M., Schill, S. R., & Macchiaverna, F. (2000). A GIS-assisted rail construction econometric model that incorporates LIDAR data. *Photogrammetric Engineering and Remote Sensing, 66*(11), 1323—1328.

Doneus, M., & Briese, C. (2006). Digital terrain modelling for archaeological interpretation within forested areas using full-waveform laserscanning. In *Proceedings of the 7th international conference on virtual reality, archaeology and intelligent cultural heritage* (pp. 155—162). Nicosia, Cyprus: Eurographics Association.

Doneus, M., Briese, C., Fera, M., & Janner, M. (2008). Archaeological prospection of forested areas using full-waveform airborne laser scanning. *Journal of Archaeological Science, 35*(4), 882—893.

Goepfert, J., Soergel, U., & Brzank, A. (2008). Integration of intensity information and echo distribution in the filtering process of LIDAR data in vegetated areas. In *Proceedings of the Silve Laser*.

Griffin, T. (1990). Measurements from maps. *Cartography, 19*(1), 66—67.

Guo, Q., Li, W., Yu, H., & Alvarez, O. (2010). Effects of topographic variability and lidar sampling density on several DEM interpolation methods. *Photogrammetric Engineering & Remote Sensing, 76*(6), 701—712.

Haralick, R. M., & Shapiro, L. G. (1992). *Computer and robot vision, 1*. Reading: Addison-wesley.

Hodgson, M. E., & Bresnahan, P. (2004). Accuracy of airborne lidar-derived elevation. *Photogrammetric Engineering & Remote Sensing, 70*(3), 331—339.

Hodgson, M. E., Jensen, J., Raber, G., Tullis, J., Davis, B. A., Thompson, G., & Schuckman, K. (2005). An evaluation of lidar-derived elevation and terrain slope in leaf-off conditions. *Photogrammetric Engineering & Remote Sensing, 71*(7), 817—823.

Hu, B., Gumerov, D., & Wang, J. (2011). An integrated approach to accurate DEM GENERARTION using airborne full waveform lidar data. *International Archives of the Photogrammetry, Remote Sensing and Spatial Information Sciences, 38*(5/W12), 237—241.

Hyyppä, H., Yu, X., Hyyppä, J., Kaartinen, H., Kaasalainen, S., Honkavaara, E., & Rönnholm, P. (2005). Factors affecting the quality of DTM generation in forested areas. *International Archives of Photogrammetry, Remote Sensing and Spatial Information Sciences, 36*(3/W19), 85—90.

Jin, S., Su, Y., Zhao, X., Hu, T., & Guo, Q. (2020). A point-based fully convolutional neural network for airborne LiDAR ground point filtering in forested environments. *IEEE Journal of Selected Topics in Applied Earth Observations and Remote Sensing, 13*, 3958—3974.

Kilian, J., Haala, N., & Englich, M. (1996). Capture and evaluation of airborne laser scanner data. *International Archives of Photogrammetry and Remote Sensing, 31*, 383—388.

Kobler, A., Pfeifer, N., Ogrinc, P., Todorovski, L., Oštir, K., & Džeroski, S. (2007). Repetitive interpolation: A robust algorithm for DTM generation from aerial laser scanner data in forested terrain. *Remote Sensing of Environment, 108*(1), 9—23.

Kraus, K., & Mikhail, E. M. (1972). Linear least-squares interpolation. *Photogrammetric Engineering & Remote Sensing, 38*, 016—1029.

Kraus, K., & Pfeifer, N. (1998). Determination of terrain models in wooded areas with airborne laser scanner data. *ISPRS Journal of Photogrammetry and Remote Sensing, 53*(4), 193—203.

Lai, X., & Wan, Y. (2007). *An algorithm to generate DEM in rural areas from Lidar data*. Second International Conference on Space Information Technology.

Lai, X., Zheng, X., & Wan, Y. (2005). A kind of filtering algorithms for LiDAR intensity image based on flatness terrain. In *Proceedings of International Symposium on Spatio-temporal Modeling, Spatial Reasoning, Analysis, Data Mining and Data Fusion* (pp. 1—5).

Lee, H. S., & Younan, N. H. (2003). DTM extraction of LiDAR returns via adaptive processing. *IEEE Transactions on Geoscience and Remote Sensing, 41*(9), 2063—2069.

Liu, X., Zhang, Z., Peterson, J., & Chandra, S. (2007). The effect of LiDAR data density on DEM accuracy. In *Proceedings of the international congress on modelling and simulation (MODSIM07)*.

Meng, X., Currit, N., & Zhao, K. (2010). Ground filtering algorithms for airborne LiDAR data: A review of critical issues. *Remote Sensing, 2*(3), 833−860.

Mongus, D., & Žalik, B. (2012). Parameter-free ground filtering of LiDAR data for automatic DTM generation. *ISPRS Journal of Photogrammetry and Remote Sensing, 67*, 1−12.

Persson, Å., Holmgren, J., Söderman, U., & Olsson, H. (2004). Tree species classification of individual trees in Sweden by combining high resolution laser data with high resolution near-infrared digital images. *International Archives of Photogrammetry, Remote Sensing and Spatial Information Sciences, 36*(8), 204−207.

Pingel, T. J., Clarke, K. C., & McBride, W. A. (2013). An improved simple morphological filter for the terrain classification of airborne LIDAR data. *ISPRS Journal of Photogrammetry and Remote Sensing, 77*, 21−30.

Popescu, S. C., & Wynne, R. H. (2004). Seeing the trees in the forest. *Photogrammetric Engineering & Remote Sensing, 70*(5), 589−604.

Sánchez-Lopera, J., & Lerma, J. L. (2014). Classification of lidar bare-earth points, buildings, vegetation, and small objects based on region growing and angular classifier. *International Journal of Remote Sensing, 35*(19), 6955−6972.

Schmohl, S., & Sörgel, U. (2019). Submanifold sparse convolutional networks for semantic segmentation of large-scale ALS point clouds. *ISPRS Annals of the Photogrammetry, Remote Sensing and Spatial Information Sciences, 4*, 77−84.

Sithole, G. (2005). *Segmentation and classification of airborne laser scanner data* (p. 59). Publications on Geodesy.

Sithole, G., & Vosselman, G. (2004). Experimental comparison of filter algorithms for bare-Earth extraction from airborne laser scanning point clouds. *ISPRS Journal of Photogrammetry and Remote Sensing, 59*(1−2), 85−101.

Su, J., & Bork, E. (2006). Influence of vegetation, slope, and lidar sampling angle on DEM accuracy. *Photogrammetric Engineering & Remote Sensing, 72*(11), 1265−1274.

Susaki, J. (2012). Adaptive slope filtering of airborne LiDAR data in urban areas for digital terrain model (DTM) generation. *Remote Sensing, 4*(6), 1804−1819.

Tóvári, D., & Pfeifer, N. (2005). Segmentation based robust interpolation-a new approach to laser data filtering. *International Archives of Photogrammetry, Remote Sensing and Spatial Information Sciences, 36*(3/19), 79−84.

Vosselman, G. (2000). Slope based filtering of laser altimetry data. *International Archives of Photogrammetry and Remote Sensing, 33*(B3/2; PART 3), 935−942.

Wagner, W., Ullrich, A., Ducic, V., Melzer, T., & Studnicka, N. (2006). Gaussian decomposition and calibration of a novel small-footprint full-waveform digitising airborne laser scanner. *ISPRS Journal of Photogrammetry and Remote Sensing, 60*(2), 100−112.

Wang, C., & Glenn, N. F. (2009). Integrating LiDAR intensity and elevation data for terrain characterization in a forested area. *IEEE Geoscience and Remote Sensing Letters, 6*(3), 463−466.

Wang, C.-K., & Tseng, Y.-H. (2010a). *DEM generation from airborne LiDAR data by an adaptive dual-directional slope filter*. na.

Wang, M., & Tseng, Y. H. (2010b). Automatic segmentation of LiDAR data into coplanar point clusters using an octree-based split-and-merge algorithm. *Photogrammetric Engineering & Remote Sensing, 76*(4), 407−420.

Yousefhussien, M., Kelbe, D. J., Ientilucci, E. J., & Salvaggio, C. (2018). A multi-scale fully convolutional network for semantic labeling of 3D point clouds. *ISPRS Journal of Photogrammetry and Remote Sensing, 143*, 191−204.

Zhang, K., Chen, S.-C., Whitman, D., Shyu, M.-L., Yan, J., & Zhang, C. (2003). A progressive morphological filter for removing nonground measurements from airborne LIDAR data. *IEEE Transactions on Geoscience and Remote Sensing, 41*(4), 872—882.

Zhang, J., & Lin, X. (2013). Filtering airborne LiDAR data by embedding smoothness-constrained segmentation in progressive TIN densification. *ISPRS Journal of Photogrammetry and Remote Sensing, 81,* 44—59.

Zhao, X., Guo, Q., Su, Y., & Xue, B. (2016). Improved progressive TIN densification filtering algorithm for airborne LiDAR data in forested areas. *ISPRS Journal of Photogrammetry and Remote Sensing, 117,* 79—91.

CHAPTER 7

Forest Structural Attribute Extraction

Contents

LiDAR Principles, Processing and Applications in Forest Ecology
ISBN 978-0-12-823894-3
https://doi.org/10.1016/B978-0-12-823894-3.00007-4

Forest structural attributes, particularly three-dimensional attributes such as the vertical foliage profile, tree height, and crown size, play an important role in forest biomass estimation, forest management and operation, and biodiversity research and protection. Thus, extracting forest structural attributes is an essential prerequisite for quantifying forest processes and functions. Traditional methods of measuring forest structures rely primarily on field surveys and two-dimensional remote sensing images. However, traditional methods have limited accuracy. Light detection and ranging (LiDAR), an active remote sensing technology, has revolutionized the measuring of forest structural attributes, especially with the recent, rapid development of near-surface LiDAR platforms. This chapter introduces the concepts, methods, and case studies for extracting forest structural attributes from the stand level to the organ level using LiDAR data from various platforms.

7.1 Stand-Level Structural Attribute Extraction

Stand-level structural attributes are an important element in understanding the status and dynamics of a forest. Light detection and ranging (LiDAR) has become an effective tool for accurately and efficiently characterizing three-dimensional (3-D) forest information. Currently, many structural attributes can be accurately extracted from LiDAR data at the stand level, including canopy height, forest gap attributes, and intensity profiles.

7.1.1 Canopy Height and Canopy Gap Detection

A canopy height model (CHM) represents the absolute height difference between the ground and the surface of forest canopies and is the most widely used raster product derived from LiDAR data. It can be calculated in several ways. One way to derive a CHM is to subtract a digital elevation model (DEM) from the corresponding digital surface model (DSM). DEM can be generated by interpolating the ground points using methods such as the ordinary Kriging algorithm (detailed in Chapter 6) (Guo et al., 2010; Jin, Su, Gao, Hu, et al., 2018), and DSM can be generated from the first returns (or canopy surface returns) of LiDAR data using a similar procedure. Another way to generate a CHM is to compute the DSM directly from the normalized LiDAR points. Because CHM eliminates the influence of terrain, it allows the extracted forest structural attributes (e.g., canopy height and cover) from LiDAR to be comparable at the stand level.

Despite the wide use of CHMs to extract forestry structural attributes (Hill et al., 2014; Khosravipour et al., 2015; Silva et al., 2019), empty pixels in the CHM, called pits, can hamper subsequent analyses (e.g., individual tree detection and the crown size calculation) (Khosravipour et al., 2014). Pits appear when a laser beam penetrates a tree canopy deeply before it generates its first echo; they are closely related to both data quality (e.g., point density) and postprocessing methods (e.g., strip alignment and filtering) (Khosravipour et al., 2014). The appearance of pits in a CHM adversely affects the extraction of canopy structural attributes.

Many attempts have been made to generate pit-free CHMs by smoothing standard CHMs using mean, median, or Gaussian filters (Hosoi et al., 2012). However, because these filters use a two-dimensional (2-D) kernel function (Dralle & Rudemo, 1996), they not only remove pits but also change all the pixel values of the CHM (Ben-Arie et al., 2009; Shamsoddini et al., 2013; Zhao et al., 2013). Therefore, efforts have been devoted to developing algorithms to directly produce pit-free CHMs from 3-D LiDAR points. For example, using only the highest first return in each pixel is recommended over using all first returns in the CHM generation because some of the first returns may fall inside the canopy (Leckie et al., 2003; Popescu & Wynne, 2004), and larger pixel size is suggested to minimize height changes within the canopy (Chen et al., 2006). Although these algorithms can remove most pits in CHM, they may be ineffective in handling pits caused by branches or leaves inside the canopy instead of the ground. Khosravipour et al. (2014) proposed an improved pit-free CHM generation algorithm that includes two steps. The first step establishes a standard CHM from all first returns and establishes a series of CHMs from only first returns above a given height. The second step merges all these CHMs into one CHM using the maximum value of all CHMs. Unlike smoothing the CHM with 2-D filters, this algorithm does not remove the treetops of small trees (Kaartinen et al., 2012; Solberg et al., 2006) and has proven to be significantly better than Gaussian-filtered CHMs for subsequent individual tree detection and forest structural attribute extraction.

Forest gaps refer to openings in the canopy at a certain height threshold caused by natural or anthropogenic disturbances (Silva et al., 2019). Forest gap size, distribution, and dynamics are important indicators of forest regeneration, succession, and maturity (Asner et al., 2013; Vepakomma et al., 2008). Therefore, detecting canopy gaps and quantifying their size, spatial distribution, and dynamics are exceptionally important in forest ecology studies (Dietze & Clark, 2008; Dupuy & Chazdon, 2008; Farrior et al., 2016).

LiDAR is widely used in canopy gap detection because it can characterize the vertical structure of the forest canopy with high efficiency and accuracy (Heiskanen et al., 2015; Silva et al., 2019). A canopy gap can be detected by analyzing the vertical profile of the 3-D point cloud or the CHM interpolated from the point cloud (Gaulton & Malthus, 2010). For example, Vepakomma et al. (2011) mapped canopy gaps from LiDAR-derived CHMs by finding open areas with height values lower than a given threshold. They eliminated false gaps (e.g., streams) using multi-temporal LiDAR-derived CHMs. Vepakomma et al. (2008) and Kellner and Asner (2009) distinguished canopy gaps along vertical canopy structure profiles. These detection methods enable the analysis of gap size and distribution (Koukoulas & Blackburn, 2004), and multitemporal LiDAR data further provide us with a better opportunity to understand forest canopy gap dynamics and the influence of climate change and human activities on forest dynamics (Vepakomma et al., 2008).

7.1.2 Canopy Height Profile

The vertical profile of forest canopies is a key structural attribute in community ecology. It plays an important role in studying community photosynthesis and species composition and quantifying animal and plant habitat quality. Traditional ecological studies typically draw canopy height profiles manually from the field. However, this method is time-consuming and subjective and can hardly be used to depict the 3-D structural information of forests on a large scale. The waveform and point cloud data recorded by LiDAR scanners enable direct quantification of canopy height profiles with high accuracy and efficiency (Jaskierniak et al., 2011).

Harding et al. (2001), Lefsky et al. (1999), and Lefsky et al. (2002) provided pioneer research using SLICER (scanning LiDAR imager of canopies by echo recovery) to quantify the spatial distribution of canopy height profiles from waveform data. By removing ground returns, Lefsky et al. (2002) calculated canopy cover from LiDAR point cloud data using the equation developed by MacArthur and Horn (1969) and transformed canopy cover into the distribution of canopy height profiles. Asner et al. (2011) used point cloud data to calculate the canopy height profiles for tropical forests in different regions. They found that the vertical structures of forest stands in Peru, Panama, and Madagascar differed from those in the Hawaiian region. They also showed that tropical forests have a subcanopy layer approximately 7 m below the canopy.

In addition to the structural attributes of the entire stand, LiDAR data can also be used to quantify the structural attributes of different forest layers. Using the mean shift algorithm, Ferraz et al. (2012) separated the point clouds into the forest canopy, shrub, and ground and estimated the height profiles of the different plant layers. Ferraz et al. (2012) used 44 quadrants in a Portugal forest dominated by *Eucalyptus globulus* Labill. and *Pinus pinaster* Ait. to test this algorithm and found that the canopy base height extracted by this algorithm explained approximately 70% of the field measured canopy base height. Duncanson et al. (2014) used a height frequency distribution diagram of point clouds to separate forest communities into a canopy layer and a canopy base layer. They then applied a watershed algorithm at each layer to segment individual trees and obtained the number of trees, heights, and crown sizes for both the canopy layer and the canopy base layer. This method was tested on forest quadrants in the United States within the Sierra Nevada mountain of California and the state of Maryland. Their results showed that this method detected approximately 21% of the small trees in the canopy layer.

7.1.3 Intensity Profile

Intensity is the returned energy strength (collected for each point) recorded for a laser beam. Its value is mainly determined by the reflectivity of an object scanned by a laser pulse. Intensity can be used to improve the accuracy of object detection and classification, estimate plant biochemical properties (Zhu et al., 2015), and generate intensity imagery similar to a black-and-white photograph. However, intensity values can easily be influenced by many factors, such as the scanning angle, surface composition, scan distance, roughness, multipath effect, and moisture content. Thus, intensity is commonly used as a relative value that is usually incomparable between datasets collected from different platforms at different locations. Several studies have examined how to calibrate intensity values (Zhu et al., 2015). For example, Zhu et al. (2015) calibrated the specular influence and angle effect of intensity values, using the calibrated values to estimate the leaf water content of broadleaved trees, achieving an accuracy of 74% at the leaf level.

The intensity profile of a forest stand refers to the vertical distribution of intensity values of all points, which can be used to extract variables like mean intensity, coefficient of variations of intensity, relative intensity percentiles, etc. Using these intensity variables can improve the estimation accuracy of certain forest structural attributes and may even be used to estimate certain functional traits (García et al., 2010). Moreover, through the combination with structural metrics, better performances might be achieved for individual tree segmentation and tree species classification (Budei & St-Onge, 2018; Kim et al., 2011)

7.1.4 Regression-Based Structural Attribute Extraction

Many stand-level structural attributes can be estimated using LiDAR data, such as the average height of a forest stand (Coops et al., 2007), Lorey's height (Næsset & Økland, 2002), average diameter at breast height (DBH) (Popescu et al., 2004), and canopy base height (Andersen et al., 2005; Kramer et al., 2014). Some attributes cannot be directly derived from certain LiDAR data types, such as DBH from airborne LiDAR. A common method to resolve this issue is to estimate those attributes by establishing a regression relationship between variables from point cloud heights and field measurements to estimate forest attributes in unknown study areas (Fig. 7.1).

Taking the study by Wing et al. (2012) as an example, they calculated the vegetation coverage of forest stands at the crown base layer (including shrubs, dead and dying trees, and small trees) using the intensity information of LiDAR points. They first extracted vegetation points at the crown base layer by using intensity and shrub height thresholds. Then, they built

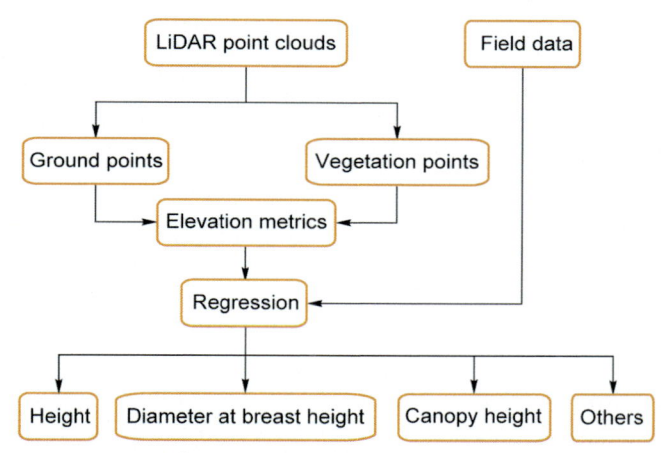

Figure 7.1 Entire process for estimating forest attributes at a stand level using the regression method.

regression models between field-measured vegetation coverage at the crown base layer and LiDAR intensity profiles. Two regression algorithms, i.e., beta regression and weighted linear regression, were tested in the Blacks Mountain Experimental Forest, California, USA. Their results showed that the R^2 values between the estimated vegetation coverage at the crown base layer and field-measured values for the two regression methods were between 0.7 and 0.8, respectively.

7.2 Individual Tree Segmentation

Individual tree segmentation identifies each tree from LiDAR point clouds, which is an important prerequisite for extracting forest structural attributes at the individual tree level. In the past decade, individual tree segmentation has been one of the main research focuses of LiDAR forestry studies. Current available individual tree segmentation methods can be divided into two categories according to the data used: segmentation based on CHMs and segmentation directly based on point clouds. They can also be grouped into three categories according to the data acquisition platform: segmentation from airborne LiDAR data (including unmanned aerial vehicles

(UAVs) and large aircraft), segmentation from mobile LiDAR data, and segmentation from terrestrial LiDAR data.

7.2.1 Canopy Height Model–Based Individual Tree Segmentation Based on Airborne LiDAR Data

Individual tree segmentation using a CHM has the longest history of use. Because a CHM is a raster image, the individual tree segmentation method using a CHM can be regarded as grayscale image processing. Some typical methods are discussed below.

Hyyppa et al. (2001) used a region-growing method for individual tree segmentation for coniferous forests. This method first processes a CHM with a low-pass convolution filter to remove noise points and then searches for the highest point within the region where its height value is greater than its 8-neighbor pixel values in the adjacent 3 × 3 window. Finally, it sets the growth rule to increase the region beginning with the highest point. The algorithm has four parameters that must be optimized based on the point features.

Popescu et al. (2002) estimated the number and the height of trees using the region-growing method. Rather than a square search window, their method uses a circular search window closer to the true shape of the canopy. The method extracts the height value of each pixel in the CHM image and then estimates the crown size based on the tree height–crown relationship obtained from ground measurements. Then, with the current pixel as the center and the crown size as the search radius, it determines whether the pixel is the local highest point within the search radius. If so, a tree is detected, and the height value of the cell is the tree height.

Koch et al. (2006) used a pouring algorithm to perform canopy segmentation of coniferous and broadleaf forests, with an overall segmentation accuracy of 62%. The algorithm first performs a local high-point search and then uses HALCON software to perform the pouring algorithm segmentation. Then, the segmentation results are optimized by combining a series of optimization conditions.

Chen et al. (2006) used the marker watershed segmentation algorithm to delineate individual trees in a savanna forest, achieving a segmentation accuracy of 64%. The marker watershed segmentation algorithm is a classic graph segmentation algorithm (Beucher, 1979), similar to the pouring algorithm used by Koch et al. (2006). Meyer and Beucher (1990) improved the watershed segmentation algorithm by adding user-defined markers to avoid oversegmentation problems. To obtain the most accurate local maximum points as markers, Chen et al. (2006) used the circular window method proposed by Popescu et al. (2002) to detect the highest point. Unlike Popescu et al. (2002), who established the relationship between tree height and crown width using ground measurements, Chen et al. (2006) measured tree height and crown width directly from CHM images and used those measurements to build the relationship. On this basis, Gaussian filtering and distance transform were performed on CHM images to achieve better segmentation accuracy. This method has been widely used to segment airborne LiDAR data of coniferous forests, but its accuracy in broadleaf forests needs improvement.

Other common CHM-based segmentation methods include morphology analysis, multiple-scale segmentation, Laplacian of Gaussian segmentation, and minimum curvature computation of CHM (Jing et al., 2012; Kaartinen et al., 2012). Detailed descriptions and comparisons of more segmentation algorithms can be found in the studies of Lu et al. (2014), Kaartinen et al. (2012), and Yang et al. (2019). Because CHM only reflects the height of the canopy surface, CHM-based individual tree segmentation is inferior at detecting understory vegetation and thus may affect the accuracy of tree numbers and height estimations.

The *skimage* library in Python provides a watershed segmentation algorithm. The following code shows how to use the watershed segmentation algorithm to perform individual tree segmentation with CHM images.

7.2.2 Point Cloud-Based Individual Tree Segmentation Based on Airborne LiDAR Data

Individual point cloud-based tree segmentation algorithms are more direct than CHM-based algorithms and can thus avoid the errors introduced by interpolating point clouds into CHMs. However, the efficiency of point

```python
# The watershed segmentation algorithm is provided by the Skimage document library in Python, the
following codes have displayed how CMH data-based watershed algorithm is utilized to proceed with
individual tree segmentation.
# -*- coding: UTF-8 -*-

import numpy as np

from osgeo import gdal

import numpy as np

import matplotlib.pyplot as plt

from scipy import ndimage as ndi

from skimage.morphology import watershed

from skimage.feature import peak_local_max

from scipy.ndimage.filters import gaussian_filter
#reading CHM

raster = gdal.Open("./chm.tif")

banddataraster = raster.GetRasterBand(1)

dataraster = banddataraster.ReadAsArray()
#applying Gaussian fitering to CHM，smoothen data

dataraster_gau = gaussian_filter(dataraster,sigma=1)
#finding the local maximum value in CHM as mark points for watershed.

local_maxi = peak_local_max(dataraster_gau, indices=False)

markers = ndi.label(local_maxi)[0]
#utilizing the watershed algorithm to proceed segmentation, each segmentation result is being stored in label
variables

labels = watershed(-dataraster_gau, markers, mask= dataraster_gau[:]>5)
#utilizing matplotlibto produce images to check for preliminary results

fig, axes = plt.subplots(ncols=3, figsize=(9, 3), sharex=True, sharey=True,

subplot_kw={'adjustable': 'box-forced'}ax = axes.ravel()
#drawing CHM
```

```
ax[0].imshow(-dataraster, cmap=plt.cm.spectral)

ax[0].set_title('CHM')
#CHM after the Gaussian filtering is being drawn

ax[1].imshow(dataraster_gau, cmap=plt.cm.spectral)

ax[1].set_title('chm_gs')
#the result of drawing segmentation

ax[2].imshow(labels, cmap=plt.cm.spectral)

ax[2].set_title('Segment')

for a in ax:

    set_axis_off()

    fig.tight_layout()

    plt.show()
```

cloud-based segmentation algorithms might be lower than CHM-based algorithms because of the computational burden that arises with a large number of points. The mainstream point cloud segmentation (PCS) methods are introduced next.

Morsdorf et al. (2004) used the K-means method to process individual tree segmentation in coniferous forests. They first detected the local highest point from the DSM and used it as the highest point of the trees. Then, they used each local highest point as the seed point for K-means clustering.

Reitberger et al. (2009) proposed a normalized conduct method to segment individual segmentation from point clouds by introducing the normalized cut of a graph-based approach. The method connects each point to form a graph, cuts off places where connectedness is weak, and produces a subgraph. This process is repeated several times until there is no weak connectedness left in the graph. Given $G = [V, E]$ for a graph, V represents nodes, and E represents the route that connects each node, then,

$$\text{Ncut}(A, B) = \frac{\text{cut}(A, B)}{\text{assoc}(A, V)} + \frac{\text{cut}(A, B)}{\text{assoc}(B, V)} \qquad (7.1)$$

where $\text{Ncut}(A,B)$ represents the normalized feature space weight between the two subgraphs, $\text{cut}(A,B)$ represents the spatial feature connection weight between subgraph A and subgraph B, assoc(A, V) represents the spatial feature connection weight of all nodes on subgraph A, and assoc(B, V) represents the spatial feature connection weight of all nodes on subgraph B. The feature space can include the 3-D coordinates of points as well as other features like the intensity information of points. The accuracy of normalized cut is considerably higher than that of other graph–based segmentation methods like minimum cut, with a notable reduction in oversegmentation errors. Reitberger et al. (2009) tested this algorithm in several coniferous and broadleaf mixed forests in Germany and compared this algorithm with the watershed algorithm. Their results showed that the normalized cut method was more accurate than the watershed segmentation algorithm. A primary drawback of the normalized cut algorithm is the extensive calculation required, even when using only the x, y, and z information in the feature space. Moreover, although the Euclidean distance between points is used in calculating the connection weight, the segmentation results may still have errors. Reitberger et al. (2009) introduced additional information like point intensity for calculating the connection weight and achieved better segmentation accuracy. At present, PCS conducted through graph–based theory is widely used in various areas (Strîmbu & Strîmbu, 2015; Tao, Wu et al., 2015).

The authors' research team proposed a PCS method called Point Cloud Segmentation (PCS), which combines region growing with a threshold value to help with segmentation (Li, Guo et al., 2012). With this algorithm, individual tree segmentation can be applied in coniferous and broadleaf mixed forests. This method uses the distance between trees as the criteria to separate them, especially the distance between different treetops. First, the algorithm locates the tallest point in the point cloud; it then applies region growing to segment an individual tree. This process repeats until all trees have been segmented. The detailed process of running the PCS algorithm includes first assuming p is the original point cloud to be segmented. Let a be a target point, Ti the tree segmented at the ith time, Ni an empty set, $dmin1$ (or $dmin2$) the minimum 2-D projection distance from all points in Ti (or Ni) to point a, and d the predefined distance threshold value. Then, the core principle of the PCS algorithm can be displayed by the following pseudocode:

While P is not empty:

    If point a is the highest point in the 3D space j with the diameter R

        If $dmin1 > d$: point $a \in Ni$;

        If $dmin1 <= d$ and $dmin1 < dmin2$: point $a \in Ti$;

        If $dmin1 <= d$ and $dmin1 > dmin2$: point $a \in Ni$;

    Else:

        If $dmin1 <= dmin2$: point $a \in Ti$;

        If $dmin1 > dmin2$: point $a \in Ni$;

    $P = P - T$.

As shown in Fig. 7.2, when the minimum distance value of the 2–D projection of all points of one tree after being separated from the target point is larger than the predefined threshold d, it can be concluded that the target point belongs to another tree Ni. If the threshold value is smaller than

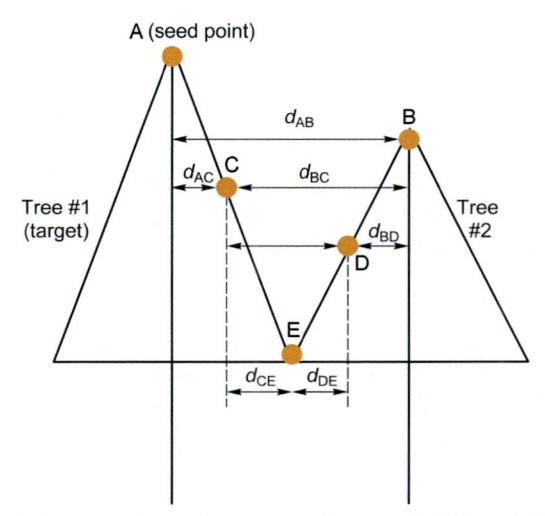

Figure 7.2 Schematic diagram of individual tree segmentation by point cloud segmentation algorithm. A represents the highest point; B, C, D, and E represent the points to be segmented; and d_{AB}, d_{AC}, d_{BC}, d_{BD}, d_{CE}, and d_{DE} represent the distance between two labeled points.

d, and the distance to *Ti* is smaller than the distance to *Ni*, it can be concluded that the target point belongs to the target tree *Ti*; otherwise, the target point belongs to another tree *Ni*. The algorithm has been tested in coniferous forests located in California, USA, and the overall accuracy reached 90% (Yang et al., 2019). Compared with other segmentation algorithms, the PCS algorithm requires fewer parameters and fewer parameter adjustments, while it provides a faster segmentation speed. Therefore, the PCS algorithm has been widely adopted in related applications (Fig. 7.3) (Ali et al., 2016). It should be noted that PCS should be performed based on normalized point clouds. Point cloud normalization is a process aiming to remove the influence of terrain elevation on lidar height measurements by subtracting their corresponding ground height measurements.

Based on the PCS algorithm, the authors' research team developed a bottom–up method for segmenting individual trees in broadleaf forests (Lu et al., 2014). This method is similar to the PCS algorithm, the major difference being that the bottom–up approach uses tree trunk points to find seed points and grow the trees in a bottom–up direction. This algorithm identifies tree trunks by including normalized intensity information. Usually, the normalized intensity value of tree trunks shows an obvious difference from that of tree branches and leaves. Thus, using a predefined threshold intensity value, a preliminary classification can be initiated to obtain the trunk points. Individual trees can then be segmented, starting from the trunk points, using a distance threshold with the region-growing method. This algorithm was tested in a broadleaf deciduous forest located in the Shaver's Creek Watershed region, Pennsylvania, USA, and an

(a)   (b)

Figure 7.3 An example of individual tree segmentation of unmanned aerial vehicle point clouds using the PCS algorithm. (a) Original point cloud and (b) the point cloud after individual tree segmentation. (Data acquisition date: August 2016; Location: Changbai Mountains, Jilin Province, China.)

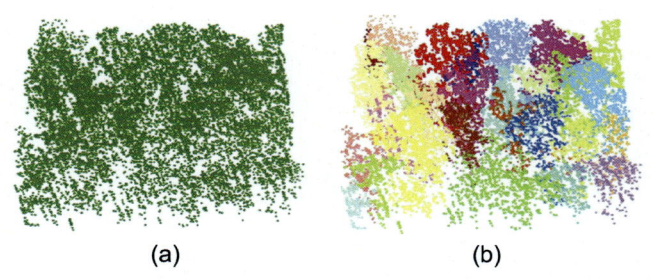

(a)   (b)

Figure 7.4 An example of individual tree segmentation from airborne light detection and ranging (LiDAR) data in a broadleaf forest using the bottom-up point cloud segmentation algorithm. (Data acquisition time: 2009; Location: Shaver's Creek Watershed, Pennsylvania, USA).

individual tree detection rate of 84% was achieved, among which 94% of trees were correctly segmented (Fig. 7.4).

7.2.3 Comparison Between Point Cloud-Based and Canopy Height Model—Based Individual Tree Segmentation Algorithms

Numerous individual tree segmentation methods have been proposed for various situations and can be grouped into two types, CHM-based and point cloud-based (Table 7.1). However, comparing their accuracy is difficult because LiDAR datasets are acquired under a variety of forest conditions, and accuracy assessment methods often differ (Table 7.1). For example, Chen (2007) evaluated the absolute accuracy for tree isolation using manually delineated benchmark data, whereas Li, Guo et al. (2012) compared segmented trees with reference trees using global navigation satellite system (GNSS) measurement and the F-score to evaluate segmentation accuracy (Sokolova et al., 2006). A systematic evaluation of different individual tree segmentation methods is needed.

To compare the performance of CHM-based and point cloud-based algorithms in extracting individual trees from airborne point data, the authors' research team compared the performances of different individual tree segmentation methods—i.e., CHM-based, pit-free CHM-based (PFCHM), PCS, and layer stacking seed point-based (LSS)—using a consistent airborne dataset and evaluation metrics (i.e., F-score and OA). This work was conducted at 120 plots covering three vegetation types (i.e., coniferous forest, broadleaf forest, and coniferous and broadleaf mixed forests) in California and Pennsylvania, USA (Fig. 7.5). A total of 40 circular

Table 7.1 Accuracies of individual tree segmentation methods from the literature.

Algorithm	Reference	Vegetation type	Point density (pts/m^2)	Accuracy (%)	Evaluation method
Region growing[a]	Hyyppa et al. (2001)	Conifer	8–10	—	—
Pouring[a]	Koch et al. (2006)	Conifer, broadleaf	5/10	62	MARA
Watershed[a]	Jing et al. (2012)	Conifer, broadleaf	45	69	C, O
Marker-controlled watershed[a]	Chen et al. (2006)	Conifer	9.5	64	AATI
Local maxima[a]	Smits et al. (2012)	Conifer, broadleaf	9	87.50	D
Pit-free canopy height model (PFCHM)[a]	Khosravipour et al. (2014)	Conifer, broadleaf	160	74	AI
Normalized cut[b]	Reitberger et al. (2009)	Conifer, broadleaf	25/10	66	—
Point cloud segmentation[b]	Li, Guo et al. (2012)	Conifer	6	90	F-score
Bottom-up region growing[b]	Lu et al. (2014)	Broadleaf	10.28	84	F-score
Region growing[b]	Hamraz et al. (2017)	Broadleaf	25/1.5	—	—
Layer stacking[b]	Ayrey et al. (2017)	Conifer, broadleaf	21/6/5	72	D, C, O
Iterative watershed[b]	Duncanson et al. (2014)	Broadleaf	18	70	—
Watershed + K-means[b]	Tochon et al. (2015)	Conifer, broadleaf	—	69.86	D, US, M, OS
Hierarchical approach[b]	Paris et al. (2016)	Conifer (multilayer)	50/8	92–97	D, C, O

Note that "—" means that information was not available. [a] refers to a segmentation method that is CHM-based, while [b] refers to a segmentation method that is point cloud-based. *AATI*, comparison of the overlay areas of the segment crown and reference crown polygons; *AI*, accuracy index = ((N−O + C)/N) × 100 (Pouliot et al., 2005); *C*, commission; *D*, detection; *M*, missed; *MARA*, manual to automated recognition accuracy (Leckie et al., 2003); *N*, the number of trees; *O*, omission; *OA*, over accuracy; *OS*, oversegmented; *US*, undersegmented.

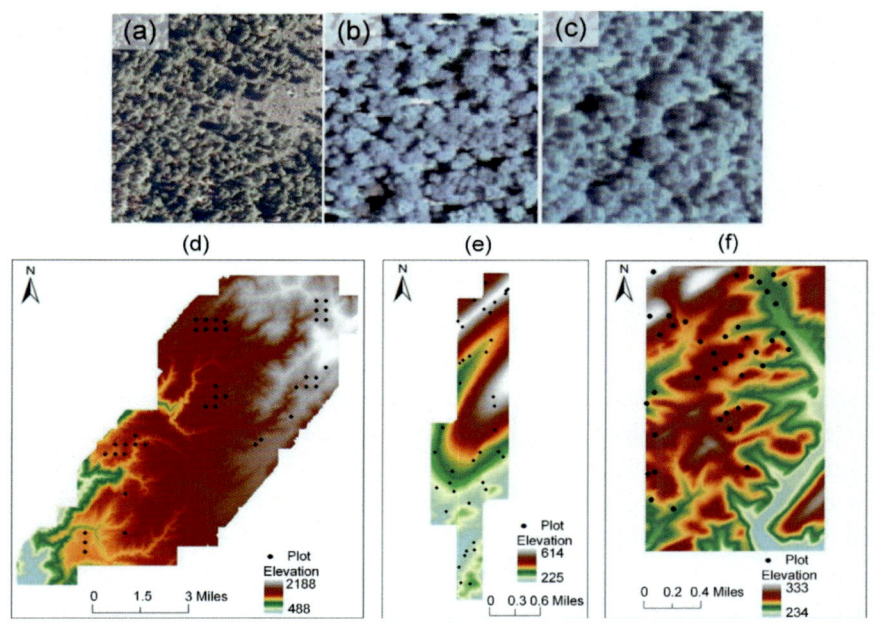

Figure 7.5 Aerial image examples from the national agriculture imagery program at 1 m resolution of a (a) coniferous forest in California, (b) broadleaf forest in Pennsylvania, and (c) mixed forest in Pennsylvania and the plot distributions of a (d) coniferous forest site, (e) broadleaf forest site, and (f) mixed forest site overlaid by elevation maps of the study areas.

plots were selected for each forest type (Fig. 7.5). The radii of coniferous, broadleaf, and mixed forest plots were 12.62, 15, and 15 m, respectively.

The results for the four segmentation methods were evaluated at the individual tree level. If a tree existed in the ground truth data and was segmented correctly, it was called a true positive (TP). If a tree did not exist in the ground truth but was segmented, it was called a false positive (FP). If a tree existed in the ground truth but was not segmented, it was called a false negative (FN). Higher TP, lower FP, and lower FN indicated better segmentation results. The recall (r), precision (p), F–score (F), and overall accuracy (OA) for each plot were calculated with Eqs. (7.2)–(7.5) (Csillik et al., 2018; Goutte & Gaussier, 2005; Sokolova et al., 2006). Recall is the tree detection rate, precision is the accuracy rate, F–score is the weighted harmonic average of precision and recall, and OA is the overall accuracy calculated as the percentage of trees segmented correctly within each plot.

$$F = 2 \times \frac{r \times p}{r + p} \tag{7.2}$$

$$r = \frac{\mathrm{TP}}{\mathrm{TP} + \mathrm{FN}} \tag{7.3}$$

$$p = \frac{\mathrm{TP}}{\mathrm{TP} + \mathrm{FP}} \tag{7.4}$$

$$\mathrm{OA} = \frac{\mathrm{TP}}{N} \tag{7.5}$$

where N represents the number of trees in each plot.

We compared the accuracy of the four segmentation methods for three forest types (Fig. 7.6). The quantitative accuracy assessment results showed that for the forest type, all segmentation methods had good accuracy (the highest F and OA) and stability (the lowest standard deviation of F and OA) in coniferous forests, followed by mixed forest and broadleaf forests. In addition, the recall of all segmentation methods was larger than the precision in all three vegetation types, especially the broadleaf forest. The differences between precision and recall were also larger with point cloud-based segmentations (PCS, LSS) than with the CHM-based segmentation and PFCHM, indicating that the point cloud-based methods were more sensitive to different vegetation types than CHM-based methods. The differences among the segmentation algorithms mainly arose from differences in the precision of the segmentation rather than from differences in the recall of the segmentation.

Out of the four segmentation methods, the PFCHM performed the best in all forest types (Fig. 7.7 and Table 7.2). Specifically, in the coniferous forest, the F-score and OA of the PFCHM were 0.88 and 0.9, respectively,

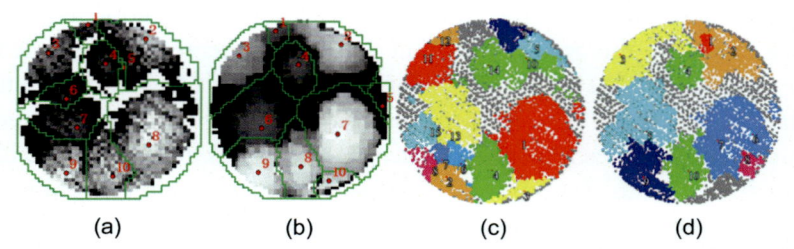

| (a) | (b) | (c) | (d) |

Figure 7.6 Individual tree segmentation results of (a) canopy height model (CHM)-based segmentation, (b) pit-free CHM-based segmentation, (c) point cloud segmentation, and (d) layer stacking seed point-based segmentation in a coniferous plot.

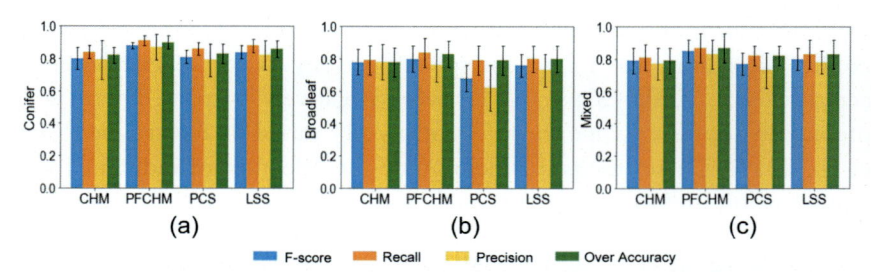

Figure 7.7 Accuracy comparison of the four segmentation methods in (a) coniferous, (b) broadleaf, and (c) mixed forests. *CHM* is short for CHM-based segmentation.

followed by LSS-, PCS-, and CHM-based segmentation. In the broadleaf forest, the F-score and OA of the PFCHM were 0.8 and 0.83, respectively, followed by CHM-, LSS-, and PCS-based segmentation. In the mixed forest, the F-score and OA of the PFCHM were 0.85 and 0.87, respectively, followed by LSS-, CHM-, and PCS-based segmentation. In addition, the PFCHM and LSS exhibited the best stability in coniferous and broadleaf forests, respectively.

The influence of different forest characteristics on segmentation accuracy has rarely been studied. A few studies have suggested that tree density, vegetation type, and canopy cover affect the performance of individual segmentation (Peuhkurinen et al., 2011). Therefore, it is necessary to analyze the influence of vegetation characteristics on different segmentation methods in different forest types (Forzieri et al., 2009). Along with the individual algorithm comparison study, the authors' research team analyzed the influence of different vegetation conditions (i.e., forest type, leaf area index (LAI), canopy cover (CC), tree density (TD), and the coefficient of variation of tree height (CVTH)) on different segmentation methods. According to the experiments, the main forest characteristics influencing

Table 7.2 Accuracy of four segmentation methods in different forest types. *CHM* is short for CHM-based segmentation.

Forest type	CHM		PFCHM		PCS		LSS	
	F-score	OA	F-score	OA	F-score	OA	F-score	OA
Conifer	0.80	0.82	0.88	0.90	0.82	0.83	0.84	0.86
Broadleaf	0.78	0.78	0.80	0.83	0.68	0.79	0.76	0.80
Mixed	0.79	0.79	0.85	0.87	0.77	0.82	0.80	0.83

segmentation methods were LAI and CC, LAI and TD, and CVTH in conifer, broadleaf, and mixed forests, respectively. Most vegetation characteristics (i.e., LAI, CC, and TD) were negatively correlated with segmentation accuracy, while the effect of CVTH varied by forest type (Yang et al., 2019). These results can help guide the selection of individual tree segmentation methods.

7.2.4 Individual Tree Segmentation Based on Mobile and Terrestrial LiDAR Data

Mobile LiDAR is mostly used for scanning urban trees or forests that can be accessed by vehicles. Trees in this environment tend to be evenly planted, which makes them less difficult to segment. Livny et al. (2010) started segmentation by locating the regional highest point. Each local highest point is assumed to be the highest point of a tree. Then, all points located within a distance threshold to each local highest point are deemed as points belonging to the same tree. Rutzinger et al. (2010) used the graph-based connected-region ideology, and Wu, Yu et al. (2013) used grids and a proximity research method to segment landscape trees along urban roads.

Compared with airborne LiDAR data, terrestrial LiDAR data have much higher position accuracy and point density and can record clear trunk information. Therefore, the individual tree segmentation from TLS point clouds usually has a higher accuracy requirement to ensure the success of following fine forest structural attribute extraction steps. However, tree canopies are usually displayed in irregular patterns with overlaps, which occurs even more commonly in broadleaf forests with high canopy cover. Therefore, accurate segmentation from terrestrial LiDAR data is still a difficult task, especially delineating individual tree crowns, which has a very high requirement for the individual tree segmentation algorithm.

The authors' research team proposed a segmentation method based on ecological theories. The basic concept of this algorithm is that trees tend to allocate their water and nutrients using the shortest transport path to optimize the allocation of resources (Tao, Wu et al., 2015). This method starts by identifying trunk points by slicing point clouds at the layer of the breast height. Each tree trunk can be recognized from the sliced point clouds using the density-based spatial clustering of applications with noise (DBSCAN) algorithm or a circular detection method. Then, all points are connected to form a graph, the shortest-path distance from each point to its closest trunk is calculated, and points are assigned to trees with the shortest path distance to their trunks. For trees with large differences in size, this

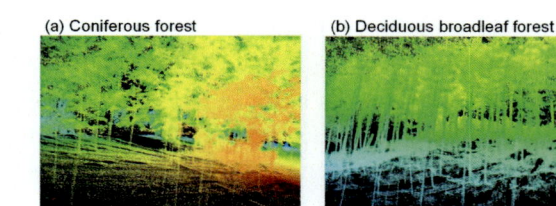

(a) Coniferous forest

(b) Deciduous broadleaf forest

(b) Evergreen broadleaf forest

Figure 7.8 Example LiDAR point cloud data collected in different types of forests by the authors' research team using a RIEGL VZ-400 scanner (RIEGL Laser Measurement System GmbH, Horn, Austria): (a) Coniferous forests, (b) Deciduous broadleaf forest, and (c) Evergreen broadleaf forest.

algorithm uses ecological metabolism theory to normalize the shortest path according to the size of the DBH and then compares the normalized shortest path. This method has achieved relatively high segmentation accuracy with different test data collected from 16 sites across China, including the broadleaf deciduous forest in the Beijing Botanical Garden, coniferous forests on Dongling Mountain, and evergreen broadleaf forests in Anhui province. Example point clouds of the plots are displayed in Fig. 7.8, and examples of segmentation results are displayed in Fig. 7.9. The segmentation accuracy quantified at the point-level by the kappa coefficient was 0.83–0.93 for the various plots.

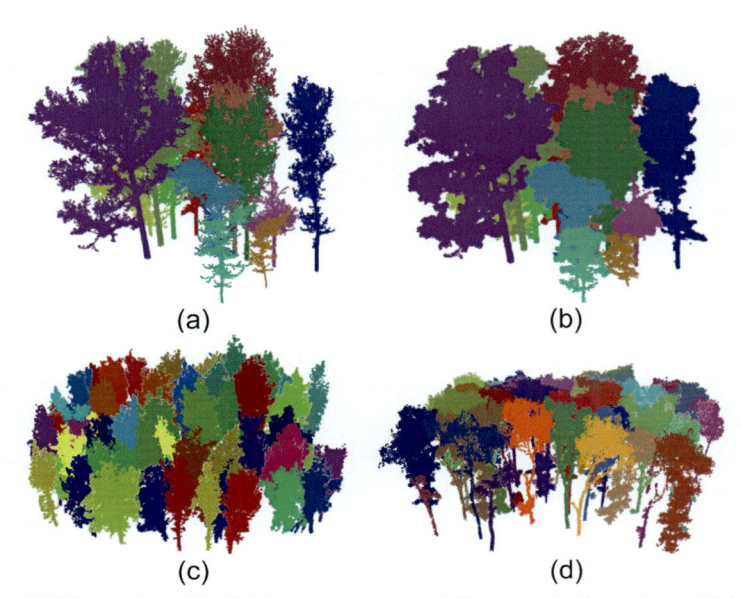

(a)

(b)

(c)

(d)

Figure 7.9 Examples of individual tree segmentation results from terrestrial LiDAR data. A *Carya illinoinensis* forest plot during (a) the leaf-off condition and (b) the leaf-on condition, (c) a *Pinus tabuliformis* forest plot, and (d) a *Castanopsis sclerophylla* forest plot.

7.2.5 Individual Tree Segmentation Using Deep Learning

Individual tree segmentation using deep learning is a novel research direction that has arisen because of the popularity of artificial intelligence, the maturity of computation resources, and the accumulation of big data. Deep learning, which is a new area of machine learning, can extract complex structural information from high-dimensional and massive data (Guo et al., 2020; LeCun et al., 2015) and has achieved notable results in object classification (Guan et al., 2015), detection (Jin, Su, Gao, Wu, et al., 2018), and segmentation (Jin et al., 2019).

The essence of individual tree segmentation using deep learning is 3-D target detection. Currently, there are two main ways to perform 3-D detection: one is detecting the object in 2-D space using multiview image-based deep learning methods, and the other is detecting the object directly in 3-D space using point-based deep learning methods.

The image-based deep learning method for individual tree segmentation is indirect but mature. For example, Jin, Su, Gao, Wu, et al. (2018) proposed an indirect method that used 2-D Faster R−CNN (region-based convolutional neural networks) to detect objects in 2-D images compressed from 3-D points. Then, the detected objects in 2-D images were mapped to 3-D points, and these were further used as seed points for the region-growing method proposed by Tao, Guo et al., (2015). This method uses the 2-D CNN method to avoid the high computational and storage cost of 3-D CNNs. In addition, the region-growing method can maintain the geometric relationship of individual plants. This indirect method achieved promising r, p, and F-score values, which were all higher than 0.9 at three testing sites with different tree densities. The algorithm was also used by Wang et al. (2019) and Windrim and Bryson (2020) to explore the segmentation of individual trees and stems from terrestrial and airborne LiDAR data. They also found that the individual tree/stem segmentation accuracy could be greater than 90%.

Compared with the image-based method, the point-based method is direct but computationally-intensive. The direct point-cloud-based method cannot fully consider the spatial structure of the point clouds and does not converge easily (Qi et al., 2016, 2017). Therefore, some studies have converted points into voxels, octrees, and graphs to enhance the structural relationship between points and reduce the computational cost. For example, Wang (2020) constructed a superpoint graph from terrestrial LiDAR points. The superpoint graph can be theoretically analyzed using graph-based deep learning methods to achieve instance/individual segmentation (Landrieu & Simonovsky, 2017). However, current point-based deep learning methods for individual tree segmentation are still rare. The main challenges include the overlapping of tree crowns, the complex vertical layer structure of forests, and the computational cost in large-scale forest scenes.

7.3 Wood-Leaf Separation

Wood-leaf separation is used to separate wood and leaf points from the original point cloud, an essential prerequisite for measuring individual and organ level attributes (Tao, Guo et al., 2015). Among them, wood points are needed to calculate the number of branches, length and radius of each branch, total wood volume of the tree, etc. (Kelbe et al., 2013; Lefsky & McHale, 2008), and leaf points are needed to calculate LAI, leaf area density, leaf inclination angles, etc. For better modeling of the 3-D radiative transfer properties of vegetation canopies, the wood and leaf parts of a tree should also be separated and assigned different spectral reflectances (Widlowski et al., 2014). In addition, computer visualization of trees requires wood points and leaf points to be separated and visualized using different techniques because the shapes of branches and leaves differ greatly (Xu et al., 2007). Applications of these techniques in forestry would improve our knowledge in a range of areas, from exploring the allometric relationships hidden in tree branching networks to quantifying the structural complexity of forests (Bentley et al., 2013).

At present, two kinds of methods are commonly used to separate wood and leave points: manual separation and automatic separation using various algorithms. Hosoi and Omasa (2006) separated branch points from leaf points manually, which proved to be relatively accurate but time-consuming. The automatic approach relies on the different geometric or physical attributes of leaf and wood points. Compared with the manual approach, automatic separation is faster, but its accuracy depends on the quality of the point cloud. Existing algorithms can be generally divided into three groups: threshold-based (e.g., intensity, multiwavelength, and waveform threshold), geometry-based (e.g., point-based), and machine learning-based methods.

7.3.1 Threshold-Based Methods

The threshold-based methods rely on a threshold of intensity or waveform width (Wu, Cawse-Nicholson et al., 2013) to distinguish wood and leaf points. This approach assumes significant differences among the optical properties of different tree components at the operating wavelength of a LiDAR system (Tao, Guo et al., 2015). For example, Côté et al. (2009) and Côté et al. (2011) used the reflectance difference of branches and leaves in the infrared band; that is, higher reflectance from branches than leaves, to realize the separation of branch points and leaf points. Yao et al. (2011) and Yang et al. (2013) used the differences in the width and height of the echo information of branches and leaves, respectively. However, the wavelengths and powers of different LiDAR systems are inconsistent (Lefsky et al., 2002). Therefore, successful intensity-based wood-leaf separation for data collected by one LiDAR system does not guarantee successful separation for data collected from other LiDAR systems. Furthermore, trees of different species have varied responses to LiDAR signals at different wavelengths, which means that the intensity approach cannot be used for certain tree species. Recently, a multiwavelength approach for wood-leaf separation was reported (Li et al., 2013). They merged data obtained using the shortwave-infrared (SWIR) laser in the Dual-Wavelength Echidna LiDAR instrument with data from the near-infrared (NIR) laser in the Echidna Validation Instrument. They found that leaf reflectance intensities were lower in SWIR than in NIR, and trunks were approximately twice as bright as leaves in SWIR. These results suggest that multiwavelength information can be used for wood—leaf separation. However, the development of multiwavelength LiDAR is still in the early stages (Chen et al., 2010; Kaasalainen et al., 2007). In addition to intensity and multiwavelength approaches, Yao et al. (2011) and Yang et al. (2013) classified trunk/branch or foliage points by thresholding the relative widths of the return waveforms. However, full-waveform data can only be produced using a few LiDAR platforms; thus, this approach has not been widely adopted.

In summary, threshold-based methods are empirical and less popular because they often require multiband or full waveform information, which is highly uncertain because of different object properties and scanning settings.

7.3.2 Geometry-Based Methods

Geometry-based methods use the geometric feature difference between branches and leaves. Xu et al. (2007) treated points with the shortest path distance from the tree root node less than a certain empirical value as branch points and other points as leaf points. However, because the focus of their work was leafy tree modeling, wood points were loosely defined as having a geodesic distance to the root that was less than a threshold (approximately two-thirds of the tree height), which led to the false classification of a large number of wood points as leaf points.

The authors' research team developed a geometric method for separating wood points from leaf points, using only 3-D information from terrestrial LiDAR point clouds (Tao, Guo et al., 2015). This method was inspired by the observation that trunk/branch boundaries appear as circles or circle-like shapes (such as arcs or incomplete circles) in point slices, whereas leaf clusters can be abstracted as line segments. They identified these geometric forms using common algorithms such as the shortest-path algorithm for graph analysis (Côté et al., 2009; Livny et al., 2010; Xu et al., 2007) and the medial axis transform (Verroust & Lazarus, 1999) for simplifying complex shapes. This algorithm was tested using two broadleaf trees and one virtual tree, and it produced Cohen's kappa coefficients ranging from 0.80 to 0.90 (Fig. 7.10).

In summary, geometric methods are more intensely studied for wood—leaf separation because the 3-D coordinates of each point are the most fundamental information acquired by a LiDAR system. However, it is challenging to define an optimal and universal geometric feature for differentiating branches and leaves among various tree species.

7.3.3 Machine Learning-Based Methods

Machine learning-based methods, especially the recently developed deep learning-based methods, are tailored to learn complex features from large amounts of high-dimensional data automatically (LeCun et al., 2015). These methods have the potential to outperform traditional methods (Xi et al., 2018) that need to define separation thresholds or geometry descriptors. Moreover, machine learning-based methods can combine various types of information instead of relying solely on certain attribute information from the tree point cloud.

In machine learning-based methods, different geometric (e.g., linear, surface, and random distribution) and spectral features are extracted and

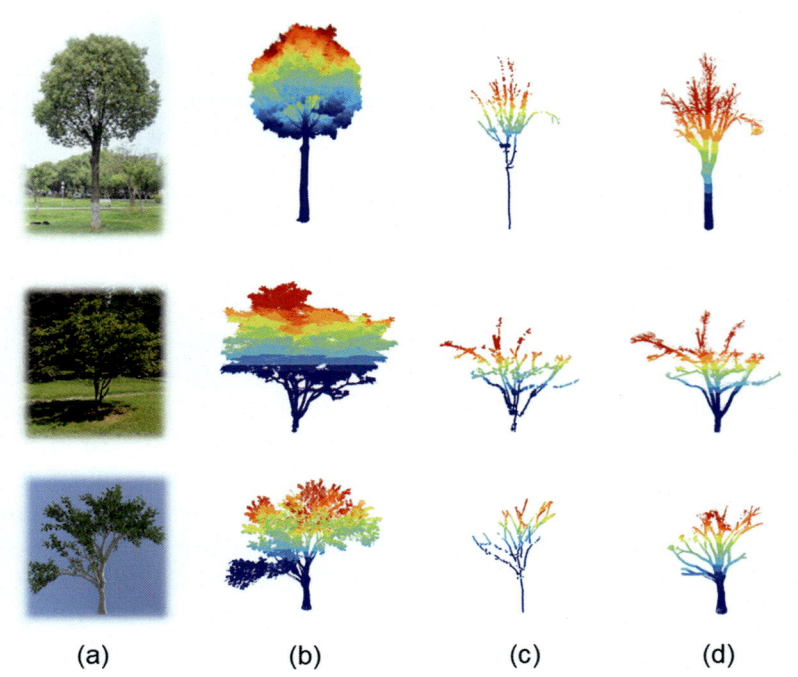

(a)   (b)   (c)   (d)

Figure 7.10 Examples of the wood—leaf separation using the geometric method proposed by the authors' research team. The first row is a camphor tree, the second row is a magnolia tree, and the third row is a virtual tree. (a) Photos, (b) raw point clouds, (c) skeleton of wood points after detecting lines and circles, and (d) final wood points.

combined to separate wood and leaf points (Lalonde et al., 2006; Ma et al., 2015; Moorthy et al., 2019). Hierarchical local feature extraction is especially beneficial (Moorthy et al., 2019). Many machine learning algorithms have been successfully applied, including unsupervised machine learning algorithms such as the DBSCAN algorithm (Ferrara et al., 2018), the LeWoS model (Wang et al., 2020), graph-base method (Wang, 2020), and the Gaussian mixed model (Wang et al., 2017) and supervised machine learning algorithms such as XGBoost (Moorthy et al., 2019), support vector machine (Yun et al., 2016), and random forest (Zhu et al., 2018).

Recent efforts have proved that deep learning has great potential in discriminating wood and leaf points in complex forests (Ayrey & Hayes, 2018; Jin et al., 2019). For example, Wu et al. (2020) developed a CNN-based model to separate wood and leaves from terrestrial LiDAR data.

Their method combined the geometric and laser return intensity information for local point sets, and the results showed a promising accuracy in broadleaf (OA > 95.05%), coniferous (OA > 93.46%), and mixed forests (OA > 84.26%). In the future, combining expert knowledge of plants with multisource data (e.g., images and multispectral data) will be interesting and perhaps beneficial for wood and leaf separation.

7.4 Individual Tree- and Organ-Level Structural Attribute Extraction

7.4.1 Basic Forest Inventory Attributes

7.4.1.1 Tree Height

Tree height is typically calculated by the height difference between the highest point of a tree and the ground point, which is one of the most common individual attributes that can be extracted by LiDAR. Owing to the setting of scanning frequency, scanning angle, and scanning mode, sometimes the highest area of a tree may not be scanned, leading to an underestimation of tree height. Fortunately, this type of error is usually at the centimeter level, which may not have a substantial influence considering the meter-level tree height. Typically, tree height can be obtained directly after individual tree segmentation (Van Leeuwen & Nieuwenhuis, 2010). The authors' research team compared tree heights obtained by CHM and PCS segmentation methods to field-measured tree heights in coniferous forests in the Sierra Nevada Mountains, California, USA. CHM segmentation-based tree height estimates were derived from the local maximum points in CHMs, while PCS-based tree height estimates were calculated using the highest point of the tree. The field survey of this study was conducted in the summer of 2008, and locations and heights of all trees with a DBH larger than 5 cm were recorded. Through careful visual inspection, the positions of trees recorded by GNSS were matched with segmented individual trees. The LiDAR- and field-measured heights of these matched trees were then compared. The results showed that PCS-based tree height more consistently agreed with the field-measured tree height than CHM segmentation-based tree height (Figs. 7.11 and 7.12), and the CHM segmentation-based method tends to underestimate tree height (Fig. 7.13).

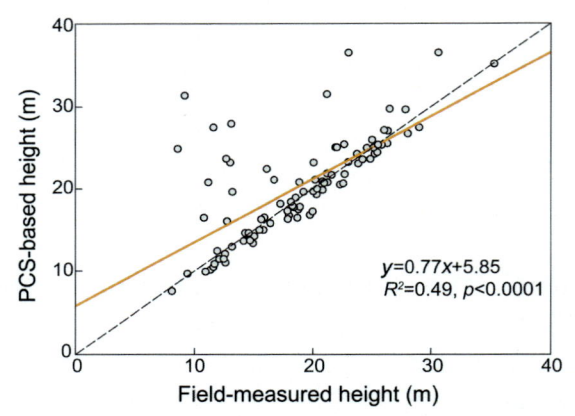

Figure 7.11 Comparison between field-measured tree heights and PCS-based tree height estimates. Each dot represents a tree.

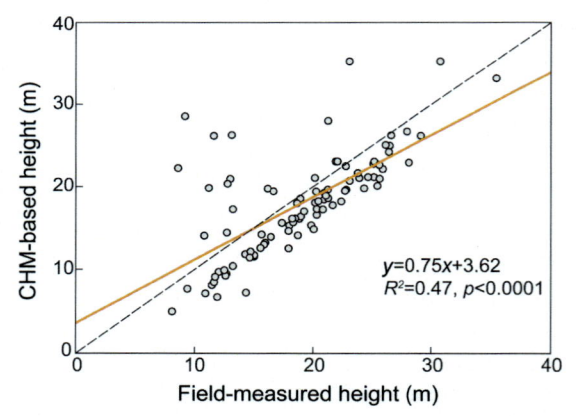

Figure 7.12 Comparison between field-measured tree heights and CHM segmentation-based tree height estimates. Each dot represents a tree.

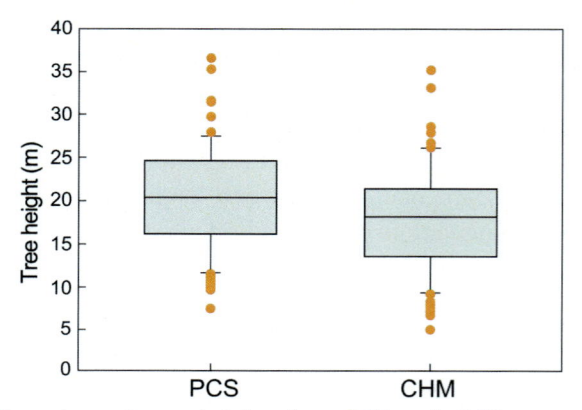

Figure 7.13 Box plots of tree heights from PCS and CHM segmentation-based methods.

7.4.1.2 Diameter at Breast Height

Since airborne LiDAR may miss trunk information, DBH is derived from airborne LiDAR data typically through the relationships between tree height and DBH (Tao et al., 2014). In contrast, terrestrial LiDAR data can capture detailed trunk information and directly estimate DBH using circular fitting methods such as the Hough transform method (Li, Pang et al., 2012; Simonse et al., 2003). Li, Pang et al. (2012) successfully used Hough transform to achieve high-accuracy DBH estimation in a mixed forest in Liangshui Reserve (located in Heilongjiang Province, China). In addition, some studies used classification techniques to extract DBH. For instance, Brolly and Kiraly (2009) calculated DBH based on identified individual trunks using the K-means clustering method.

The authors' research team developed a DBSCAN-based method to identify trunk points for DBH estimation. Compared with the K-means clustering method, the advantage of DBSCAN is that it does not require a predefined number of classes. This method was tested using a terrestrial LiDAR dataset acquired in the Institute of Botany, the Chinese Academy of Sciences, Beijing, China (Fig. 7.14). This dataset contains 14 trees, and the DBH of each tree is measured simultaneously with LiDAR data acquisition. Overall, there were three main steps of the developed DBSCAN-based DBH estimation method. First, normalization was applied to the point cloud data to remove the impact of topographical variations. Then, points around 1.3 m from the ground were fed to the process of DBSCAN-based trunk identification. Finally, a circle-fitting method was used to estimate DBH. As shown in Fig. 7.15, the DBH estimates from the proposed method had a high correlation with ground truth measurements.

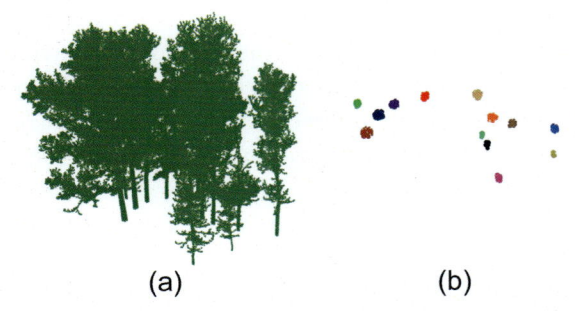

(a)   (b)

Figure 7.14 An example of trunk identification. (a) Raw terrestrial LiDAR data and (b) identified trunk through density-based spatial clustering of applications with noise (DBSCAN) algorithm.

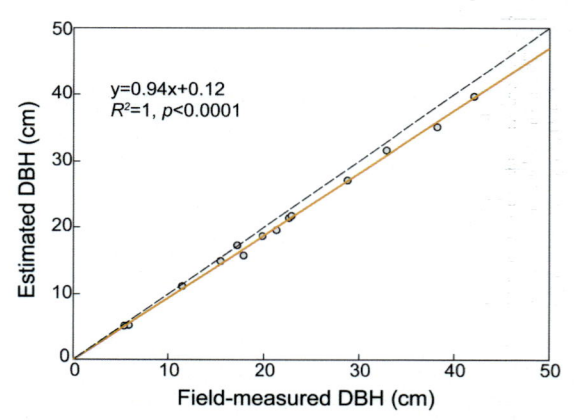

Figure 7.15 Scatter plot between diameter at breast height (DBH) estimated from terrestrial LiDAR data and field-measured DBH. DBH estimates from terrestrial LiDAR data were derived using the DBSCAN-based method.

7.4.1.3 Crown Base Height

Crown base height (also known as height to crown, or height to live crown base; CBH) is an essential biophysical tree parameter for various applications such as forest management, forest fuel treatment, wildfire modeling, ecosystem modeling, and global climate change studies (Stephens, 1998). It refers to the height from the bottom of the canopy to the base of the trunk. Currently, methods of CBH estimation can be divided into regression-based methods (Holmgren & Persson, 2004; Næsset & Økland, 2002), curve fitting methods (Dean et al., 2009), and layer slicing-based methods (Kato et al., 2009; Popescu & Zhao, 2008).

Regression-based methods usually used LiDAR-derived height metrics as prediction variables of CBH. For example, Næsset and Økland (2002) estimated the CBH of a coniferous forest in the Østmarka Nature Reserve, Norway, by extracting quantile height metrics from point clouds, and their results explained 53% of the variability in ground-truth measurements. Holmgren and Persson (2004) predicted CBH using percentile height variables, and R^2 between the estimation value and the ground truth was 0.71. The disadvantages of regression-based methods are that they need the support of field measurements and cannot be used to estimate CBH at the individual tree-level.

Curve fitting methods are also designed for stand-level CBH estimation. Dean et al. (2009) sliced the point cloud of a forest stand into 40 vertical layers and calculated a frequency distribution curve using the number of points in each layer. Then, they used the Weibull probability density function to fit the distribution curve and extracted stand–level CBH based on the fitted curve. Their estimated CBH was approximately 0.6 m higher than ground truth measurements.

Like curve fitting methods, layer slicing-based methods slice point clouds into different layers, but they target individual trees. Since the number of points gradually increases from the top to the bottom of the canopy, and point density may suddenly increase at the height of CBH, the detecting discontinuity in point density can be used to estimate CBH. For example, Popescu and Zhao (2008) estimated the CBH of 94 conifers and 23 broadleaf trees in the southeastern United States by first using the wavelet and Fourier transforms to smooth the point density curve in the vertical direction and then performing a polynomial fitting on the smoothed curve to find heights with sudden changes in point density. The R^2 between their estimated values and ground truth measurements reached 0.8, indicating the effectiveness of the layer slice-based methods. Kato et al. (2009) discussed the possibility of using a point density change threshold along the vertical direction to estimate CBH. Their results showed that the R^2 and RMSE estimates for conifer forests were 0.92 and 1.62 m; for broadleaf forests, they were 0.53 and 2.23 m.

The authors' research team also proposed a layer slicing-based method for directly estimating individual-tree CBH from airborne LiDAR data (Luo et al., 2018). The method involves two main steps: (1) removing noise and understory vegetation for each tree and (2) estimating the CBH by generating a percentile ranking profile for each tree and using a spline curve to identify its inflection points (Luo et al., 2018) (Fig. 7.16). The proposed method was tested in a mixed conifer forest in the Sierra Nevada,

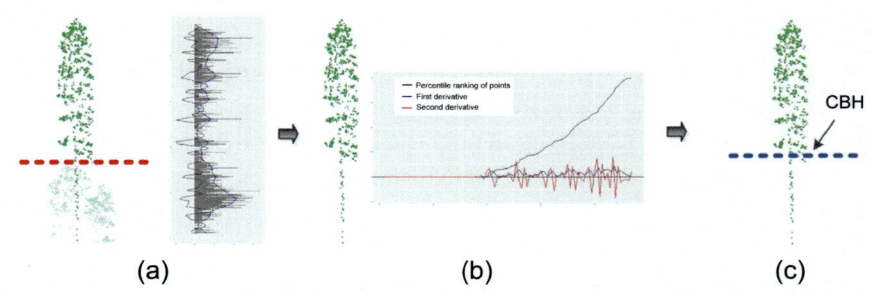

(a)   (b)   (c)

Figure 7.16 An example of estimating individual tree canopy base height (CBH) from airborne LiDAR data. (a) Identify understory points beneath a target tree and remove them, (b) generate a percentile ranking profile for the target tree, and (c) estimate CBH for the target tree.

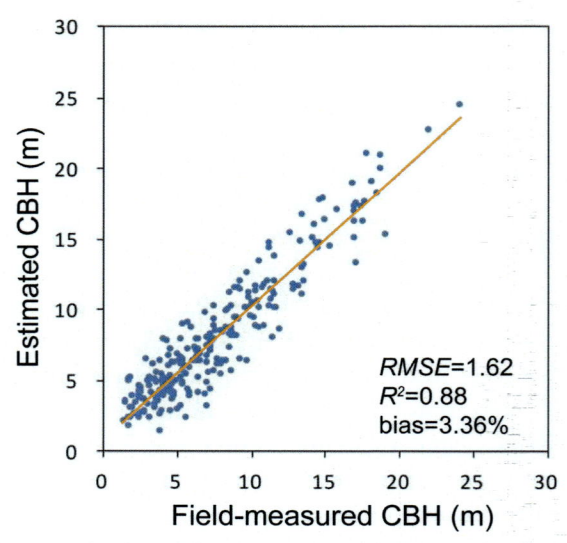

Figure 7.17 Scatter plot between estimated individual tree canopy base heights (CBHs) from LiDAR point clouds and field-measured CBHs.

California, and was validated by field measurements. The results showed that our method could directly estimate the CBH at the individual tree level with an RMSE of 1.62 m, an R^2 of 0.88, and a relative bias of 3.36% (Fig. 7.17).

Although layer slicing-based methods have the advantages of being highly accurate and not needing field measurements. However, its performance is highly dependent on individual tree segmentation, and point density change along the vertical direction in structurally complex forests may not follow the abovementioned rules.

7.4.2 Crown Size Attributes

The shape and size of tree crowns can be acquired using airborne, mobile, or terrestrial LiDAR. There are two commonly used methods. (1) CHM-based: Popescu et al. (2003) segmented each tree from a CHM, then fitted a curve to estimate the portion of CHM from the top of a tree to its crown. In this way, the edge location of the crown can be acquired, and the crown diameter can then be calculated. (2) Point-based: Tao et al. (2014) projected the point cloud of an individual tree on a plane and calculated the convex hull and concave hull in the projection region to acquire the crown projection area and calculate the crown width (Fig. 7.18). For terrestrial and

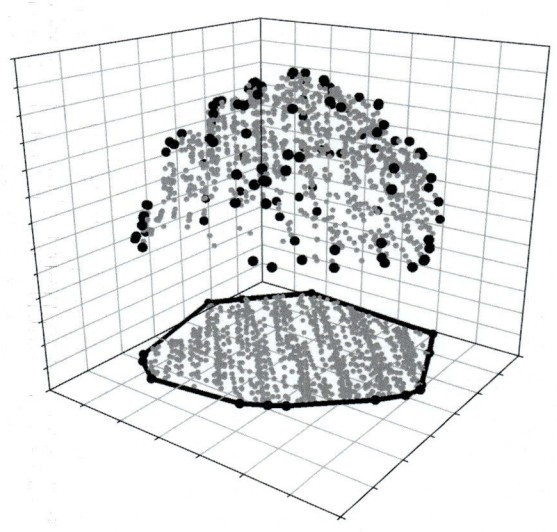

Figure 7.18 Schematic diagram of crown size estimation using the point-based method.

mobile LiDAR, the accuracy of crown extraction is higher because they have higher point density.

Crown volume has increasingly been used to calculate urban greenness (Rafiee et al., 2013). Based on CBH estimates, LiDAR data can be used to estimate crown volume. The commonly used methods comprise the following steps: (1) applying a 3-D convex hull to a region above the height of the CBH, (2) further flattening the 3-D convex hull points, and finally, (3) calculating the volume of the space surrounded by the convex hull (Kato et al., 2009; Tao et al., 2014) (Fig. 7.19). The accuracy of this method is affected by CBH estimation accuracy and the size of the blank region between branches, which was included in the crown volume calculation.

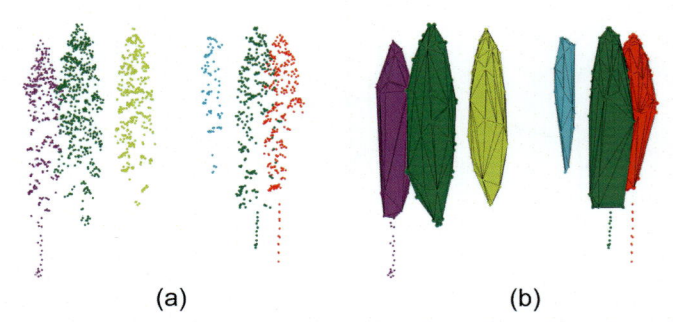

(a)   (b)

Figure 7.19 Schematic diagram of crown volume estimation using point clouds. (a) Original point clouds, and (b) calculation of crown volume based on three-dimensional (3-D) convex hulls.

7.4.3 Branch Architecture Attributes

Branch architecture, defined as the size, shape, and 3-D arrangement of tree branches, directly influences the biological and physical processes of vegetation and can be used to reveal vegetation dynamics during climate change (Li et al., 2020; Su et al., 2020). However, tree branch architecture has rarely been quantitatively described because it is extremely time-consuming and labor-intensive to quantify tree branch architecture by traditional approaches (Bentley et al., 2013; Dassot et al., 2012; Henning & Radtke, 2006; Van der Zande et al., 2006).

Terrestrial LiDAR has proved to be a promising tool to quantitatively describe the branch architecture attributes. Extracting information on the tree skeleton and establishing the topological relationship of branch systems are two critical steps in retrieving branch structural attributes from terrestrial LiDAR data. Current skeletonization algorithms have been mostly adopted from the field of computer graphics and computer vision, such as volumetric methods (Gorte & Pfeifer, 2004), octree structure methods (Bucksch & Lindenbergh, 2008), Voronoi diagram methods (Dey & Sun, 2006), Reeb graph-based methods (Lazarus & Verroust, 1999), segmentation methods (Dai et al., 2009; Dai et al., 2010), and Laplacian-based methods (Cao et al., 2010).

The authors' research team improved the Laplacian-based contraction skeletonization algorithm using the Dijkstra algorithm, developed a new path discrimination method to identify and encode branch orders (Fig. 7.20) and retrieved branch architecture parameters using branch order

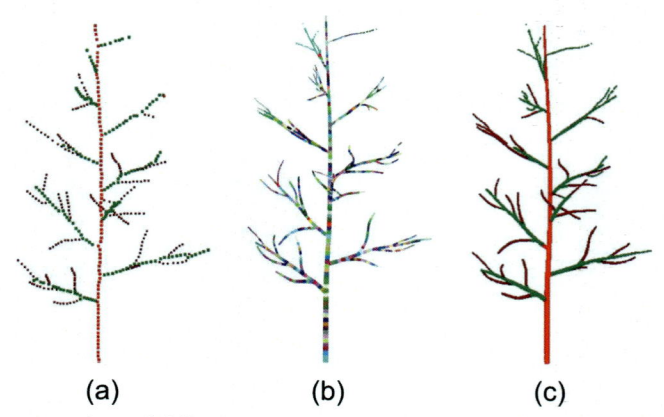

(a)   (b)   (c)

Figure 7.20 Encoding of different branch orders. (a) The encoding of skeleton points, (b) raw point cloud data, where different colors represent different segmented sections, and (c) the encoding of raw point cloud data. Red represents first-order branches, green represents second-order branches, and brown represents third-order branches.

and topology information. Details of the improved skeletonization algorithm can be found in (Li et al., 2020). After labeling the branch ID, the raw point cloud can be distinguished as first-order, second-order, and third-order branches, and so on (Fig. 7.20c) according to the one-to-one match between skeleton points and the raw point cloud (Fig. 7.20a and b). The branch architecture attributes can then be retrieved through the following steps:

7.4.3.1 Branch Angle

The branch angle was defined as the angle θ at the junction of two branches (Fig. 7.21) and is calculated using Eq. (7.6):

$$\theta_{H,N} = \arccos \left(\frac{\left(\overrightarrow{AB}, \overrightarrow{AC} \right)}{\left| \overrightarrow{AB} \right| * \left| \overrightarrow{AC} \right|} \right) \tag{7.6}$$

where H represents the branch order, N is the number of the same branch order (from the bottom to up), A is the starting point of branch level H, B is the nearest skeleton point of branch level H-1 above A, and C is the nearest skeleton point of branch level H.

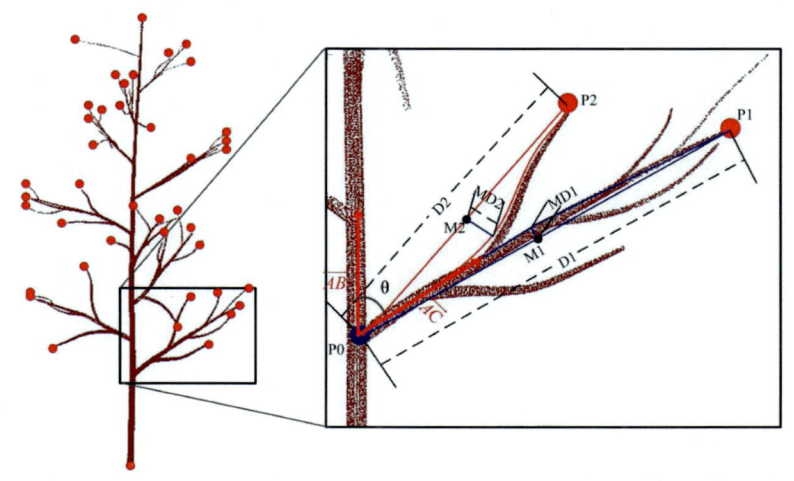

Figure 7.21 Illustration of skeleton order identification. Red points are skeleton endpoints, blue points are the starting point, and black points are the midpoint of two endpoints.

7.4.3.2 Branch Diameter and Length

The relationship between skeleton points and raw point clouds was established during point cloud contraction, and each branch was segmented into s sections according to the skeleton points (Fig. 7.20b). We constructed the covariance matrix for each section point cloud using Eqs. (7.7)–(7.9) to retrieve the branch diameter and length:

$$M = \begin{pmatrix} \text{cov}(x,x) & \text{cov}(x,y) & \text{cov}(x,z) \\ \text{cov}(y,x) & \text{cov}(y,y) & \text{cov}(y,z) \\ \text{cov}(z,x) & \text{cov}(z,y) & \text{cov}(y,z) \end{pmatrix} \tag{7.7}$$

$$o_i = (1/n) \sum_{i=1}^{n} p_i \tag{7.8}$$

$$\text{cov}(i,j) = \frac{\sum_{1}^{n}(p_i - o_i)\left(p_j - o_j\right)^{\text{T}}}{n-1} \tag{7.9}$$

where M is the covariance matrix, (x, y, z) are the variables corresponding to the 1st, 2nd, and 3rd columns of the point cloud data, o_i is the average value of all elements of the ith column, p_i and p_j are all the elements of the ith and jth column, respectively, and n is the number of points. The eigenvectors (e_1, e_2, e_3) and corresponding eigenvalues $(\lambda_1, \lambda_2, \lambda_3)$ of the covariance matrix are extracted. Assuming $\lambda_1 \approx \lambda_2$, eigenvectors e_1 and e_2 would be considered two vectors in the cross-sectional plane of each branch, with eigenvector e_3 representing the orientation of the branch. The product of points set P and eigenvector e_i ($i = 1$, 2, or 3) represents the projection of p in the direction of eigenvectors e_i. The branch diameter D is obtained by calculating the average value of the difference of maximum and minimum values of $p \times e_1$ and $p \times e_1$. Each branch section length L is the difference between the maximum and minimum value of $p \times e_3$, and the whole branch length is the accumulation of each section length L.

7.4.3.3 Branch Volume

Each segmented point cloud section can be regarded as a cylinder, to which volume V can be retrieved using Eq. (7.10):

$$V_{H,N} = \sum_{1}^{s} (\pi D^2 L)/4 \tag{7.10}$$

To assess the influence of branching complexity and branching pattern on the estimation accuracy, we scanned 15 magnolia trees of different sizes without a leading stem, simulating 10 trees with leading stems of different sizes. Results showed that the overall branch order identification and parameter retrieval accuracy for trees with a leading stem were higher than for trees without a leading stem. The identification accuracy of the branching order decreased with increasing branch numbers and branching complexity. The estimated branch architecture attributes agreed well with ground truth measurements (R^2 up to 0.99), except for the second- and third-order branch volume. Compared with branch angle and diameter, branch length showed the best correlations with manually measured values (0.14 m vs. 0.002 degree and 8.48 m in RMSE; 0.99 vs. 0.99 and 0.78 in R^2). The second-and third-order branch volume estimations were highly underestimated compared with the ground truth values ($R^2 = 0.53$, RMSE $= 0.0239$ and $R^2 = 0.70$, RMSE $= 0.0257$, respectively).

7.4.4 Timber-Related Attributes

Timber-related attributes, such as stem straightness and branchiness, are important for determining tree quality and the possible amount of logs (Vepakomma & Cormier, 2019), which is thus important for forest management. Stem straightness and branchiness in standing trees are mainly quantified by three methods—visual assessment, photogrammetric measurement, and LiDAR measurement. The visual assessment method is the fastest and simplest, but it is subjective and error-prone. The photogrammetric measurement method requires 3-D reconstruction through multiview images, which is subjective and technically demanding and can be easily affected by crown density, branch overlapping, and light conditions. By contrast, LiDAR measurement is the most applicable method, especially proximal LiDAR methods, because they can provide highly accurate 3-D information within forest canopies.

Calculating stem straightness from LiDAR points is usually straightforward, including steps of extracting and reconstructing tree stems. First, points of a clear stem should be sliced into several vertical layers. Second, a circle is fitted at each vertical layer to extract the stem diameter of each layer and reconstruct the diameter profile along the stem. Finally, each stem section is represented by a cylinder, and the stem straightness is estimated by the sinuosity between the centroids of circles (Vepakomma & Cormier, 2019).

After fitting the stem cylinder, branches are estimated by connecting the voxels (cubic grids containing tree points) attached to the stem cylinder using a neighbor tracing algorithm (Vepakomma & Cormier, 2019). The extracted branch voxels are used to calculate stem branchiness (the proportion of space filled with branches). Many other timber-related attributes, such as timber volume, can also be extracted from the fitted stem cylinder or regression methods, and readers are referred to (Maltamo et al., 2004; Steinmann et al., 2013; Tonolli et al., 2011) for more details.

7.4.5 Leaf-Related Attributes

Leaf-related attributes are important for understanding the transmission of radiation within the vegetation canopy and the distribution of incident photosynthetically active radiation (Asner, 1998; Houborg et al., 2007; Ross, 2012). However, extracting individual leaf-level attributes is challenging owing to the small size of the leaves. Only a few studies have tried to extract leaf properties, such as leaf inclination and azimuthal angle (Li et al., 2018).

The authors' research team developed a new method to accurately estimate leaf inclination and azimuthal angles. The method comprises three steps: (1) voxelization; (2) leaf segmentation and filtration; and (3) leaf inclination and azimuthal angle estimation.

7.4.5.1 Voxelization

After data preprocessing (e.g., outlier removal), the leaf points of each tree were first translated. Specifically, the central point of each leaf point cloud was transformed into the origin point of the Cartesian coordinate system. Then, we voxelized the leaf points using the voxel-based canopy profiling method (Hosoi & Omasa, 2006). The width and length of a voxel were defined as twice the maximum leaf length of each tree, and the height of a voxel was determined by the difference between the minimum and maximum values of the z-coordinates (Fig. 7.22). The leaf point cloud data were divided into a finite number of small parts through voxelization, and each voxel was regarded as a basic computing unit.

7.4.5.2 Leaf Segmentation and Filtration

Accurate leaf segmentation is an important prerequisite for retrieving leaf inclination and azimuthal angles. After voxelization, we adopted the clustering algorithm proposed by Ester et al. (1996), namely DBSCAN, for leaf segmentation. After the clustering, leaf points in the same voxel were

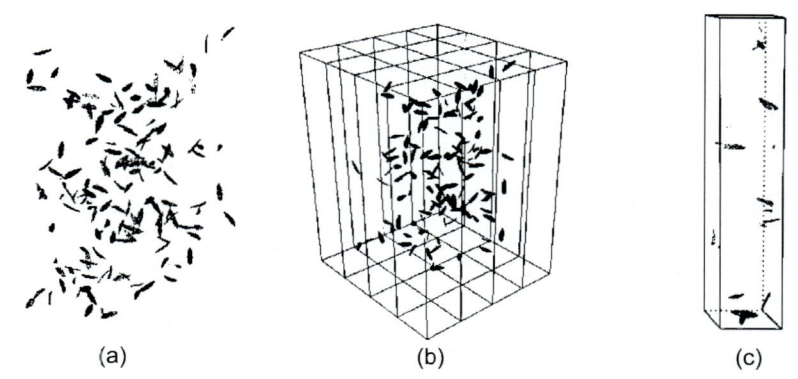

Figure 7.22 Illustration for point cloud voxelization: (a) Leaf point cloud data, (b) voxelization, and (c) leaf point cloud data in a single voxel.

segmented into different clusters (Figs. 7.22c and 7.23a). We constructed a 3-D convex hull for each cluster (Fig. 7.23b), and one-half of the total surface area of the 3-D convex hull was regarded as the surface area of a cluster in reality (Zheng & Moskal, 2012). The cluster where one-half of the surface area was greater than two-thirds of the manually measured single leaf area and smaller than the manually measured single leaf area (leaves 4, 5, 14, and 15 in Fig. 7.23) was selected as the point cloud data of a complete

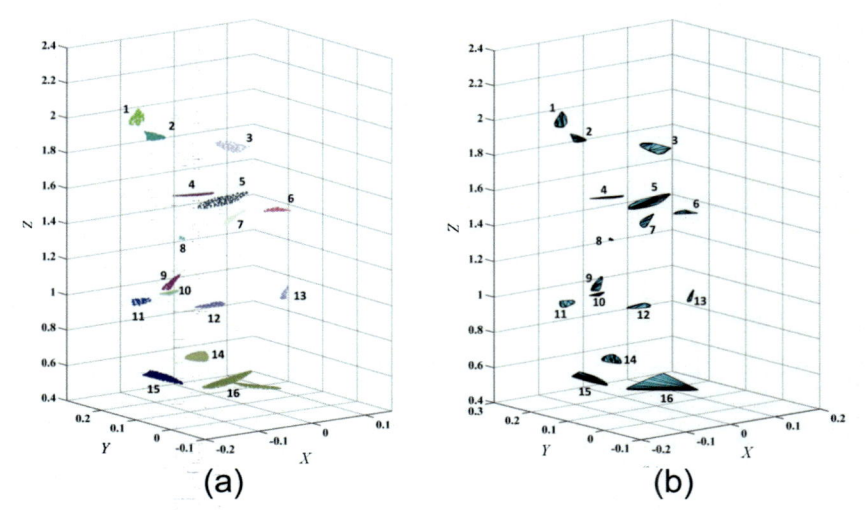

Figure 7.23 Leaf point cloud clustering analysis and 3-D convex hull construction. (a) Final clustering results, and (b) 3-D convex hull for each cluster. The numbers 1–16 represent the different clusters.

leaf and removed from the raw point cloud data. True leaves were collected in the field, and the leaf area was measured in the laboratory. The clusters with incomplete leaves (leaves 1−3 and 6−13 in Fig. 7.23), as well as several other leaves (leaf 16 in Fig. 7.23), were retained. In addition, to obtain more point cloud data for complete leaves, we voxelized the remaining leaf point clouds and repeated the abovementioned process after data rotation until no more complete leaves could be found. Visual inspection was used to judge whether the segmentation was correct.

7.4.5.3 Leaf Inclination and Azimuthal Angle Estimation

After leaf segmentation and filtration, we constructed the covariance matrix for each complete leaf point cloud data using Eqs. (7.7)−(7.9) to retrieve the leaf inclination and azimuthal angles. The leaf inclination angle was calculated as the angle between the eigenvector e_3 and the vector (0, 0, 1). The eigenvector e_1 represents the direction of the principal leaf axis under the assumption that the leaf length is greater than the leaf width, and the leaf azimuthal angle is calculated as the angle between the eigenvector e_1 and the vector representing the north direction.

We simulated 10 trees with different leaf numbers (Table 7.3: Tree IDs 1−10) using the stochastic L-system and scanned three magnolia trees

Table 7.3 Total leaf number, number of segmented leaves, number of true segmented leaves, and corresponding recognition ratio and correction ratio for each tree. Tree identification numbers (IDs) 1−10 represent simulated trees, and tree IDs 11−13 represent scanned magnolia trees.

Tree ID	No. of leaves	No. of segmented leaves	No. of true segmented leaves	Recognition ratio (%)	Correction ratio (%)
1	15	15	15	100	100
2	30	28	28	93	100
3	45	43	43	95	100
4	60	54	54	90	100
5	75	67	67	89	100
6	90	82	82	91	100
7	105	94	94	89	100
8	120	102	102	85	100
9	135	117	117	86	100
10	150	120	120	80	100
11	321	102	35	32	34
12	617	172	37	28	22
13	911	212	28	23	13

(*Magnolia denudata* Desr.) with different leaf numbers using a RIEGL VZ-400 terrestrial LiDAR (RIEGL Laser Measurement System GmbH, Horn, Austria) mounted on a tripod approximately 1.5 m above the ground. Our method provided accurate leaf inclination and azimuthal angles (leaf inclination angle: $R^2 = 0.98$, RMSE = 2.41° and leaf azimuthal angle: $R^2 = 0.99$, RMSE = 3.44°).

7.5 Structural Attributes Extracted Through the Fusion of Multiplatform LiDAR Data

Structural attributes can be extracted from various LiDAR systems (Liang et al., 2014, 2016; Su et al., 2018; Wallace et al., 2016). However, each of these LiDAR platforms has its limitations. The downward-looking LiDAR systems (e.g., UAVs and airborne LiDAR systems) can provide highly accurate tree canopy information but lack tree trunk information (Paris et al., 2016). In contrast, mobile and backpack LiDAR systems can provide detailed tree trunk information, but the limited vertical field of view and measurement range may result in missed upper canopy information in the point cloud (Hilker et al., 2012). Single-location terrestrial LiDAR scans suffer from the occlusion effect of branches and leaves, and the registration of multiple terrestrial LiDAR scans can be time-consuming (Theiler et al., 2015; Yan et al., 2017). The fusion of multiplatform LiDAR data has the potential to address the limitations of each LiDAR platform and thus obtain a more complete and accurate estimation of forest structural attributes (Hämmerle et al., 2017). Consequently, multiplatform LiDAR data fusion gradually attracted the attention of many researchers and has become an interesting and arduous task.

Early multiplatform LiDAR data fusion mainly relies on the assistance of exterior data or targets. Positioning information from the GNSS and marker points identified from high-reflectance referencing targets are usually used to register point clouds collected from different platforms. However, this information is unavailable, inaccurate, or hard to acquire in forests (Guan, Su, Hu et al., 2020). For example, GNSS signals can be easily blocked or influenced by the multipath effect under the forest canopy, and arranging referencing targets can be very time-consuming. Furthermore, the occlusion effects of forests and different viewing angles of platforms make it hard to acquire corresponding marker points from referencing targets. To address this issue, many researchers attempt to use tie points/lines/polygons identified within point clouds to register multiplatform

LiDAR data automatically. Although these methods perform well in urban areas, most are failed in forested areas since the complexity and irregularity of forest environments give rise to the absence of repeatable and unambiguous features. Recently, numerous efforts have been put forth in developing marker-free methods to register multiplatform LiDAR data in forest environments. These methods typically used tree attributes (e.g., tree location, tree height, and DBH) as the required exterior features for registration (Guan, Su, Sun et al., 2020). It has been shown that using tree attributes has the advantages of high efficiency and robustness of multiplatform LiDAR data fusion in forested areas, thereby improving the efficiency and accuracy of LiDAR forestry applications. The detailed principle of the above method can be found in Chapter 5. Once with the fused point clouds from multiple LiDAR platforms, the procedures for extracting forest structural attributes from them are similar to those for point clouds of a single LiDAR platform. Studies have shown that when a single platform scans incompletely, fusing lidar data from different platforms can effectively improve the accuracy of LiDAR forestry applications. For example, if tree height and DBH can be simultaneously extracted from UAV LiDAR and terrestrial LiDAR, respectively, the accuracy of forest aboveground biomass estimation is expected to be largely improved (Graves et al., 2018).

7.6 Chapter Summary

Estimating forest structural attributes is crucial for the sustainable management of forests and helps deepen our understanding of forest ecosystems. In recent decades, LiDAR has become a promising technology for estimating forest structural attributes. The extraction methods of stand-level attributes (e.g., canopy height, canopy height profile, and canopy gap), individual tree-level attributes (e.g., tree height, DBH, crown base height, and trunk volume), and organ-level attributes (branch architecture attributes, timber-related attributes, and leaf-related attributes) using LiDAR data from various platforms were introduced in this chapter. The prerequisite of individual tree-level and organ-level forest structural attribute extraction is individual tree segmentation and wood and leaf separation, respectively. Various CHM-based or point cloud-based individual tree segmentation methods have been proposed in recent years. In practice, the selection of individual tree segmentation methods should consider factors such as the LiDAR platforms, forest types, and terrain conditions. In addition to the choice of method, the completeness of the LiDAR scan of the forest can have a significant impact on the accuracy of forest structural attribute extraction. Each LiDAR platform has its scanning angle limitations in forested environments. Multiplatform LiDAR fusion has become a potential solution to address the limitations of each single LiDAR platform and thus obtain a more complete and accurate estimation of forest structural attributes.

References

Ali, A. M., Darvishzadeh, R., Skidmore, A. K., van Duren, I., Heiden, U., & Heurich, M. (2016). Estimating leaf functional traits by inversion of PROSPECT: Assessing leaf dry matter content and specific leaf area in mixed mountainous forest. *International Journal of Applied Earth Observation and Geoinformation, 45*, 66—76.

Andersen, H.-E., McGaughey, R. J., & Reutebuch, S. E. (2005). Estimating forest canopy fuel parameters using LIDAR data. *Remote Sensing of Environment, 94*(4), 441—449.

Asner, G. P. (1998). Biophysical and biochemical sources of variability in canopy reflectance. *Remote Sensing of Environment, 64*(3), 234—253.

Asner, G. P., Hughes, R. F., Mascaro, J., Uowolo, A. L., Knapp, D. E., Jacobson, J., Kennedy-Bowdoin, T., & Clark, J. K. (2011). High-resolution carbon mapping on the million-hectare Island of Hawaii. *Frontiers in Ecology the Environment, 9*(8), 434—439.

Asner, G. P., Kellner, J. R., Kennedy-Bowdoin, T., Knapp, D. E., Anderson, C., & Martin, R. E. (2013). Forest canopy gap distributions in the southern Peruvian Amazon. *PloS One, 8*(4), e60875.

Ayrey, E., Fraver, S., Kershaw, J. A., Jr., Kenefic, L. S., Hayes, D., Weiskittel, A. R., & Roth, B. E. (2017). Layer stacking: A novel algorithm for individual forest tree segmentation from LiDAR point clouds. *Canadian Journal of Remote Sensing, 43*(1), 16—27.

Ayrey, E., & Hayes, D. J. (2018). The use of three-dimensional convolutional neural networks to interpret LiDAR for forest inventory. *Remote Sensing, 10*(4), 649.

Ben-Arie, J. R., Hay, G. J., Powers, R. P., Castilla, G., & St-Onge, B. (2009). Development of a pit filling algorithm for LiDAR canopy height models. *Computers & Geosciences, 35*(9), 1940—1949.

Bentley, L. P., Stegen, J. C., Savage, V. M., Smith, D. D., von Allmen, E. I., Sperry, J. S., Reich, P. B., & Enquist, B. J. (2013). An empirical assessment of tree branching networks and implications for plant allometric scaling models. *Ecology Letters, 16*(8), 1069—1078.

Beucher, S. (1979). Use of watersheds in contour detection. In *Proceedings of the International Workshop on Image Processing*.

Brolly, G., & Kiraly, G. (2009). Algorithms for stem mapping by means of terrestrial laser scanning. *Acta Silvatica et Lignaria Hungarica, 5*, 119—130.

Bucksch, A., & Lindenbergh, R. (2008). CAMPINO—a skeletonization method for point cloud processing. *ISPRS Journal of Photogrammetry Remote Sensing, 63*(1), 115—127.

Budei, B. C., & St-Onge, B. (2018). Variability of multispectral lidar 3D and intensity features with individual tree height and its influence on needleleaf tree species identification. *Canadian Journal of Remote Sensing, 44*(4), 263—286.

Cao, J., Tagliasacchi, A., Olson, M., Zhang, H., & Su, Z. (2010). *Point cloud skeletons via laplacian based contraction. 2010 Shape Modeling International Conference*.

Chen, Q. (2007). Airborne lidar data processing and information extraction. *Photogrammetric Engineering Remote Sensing, 73*(2), 109.

Chen, Q., Baldocchi, D., Gong, P., & Kelly, M. (2006). Isolating individual trees in a savanna woodland using small footprint lidar data. *Photogrammetric Engineering Remote Sensing, 72*(8), 923–932.

Chen, Y., Räikkönen, E., Kaasalainen, S., Suomalainen, J., Hakala, T., Hyyppä, J., & Chen, R. (2010). Two-channel hyperspectral LiDAR with a supercontinuum laser source. *Sensors, 10*(7), 7057–7066.

Coops, N. C., Hilker, T., Wulder, M. A., St-Onge, B., Newnham, G., Siggins, A., & Trofymow, J. T. (2007). Estimating canopy structure of Douglas-fir forest stands from discrete-return LiDAR. *Trees, 21*(3), 295.

Côté, J.-F., Fournier, R. A., & Egli, R. (2011). An architectural model of trees to estimate forest structural attributes using terrestrial LiDAR. *Environmental Modelling Software, 26*(6), 761–777.

Côté, J.-F., Widlowski, J.-L., Fournier, R. A., & Verstraete, M. M. (2009). The structural and radiative consistency of three-dimensional tree reconstructions from terrestrial lidar. *Remote Sensing of Environment, 113*(5), 1067–1081.

Csillik, O., Cherbini, J., Johnson, R., Lyons, A., & Kelly, M. (2018). Identification of citrus trees from unmanned aerial vehicle imagery using convolutional neural networks. *Drones, 2*(4), 39.

Dai, M., Li, H., & Zhang, X. (2010). Tree modeling through range image segmentation and 3D shape analysis. In *Advances in neural network research and applications* (pp. 413–422). Springer.

Dai, M., Zhang, X., Zhang, Y., & Jaeger, M. (2009). *Segmentation of point cloud scanned from trees. Workshop on community based 3D content and its applications in mobile internet environments, ACCV.*

Dassot, M., Colin, A., Santenoise, P., Fournier, M., & Constant, T. (2012). Terrestrial laser scanning for measuring the solid wood volume, including branches, of adult standing trees in the forest environment. *Computers Electronics in Agriculture, 89*, 86–93.

Dean, T. J., Cao, Q. V., Roberts, S. D., & Evans, D. L. (2009). Measuring heights to crown base and crown median with LiDAR in a mature, even-aged loblolly pine stand. *Forest Ecology Management, 257*(1), 126–133.

Dey, T. K., & Sun, J. (2006). Defining and computing curve-skeletons with medial geodesic function. *Symposium on Geometry Processing, 6*, 143–152.

Dietze, M. C., & Clark, J. S. (2008). Changing the gap dynamics paradigm: Vegetative regeneration control on forest response to disturbance. *Ecological Monographs, 78*(3), 331–347.

Dralle, K., & Rudemo, M. (1996). Stem number estimation by kernel smoothing of aerial photos. *Canadian Journal of Forest Research, 26*(7), 1228–1236.

Duncanson, L., Cook, B., Hurtt, G., & Dubayah, R. (2014). An efficient, multi-layered crown delineation algorithm for mapping individual tree structure across multiple ecosystems. *Remote Sensing of Environment, 154*, 378–386.

Dupuy, J. M., & Chazdon, R. L. (2008). Interacting effects of canopy gap, understory vegetation and leaf litter on tree seedling recruitment and composition in tropical secondary forests. *Forest Ecology Management, 255*(11), 3716–3725.

Ester, M., Kriegel, H.-P., Sander, J., & Xu, X. (1996). *A density-based algorithm for discovering clusters in large spatial databases with noise.* Kdd.

Farrior, C., Bohlman, S., Hubbell, S., & Pacala, S. W. (2016). Dominance of the suppressed: Power-law size structure in tropical forests. *Science, 351*(6269), 155–157.

Ferrara, R., Virdis, S. G., Ventura, A., Ghisu, T., Duce, P., & Pellizzaro, G. (2018). An automated approach for wood-leaf separation from terrestrial LIDAR point clouds using the density based clustering algorithm DBSCAN. *Agricultural and Forest Meteorology, 262*, 434–444.

Ferraz, A., Bretar, F., Jacquemoud, S., Gonçalves, G., Pereira, L., Tomé, M., & Soares, P. (2012). 3-D mapping of a multi-layered Mediterranean forest using ALS data. *Remote Sensing of Environment, 121*, 210−223.

Forzieri, G., Guarnieri, L., Vivoni, E. R., Castelli, F., & Preti, F. (2009). Multiple attribute decision making for individual tree detection using high-resolution laser scanning. *Forest Ecology Management, 258*(11), 2501−2510.

García, M., Riaño, D., Chuvieco, E., & Danson, F. M. (2010). Estimating biomass carbon stocks for a Mediterranean forest in central Spain using LiDAR height and intensity data. *Remote Sensing of Environment, 114*(4), 816−830.

Gaulton, R., & Malthus, T. J. (2010). LiDAR mapping of canopy gaps in continuous cover forests: A comparison of canopy height model and point cloud based techniques. *International Journal of Remote Sensing, 31*(5), 1193−1211.

Gorte, B., & Pfeifer, N. (2004). Structuring laser-scanned trees using 3D mathematical morphology. *International Archives of Photogrammetry Remote Sensing, 35*(B5), 929−933.

Goutte, C., & Gaussier, E. (2005). A probabilistic interpretation of precision, recall and F-score, with implication for evaluation. In *European conference on information retrieval*.

Graves, S. J., Caughlin, T. T., Asner, G. P., & Bohlman, S. A. (2018). A tree-based approach to biomass estimation from remote sensing data in a tropical agricultural landscape. *Remote Sensing of Environment, 218*, 32−43.

Guan, H., Su, Y., Hu, T., Wang, R., Ma, Q., Yang, Q., Sun, X., Li, Y., Jin, S., Zhang, J., Ma, Q., Liu, M., Wu, F., & Guo, Q. (2020). A novel framework to automatically fuse multiplatform LiDAR data in forest environments based on tree locations. *IEEE Transactions on Geoscience and Remote Sensing, 58*(3), 2165−2177.

Guan, H., Su, Y., Sun, X., Xu, G., Li, W., Ma, Q., Wu, X., Wu, J., Liu, L., & Guo, Q. (2020). A marker-free method for registering multi-scan terrestrial laser scanning data in forest environments. *Isprs Journal of Photogrammetry and Remote Sensing, 166*, 82−94.

Guan, H., Yu, Y., Ji, Z., Li, J., & Zhang, Q. (2015). Deep learning-based tree classification using mobile LiDAR data. *Remote Sensing Letters, 6*(11), 864−873.

Guo, Q., Jin, S., Li, M., Yang, Q., Xu, K., Ju, Y., Zhang, J., Xuan, J., Liu, J., Su, Y., Xu, Q., & Liu, Y. (2020). Application of deep learning in ecological resource research: Theories, methods, and challenges. *Science China Earth Sciences, 63*(10), 1457−1474.

Guo, Q., Li, W., Yu, H., & Alvarez, O. (2010). Effects of topographic variability and lidar sampling density on several DEM interpolation methods. *Photogrammetric Engineering & Remote Sensing, 76*(6), 701−712.

Hämmerle, M., Lukač, N., Chen, K.-C., Koma, Z., Wang, C.-K., Anders, K., & Höfle, B. (2017). Simulating various terrestrial and UAV lidar scanning configurations for understory forest structure modelling. *ISPRS Annals of Photogrammetry, Remote Sensing Spatial Information Sciences, IV-2/W4*, 59−65.

Hamraz, H., Contreras, M. A., & Zhang, J. (2017). Vertical stratification of forest canopy for segmentation of understory trees within small-footprint airborne LiDAR point clouds. *ISPRS Journal of Photogrammetry Remote Sensing, 130*, 385−392.

Harding, D., Lefsky, M., Parker, G., & Blair, J. (2001). Laser altimeter canopy height profiles: Methods and validation for closed-canopy, broadleaf forests. *Remote Sensing of Environment, 76*(3), 283−297.

Heiskanen, J., Korhonen, L., Hietanen, J., & Pellikka, P. K. (2015). Use of airborne lidar for estimating canopy gap fraction and leaf area index of tropical montane forests. *International Journal of Remote Sensing, 36*(10), 2569−2583.

Henning, J. G., & Radtke, P. J. (2006). Detailed stem measurements of standing trees from ground-based scanning lidar. *Forest Science, 52*(1), 67−80.

Hilker, T., Coops, N. C., Newnham, G. J., van Leeuwen, M., Wulder, M. A., Stewart, J., & Culvenor, D. S. (2012). Comparison of terrestrial and airborne LiDAR in describing stand structure of a thinned lodgepole pine forest. *Journal of Forestry, 110*(2), 97−104.

Hill, A., Breschan, J., & Mandallaz, D. (2014). Accuracy assessment of timber volume maps using forest inventory data and LiDAR canopy height models. *Forests, 5*(9), 2253–2275.

Holmgren, J., & Persson, Å. (2004). Identifying species of individual trees using airborne laser scanner. *Remote Sensing of Environment, 90*(4), 415–423.

Hosoi, F., Matsugami, H., Watanuki, K., Shimizu, Y., & Omasa, K. (2012). Accurate detection of tree apexes in coniferous canopies from airborne scanning light detection and ranging images based on crown-extraction filtering. *Journal of Applied Remote Sensing, 6*(1), 063502.

Hosoi, F., & Omasa, K. (2006). Voxel-based 3-D modeling of individual trees for estimating leaf area density using high-resolution portable scanning lidar. *IEEE Transactions on Geoscience Remote Sensing, 44*(12), 3610–3618.

Houborg, R., Soegaard, H., & Boegh, E. (2007). Combining vegetation index and model inversion methods for the extraction of key vegetation biophysical parameters using Terra and Aqua MODIS reflectance data. *Remote Sensing of Environment, 106*(1), 39–58.

Hyyppa, J., Kelle, O., Lehikoinen, M., & Inkinen, M. (2001). A segmentation-based method to retrieve stem volume estimates from 3-D tree height models produced by laser scanners. *IEEE Transactions on Geoscience Remote Sensing, 39*(5), 969–975.

Jaskierniak, D., Lane, P. N., Robinson, A., & Lucieer, A. (2011). Extracting LiDAR indices to characterise multilayered forest structure using mixture distribution functions. *Remote Sensing of Environment, 115*(2), 573–585.

Jing, L., Hu, B., Li, J., & Noland, T. (2012). Automated delineation of individual tree crowns from LiDAR data by multi-scale analysis and segmentation. *Photogrammetric Engineering Remote Sensing, 78*(12), 1275–1284.

Jin, S., Su, Y., Gao, S., Wu, F., Hu, T., Liu, J., Li, W., Wang, D., Chen, S., Jiang, Y., Pang, S., & Guo, Q. (2018). Deep learning: Individual maize segmentation from terrestrial lidar data using faster R-CNN and regional growth algorithms [original research]. *Frontiers in Plant Science, 9*, 866–875.

Jin, S., Su, Y., Gao, S., Hu, T., Liu, J., & Guo, Q. (2018). The transferability of random forest in canopy height estimation from multi-source remote sensing data. *Remote Sensing, 10*(8), 1183–1203.

Jin, S., Su, Y., Gao, S., Wu, F., Ma, Q., Xu, K., Ma, Q., Hu, T., Liu, J., Pang, S., Guan, H., Zhang, J., & Guo, Q. (2019). Separating the structural components of maize for field phenotyping using terrestrial lidar data and deep convolutional neural networks. *IEEE Transactions on Geoscience and Remote Sensing, 58*(4), 2644–2658.

Kaartinen, H., Hyyppä, J., Yu, X., Vastaranta, M., Hyyppä, H., Kukko, A., Holopainen, M., Heipke, C., Hirschmugl, M., & Morsdorf, F. (2012). An international comparison of individual tree detection and extraction using airborne laser scanning. *Remote Sensing, 4*(4), 950–974.

Kaasalainen, S., Lindroos, T., & Hyyppa, J. (2007). Toward hyperspectral lidar: Measurement of spectral backscatter intensity with a supercontinuum laser source. *IEEE Geoscience Remote Sensing Letters, 4*(2), 211–215.

Kato, A., Moskal, L. M., Schiess, P., Swanson, M. E., Calhoun, D., & Stuetzle, W. (2009). Capturing tree crown formation through implicit surface reconstruction using airborne lidar data. *Remote Sensing of Environment, 113*(6), 1148–1162.

Kelbe, D., Romanczyk, P., van Aardt, J., & Cawse-Nicholson, K. (2013). Reconstruction of 3D tree stem models from low-cost terrestrial laser scanner data. *Laser Radar Technology and Applications XVIII, 8731*, 44–55.

Kellner, J. R., & Asner, G. P. (2009). Convergent structural responses of tropical forests to diverse disturbance regimes. *Ecology Letters, 12*(9), 887–897.

Khosravipour, A., Skidmore, A. K., Isenburg, M., Wang, T., & Hussin, Y. A. (2014). Generating pit-free canopy height models from airborne lidar. *Photogrammetric Engineering & Remote Sensing, 80*(9), 863–872.

Khosravipour, A., Skidmore, A. K., Wang, T., Isenburg, M., & Khoshelham, K. (2015). Effect of slope on treetop detection using a LiDAR canopy height model. *ISPRS Journal of Photogrammetry Remote Sensing, 104*, 44−52.

Kim, S., Hinckley, T., & Briggs, D. (2011). Classifying individual tree genera using stepwise cluster analysis based on height and intensity metrics derived from airborne laser scanner data. *Remote Sensing of Environment, 115*(12), 3329−3342.

Koch, B., Heyder, U., & Weinacker, H. (2006). Detection of individual tree crowns in airborne lidar data. *Photogrammetric Engineering Remote Sensing, 72*(4), 357−363.

Koukoulas, S., & Blackburn, G. A. (2004). Quantifying the spatial properties of forest canopy gaps using LiDAR imagery and GIS. *International Journal of Remote Sensing, 25*(15), 3049−3072.

Kramer, H. A., Collins, B. M., Kelly, M., & Stephens, S. L. (2014). Quantifying ladder fuels: A new approach using LiDAR. *Forests, 5*(6), 1432−1453.

Lalonde, J. F., Vandapel, N., Huber, D. F., & Hebert, M. (2006). Natural terrain classification using three-dimensional ladar data for ground robot mobility. *Journal of Field Robotics, 23*(10), 839−861.

Landrieu, L., & Simonovsky, M. (2017). *Large-scale point cloud semantic segmentation with superpoint graphs. arXiv preprint arXiv:1711.09869.*

Lazarus, F., & Verroust, A. (1999). Level set diagrams of polyhedral objects. In *Proceedings of the Fifth ACM Symposium on Solid Modeling and Applications.*

Leckie, D., Gougeon, F., Hill, D., Quinn, R., Armstrong, L., & Shreenan, R. (2003). Combined high-density lidar and multispectral imagery for individual tree crown analysis. *Canadian Journal of Remote Sensing, 29*(5), 633−649.

LeCun, Y., Bengio, Y., & Hinton, G. (2015). Deep learning. *Nature, 521*(7553), 436−444.

Lefsky, M. A., Cohen, W. B., Parker, G. G., & Harding, D. J. J. B. (2002). Lidar remote sensing for ecosystem studies: Lidar, an emerging remote sensing technology that directly measures the three-dimensional distribution of plant canopies, can accurately estimate vegetation structural attributes and should be of particular interest to forest, landscape, and global ecologists. *BioScience, 52*(1), 19−30.

Lefsky, M. A., Harding, D., Cohen, W., Parker, G., & Shugart, H. (1999). Surface lidar remote sensing of basal area and biomass in deciduous forests of eastern Maryland, USA. *Remote Sensing of Environment, 67*(1), 83−98.

Lefsky, M. A., & McHale, M. R. (2008). Volume estimates of trees with complex architecture from terrestrial laser scanning. *Journal of Applied Remote Sensing, 2*(1), 023521.

Li, D., Pang, Y., Yue, C. R., Zhao, D., & Xu, G. C. J. J.o. B. F. U. (2012). *Extraction of individual tree DBH and height based on terrestrial laser scanner data.*

Li, W., Guo, Q., Jakubowski, M. K., & Kelly, M. (2012). A new method for segmenting individual trees from the lidar point cloud. *Photogrammetric Engineering & Remote Sensing, 78*(1), 75−84.

Liang, X., Hyyppä, J., Kukko, A., Kaartinen, H., Jaakkola, A., & Yu, X. (2014). The use of a mobile laser scanning system for mapping large forest plots. *IEEE Geoscience Remote Sensing Letters, 11*(9), 1504−1508.

Liang, X., Kankare, V., Hyyppä, J., Wang, Y., Kukko, A., Haggrén, H., Yu, X., Kaartinen, H., Jaakkola, A., & Guan, F. (2016). Terrestrial laser scanning in forest inventories. *ISPRS Journal of Photogrammetry Remote Sensing, 115*, 63−77.

Li, Z., Douglas, E., Strahler, A., Schaaf, C., Yang, X., Wang, Z., Yao, T., Zhao, F., Saenz, E. J., & Paynter, I. (2013). Separating leaves from trunks and branches with dual-wavelength terrestrial LiDAR scanning. In *2013 IEEE International Geoscience and Remote Sensing Symposium-IGARSS.*

Li, D., Pang, Y., Yue, C., Zhao, D., & Xu, G. (2012). Extraction of individual tree DBH and height based on terrestrial laser scanner data. *Journal of Beijing Forestry University, 34*, 79−86.

Li, Y., Su, Y., Hu, T., Xu, G., & Guo, Q. (2018). Retrieving 2-D leaf angle distributions for deciduous trees from terrestrial laser scanner data. *IEEE Transactions on Geoscience and Remote Sensing, 56*(8), 4945–4955.

Li, Y., Su, Y., Zhao, X., Yang, M., Hu, T., Zhang, J., Liu, J., Liu, M., & Guo, Q. (2020). Retrieval of tree branch architecture attributes from terrestrial laser scan data using a Laplacian algorithm. *Agricultural and Forest Meteorology, 284*, 107874.

Livny, Y., Yan, F., Olson, M., Chen, B., Zhang, H., & El-Sana, J. (2010). Automatic reconstruction of tree skeletal structures from point clouds. In *ACM SIGGRAPH Asia 2010 papers* (pp. 1–8).

Lu, X., Guo, Q., Li, W., & Flanagan, J. (2014). A bottom-up approach to segment individual deciduous trees using leaf-off lidar point cloud data. *ISPRS Journal of Photogrammetry Remote Sensing, 94*, 1–12.

Luo, L., Zhai, Q., Su, Y., Ma, Q., Kelly, M., & Guo, Q. (2018). Simple method for direct crown base height estimation of individual conifer trees using airborne LiDAR data. *Optics Express, 26*(10), A562–A578.

MacArthur, R. H., & Horn, H. S. (1969). Foliage profile by vertical measurements. *Ecology Letters, 50*(5), 802–804.

Maltamo, M., Eerikainen, K., Pitkanen, J., Hyyppa, J., & Vehmas, M. (2004). Estimation of timber volume and stem density based on scanning laser altimetry and expected tree size distribution functions. *Remote Sensing of Environment, 90*(3), 319–330.

Ma, L., Zheng, G., Eitel, J. U., Moskal, L. M., He, W., & Huang, H. (2015). Improved salient feature-based approach for automatically separating photosynthetic and non-photosynthetic components within terrestrial lidar point cloud data of forest canopies. *IEEE Transactions on Geoscience Remote Sensing, 54*(2), 679–696.

Meyer, F., & Beucher, S. (1990). Morphological segmentation. *Journal of Visual Communication Image Representation, 1*(1), 21–46.

Moorthy, S. M. K., Calders, K., Vicari, M. B., & Verbeeck, H. (2019). Improved supervised learning-based approach for leaf and wood classification from LiDAR point clouds of forests. *IEEE Transactions on Geoscience and Remote Sensing, 58*(5), 3057–3070.

Morsdorf, F., Meier, E., Kötz, B., Itten, K. I., Dobbertin, M., & Allgöwer, B. (2004). LIDAR-based geometric reconstruction of boreal type forest stands at single tree level for forest and wildland fire management. *Remote Sensing of Environment, 92*(3), 353–362.

Næsset, E., & Økland, T. (2002). Estimating tree height and tree crown properties using airborne scanning laser in a boreal nature reserve. *Remote Sensing of Environment, 79*(1), 105–115.

Paris, C., Valduga, D., & Bruzzone, L. (2016). A hierarchical approach to three-dimensional segmentation of LiDAR data at single-tree level in a multilayered forest. *IEEE Transactions on Geoscience Remote Sensing, 54*(7), 4190–4203.

Peuhkurinen, J., Mehtätalo, L., & Maltamo, M. (2011). Comparing individual tree detection and the area-based statistical approach for the retrieval of forest stand characteristics using airborne laser scanning in Scots pine stands. *Canadian Journal of Forest Research, 41*(3), 583–598.

Popescu, S. C., & Wynne, R. H. (2004). Seeing the trees in the forest. *Photogrammetric Engineering Remote Sensing, 70*(5), 589–604.

Popescu, S. C., Wynne, R. H., & Nelson, R. F. (2002). Estimating plot-level tree heights with lidar: Local filtering with a canopy-height based variable window size. *Computers Electronics in Agriculture, 37*(1–3), 71–95.

Popescu, S. C., Wynne, R. H., & Nelson, R. F. (2003). Measuring individual tree crown diameter with lidar and assessing its influence on estimating forest volume and biomass. *Canadian Journal of Remote Sensing, 29*(5), 564–577.

Popescu, S. C., Wynne, R. H., & Scrivani, J. A. (2004). Fusion of small-footprint lidar and multispectral data to estimate plot-level volume and biomass in deciduous and pine forests in Virginia, USA. *Forest Science, 50*(4), 551−565.

Popescu, S. C., & Zhao, K. (2008). A voxel-based lidar method for estimating crown base height for deciduous and pine trees. *Remote Sensing of Environment, 112*(3), 767−781.

Pouliot, D., King, D., & Pitt, D. (2005). Development and evaluation of an automated tree detection delineation algorithm for monitoring regenerating coniferous forests. *Canadian Journal of Forest Research, 35*(10), 2332−2345.

Qi, C. R., Su, H., Mo, K., & Guibas, L. J. (2016). Pointnet: Deep learning on point sets for 3d classification and segmentation. In *Proceedings of the IEEE conference on computer vision and pattern recognition*.

Qi, C. R., Yi, L., Su, H., & Guibas, L. J. (2017). PointNet++: Deep hierarchical feature learning on point sets in a metric space. *Advances in Neural Information Processing Systems, 30*, 5105−5114.

Rafiee, A., Dias, E., & Koomen, E. (2013). Between green and grey: Towards a new green volume indicator for cities. In *Proceedings of the 13th International Conference on Computers in Urban Planning and Urban Management (CUPUM)*.

Reitberger, J., Schnörr, C., Krzystek, P., & Stilla, U. (2009). 3D segmentation of single trees exploiting full waveform LIDAR data. *ISPRS Journal of Photogrammetry Remote Sensing, 64*(6), 561−574.

Ross, J. (2012). *The radiation regime and architecture of plant stands* (Vol 3). Springer Science & Business Media.

Rutzinger, M., Pratihast, A. K., Oude Elberink, S., & Vosselman, G. (2010). Detection and modelling of 3D trees from mobile laser scanning data. *International Archives of Photogrammetry Remote Sensing and Spatial Information Sciences, 38*, 520−525.

Shamsoddini, A., Turner, R., & Trinder, J. (2013). Improving lidar-based forest structure mapping with crown-level pit removal. *Journal of Spatial Science, 58*(1), 29−51.

Silva, C. A., Valbuena, R., Pinagé, E. R., Mohan, M., de Almeida, D. R., North Broadbent, E., Jaafar, W. S. W. M., de Almeida Papa, D., Cardil, A., & Klauberg, C. (2019). ForestGapR: An r Package for forest gap analysis from canopy height models. *Methods in Ecology and Evolution, 10*(8), 1347−1356.

Simonse, M., Aschoff, T., Spiecker, H., & Thies, M. (2003). Automatic determination of forest inventory parameters using terrestrial laser scanning. In *Proceedings of the Scandlaser Scientific Workshop on Airborne Laser Scanning of Forests*.

Smits, I., Prieditis, G., Dagis, S., & Dubrovskis, D. (2012). Individual tree identification using different LIDAR and optical imagery data processing methods. *Biosystems Inform Tech, 1*, 19−24.

Sokolova, M., Japkowicz, N., & Szpakowicz, S. (2006). Beyond accuracy, F-score and ROC: A family of discriminant measures for performance evaluation. In *Australasian joint conference on artificial intelligence*. Springer.

Solberg, S., Naesset, E., & Bollandsas, O. M. (2006). Single tree segmentation using airborne laser scanner data in a structurally heterogeneous spruce forest. *Photogrammetric Engineering Remote Sensing, 72*(12), 1369−1378.

Steinmann, K., Mandallaz, D., Ginzler, C., & Lanz, A. (2013). Small area estimations of proportion of forest and timber volume combining Lidar data and stereo aerial images with terrestrial data. *Scandinavian Journal of Forest Research, 28*(4), 373−385.

Stephens, S. L. (1998). Evaluation of the effects of silvicultural and fuels treatments on potential fire behaviour in Sierra Nevada mixed-conifer forests. *Forest Ecology Management, 105*(1−3), 21−35.

Strîmbu, V. F., & Strîmbu, B. M. (2015). A graph-based segmentation algorithm for tree crown extraction using airborne LiDAR data. *ISPRS Journal of Photogrammetry Remote Sensing, 104*, 30−43.

Su, Y., Guan, H., Hu, T., & Guo, Q. (2018). The integration of uavand backpack LiDAR systems for forest inventory. In *IGARSS 2018-2018 IEEE International Geoscience and Remote Sensing Symposium.*

Su, Y., Hu, T., Wang, Y., Li, Y., Dai, J., Liu, H., Jin, S., Ma, Q., Wu, J., Liu, L., Fang, J., & Guo, Q. (2020). Large-scale geographical variations and climatic controls on crown architecture traits. *Journal of Geophysical Research: Biogeosciences, 125*(2). e2019JG005306.

Tao, S., Guo, Q., Li, L., Xue, B., Kelly, M., Li, W., Xu, G., & Su, Y. (2014). Airborne lidar-derived volume metrics for aboveground biomass estimation: A comparative assessment for conifer stands. *Agricultural Forest Meteorology, 198*, 24—32.

Tao, S., Guo, Q., Xu, S., Su, Y., Li, Y., & Wu, F. (2015). A geometric method for wood-leaf separation using terrestrial and simulated lidar data. *Photogrammetric Engineering & Remote Sensing, 81*(10), 767—776.

Tao, S., Wu, F., Guo, Q., Wang, Y., Li, W., Xue, B., Hu, X., Li, P., Tian, D., & Li, C. (2015). Segmenting tree crowns from terrestrial and mobile LiDAR data by exploring ecological theories. *ISPRS Journal of Photogrammetry Remote Sensing, 110*, 66—76.

Theiler, P. W., Wegner, J. D., & Schindler, K. (2015). Globally consistent registration of terrestrial laser scans via graph optimization. *ISPRS Journal of Photogrammetry Remote Sensing, 109*, 126—138.

Tochon, G., Feret, J.-B., Valero, S., Martin, R. E., Knapp, D. E., Salembier, P., Chanussot, J., & Asner, G. P. (2015). On the use of binary partition trees for the tree crown segmentation of tropical rainforest hyperspectral images. *Remote Sensing of Environment, 159*, 318—331.

Tonolli, S., Dalponte, M., Neteler, M., Rodeghiero, M., Vescovo, L., & Gianelle, D. (2011). Fusion of airborne LiDAR and satellite multispectral data for the estimation of timber volume in the Southern Alps. *Remote Sensing of Environment, 115*(10), 2486—2498.

Van Leeuwen, M., & Nieuwenhuis, M. (2010). Retrieval of forest structural parameters using LiDAR remote sensing. *European Journal of Forest Research, 129*(4), 749—770.

Van der Zande, D., Hoet, W., Jonckheere, I., van Aardt, J., & Coppin, P. (2006). Influence of measurement set-up of ground-based LiDAR for derivation of tree structure. *Agricultural Forest Meteorology, 141*(2—4), 147—160.

Vepakomma, U., & Cormier, D. J. I. A. P. R. S. S. I. S. (2019). Valuing forest stand at a glance with UAV based Lidar. *The International Archives of Photogrammetry, Remote Sensing and Spatial Information Sciences, 42*, 643—647.

Vepakomma, U., St-Onge, B., & Kneeshaw, D. (2008). Spatially explicit characterization of boreal forest gap dynamics using multi-temporal lidar data. *Remote Sensing of Environment, 112*(5), 2326—2340.

Vepakomma, U., St-Onge, B., & Kneeshaw, D. (2011). Response of a boreal forest to canopy opening: Assessing vertical and lateral tree growth with multi-temporal lidar data. *Ecological Applications, 21*(1), 99—121.

Verroust, A., & Lazarus, F. (1999). Extracting skeletal curves from 3D scattered data. In *Proceedings Shape Modeling International' Vol 99. International conference on shape modeling and applications.*

Wallace, L., Lucieer, A., Malenovský, Z., Turner, D., & Vopěnka, P. (2016). Assessment of forest structure using two UAV techniques: A comparison of airborne laser scanning and structure from motion (SfM) point clouds. *Forests, 7*(3), 62.

Wang, D. (2020). Unsupervised semantic and instance segmentation of forest point clouds. *ISPRS Journal of Photogrammetry and Remote Sensing, 165*, 86—97.

Wang, J., Chen, X., Cao, L., An, F., Chen, B., Xue, L., & Yun, T. (2019). Individual rubber tree segmentation based on ground-based LiDAR data and faster R-CNN of deep learning. *Forests, 10*(9), 793—812. Article 793.

Wang, D., Hollaus, M., Pfeifer, N. J. I. A.o. P., & Remote Sensing. (2017). Feasibility of machine learning methods for separating wood and leaf points from terrestrial laser scanning data. *ISPRS Annals of Photogrammetry, Remote Sensing Spatial Information Sciences, 4.*

Wang, D., Momo Takoudjou, S., & Casella, E. (2020). LeWoS: A universal leaf-wood classification method to facilitate the 3D modelling of large tropical trees using terrestrial LiDAR. *Methods in Ecology Evolution, 11*(3), 376−389.

Widlowski, J.-L., Côté, J.-F., & Béland, M. (2014). Abstract tree crowns in 3D radiative transfer models: Impact on simulated open-canopy reflectances. *Remote Sensing of Environment, 142*, 155−175.

Windrim, L., & Bryson, M. (2020). Detection, segmentation, and model fitting of individual tree stems from airborne laser scanning of forests using deep learning. *Remote Sensing, 12*(9), 1469.

Wing, B. M., Ritchie, M. W., Boston, K., Cohen, W. B., Gitelman, A., & Olsen, M. J. (2012). Prediction of understory vegetation cover with airborne lidar in an interior ponderosa pine forest. *Remote Sensing of Environment, 124*, 730−741.

Wu, J., Cawse-Nicholson, K., & van Aardt, J. (2013). 3D Tree reconstruction from simulated small footprint waveform lidar. *Photogrammetric Engineering Remote Sensing, 79*(12), 1147−1157.

Wu, B., Yu, B., Yue, W., Shu, S., Tan, W., Hu, C., Huang, Y., Wu, J., & Liu, H. (2013). A voxel-based method for automated identification and morphological parameters estimation of individual street trees from mobile laser scanning data. *Remote Sensing, 5*(2), 584−611.

Wu, B., Zheng, G., & Chen, Y. (2020). An improved convolution neural network-based model for classifying foliage and woody components from terrestrial laser scanning data. *Remote Sensing, 12*(6), 1010−1021.

Xi, Z., Hopkinson, C., & Chasmer, L. (2018). Filtering stems and branches from terrestrial laser scanning point clouds using deep 3-D fully convolutional networks. *Remote Sensing, 10*(8), 1215.

Xu, H., Gossett, N., & Chen, B. J. A. T.o. G. (2007). Knowledge and heuristic-based modeling of laser-scanned trees. *ACM Transactions on Graphics, 26*(4), 19-es.

Yang, X., Strahler, A. H., Schaaf, C. B., Jupp, D. L., Yao, T., Zhao, F., Wang, Z., Culvenor, D. S., Newnham, G. J., & Lovell, J. L. (2013). Three-dimensional forest reconstruction and structural parameter retrievals using a terrestrial full-waveform lidar instrument (Echidna®). *Remote Sensing of Environment, 135*, 36−51.

Yang, Q., Su, Y., Jin, S., Kelly, M., Hu, T., Ma, Q., Li, Y., Song, S., Zhang, J., Xu, G., Wei, J., & Guo, Q. (2019). The influence of vegetation characteristics on individual tree segmentation methods with airborne LiDAR data. *Remote Sensing, 11*(23), 2880−2897.

Yan, L., Tan, J., Liu, H., Xie, H., & Chen, C. (2017). Automatic registration of TLS-TLS and TLS-MLS point clouds using a genetic algorithm. *Sensors, 17*(9), 1979.

Yao, T., Yang, X., Zhao, F., Wang, Z., Zhang, Q., Jupp, D., Lovell, J., Culvenor, D., Newnham, G., & Ni-Meister, W. (2011). Measuring forest structure and biomass in New England forest stands using Echidna ground-based lidar. *Remote Sensing of Environment, 115*(11), 2965−2974.

Yun, T., An, F., Li, W., Sun, Y., Cao, L., & Xue, L. J. R. S. (2016). A novel approach for retrieving tree leaf area from ground-based LiDAR. *Remote Sensing, 8*(11), 942.

Zhao, D., Pang, Y., Li, Z., & Sun, G. (2013). Filling invalid values in a lidar-derived canopy height model with morphological crown control. *International Journal of Remote Sensing, 34*(13), 4636−4654.

Zheng, G., & Moskal, L. M. (2012). Computational-geometry-based retrieval of effective leaf area index using terrestrial laser scanning. *IEEE Transactions on Geoscience Remote Sensing, 50*(10), 3958−3969.

Zhu, X., Skidmore, A. K., Darvishzadeh, R., Niemann, K. O., Liu, J., Shi, Y., & Wang, T. (2018). Foliar and woody materials discriminated using terrestrial LiDAR in a mixed natural forest. *International Journal of Applied Earth Observation Geoinformation, 64*, 43–50.

Zhu, X., Wang, T., Darvishzadeh, R., Skidmore, A. K., & Niemann, K. O. (2015). 3D leaf water content mapping using terrestrial laser scanner backscatter intensity with radiometric correction. *Isprs Journal of Photogrammetry and Remote Sensing, 110*, 14–23.

CHAPTER 8

Estimation of Forest Functional Attributes

Contents

Forest ecosystems, one of the most important components of the terrestrial biosphere, account for approximately 70% of terrestrial ecosystem net productivity. Their role in regulating the carbon cycle and mitigating global climate change is irreplaceable. Light detection and ranging (LiDAR) has tremendous advantages over traditional forest surveys in extracting three-dimensional forest structural attributes such as vertical foliage profile, tree height, and crown size and forest functional attributes such as canopy cover, leaf area index, and biomass. This chapter introduces the principles, methods, and applications of LiDAR for extracting those functional attributes of forest ecosystems, as well as tree species classification using LiDAR data. These functional attributes are critical for forest ecosystem studies and various applications.

8.1 Canopy Cover and Closure

Canopy cover and closure are important attributes in forest surveys. They reflect the forest structure and environment, but the two concepts can be easily confused in field surveys. Canopy cover is the ratio of the vertical projection of

LiDAR Principles, Processing and Applications in Forest Ecology
ISBN 978-0-12-823894-3
https://doi.org/10.1016/B978-0-12-823894-3.00008-6

tree crowns to the forest area (Jennings et al., 1999). Meanwhile, canopy closure is the ratio of the sky hemisphere blocked by vegetation when viewed from a single point, usually on the ground under the canopy (Jennings et al., 1999) (Fig. 8.1). Canopy cover is obtained by vertical projection, with the values closely related to the projected area corresponding to tree crowns. In contrast, canopy closure is obtained by central projection, with the values closely related to the observation location and angles during measurement. Canopy cover and closure are highly correlated, as well as the gap fraction and leaf area index (LAI) (Zheng et al., 2015). Canopy cover is adopted more often than canopy closure in forest management practices because it is easier to measure. Canopy cover and closure are important forest management attributes for planning forest thinning and estimating forest volume. They are also key indicators for classifying land cover as a forest. For example, if the canopy closure of an area is greater than 20%, and its area is larger than 0.67 ha, it is defined as forest land in China according to forest inventory guidelines. Forest land with a canopy closure greater than 10% is defined as a forest. The traditional canopy cover and closure measurements use visual estimation, remote sensing image interpretation, canopy projection, line transect, point count, observation tube, and photo methods (Zou & Zhuge, 2011).

Light detection and ranging (LiDAR) provides a new way to accurately measure forest canopy cover and closure. Compared with traditional measurement methods, LiDAR can penetrate forest canopies and measure vertical forest structures for various forest types. Studies have found that laser returns from forest canopies are closely related to tree height and canopy closure (Kato et al., 2009; Luo, 2012, pp. 16–20; Tang, 2013, pp. 4–25). Small- and large-footprint LiDAR has been successfully applied to estimate forest canopy cover and closure. With small-footprint LiDAR, canopy cover and closure can be

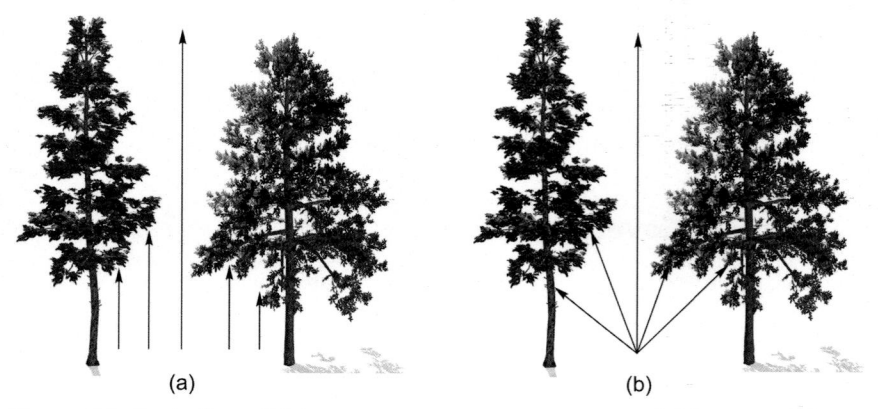

(a)   (b)

Figure 8.1 Illustration of (a) canopy cover measurement and (b) canopy closure measurement.

estimated from the ratio of vegetation returns to ground returns (Korhonen et al., 2011; Solberg et al., 2009). With large-footprint LiDAR, these two attributes can be calculated from the ratio of vegetation returns energy to total returns energy through waveform analysis (Farid et al., 2008; Solberg et al., 2007).

Airborne laser scanning (ALS) can measure both large-scale and fine-resolution canopy cover. ALS emits laser pulses from the aircraft above vegetation canopies to the ground at a narrow angle and records the height of the land surface, which can be used to classify the vegetation on the ground (Coops et al., 2007). The two main methods currently used to estimate canopy cover based on airborne LiDAR data are the three-dimensional (3-D) point cloud-based method and the two-dimensional (2-D) canopy height model (CHM)-based method. In point cloud-based methods, canopy cover is obtained from the proportion of canopy echoes, such as the first echo cover index (FCI) and all echo cover index (ACI) (Hopkinson & Chasmer, 2009; Korhonen et al., 2011). The equation for FCI is as follows:

$$FCI = \frac{\sum Single_{Canopy} + \sum First_{Canopy}}{\sum Single_{All} + \sum First_{All}} \tag{8.1}$$

where FCI approximates vertical canopy cover (VCC); $Single_{Canopy}$ is the single canopy echoes; $First_{Canopy}$ is the first canopy echoes; $Single_{All}$ is all single echoes; and $First_{All}$ is all first echoes. For consistency with traditional observations, a height threshold is always required (usually set to $1-2$ m) to remove the vegetation points below the height threshold. The ACI is expressed as

$$ACI = \frac{\sum ALL_{Canopy}}{\sum ALL}$$
$$= \frac{\sum \left(Single_{Canopy} + First_{Canopy} + Intermediate_{Canopy} + Last_{Canopy} \right)}{\sum ALL} \tag{8.2}$$

Like FCI, ACI is commonly used to calculate canopy cover. In the equation above, $Last_{Canopy}$ is the last canopy echoes; $Intermediate_{Canopy}$ is intermediate canopy echoes that are neither the first nor the last; ALL_{Canopy} is all canopy echoes; and ALL is all echoes. Estimations of canopy cover based on the ACI are usually lower than those based on the FCI because the first echo is generally higher than subsequent echoes. However, the difference between the two estimates is small, usually within 3% (Ma et al., 2017).

The principle of the CHM-based method is similar to that of the point cloud-based method. The key is to count the number of CHM pixels with a tree height value greater than the threshold (usually 2 m) within the canopy cover estimation resolution, which should be coarser than the CHM resolution. The finer the CHM resolution is, the more accurate the canopy cover estimation will be. The CHM resolution must generally be finer than 1 m to obtain an accurate canopy cover map. Therefore, canopy cover estimated by the CHM-based method tends to be higher than that of the point cloud-based method because some ground returns under the canopy within the same CHM pixel can be classified into canopy pixels after point cloud rasterization. Studies have found that canopy cover estimated from the CHM-based method was approximately 5% higher than that from the ACI-based method (Ma et al., 2017).

Although ALS emits a laser beam that is close to vertical, there is still a certain angle that may bias canopy cover estimation. Ma et al. (2017) showed that bias increases with ALS scan angle. When the scan angle is greater than 12°, canopy cover estimation error can be as high as 20% in mountainous forests. To eliminate errors caused by scan angle, Korhonen et al. (2011) proposed a corrected method employing the following equations:

$$\begin{aligned} VCC &= FCI - 0.623 \times \theta_{scan} \\ VCC &= FCI - 0.0253 \times \theta_{scan} \times F_{max} \end{aligned} \tag{8.3}$$

where VCC is vertical canopy cover; θ_{scan} is the average scan angle recorded by an ALS; and F_{max} is the distance from the highest return to the ground.

Point density also affects canopy cover estimation accuracy. High point density data can provide more detailed vegetation structure information, which improves the accuracy of canopy cover estimates, although there are additional data acquisition and processing costs. An average density of 1 pt/m^2 can achieve an accuracy (more than 90%) that is satisfactory for estimating forest canopy cover. After the point density reaches 10 pt/m^2, the accuracy of canopy cover estimation no longer increases with the point density (Jakubowski et al., 2013; Ma et al., 2017).

Compared with canopy cover, it is more difficult to estimate canopy closure from ALS data; however, it can be approximated from several indices. For example, Solberg et al. (2009) proposed Solberg's cover index (SCI):

$$SCI = \frac{\sum Single_{Canopy} + a\left(\sum First_{Canopy} + \sum Last_{Canopy}\right)}{\sum Single_{All} + a\left(\sum First_{All} + \sum Last_{All}\right)} \tag{8.4}$$

In this equation, the value of a is related to the properties of the laser sensor and often set at 0.5.

Hopkinson and Chasmer (2009) proposed another indicator, the intensity cover index (ICI), which considers the energy loss of a laser beam during its travel through the canopy:

$$ICI = 1 - \frac{\left(\dfrac{\sum I_{Ground\ single}}{\sum I_{Total}}\right) + \sqrt{\dfrac{\sum I_{Ground\ last}}{\sum I_{Total}}}}{\left(\dfrac{\sum I_{First} + \sum I_{Single}}{\sum I_{Total}}\right) + \sqrt{\dfrac{\sum I_{Intermediate} + \sum I_{Last}}{\sum I_{Total}}}} \tag{8.5}$$

where I is the intensity of returns; $I_{\text{Ground single}}$ is the intensity of single ground returns; $I_{\text{Ground last}}$ is the intensity of the last ground returns; I_{First} is the intensity of all first returns; I_{Single} is the intensity of all single returns; $I_{\text{Intermediate}}$ is the intensity of intermediate returns; I_{Last} is the intensity of all last returns; and I_{Total} is the intensity of all returns. Intensity is highly correlated with sensor height and must be corrected before the calculation. The following code is for a point cloud-based method to calculate canopy cover implemented in Python.

```python
# The following code is a point cloud-based method to calculate canopy cover
# -*- coding: UTF-8 -*-
import laspy
from osgeo import ogr
from osgeo import osr
from osgeo import gdal
import numpy as np
# Read las file
lasfile = "./canopycover.las"
inFile = laspy.file.File(lasfile, mode = "r")
# Get point cloud coordinates and attribute information
x,y,z = inFile.x,inFile.y,inFile.z
classfication = inFile.raw_classification
return_number = inFile.return_num
# Read DEM corresponding to the point cloud
ds_dem = gdal.Open( "sample_dem.tif" )
rows = ds_dem.RasterYSize
cols = ds_dem.RasterXSize
bands = ds_dem.RasterCount
# Get the origin and resolution information of the DEM raster data
transform = ds_dem.GetGeoTransform()
xOrigin = transform[0]
yOrigin = transform[3]
pixelWidth = transform[1]
pixelHeight = transform[5]
# Calculate the DEM grid position for each point cloud
xOffset = (x - xOrigin) / pixelWidth
xOffset= xOffset.astype(int)
yOffset = (y - yOrigin) / pixelHeight
yOffset= yOffset.astype(int)
# Create an empty DEM array corresponding to the stored point cloud
point_dem = np.empty(len(x))
# Read DEM corresponding to the point cloud
band = ds_dem.GetRasterBand(1)
band1Array = band.ReadAsArray()
for i in range(0,len(x)):
```

```
        point_dem[i] = band.ReadAsArray(xOffset[i], yOffset[i], 1, 1)
# Normalized the Z value of the point cloud
normilze_z = z - point_dem
# Get the range of point cloud
x_max,y_max = inFile.header.max[0:2]
x_min,y_min = inFile.header.min[0:2]
# Set the resolution to generate coverage
pixelWidth , pixelHeight = 10,-10
# Set the origin, row number of the coverage raster
xOrigin = x_min
yOrigin = y_max
cols = int(round((x_max - x_min) / pixelWidth))
rows = int(round((y_max - y_min) / abs(pixelHeight)))
# Calculate the coverage raster position for each point cloud
xOffset = (x - xOrigin) / pixelWidth
xOffset= xOffset.astype(int)
yOffset = (y - yOrigin) / pixelHeight
yOffset= yOffset.astype(int)
# Set the variable that stores the coverage raster
cc = np.zeros((cols, rows))
for j in range(0, rows):
    for i in range(0,cols):
        vp = len( np.where((xOffset==i) & (yOffset==j) & (normilze_z >2) & (return_number ==1))[0])
        ttp = len(np.where((xOffset==i) & (yOffset==j) & (return_number ==1))[0])
        cc[i,j] =float(vp)/float(ttp)
# Output into geoTIFF
driver = gdal.GetDriverByName('GTiff')
outRaster = driver.Create('canopy_cover.tif', cols, rows, 1, gdal.GDT_Float32)
outRaster.SetGeoTransform((xOrigin, pixelWidth, 0, yOrigin, 0, pixelHeight))
outband = outRaster.GetRasterBand(1)
outband.WriteArray(np.transpose(cc))
# Set projection information
outRasterSRS = osr.SpatialReference()
outRasterSRS.ImportFromProj4("+proj=utm +zone=10 +datum=NAD83 +units=m +no_defs")
outRaster.SetProjection(outRasterSRS.ExportToWkt())
outband.FlushCache()
outRaster = None
```

8.2 Leaf Area Index

LAI is one of the most essential functional attributes for characterizing vegetation canopy structure. The index plays a crucial role in understanding the interactions between the land surface and the atmosphere through the biological and physical processes of vegetation, such as photosynthesis, respiration, and transpiration (Boegh et al., 2002; Luo et al., 2013; Parker, 1995). The LAI is defined as the sum of the plant photosynthesis area per unit of ground surface area. It was first proposed by the British agroecologist Watson in the mid-1940s (Wang et al., 2005; Wu et al., 2007). There are various definitions of LAI in different fields for different applications and measurement methods. Chen et al. (2002) defined LAI as one-half of the total green leaf area per unit of ground surface area, which is currently the most widely used and recognized definition (Chen & Black, 1991; Chen et al., 2002; Fassnacht et al., 1994; Lang et al., 1991; Stenberg et al., 1994).

The two primary means of estimating LAI are the direct method and the indirect method. Direct methods include destructive sampling, allometric equations, and litter collection using traps (Ryu et al., 2010). Litter collection is widely used in deciduous forest ecosystems because it is nondestructive. However, all these direct methods are extremely labor-intensive, leading to limited measurements and making them unsuitable for large-scale LAI estimation. Therefore, indirect methods have been developed to overcome these disadvantages by estimating LAI from other forest traits or using optical instruments (such as tracing radiation and architecture of canopies, LAI-2000 Plant Canopy Analyzer and hemispherical photography) (Jonckheere et al., 2004; Pierce & Running, 1988). Hemispherical photography has been widely applied because of its low cost and convenience. The emergence of high-resolution single-lens reflex cameras also advanced its development. However, LAI tends to be underestimated by optical instruments by up to 25%−50%, depending on the type of forest (Breda, 2003). With recent developments in remote sensing technology, studies have focused on estimating LAI quantitatively using optical remote sensing data. Approaches can be divided into two categories: statistical models and optical models (Fang et al., 2003). Statistical models usually rely on the correlation between LAI and band reflectance or various vegetation indices (VIs), such as the normalized difference vegetation index (NDVI) (Rouse et al., 1974). Optical models, usually based on the Beer−Lambert law or Miller's equation, are introduced in the following section. The LAI derived by optical remote sensing data also suffers from the saturation effect of VIs, i.e., their values no longer increase with LAI values, especially when the LAI value is higher than 3 or 4 (Baret & Guyot, 1991; Spanner et al., 1990; Wulder et al., 1998). Currently, LiDAR is widely used to retrieve

LAI and can overcome some limitations of LAI estimations using hemispherical photography and optical remote sensing data.

8.2.1 Theoretical Basis

As shown in Fig. 8.2, the estimation of LAI using traditional optical instruments, remote sensing, and LiDAR technology is often based on three classic models. The principles of these models are shown in the following paragraphs.

(1) Beer—Lambert law

$$L(z) = -\frac{\cos\theta}{G(\theta)\Omega}\log P_{\text{gap}}(\theta, z) \tag{8.6}$$

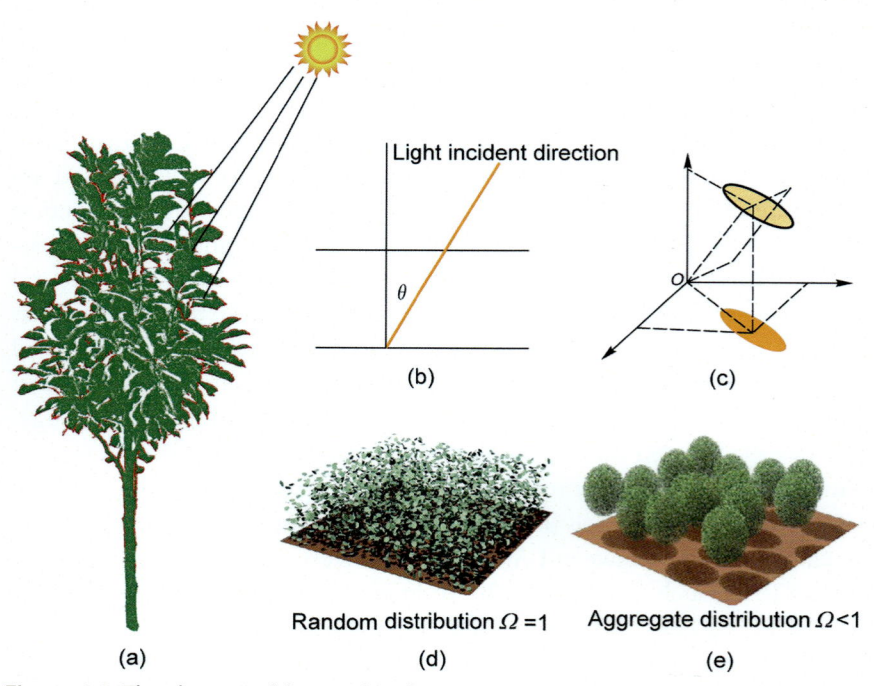

Figure 8.2 The theoretical basis of leaf area index estimation: (a) light incident on vegetation; (b) zenith angle (refers to the angle between the incident ray and the vertical direction); (c) extinction coefficient (affected by the projection of plant leaves in the direction of the incident ray); (d) randomly distributed leaves with a clumping index of one; and (e) aggregated leaves with a clumping index lower than 1 (Revised from Pinty et al. (2001).).

where L is the true LAI; θ is the zenith angle of the incident light; z is the height; P_{gap} is the gap fraction; $G(\theta)$ is the extinction coefficient or the Ross–Nilson G-function (the mean projection of a unit foliage area on the plane perpendicular to the direction of the incident light) (Nilson, 1999); and Ω is the clumping index. In most studies, $G(\theta)$ is set to a constant value of 0.5, which is based on the spherical leaf inclination distribution and $\theta = 57.5°$ as the effect of leaf inclination distribution on the extinction coefficient is the smallest at this zenith angle.

(2) Miller's formula

According to Miller's formula Eqs. (8.7) (Nilson, 1999),

$$\int_0^{\frac{\pi}{2}} G(\theta)\sin\theta d\theta = \frac{1}{2} \tag{8.7}$$

and it can be derived that

$$L(z) = -2\int_0^{\frac{\pi}{2}} \log P_{\text{Gap}}(\theta, z)\cos\theta\sin\theta d\theta \tag{8.8}$$

The equation considers all zenith angles based on the Ross–Nilson G-function (Nilson, 1999; Zhao & Popescu, 2009). In practical applications, zenith angles are often selected at several intervals (e.g., 7°, 23°, 38°, 53° and 68°) to analyze the canopy gap fraction.

(3) Jupp's method

The third algorithm assumes that $G(\theta, \chi)$ is a specific function of θ and χ, and χ is an unknown parameter. Jupp et al. (2005) assumed that

$$G(\theta, \chi) = \chi\cos\theta + (1 - \chi)\frac{2}{\pi}\sin(\theta) \tag{8.9}$$

and the following equations can be derived from Eq. (8.9):

$$-\log P_{\text{gap}}(\theta, z) = A + B\frac{2}{\pi}\tan\theta \tag{8.10}$$

$$A = L(z)\chi$$

$$B = L(z)(1 - \chi)$$

$$L(z) = A + B$$

where parameters A and B can be derived from the regression analysis between $-\log P_{\text{gap}}(\theta, z)$ and $\frac{2}{\pi}\tan\theta$, and LAI can be obtained by adding A and B (Jupp et al., 2005).

The abovementioned theories have formed the basis for methods estimating LAI from LiDAR data acquired by different platforms. The applications of LiDAR technology in LAI extraction are introduced for two LiDAR systems: terrestrial laser scanning (TLS) and ALS.

8.2.2 Leaf Area Index Extraction from Terrestrial LiDAR Data

In recent years, the application of TLS in LAI extraction has developed rapidly because it can provide detailed structural information with millimeter-level positioning precision. TLS can be applied to extract the LAI at different scales, such as the individual tree level, stand level, and ecosystem level. At present, the format of TLS data can be divided into two categories: point cloud data and full waveform data. The most commonly used algorithm for LAI extraction from a TLS point cloud is the voxel-based method (Hosoi & Omasa, 2006; Zheng et al., 2013). First, according to the boundary of the point cloud data, the point cloud is divided into $i*j*s$ grids of size $(\Delta i)*(\Delta j)*(\Delta s)$ using Eq. (8.11) and Fig. 8.3:

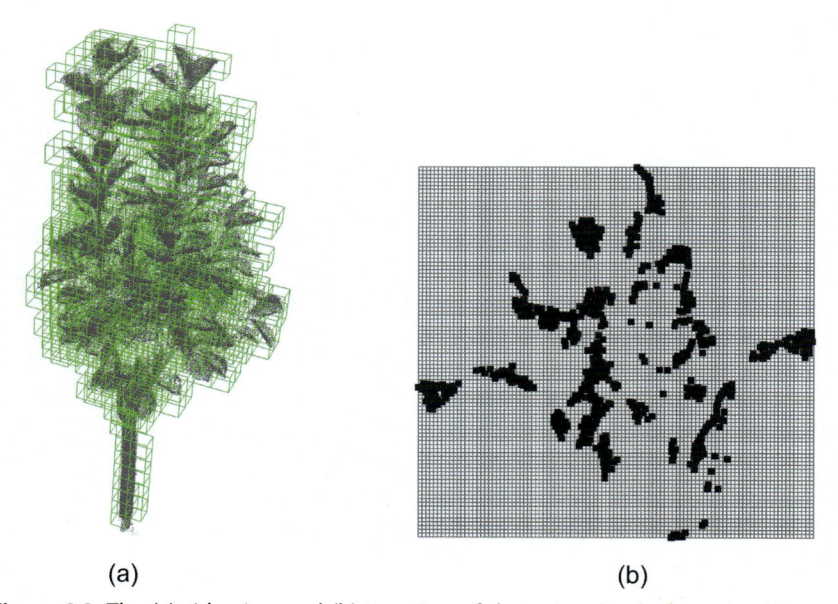

(a) (b)

Figure 8.3 The (a) side view and (b) top view of the point cloud of a tree with voxel grids.

$$i = \text{int}\left(\frac{x_{\max} - x_{\min}}{\Delta i}\right)$$

$$j = \text{int}\left(\frac{y_{\max} - y_{\min}}{\Delta j}\right) \qquad (8.11)$$

$$s = \text{int}\left(\frac{z_{\max} - z_{\min}}{\Delta s}\right)$$

Second, all voxels are checked and a voxel is assigned a value of "1" when that voxel contains one or more laser points; otherwise, the voxel is assigned a value of "0", indicating gaps in the forest canopy. Then, three methods can be applied to estimate LAI. The first one is based on the cumulative leaf area density profiles of trees, using the following equations:

$$\text{LAD}(h, \Delta H) = 1.1 \times \frac{1}{\Delta H} \sum_{k=m_h}^{m_h + \Delta H} \frac{n_I(k)}{n_T(k)} \qquad (8.12)$$

$$\text{LAI} = 1.1 \times \sum_{k} \frac{n_I(k)}{n_T(k)} \qquad (8.13)$$

where $n_I(k)$ is the number of voxel grids with attribute "1" for the kth horizontal layer; $n_T(k)$ is the total number of voxel grids for the kth horizontal layer; and ΔH is the horizontal layer thickness.

The second LAI estimation method calculates the ratio of the number of voxel grids without points to the total number of voxel grids in one layer, i.e., the gap fraction $P(\theta)$ (Eq. 8.14). The LAI of a layer can be calculated using Eqs. (8.6)–(8.10), and then the cumulative LAI of all layers can be obtained:

$$P(\theta) = n_{\text{empty}}/n_k \qquad (8.14)$$

Another method for estimating the LAI is converting 3-D point cloud data to 2-D raster images, similar to hemispherical photographs. This method uses corresponding processing software (e.g., Gap Light Analyzer, Hemiview) to estimate LAI. Zheng et al. (2013) converted 3-D point cloud data into 2-D raster images using four different projection systems (i.e., polar projection, orthogonal projection, Lambert azimuthal equal-area projection, and stereographic projection) and compared the effects of different projections on LAI estimates. The results show that the best projection varies among forest types, and it is necessary to choose a suitable projection according to different forest types in practice. The following Python code is for extracting LAI from TLS data.

```python
# The following Python code is the LAI extraction from terrestrial lidar data
# -*- coding: UTF-8 -*-
import numpy as np
import pandas as pd
# Read point cloud data
data = np.loadtxt(open("./TLS_LAI.txt","rb"),delimiter=" ",usecols=(0,1,2), dtype=float,skiprows=1)
# Set the voxel size, usually set to 1.5 times of the average spacing
sizeofvoxel = [0.003,0.003,0.003]
x = data[:, 0];y = data[:, 1];z = data[:, 2]
# Get the maximum and minimum of x, y, z
xmin=min(x)
xmax=max(x)
ymin=min(y)
ymax=max(y)
zmin=min(z)
zmax=max(z)
# Length, width, and height of voxels
pixelWidth=sizeofvoxel[0]
pixeLong=sizeofvoxel[1]
pixelHeight=sizeofvoxel[2]
# Calculate the offset of each point in the x, y, and z directions
xOffset = np.ceil((x - xmin)/pixelWidth)
yOffset = np.ceil((y - ymin)/pixeLong)
zOffset = np.ceil((z - zmin)/pixelHeight)
# Calculate the number of voxels in the three directions of length, width and height of point cloud data
rows=np.ceil((ymax-ymin)/pixelWidth)
cols=np.ceil((xmax-xmin)/pixeLong)
heis = np.ceil((zmax-zmin)/pixelHeight)
# Number of voxels per layer
one_layer=rows*cols
# Offset matrix in three directions
points = np.column_stack([xOffset,yOffset,zOffset])
points = np.array(points)
```

```
# Matrix deduplication
a =np.unique(points.axis =0)
# Grouped by offset in the z direction
pts = pd.DataFrame(a.columns=['Off_x','Off_y','Off_z'])
groups = pts.groupby(pts['Off_z'])
# Calculate the ratio of the number of voxels with point cloud data and the total number of voxels in
each layer
    layers = groups.size()/one_layer
    #LAI estimation
    LAI=sum(layers)*1.1
```

Full-waveform data can also be used to estimate the LAI. Zhao et al. (2011) proposed an algorithm for calculating gap fraction using full-waveform data provided by the Echidna validation instrument (Eqs. 8.15 and 8.16). Then, LAI can be estimated based on Eq. (8.6) described in the previous section.

$$\rho_{App}(r) = \frac{r^2 E(r) - e}{C(r) T_A^2 E_0} = \rho_V P_{FC}(r) = -\rho_V \frac{dP_{Gap}}{dr}(r) \tag{8.15}$$

$$P_{Gap}(\theta, r) = 1 - \frac{1}{\rho_A} \int \rho_{App} r' dr' \tag{8.16}$$

where r is the range; $C(r)$ is the optics calibration factor for the target at range r; $E(r)$ is the measured energy at range r; E_0 is the signal energy at the laser emitter; T_A is the atmospheric transmission; e is the background signal power; ρ_V is the effective emissivity; $P_{FC}(r)$ is the probability of first touching the target when the scan distance is r; $P_{Gap}(r)$ is the probability of a gap between the laser source and a point at range r; and ρ_{App} is the apparent reflectance of a facet along its normal direction.

8.2.3 Leaf Area Index Extraction from Airborne LiDAR Data

ALS is one of the most mature laser scanning technologies. Currently, two approaches are used to estimate the LAI from ALS point cloud data: indirectly using regression analysis of the related LiDAR-derived variables or directly using models based on the Beer—Lambert law. For example, Lim et al. (2003) and Roberts et al. (2005) used field-measured LAI data to build a linear regression model with various parameters derived from ALS data to estimate LAI, such as height quantiles, tree height, and crown width.

Nevertheless, most studies still prefer the direct method based on the Beer–Lambert law to extract LAI from ALS data. In this method, the most critical variable is the gap fraction (see Section 8.1, where the concept and calculation of canopy cover and closure were introduced). The gap fraction is linked to canopy closure through $P_{Gap} = 1 - $ canopy closure. The following is a Python script for extracting LAI from ALS data.

```python
# The following code is used to extract leaf area index from ALS data
# -*- coding: UTF-8 -*-
import laspy
from osgeo import ogr
from osgeo import osr
from osgeo import gdal
import numpy as np
# Read las file
lasfile = "./ALS_LAI.las"
inFile = laspy.file.File(lasfile, mode = "r")
# Get point cloud coordinates and attributes
x,y,z = inFile.x,inFile.y,inFile.z
classfication = inFile.raw_classification
return_number = inFile.return_num
scan_angle = inFile.scan_angle
# Read the corresponding DEM data
ds_dem = gdal.Open( "sample_dem.tif" )
rows = ds_dem.RasterYSize
cols = ds_dem.RasterXSize
bands = ds_dem.RasterCount
# Get the origin and resolution information of the DEM raster data
transform = ds_dem.GetGeoTransform()
xOrigin = transform[0]
yOrigin = transform[3]
pixelWidth = transform[1]
pixelHeight = transform[5]
# Calculate the DEM grid position for each point cloud
xOffset = (x - xOrigin) / pixelWidth
xOffset= xOffset.astype(int)
yOffset = (y - yOrigin) / pixelHeight
yOffset= yOffset.astype(int)
# Read DEM data
point_dem = np.empty(len(x))
band = ds_dem.GetRasterBand(1)
band1Array = band.ReadAsArray()
for i in range(0,len(x)):
```

```
        point_dem[i] = band.ReadAsArray(xOffset[i], yOffset[i], 1, 1)
# Point cloud Z value normalization
normilze_z = z - point_dem
# Get the range of the point cloud data
x_max,y_max = inFile.header.max[0:2]
x_min,y_min = inFile.header.min[0:2]
# Set the resolution of the generated LAI
pixelWidth , pixelHeight = 10,-10
# Set the origin and rank number of the LAI raster
xOrigin = x_min
yOrigin = y_max
cols = int(round((x_max - x_min) / pixelWidth))
rows = int(round((y_max - y_min) / abs(pixelHeight)))
# Calculate the position of the LAI grid corresponding to each point cloud
xOffset = (x - xOrigin) / pixelWidth
xOffset= xOffset.astype(int)
yOffset = (y - yOrigin) / pixelHeight
yOffset= yOffset.astype(int)
# Set variables for storing void ratios and LAI rasters
gp = np.zeros((cols, rows))
lai = np.zeros((cols, rows))
# Calculate the LAI value of each grid
for j in range(0, rows):
    for i in range(0,cols):
        vp = len( np.where((xOffset==i) & (yOffset==j) & (normilze_z >2))[0])
        ttp = len(np.where((xOffset==i) & (yOffset==j))[0])
        angles = np.mean(np.abs(scan_angle[np.where((xOffset==i) & (yOffset==j))]))
        gp[i,j] = 1 - float(vp)/float(ttp)
        lai[i,j] =   -1 * np.cos(np.deg2rad(angles)) * np.log(gp[i,j])/0.5
# Output into geoTIFF
driver = gdal.GetDriverByName('GTiff')
outRaster = driver.Create('lai.tif', cols, rows, 1, gdal.GDT_Float32)
outRaster.SetGeoTransform((xOrigin, pixelWidth, 0, yOrigin, 0, pixelHeight))
outband = outRaster.GetRasterBand(1)
outband.WriteArray(np.transpose(lai))
# Set projection information
outRasterSRS = osr.SpatialReference()
outRasterSRS.ImportFromProj4("+proj=utm +zone=10 +datum=NAD83 +units=m +no_defs")
outRaster.SetProjection(outRasterSRS.ExportToWkt())
outband.FlushCache()
outRaster = None
```

8.3 Growing Stock and Biomass

Forest growing stock refers to the total volume of tree trunks over a certain forest area. It is one of the basic indicators of forest resources in a country or region. It is also an important indicator of the ecological environment. Forest biomass comprises the biomass of trees (the total weight of roots, stems, leaves, flowers, fruits, seeds, and litter) and the biomass of understory vegetation. These are often calculated as the dry matter mass or energy per unit area or unit time. Forest growing stock and biomass are important biophysical attributes that describe the function and productivity of forest ecosystems (Dubayah & Drake, 2000) and can be used to assess the carbon sink potential of forests. Hence, rapid and accurate estimation of forest growing stock and biomass is essential for studies on forest ecosystems and global climate change. Methods for estimating forest growing stock and biomass can be divided into two main categories: traditional field surveys and remote sensing-based methods. Traditional field surveys are time-consuming, labor-intensive, and destructive. Moreover, it is difficult to obtain information about each tree in the forest (He et al., 2007; Wang et al., 2010).

The rapid development of remote sensing technology provides a more accurate and faster way to monitor forest growing stock and biomass in the long term. LiDAR can be used to accurately obtain the 3-D structural information of forests and thus monitor forest stock and biomass changes at large scales. The current methods for estimating forest biomass and growing stock using LiDAR can be divided into two types, direct and indirect. Direct methods establish relationships between response signals of forests received by LiDAR and forest growing stock and biomass through regression models, such as multiple regression, k-nearest neighbor regression, neural networks, and Bayes models. Indirect methods generally use statistical analysis to establish relationships between field-measured biomass or growing stock and LiDAR-derived forest attributes, such as tree height, crown width, canopy closure, tree species, and forest type (Cao et al., 2013; Cháidez, 2008). LiDAR can be divided into large- and small-footprint LiDAR. The footprint size affects the accuracy of the acquired forest attributes. Different footprint sizes are suitable for estimating biomass and growing stock at different scales. Overall, large-footprint LiDAR (usually larger than 5 m) records the features of an object through waveforms and is more suitable for large-scale estimation of forest growing stock and biomass. In contrast, small-footprint LiDAR (usually smaller than 1 m) detects an object using pulses or waveforms, which is suitable for small-scale estimation of forest growing stock and biomass (Liu et al., 2012).

8.3.1 Growing Stock and Biomass Estimation at the Individual Tree Level

Individual tree segmentation is an important prerequisite for estimating forest growing stock and biomass at the individual tree level. Therefore, the accuracy of individual tree segmentation significantly affects the accuracy of forest growing stock and biomass estimation. Yu et al. (2006) applied several individual tree segmentation methods to estimate forest growing stock and biomass. Their results showed that the estimation accuracy increased with the accuracy of individual tree segmentation. At present, individual tree segmentation based on LiDAR point cloud data can be either manual or automatic. Manual segmentation mainly uses visual identification of treetops and crown boundaries through software (Andersen et al., 2006). Then the canopy structure and attribute information of each tree (e.g., tree height, crown width, and crown height) are automatically extracted using computer algorithms (Popescu & Wynne, 2004). Automatic segmentation relies on the existing individual tree segmentation algorithms. The related algorithms can be divided into two types: the segmentation algorithm based on CHM image and that based on point clouds, which are described in detail in Chapter 7. The structural attributes of individual trees from segmentation results are used to estimate growing stock and biomass either directly by linking to field-measured biomass or indirectly through existing allometric equations.

8.3.2 Growing Stock and Biomass Estimation at the Community Level

Community-level growing stock and biomass can be obtained by directly summarizing the individual tree-level estimates. However, when the LiDAR point density is too low to segment individual trees accurately, regression methods are more suitable for estimating forest growing stock and biomass at the community level, which include two main steps. First, canopy structural attributes are directly estimated from LiDAR point clouds, such as average tree height, crown width, canopy height, and the number of sample trees. Then, regression models between field-measured plot-scale growing stock or biomass and LiDAR-derived structural metrics are established, which are then used to estimate forest growing stock and biomass at the community level. However, LiDAR data do not provide the rich spectral information of objects and thus are of limited use for differentiating tree species compositions (Hill & Thomson, 2005; Su, Guo, Fry, et al., 2016). Complex forest stands are usually composed of various tree species differing in wood density, and species-induced uncertainty can account for 25% of the total uncertainty of LiDAR-derived forest aboveground biomass (AGB) (Rodig et al., 2019). The combination of optical imagery and LiDAR data can provide both spectral and structural information and can thus improve forest AGB estimation accuracy (de Almeida et al., 2019; Laurin et al., 2014; Li et al., 2021).

Currently, most forest AGB estimation methods based on fusing multisource data simply feed various spectral and structural attributes into regression models (de Almeida et al., 2019; Ma et al., 2019). A substantial number of field measurements are usually required to train regression models to learn the links between forest AGB and its spectral and structural attributes extracted from remote-sensing images (Zeide, 1993). It has been well documented that feeding more variables into regression models may not necessarily lead to higher AGB estimation accuracy (Poorazimy et al., 2020). Yang et al. (2022) proposed an approach that used a VI metric as the power of a canopy height (TH) metric at the community level to build a physically meaningful regression model for forest AGB estimation (Eqs. 8.17 and 8.18). Compared with methods using structural attributes alone, the new approach combining structural and spectral information can improve the estimation accuracy of AGB, with the coefficient of determination (R^2) increased by about 10% and the root-mean-square error reduced by about 22% (Fig. 8.4):

$$AGB = \left(aTH^{b}\right)^{VI} \tag{8.17}$$

$$\log AGB = aVI * \log(TH) + b \tag{8.18}$$

where TH is a canopy height metric in each plot, VI is a vegetation index in each plot, and a and b are fitted model parameters.

With the continuous development of satellite remote sensing technology, the recently launched spaceborne LiDAR missions, Global Ecosystem Dynamics Investigation LiDAR and Ice, Cloud, and land Elevation Satellite 2, provide nearly global coverage and high-density ground sampling footprints, which makes it possible to map global forest height and biomass precisely. In addition, existing optical remote sensing satellites, such as Sentinel-2 and the upcoming hyperspectral mission (e.g., the Environmental Mapping and Analysis Program), can better capture the spectral information of

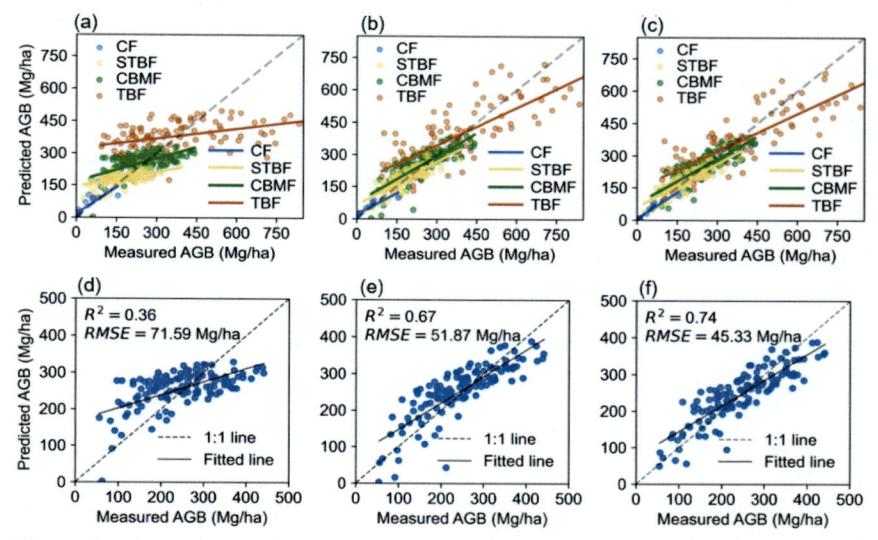

Figure 8.4 Comparisons between measured aboveground biomass (AGB) and predicted AGB based on different models: (a) only using the mean normalized difference vegetation index ($NDVI_{mean}$) data, (b) only using the mean canopy height (TH_{mean}) data, and (c) combining $NDVI_{mean}$ with TH_{mean} using the proposed method; (d–f) are the measured AGB and predicted AGB based on those models for CBMF. *CBMF*, coniferous and broadleaf mixed forest; *CF*, conifer forest; *STBF*, subtropical broadleaf forest; *TBF*, tropical broadleaf forest.

heterogeneous forest canopies (Rodig et al., 2019). In the future, we can try to integrate these new remote sensing data with higher spatial and spectral resolutions to map forest AGB at regional and global scales.

8.4 Tree Species Classification

Information about tree species composition and distribution is essential for precision forestry, biodiversity assessment and monitoring, and wildlife habitat mapping. Moreover, tree species are the basic input for tree growth models, yield models, and allometric equations. Understanding how to accurately and efficiently identify tree species in the forest is crucial for forest managers and researchers. Traditionally, tree species composition is obtained from forest inventories, which are highly time- and labor-consuming and have limited coverage. Because different tree species have various spectral reflectance features, studies have tried to use optical remote sensing data to classify tree species at different scales. However, since different species may share similar reflectance characteristics, and the same species may have different reflectance characteristics because of various background effects in remote sensing images with narrow spectral bands, accurate species classification from multispectral satellite imagery is still a challenge. LiDAR collects 3-D information from forests and thus can provide a new perspective to classify tree species.

8.4.1 Tree Species Classification at the Individual Tree Level

Both ALS and TLS can identify tree species at the individual tree level. Studies have used the structural information of segmented individual trees from point cloud data combined with optical images, especially hyperspectral images, to build classification models for tree species identification. Crown and branch structural features are the basic elements used to identify tree species using LiDAR data. Othmani et al. (2013) used the 3-D geometric texture of the bark to identify five tree species (hornbeam, oak, spruce, beech, and pine) among 75 trees using the random forest algorithm. Lin and Herold (2016) used the support vector machine algorithm to classify four species from 40 trees based on tree structural attributes (such as tree height, stem DBH, crown height, branch angle, and crown length). Akerblom et al. (2017) provided a framework to recognize tree species with features derived from quantitative structure models. Using this method, they classified five forest plots with over 1200 trees using 15 geometry and topology features derived from a quantitative structure model. Terryn et al. (2020) also used the framework to classify five species with 15 features derived from quantitative structure models. The species classification accuracy of these studies varied from 77.5% to 93%. However, the number of recognized tree species was limited to five or fewer. Individual tree species classification based on LiDAR data is still a challenge. Fusing LiDAR data with optical remote sensing data can substantially improve the accuracy of tree species classification. Xu et al. (2020) used digital aerial photogrammetry point clouds from unmanned aerial systems and multispectral imagery to extract the spectral, textural, and structural metrics of subtropical natural forests and successfully classify eight tree species. Zhao et al. (2020) combined airborne LiDAR and hyperspectral data to classify tree species using a support vector machine and spectral angle mapper classifiers in a natural mixed forest in Northeast China. They successfully identified six species with a classification accuracy greater than 90%. However, most previous methods only work for a limited number of tree species over a small region, and tree segmentation accuracy strongly affects classification accuracy. Many researchers have adopted deep learning methods to identify tree species using preclassified tree species photographs and a preclassified individual tree point cloud database, which has the potential to improve individual tree species classification accuracy as the data accumulate in the future.

8.4.2 Tree Species Classification at the Community Level

Tree species classification at the community level is also called vegetation type classification. The vegetation type is essential information for forest management and biodiversity conservation. The 3-D structural information provided by LiDAR data combined with spectral information from optical remote sensing imagery has been widely used to classify vegetation types. Su, Guo, Fry, et al. (2016) developed an unsupervised classification method based on the Bayes and k-means clustering algorithms. Results showed that the classification results from integrating LiDAR data with aerial photographs had higher accuracy and could generate more diverse vegetation types than traditional classification methods (Fig. 8.5). Studies have also found that the structural information provided by LiDAR substantially improved the performance of vegetation type classification. Voss and Sugumaran (2008) combined LiDAR data with the Airborne Imaging Spectrometer for Applications (AISA) hyperspectral data to classify four deciduous species and three evergreen species. With the help of LiDAR data, the shadowing effect in the hyperspectral data was successfully corrected. Also, using combined LiDAR and AISA data accurately differentiated species with tall and short trees, thus improving the classification accuracy. Sugumaran and Voss (2007) used object–oriented and regression analysis methods by adding the intensity of laser returns and elevation information into the image segmentation of optical imagery and improved the classification accuracy by approximately 12%—24%.

8.5 Chapter Summary

At present, LiDAR data with different footprint sizes acquired from various platforms have been widely applied to estimate ecosystem functional attributes, including canopy cover, canopy closure, LAI, growing stock and biomass, and tree species. Different LiDAR data have specific advantages, and various methods have been developed to estimate ecosystem functional attributes. Integrating LiDAR data from a variety of sources is a future

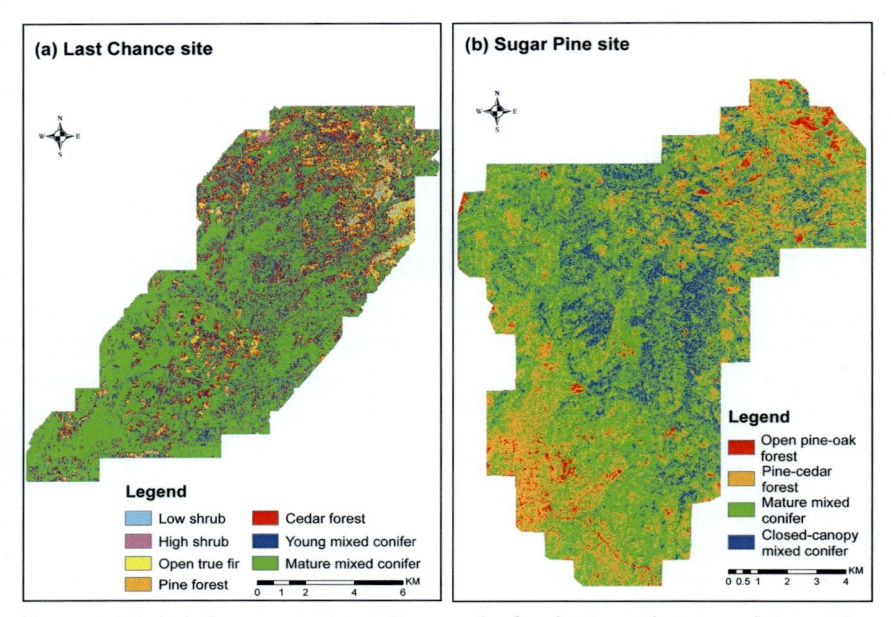

Figure 8.5 Labeled vegetation mapping results for the Last Chance and Sugar Pine sites in the Sierra Nevada Mountains, California, USA.

direction for improving the accuracy of ecosystem functional parameter estimation. Merging LiDAR data with other data types, such as passive optical remote sensing data, can improve the accuracy of large-scale forest functional parameter inversion. This will become one of the main directions for future forest ecosystem applications.

References

Akerblom, M., Raumonen, P., Makipaa, R., & Kaasalainen, M. (2017). Automatic tree species recognition with quantitative structure models. *Remote Sensing of Environment, 191*, 1–12.

de Almeida, C. T., Galvao, L. S., Aragao, L., Ometto, J., Jacon, A. D., Pereira, F. R. D., Sato, L. Y., Lopes, A. P., Graca, P., Silva, C. V. D., Ferreira-Ferreira, J., & Longo, M. (2019). Combining LiDAR and hyperspectral data for aboveground biomass modeling in the Brazilian Amazon using different regression algorithms. *Remote Sensing of Environment, 232*. Article 111323.

Andersen, H. E., Reutebuch, S. E., & McGaughey, R. J. (2006). A rigorous assessment of tree height measurements obtained using airborne lidar and conventional field methods. *Canadian Journal of Remote Sensing, 32*(5), 355–366.

Baret, F., & Guyot, G. (1991). Potentials and limits of vegetation indices for LAI and APAR assessment. *Remote Sensing of Environment, 35*(2–3), 161–173.

Boegh, E., Soegaard, H., Broge, N., Hasager, C. B., Jensen, N. O., Schelde, K., & Thomsen, A. (2002). Airborne multispectral data for quantifying leaf area index, nitrogen concentration, and photosynthetic efficiency in agriculture. *Remote Sensing of Environment, 81*(2–3), 179–193.

Breda, N. J. J. (2003). Ground-based measurements of leaf area index: A review of methods, instruments and current controversies. *Journal of Experimental Botany, 54*(392), 2403–2417.

Cao, L., Yu, G., & Dai, J. (2013). Status and prospects of the LiDAR-based forest biomass estimation. *Journal of Nanjing Forestry University, 37,* 163–169 (in Chinese).

Cháidez, J. D. J. N. (2008). Allometric equations and expansion factors for tropical dry trees of eastern Sinaloa, Mexico. *Tropical and Subtropical Agroecosystems, 10*(1), 45–52.

Chen, J., & Black, T. (1991). Measuring leaf area index of plant canopies with branch architecture. *Agricultural and Forest Meteorology, 57*(1–3), 1–12.

Chen, J. M., Pavlic, G., Brown, L., Cihlar, J., Leblanc, S. G., White, H. P., Hall, R. J., Peddle, D. R., King, D. J., Trofymow, J. A., Swift, E., Van der Sanden, J., & Pellikka, P. K. E. (2002). Derivation and validation of Canada-wide coarse-resolution leaf area index maps using high-resolution satellite imagery and ground measurements. *Remote Sensing of Environment, 80*(1), 165–184.

Coops, N. C., Hilker, T., Wulder, M. A., St-Onge, B., Newnham, G., Siggins, A., & Trofymow, J. A. (2007). Estimating canopy structure of Douglas-fir forest stands from discrete-return LiDAR. *Trees-Structure and Function, 21*(3), 295–310.

Dubayah, R. O., & Drake, J. B. (2000). Lidar remote sensing for forestry. *Journal of Forestry, 98*(6), 44–46.

Fang, H., Liang, S., & Kuusk, A. (2003). Retrieving leaf area index using a genetic algorithm with a canopy radiative transfer model. *Remote Sensing of Environment, 85*(3), 257–270.

Farid, A., Goodrich, D. C., Bryant, R., & Sorooshian, S. (2008). Using airborne lidar to predict Leaf Area Index in cottonwood trees and refine riparian water-use estimates. *Journal of Arid Environments, 72*(1), 1–15.

Fassnacht, K. S., Gower, S. T., Norman, J. M., & McMurtric, R. E. (1994). A comparison of optical and direct methods for estimating foliage surface area index in forests. *Agricultural and Forest Meteorology, 71*(1–2), 183–207.

He, H., Guo, Z., & Xiao, W. (2007). Application of remote sensing in forest aboveground biomass estimation. *Chinese Journal of Ecology, 169*(08), 1317–1322 (in Chinese).

Hill, R., & Thomson, A. (2005). Mapping woodland species composition and structure using airborne spectral and LiDAR data. *International Journal of Remote Sensing, 26*(17), 3763–3779.

Hopkinson, C., & Chasmer, L. (2009). Testing LiDAR models of fractional cover across multiple forest ecozones. *Remote Sensing of Environment, 113*(1), 275–288.

Hosoi, F., & Omasa, K. (2006). Voxel-based 3-D modeling of individual trees for estimating leaf area density using high-resolution portable scanning lidar. *Ieee Transactions on Geoscience and Remote Sensing, 44*(12), 3610–3618.

Jakubowski, M. K., Guo, Q. H., & Kelly, M. (2013). Tradeoffs between lidar pulse density and forest measurement accuracy. *Remote Sensing of Environment, 130,* 245–253.

Jennings, S. B., Brown, N. D., & Sheil, D. (1999). Assessing forest canopies and understorey illumination: Canopy closure, canopy cover and other measures. *Forestry, 72*(1), 59–73.

Jonckheere, I., Fleck, S., Nackaerts, K., Muys, B., Coppin, P., Weiss, M., & Baret, F. (2004). Review of methods for in situ leaf area index determination—part I. Theories, sensors and hemispherical photography. *Agricultural and Forest Meteorology, 121*(1–2), 19–35.

Jupp, D. L., Culvenor, D., Lovell, J. L., & Newnham, G. (2005). Evaluation and validation of canopy laser radar (LIDAR) systems for native and plantation forest inventory. *Final*

Report Prepared for the Forest and Wood Products Research and Development Corporation (FWPRDC: PN 02.2902) by CSIRO, 20, 150.

Kato, A., Moskal, L. M., Schiess, P., Swanson, M. E., Calhoun, D., & Stuetzle, W. (2009). Capturing tree crown formation through implicit surface reconstruction using airborne lidar data. *Remote Sensing of Environment, 113*(6), 1148—1162.

Korhonen, L., Korpela, I., Heiskanen, J., & Maltamo, M. (2011). Airborne discrete-return LIDAR data in the estimation of vertical canopy cover, angular canopy closure and leaf area index. *Remote Sensing of Environment, 115*(4), 1065—1080.

Lang, A. R. G., McMurtrie, R. E., & Benson, M. L. (1991). Validity of surface area indices of Pinus radiata estimated from transmittance of the sun's beam. *Agricultural and Forest Meteorology, 57*(1), 157—170.

Laurin, G. V., Chen, Q., Lindsell, J. A., Coomes, D. A., Del Frate, F., Guerriero, L., Pirotti, F., & Valentini, R. (2014). Above ground biomass estimation in an African tropical forest with lidar and hyperspectral data. *Isprs Journal of Photogrammetry and Remote Sensing, 89,* 49—58.

Lim, K., Treitz, P., Baldwin, K., Morrison, I., & Green, J. (2003). Lidar remote sensing of biophysical properties of tolerant northern hardwood forests. *Canadian Journal of Remote Sensing, 29*(5), 658—678.

Lin, Y., & Herold, M. (2016). Tree species classification based on explicit tree structure feature parameters derived from static terrestrial laser scanning data. *Agricultural and Forest Meteorology, 216,* 105—114.

Liu, D., Fan, W., & MZ, L. (2012). Estimation of forest stand parameters and biomass by small-footprint lidar. *Journal of Northeast Forestry University, 40*(1), 39—43 (in Chinese).

Li, Q. S., Wong, F. K. K., & Fung, T. (2021). Mapping multi-layered mangroves from multispectral, hyperspectral, and LiDAR data. *Remote Sensing of Environment, 258.* Article 112403.

Luo, S. Z. (2012). *Research and application of extracting forest leaf area index using LiDAR remote sensing.* Chinese University of Geosciences for Doctoral Degree (in Chinese).

Luo, S.-Z., Wang, C., Zhang, G., Xi, X., & Li, G. (2013). Forest leaf area index (LAI) inversion using airborne LiDAR data. *Chinese Journal of Geophysics, 56*(5), 1467—1475 (in Chinese).

Ma, X. L., Mahecha, M. D., Migliavacca, M., van der Plas, F., Benavides, R., Ratcliffe, S., Kattge, J., Richter, R., Musavi, T., Baeten, L., Barnoaiea, I., Bohn, F. J., Bouriaud, O., Bussotti, F., Coppi, A., Domisch, T., Huth, A., Jaroszewicz, B., Joswig, J., ... Wirth, C. (2019). Inferring plant functional diversity from space: The potential of Sentinel-2. *Remote Sensing of Environment, 233.* Article 111368.

Ma, Q., Su, Y. J., & Guo, Q. H. (2017). Comparison of canopy cover estimations from airborne LiDAR, aerial imagery, and satellite imagery. *IEEE Journal of Selected Topics in Applied Earth Observations and Remote Sensing, 10*(9), 4225—4236.

Nilson, T. (1999). Inversion of gap frequency data in forest stands. *Agricultural and Forest Meteorology, 98—99,* 437—448.

Othmani, A., Voon, L., Stolz, C., & Piboule, A. (2013). Single tree species classification from terrestrial laser scanning data for forest inventory. *Pattern Recognition Letters, 34*(16), 2144—2150.

Parker, G. G. (1995). Structure and microclimate of forest canopies. *Forest Canopies,* 73—106.

Pierce, L. L., & Running, S. W. (1988). Rapid estimation of coniferous forest leaf area index using a portable integrating radiometer. *Ecology, 69*(6), 1762—1767.

Pinty, B., Gobron, N., Widlowski, J. L., Gerstl, S. A., Verstraete, M. M., Antunes, M., Bacour, C., Gascon, F., Gastellu, J. P., & Goel, N. (2001). Radiation transfer model intercomparison (RAMI) exercise. *Journal of Geophysical Research: Atmospheres, 106*(D11), 11937—11956.

Poorazimy, M., Shataee, S., McRoberts, R. E., & Mohammadi, J. (2020). Integrating airborne laser scanning data, space-borne radar data and digital aerial imagery to estimate aboveground carbon stock in Hyrcanian forests, Iran. *Remote Sensing of Environment, 240.* Article 111669.

Popescu, S. C., & Wynne, R. H. (2004). Seeing the trees in the forest: Using lidar and multispectral data fusion with local filtering and variable window size for estimating tree height. *Photogrammetric Engineering and Remote Sensing, 70*(5), 589–604.

Roberts, S. D., Dean, T. J., Evans, D. L., McCombs, J. W., Harrington, R. L., & Glass, P. A. (2005). Estimating individual tree leaf area in loblolly pine plantations using LiDAR-derived measurements of height and crown dimensions. *Forest Ecology and Management, 213*(1–3), 54–70.

Rödig, E., Knapp, N., Fischer, R., Bohn, F. J., Dubayah, R., Tang, H., & Huth, A. (2019). From small-scale forest structure to Amazon-wide carbon estimates. *Nature Communications, 10.* Article 5088.

Rouse, J. W., Jr., Haas, R. H., Schell, J. A., & Deering, D. W. (1974). *Monitoring vegetation systems in the great plains with erts* (Vol. 351, p. 309). NASA Special Publication.

Ryu, Y., Sonnentag, O., Nilson, T., Vargas, R., Kobayashi, H., Wenk, R., & Baldocchi, D. D. (2010). How to quantify tree leaf area index in an open savanna ecosystem: A multi-instrument and multi-model approach. *Agricultural and Forest Meteorology, 150*(1), 63–76.

Solberg, S., Brunner, A., Hanssen, K. H., Lange, H., Naesset, E., Rautiainen, M., & Stenberg, P. (2009). Mapping LAI in a Norway spruce forest using airborne laser scanning. *Remote Sensing of Environment, 113*(11), 2317–2327.

Solberg, S., Weydahl, D. J., & Næsset, E. (2007). SAR forest canopy penetration depth as an indicator for forest health monitoring based on leaf area index (LAI). In *Proceedings of the 5th international symposium on retrieval of bio-and geophysical parameters from SAR data for land applications, Bari, Italy.*

Spanner, M. A., Pierce, L. L., Peterson, D. L., & Running, S. W. (1990). Remote sensing of temperate coniferous forest leaf area index the influence of canopy closure, understory vegetation and background reflectance. *International Journal of Remote Sensing, 11*(1), 95–111.

Stenberg, P., Linder, S., Smolander, H., & Flower-Ellis, J. (1994). Performance of the LAI-2000 plant canopy analyzer in estimating leaf area index of some Scots pine stands. *Tree Physiology, 14*(7–8-9), 981–995.

Sugumaran, R., & Voss, M. (2007). Object-oriented classification of LIDAR-fused hyperspectral imagery for tree species identification in an urban environment. *Urban Remote Sensing Joint Event,* 1–6.

Su, Y., Guo, Q., Fry, D. L., Collins, B. M., Kelly, M., Flanagan, J. P., & Battles, J. J. (2016). A vegetation mapping strategy for conifer forests by combining airborne LiDAR data and aerial imagery. *Canadian Journal of Remote Sensing, 42*(1), 1–15.

Tang, X. G. (2013). *Estimation of forest aboveground biomass by integrating ICESat/GLAS waveform and TM data.* University of Chinese Academy of Sciences for Doctoral Degree (in Chinese).

Terryn, L., Calders, K., Disney, M., Origo, N., Malhi, Y., Newnham, G., Raumonen, P., Akerblom, M., & Verbeeck, H. (2020). Tree species classification using structural features derived from terrestrial laser scanning. *ISPRS Journal of Photogrammetry and Remote Sensing, 168,* 170–181.

Voss, M., & Sugumaran, R. (2008). Seasonal effect on tree species classification in an urban environment using hyperspectral data, LiDAR, and an object-oriented approach. *Sensors, 8*(5), 3020–3036.

Wang, M., Li, G., & Zhong, G. (2010). Integraating ecosystem process models, remote sensing and groundbased observations for regional-scale analysis of the carbon storage. *Forest Resources Management, 02*, 107−112 (in Chinese).

Wang, X., Ma, L., Jia, Z., & Xu, C. (2005). Research and application advances in leaf area index (LAI). *Chinese Journal of Ecology, 24*(5), 537−541 (in Chinese).

Wu, W., Hong, T., & Wang, X. (2007). Advance in ground-based LAI measurement methods. *Journal of Huazhong Agricultural University, 02*, 270−275 (in Chinese).

Wulder, M. A., LeDrew, E. F., Franklin, S. E., & Lavigne, M. B. (1998). Aerial image texture information in the estimation of northern deciduous and mixed wood forest leaf area index (LAI). *Remote Sensing of Environment, 64*(1), 64−76.

Xu, Z., Shen, X., Cao, L., Coops, N. C., Goodbody, T. R. H., Zhong, T., Zhao, W. D., Sun, Q. L., Ba, S., Zhang, Z. N., & Wu, X. Q. (2020). Tree species classification using UAS-based digital aerial photogrammetry point clouds and multispectral imageries in subtropical natural forests. *International Journal of Applied Earth Observation and Geoinformation, 92*, 102173.

Yang, Q., Su, Y., Hu, T., Jin, S., Liu, X., Niu, C., Liu, Z., Kelly, M., Wei, J., & Guo, Q. (2022). Allometry-based estimation of forest aboveground biomass combining LiDAR canopy height attributes and optical spectral indexes. *Forest Ecosystems, 9*, 100059.

Yu, X. W., Hyyppa, J., Kukko, A., Maltamo, M., & Kaartinen, H. (2006). Change detection techniques for canopy height growth measurements using airborne laser scanner data. *Photogrammetric Engineering and Remote Sensing, 72*(12), 1339−1348.

Zeide, B. (1993). Primary unit of the tree crown. *Ecology, 74*(5), 1598−1602.

Zhao, D., Pang, Y., Liu, L. J., & Li, Z. Y. (2020). Individual tree classification using airborne LiDAR and hyperspectral data in a natural mixed forest of Northeast China. *Forests, 11*(3), 303.

Zhao, K. G., & Popescu, S. (2009). Lidar-based mapping of leaf area index and its use for validating GLOBCARBON satellite LAI product in a temperate forest of the southern USA. *Remote Sensing of Environment, 113*(8), 1628−1645.

Zhao, F., Yang, X. Y., Schull, M. A., Roman-Colon, M. O., Yao, T., Wang, Z. S., Zhang, Q. L., Jupp, D. L. B., Lovell, J. L., Culvenor, D. S., Newnham, G. J., Richardson, A. D., Ni-Meister, W., Schaaf, C. L., Woodcock, C. E., & Strahler, A. H. (2011). Measuring effective leaf area index, foliage profile, and stand height in New England forest stands using a full-waveform ground-based lidar. *Remote Sensing of Environment, 115*(11), 2954−2964.

Zheng, G., Ma, L., He, W., Eitel, J. U., Moskal, L. M., & Zhang, Z. (2015). Assessing the contribution of woody materials to forest angular gap fraction and effective leaf area index using terrestrial laser scanning data. *IEEE Transactions on Geoscience and Remote Sensing, 54*(3), 1475−1487.

Zheng, G., Moskal, L. M., & Kim, S. H. (2013). Retrieval of effective leaf area index in heterogeneous forests with terrestrial laser scanning. *IEEE Transactions on Geoscience and Remote Sensing, 51*(2), 777−786.

Zou, J., & Zhuge, X. (2011). Forest canopy closure and its determination method. *Heilongjiang Science and Technology Information, 35*, 290 (in Chinese).

CHAPTER 9

Forest Structural and Functional Attribute Upscaling Using Spaceborne LiDAR Data

Contents

Global mapping of forest structural and functional attributes is essential for monitoring forest disturbances, understanding forest ecosystem processes, and developing forest management and restoration policies to mitigate global climate change. Although near-surface and airborne light detection and ranging (LiDAR) systems have revolutionized the measurement of

LiDAR Principles, Processing and Applications in Forest Ecology
ISBN 978-0-12-823894-3
https://doi.org/10.1016/B978-0-12-823894-3.00009-8

forests from individual to stand levels, as mentioned in previous chapters, it is impractical to use these systems to estimate forest structural and functional attributes at a global scale due to their limited coverage. Spaceborne LiDAR systems, measuring Earth's surface height globally from space, provide potential datasets for upscaling forest structural and functional attributes from the local to the global scale. Currently, three spaceborne LiDAR systems, including the Geoscience Laser Altimeter System aboard the Ice, Cloud, and land Elevation Satellite (ICESat), Advanced Topographic Laser Altimeter System aboard ICESat-2, and Global Ecosystem Dynamics Investigation installed on the International Space Station, are the most frequently used systems for forestry applications. This chapter introduces these three spaceborne systems, their data products, and their processing procedures. Finally, we describe methods for upscaling forest structural and functional attributes using these spaceborne LiDAR data.

9.1 Ice, Cloud, and Land Elevation Satellite and Global Ecosystem Dynamics Investigation Spaceborne LiDAR Data

Both the Geoscience Laser Altimeter System (GLAS) aboard the Ice, Cloud, and land Elevation Satellite (ICESat) and the Global Ecosystem Dynamics Investigation (GEDI) are spaceborne LiDAR systems using waveform LiDAR sensors. The sections that follow introduce data acquisition by these two systems and their associated data products and types.

9.1.1 Mission Introduction

The ICESat GLAS was the first spaceborne LiDAR system to characterize the Earth's surface along three dimensions (Schutz et al., 2005). The ICESat GLAS was launched by the US National Aeronautics and Space Administration (NASA) on January 12, 2003, with just three lasers installed (i.e., Lasers 1, 2, and 3). Although the designed life of the three lasers was 18 months, Laser 1 stopped collecting data on the 38th day of operation because of laser failure. Laser 2 immediately took over the tasks of the ICESat GLAS system but ended its normal operation earlier than planned owing to accelerated laser energy degradation caused by continuous data collection. Based on the experience of Lasers 1 and 2, the working mode of Laser 3 was adjusted to intermittent data collection. It observed the Earth's surface in orbit for approximately 91 days each year from February to March, May to June, and October to November. This data collection mode extended the life of Laser 3, but it eventually stopped working on October 11, 2009. As a result, the decommissioning of the ICESat was announced on August 14, 2010. NASA originally planned to launch the second spaceborne LiDAR satellite, ICESat-2, in 2017, but the launch was delayed to 2018 for various reasons.

ICESat GLAS operated in a nearly circular and near-polar orbit at a height of 600 km and covered most of the world's surface between 86° N and 86° S, with an orbital inclination of 94° and a revisit period of approximately 183 days. The system emitted laser pulses at a frequency of 40 Hz and detected reflected laser signals from the ground, atmosphere, and clouds. The detected signals were then recorded as waveform data after processing by a sampler (Yang et al., 2015). The footprint of the ICESat GLAS data was an ellipse with a diameter of approximately 70 m. In the same strip, the distance between the centers of adjacent laser footprints was approximately 170 m; the distance between adjacent strips decreased with an increase in latitude, and the distance between adjacent strips was approximately 15 km near the equator and 2.5 km at latitude 80° (Abshire et al., 2005). The laser in the ICESat GLAS ranging unit had two channels: an infrared laser channel with a wavelength of 1064 nm that was used for measuring surface and cloud heights; and a 532 nm green laser channel for studying the vertical distribution of clouds and aerosols. For each returned waveform at 1064 nm, there were 544 samples if the return was from the land surface, corresponding to two maximum vertical distances of 81.6 m or 150 m that can be recorded.

GEDI is the first spaceborne LiDAR satellite designed specifically for precise measurements of forest canopy structures (Dubayah et al., 2020). GEDI was successfully launched on December 5, 2018, by NASA and installed on the International Space Station, which covered the Earth's land surfaces between 51.6° N and 51.6° S. During its 2-year service period, 4% of the Earth's surface was sampled, and more than 10 billion laser footprints of cloud-free land surface observations were obtained. These observations can be gridded into regular coverages of varying resolution, such as 1 km grid cells (Fig. 9.1), which can significantly improve our ability to characterize and understand the carbon and water cycling processes, biodiversity, and wildlife habitats.

The GEDI instrument is a geodetic-class waveform LiDAR. It contains three identical near-infrared lasers with a wavelength of 1064 nm; two are full-power lasers, and the other is a coverage laser split into two weaker energy beams. Thus, GEDI can produce a total of four beams and eight parallel-ground tracks by dithering beams (Fig. 9.2). Each GEDI laser emits a beam 242 times per second and forms a footprint with an average

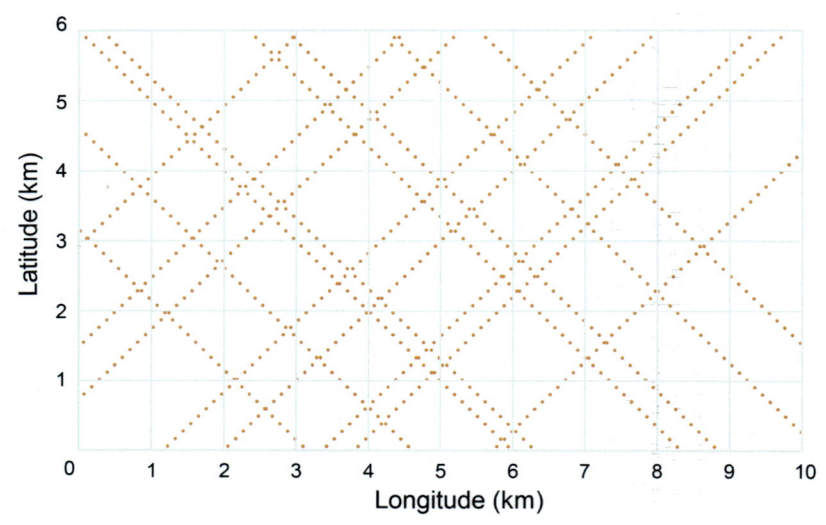

Figure 9.1 An example of Global Ecosystem Dynamics Investigation track coverage at the equator.

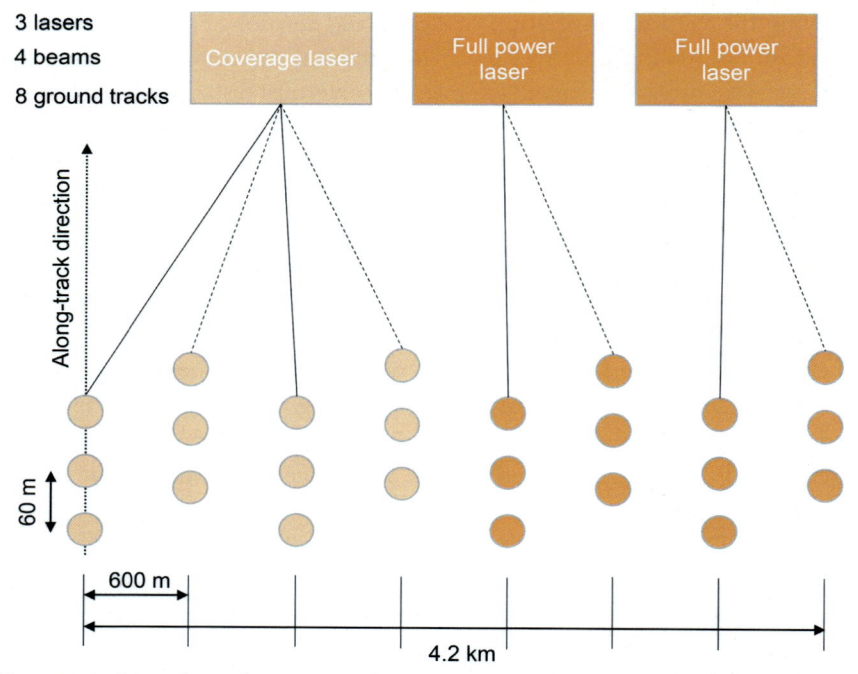

Figure 9.2 Ground sampling pattern of Global Ecosystem Dynamics Investigation. *(Revised from Dubayah, R., Blair, J. B., Goetz, S., Fatoyinbo, L., Hansen, M., Healey, S., Hofton, M., Hurtt, G., Kellner, J., & Luthcke, S. (2020). The global ecosystem dynamics investigation: High-resolution laser ranging of the Earth's forests and topography.* Science of Remote Sensing, 1, *100002.)*

diameter of 25 m. Footprints along the track are separated by 60 m, and ground tracks are separated by 600 m (Fig. 9.2). GEDI records the return laser signals from a footprint as a function of time in a 1 ns (0.15 m) interval. To prolong the service life of the instrument, the GEDI lasers are ramped down over ocean passes and only collect enough ocean data to determine the orbit and accuracy of the geographic location (Dubayah et al., 2020).

9.1.2 Introduction to Ice, Cloud, and Land Elevation Satellite and Global Ecosystem Dynamics Investigation Data Products

9.1.2.1 Introduction to Ice, Cloud, and Land Elevation Satellite Data Products

The ICESat GLAS data are processed and released by the National Snow and Ice Data Center (NSIDC). The data include 15 products at three levels: Level 1A, Level 1B, and Level 2 (Table 9.1). Level 1A (GLA01−04) is the raw data recorded by the sensor, Level 1B (GLA05−07) is the primary data product, and Level 2 (GLA08-15) is the application data product.

(1) Level 1A data products

Level 1A (GLA01−GLA04) is the raw data recorded by the ICESat GLAS. The global altimetry data (GLA01) include information about the footprint ID, time, location, and pulse energy. The GLA01 file also contains parameters (i_4nsBgMean and i_nsBgSDEV) that can be used to estimate waveform noise. The global atmospheric data (GLA02) include standardized relative backscattering data for the 532 nm and the 1064 nm channels and instrument correction data, such as laser energy correction for both channels, photon coincidence correction for the 532 nm channel, and gain correction for the probe of the 1064 nm channel. The global engineering data (GLA03) contain internal satellite data and can be used to calibrate GLA01 and GLA02 data values. The global laser positioning data (GLA04) contain the orbital information of the laser profiling array, position rate, and attitude packet.

(2) Level 1B primary data products

GLA05 data include waveform-based elevation correction parameters that can be used to characterize ground slope. The global elevation data (GLA06) provide surface elevation, surface roughness (assuming no slope), and other parameters for physical and atmospheric corrections. GLA05 and GLA06 also provide useful parameters for generating Level 2 products (e.g., GLA12−15). GLA07 is the global backscattering data, including a complete attenuation-corrected backscattering profile for

Table 9.1 Description of Ice, Cloud, and land Elevation Satellite Geoscience Laser Altimeter System (GLA01−GLA15) data.

Production level		Data description	Parameter	Main target users
L1A	GLA01	Global altimetry data	Raw waveform data	ISPIS, science teams, altimeters
	GLA02	Global atmospheric data	Sensor counting	ISPIS, science teams
	GLA03	Global engineering data	Total temperature	ISPIS, equipment teams
	GLA04	Global laser fixed point data	Attitude characteristics, orbital information	Science and equipment teams
L1B	GLA05	Global waveform altimetry correction data	Glacier/ice sheet elevation, glacier/ice sheet topography, ice sheet, surface elevation	ISPIS, science teams, altimeters
	GLA06	Global height data	Glacier/ice sheet elevation, glacier/ice sheet topography, ice roughness, ice sheet, laser reflectivity, sea level elevation	ISPIS, science teams, altimeters
	GLA07	Global backscattering data	Aerosol backscatter profile, scattering, surface elevation	ISPIS, LiDAR researchers, science teams
L2	GLA08	Global boundary and aerosol height data	Aerosol particle characteristics, boundary layer height	ISPIS, atmospheric scientists
	GLA09	Global multilayer cloud height data	The height of clouds, the vertical distribution of clouds	ISPIS, atmospheric scientists
	GLA10	Global aerosol vertical structure data	Aerosol backscattering, aerosol distribution, cloud reflectivity, transmittance	ISPIS, atmospheric scientists

Table 9.1 Description of Ice, Cloud, and land Elevation Satellite Geoscience Laser Altimeter System (GLA01–GLA15) data.—cont'd

Production level	Data description	Parameter	Main target users
GLA11	Global thin cloud/aerosol optical thickness data	Aerosol optical depth/thickness, cloud optical depth/thickness	ISPIS, atmospheric scientists
GLA12	Antarctica/Greenland ice altimetry data	Glacier/ice sheet elevation, glacier/ice sheet topography, ice sheet, laser reflectivity, reflectivity	Glaciologists
GLA13	Sea ice altimetry data	Ice roughness, laser reflectivity, sea ice elevation	Glaciologists
GLA14	Global land surface altimetry data	Laser reflectivity, surface elevation	Land scientists
GLA15	Ocean altimetry data	Laser reflectivity, sea level, sea level slope	Glaciologists and oceanographers

Note: ISPIS is the ICESat Science Investigator-led Processing System, the leading processing system for researchers using ICESat data.

the 532 and 1064 nm channels with a frequency of 5 Hz, calibration coefficients, and molecular backscattering profiles.

(3) Level 2 application data products

GLA08 is the height data of the planetary boundary and aerosol layers. The dataset contains the heights of the boundary layer, ground surface, and the top and bottom of aerosols ranging from 1.5 to 20.5 km (4 s sampling rate) and from 20.5 to 41 km (20 s sampling rate). GLA09 is the height data of the global cloud layer, including the heights of the tops and bottoms of multilayer clouds, and the sampling rate is 0.25 Hz, 1 Hz, 5 Hz, and 40 Hz. GLA10 contains global aerosol vertical structure data. It includes backscattering data that correct the attenuation of clouds and aerosols, as well as the extinction profiles of aerosols and clouds at 4 s and 1 s sampling rates, respectively. GLA11 contains thin clouds and aerosol optical thicknesses. The thin cloud layer is a layer that partially attenuates the LiDAR return signal, and its optical depth is usually less than 2. GLA12 provides surface height data for the Antarctica/Greenland ice sheets; GLA13 provides surface height data for sea ice; GLA14 provides surface height data for global land; and GLA15 provides ocean altimetry data. These four secondary altimetry datasets (GLA12, 13, 14, and 15) are calibrated by parameters from GLA05 and GLA06, including the spatial position of laser footprints, reflectance, geodetic parameters, equipment parameters, and atmospheric correction parameters needed for ranging.

9.1.2.2 Introduction to Global Ecosystem Dynamics Investigation Data Products

GEDI data are processed and released by the University of Maryland and NASA Goddard Space Flight Center. The data include primary products at five levels, including Levels 1 (L1A, L1B), 2 (L2A, L2B), 3 (L3), and 4 (L4A, L4B), and demonstrative products (Table 9.2). The Level 1 products are waveform data. The Levels 2 and 3 products are canopy structure attributes at footprint and grid levels, respectively. The Level 4 products are estimated aboveground biomass at the footprint and grid level. The demonstrative products are science products that illustrate how to use parts of GEDI data for related research, such as carbon stock monitoring, habitat mapping, and ecosystem modeling. The theories and procedures for these products can be found in the GEDI algorithm theoretical basis documents.

(1) Level 1 (L1A, L1B)

GEDI L1A data contain fundamental instrument engineering and housekeeping data, as well as raw waveform and geolocation information (Luthcke et al., 2020). The raw waveforms include transmitted and received waveforms with sampling artifacts and electronic and optical background noise. The GEDI L1B minimizes sampling artifacts by filtering the raw waveforms using a low pass filter whose shape is Gaussian and stores the corrected waveforms (Fig. 9.3) in the "txwaveform" and "rxwaveform" fields. The noise of the raw waveforms is determined according to the width of the Gaussian pulse and stored in the "noise_mean_corrected" field of the L1B product.

(2) Level 2 (L2A, L2B)

GEDI L2A data contain ground elevation and canopy height information interpreted from the waveform in L1B data. The interpretation is achieved by first decomposing the waveform into multiple Gaussian components after smoothing the corrected waveform. The lowest Gaussian component is then regarded as the corresponding ground, and the elevation of its mode is adopted as the ground elevation (field "elev_lowestmode" in GEDI L2A) (Luthcke et al., 2020). Based on the

Table 9.2 Introduction to Global Ecosystem Dynamics Investigation science data products.

Product level		Data products	Product leaders	Resolution
Level 1	L1A	Raw waveforms	Michelle Hofton Bryan Blair	25 m diameter
	L1B	Geolocated and corrected waveforms	Scott Luthcke; Tim Rebold; Taylor Thomas; Teresa Pennington	25 m diameter
Level 2	L2A	Ground elevation, canopy top height, relative height metrics	Michelle Hofton Bryan Blair	25 m diameter
	L2B	Canopy cover, canopy cover profile, plant area index (PAI), PAI profile, and foliage height diversity	Hao Tang; John Armston	25 m diameter
Level 3	L3	Gridded data of the same metrics as level 2	Scott Luthcke; Terence Sabaka; Sandra Preaux	1 km grid
Level 4	L4A	Footprint level aboveground biomass density	Jim Kellner; Laura Duncanson; John Armston	25 m diameter
	L4B	Gridded AGBD	Sean Healey; Paul Patterson	1 km grid
Demonstrative products		Prognostic ecosystem model outputs	George Hurtt	Grid size: variable
		Enhanced height/ biomass using fusion with TanDEM-X	Lola Fatoyinbo; Seung-Kuk Lee	Grid size: variable
		Enhanced height/ biomass and biomass change using fusion with landsat	Matt Hansen; Chenquan Huang	Grid size: variable
		Biodiversity/habitat model outputs	Scott Goetz; Patrick Jantz; Pat Burns	Grid size: variable

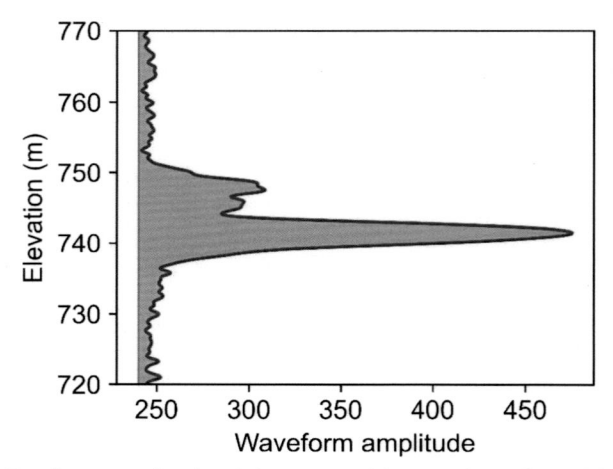

Figure 9.3 Waveform amplitude of the received laser pulse reflected by vegetation and ground in a nominal footprint.

ground elevation, GEDI L2A extracts 101 relative heights (RHs) from RH0 to RH100 with an interval of 1%. RH x represents the height of a canopy layer from which the accumulated energy of the waveform occupies $x\%$ of the total waveform energy. Thus, RH100 is the height of the highest canopy layer. GEDI L2A includes a quality flag to easily remove erroneous or lower-quality returns. The quality flag is stored in the "quality_flag" field.

GEDI L2B data contain canopy cover and its vertical profile, plant area index (PAI) and its vertical profile, as well as foliage height diversity (FHD). The canopy cover and PAI are two commonly used forest attributes in studying global environmental change and are derived from gap probability based on Beer's law and L1B waveform data (Chapter 8). The FHD measures the complexity of canopy elements and is calculated using the Shannon entropy of the PAI profile, which is important for biodiversity studies. The canopy cover, PAI, and FHD are stored in the "cover", "pai", and "fhd_normal" fields, respectively, of L2B. GEDI L2B also reports a quality flag.

(3) Level 3 (L3)

GEDI L3 data are the gridded products (in a 1 × 1 km grid) for the forest attributes in GEDI L2A and L2B data. Currently, L3 data contain the gridded mean canopy height, standard deviation of canopy height, mean ground elevation, standard deviation of ground elevation, and counts of laser footprints. The gridded mean and standard deviation of canopy height are calculated using RH100.

(4) Level 4 (L4A, L4B)

GEDI L4A contains the footprint level aboveground biomass density (AGBD; in Mg/ha). AGBD is derived using parametric models built on multiple RH metrics of L2A products. The parametric models are trained using simulated GEDI L2A RH metrics and field plot estimates of AGBD for multiple regions and plant functional types (i.e., deciduous broadleaf trees, evergreen broadleaf trees, evergreen needleleaf trees, deciduous needleleaf trees, and a combination of grasslands, shrubs, and woodlands). GEDI L4A also reports the associated uncertainty for estimated AGBD. GEDI L4B data are the gridded AGBD (in a 1 × 1 km grid) estimated from L4A using a hybrid method, which accounts for both model uncertainty and uncertainty related to the GEDI sample within a single grid cell (Patterson et al., 2019).

(5) Demonstrative products

GEDI provides demonstrative products that illustrate how to use parts of GEDI's global data for related research, such as carbon stock monitoring, habitat mapping, and ecosystem modeling. Currently, four demonstrative products are planned: prognostic ecosystem model outputs, enhanced height/biomass through fusion with TanDEM-X, biomass change through fusion with Landsat, and biodiversity and habitat model outputs. The prognostic ecosystem model produces estimates of contemporary carbon stocks, carbon fluxes, and future carbon sequestration potential by initializing the height-structured ecosystem demography model with GEDI height. Enhancing height/biomass aims to improve the resolution of GEDI gridded height and biomass products (in a 1 × 1 km grid) into a higher spatial resolution by integrating the GEDI with TanDEM-X, an interferometric synthetic aperture radar mission. Although GEDI provides precise biomass products, the short life of GEDI limits its ability to monitor biomass change. Thus, the GEDI team plans to monitor biomass change by fusing GEDI data with forest disturbance products from Landsat time series imagery. For biodiversity and habitat modeling, the GEDI team is developing habitat distribution models using GEDI vegetation structure measurements, Moderate Resolution Imaging Spectroradiometer-derived indicators of vegetation dynamics, species presence/absence observations, and databases of species traits (Dubayah et al., 2020).

9.1.3 Data Acquisition and Data Types

Users can download the data products and data reading and visualization tools for ICESat GLAS from the official website of the NSIDC, with the data products in a binary or hierarchical data file (HDF) format. The binary format file contains a standard binary file and a customized binary file— both are in big-endian byte order but with different source and naming rules. HDF is a data format developed by the National Center for Supercomputing Applications (NCSA) to meet diverse research needs for efficiently storing and distributing scientific data. Owing to its advantages of self-description, versatility, flexibility, scalability, and cross-platform applicability, HDF has been widely used since 1998. In May 2012, NCSA proposed a new generation of HDF format, the HDF5, which has been officially used for ICESat GLAS data since December 2012. Various data download tools for ICESat GLAS data, as well as their pros and cons, are listed in Table 9.3.

Table 9.3 Three common download tools for Geoscience Laser Altimeter System data.

Download tool	File format	Advantage	Disadvantage	Applicability
Data pool	Standard binary	Registration and application not required; ICESat GLAS data can be downloaded anytime and anywhere	Query and retrieval not available	Small range
Reverb	Standard binary	Provide query and retrieval	Need to register	Particular range, specific time
ICESat GLAS data subsetter	Special request file	Cutting and splicing not needed; specially targeted	Need to apply	Specific range, specific time, apply on demand

GEDI primary products are currently available for free download. Users can download lower-level data products (L1B, L2A, and L2B) in HDF5 format from the Land Processes Distributed Active Archive Center (LPDAAC) or EARTHDATA. The LPDAAC provides links for downloading the entire data (i.e., all fields) of each granule. In EARTHDATA, users can download selected fields rather than all fields. Considering that there are hundreds of fields in the L1B, L2A, and L2B products, the EARTHDATA provides a more efficient method to download GEDI data. For higher-level data products (L3, L4A, and L4B), users can directly download those products in GeoTIFF format from the Oak Ridge National Laboratory Distributed Active Archive Center.

9.2 Ice, Cloud, and Land Elevation Satellite and Global Ecosystem Dynamics Investigation Data Processing and Parameter Extraction

Both ICESat GLAS and GEDI are acquired as full waveforms (Fig. 9.4). This section introduces the appropriate methods for data processing and parameter extraction using the GLAS waveform data as an example.

9.2.1 Waveform Processing

Reading waveform data is the first step in ICESat GLAS data processing. The raw waveform data of ICESat GLAS are stored in the GLA01 products, but these products lack the location information of the laser footprint. GLA06 and GLA14 products provide geographic coordinate information and surface elevation information, respectively, for each laser footprint. GLA01, GLA06, and GLA14 products are thus joined by the unique ID of each laser beam for terrestrial ecosystem applications. NSIDC provides a GLAS data reading and visualization program based on Interactive Data Language, but it is difficult to read and process large data batches. We can effectively process data using other programming languages, such as Python and R. Raw waveform data contain errors caused by cloud cover, equipment dysfunction, and other factors. Thus, they need to be filtered before further processing. Three ideal conditions are expected after filtering: (1) data should be acquired in cloud-free conditions, (2) the signal of the laser beam is not saturated, and (3) the signal-to-noise ratio of the laser beam should be greater than a certain threshold. The saturation information of each laser beam is recorded in the GLA06 product and stored in the field "sat_corr_flg", with values lower than 2 indicating no saturation effect.

After reading and filtering data, it is necessary to convert the waveform intensity value of GLAS data (0—255) into a voltage value. This process is similar to radiometric calibration in remote sensing image processing. In theory, the converted voltage value can be used for comparative analysis among different waveforms. However, in practice, laser pulses are sensitive to environmental conditions, with errors caused by different laser energy values received in different acquisition periods. Therefore, more regulations are needed to standardize the voltage values before comparing different waveform data.

Furthermore, standard waveform data need to be smoothed to decrease background noise for determining the peak position and other waveform parameters. The received waveform should obey a Gaussian distribution in an ideal scenario where no energy is lost during the transmission since the laser pulse emitted is in a Gaussian distribution. However, energy loss is unavoidable in practice. For example, when a laser beam hits vegetation, the returned waveform data would have multiple modes that reflect the change in the vertical height of target objects (Fig. 9.4). Therefore, the returned waveform is often

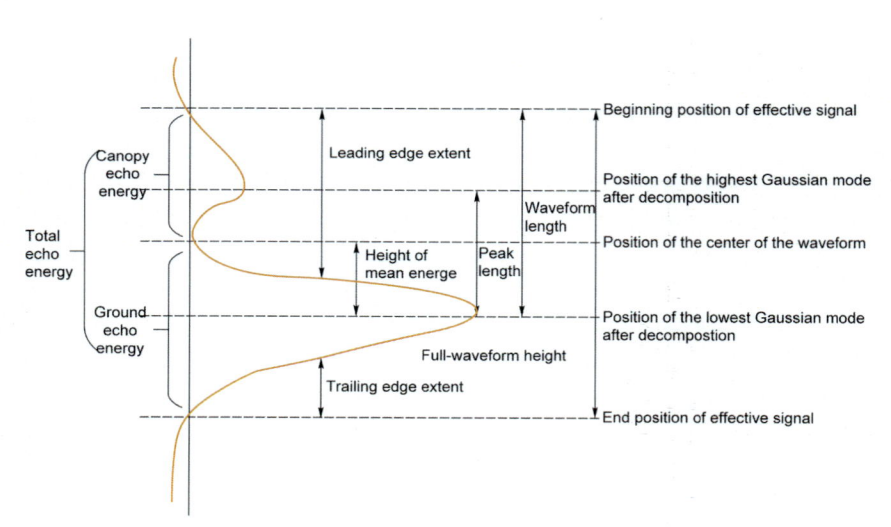

Figure 9.4 Key position parameters and waveform height and energy index of a waveform.

regarded as the composition of several different Gaussian components, which contain information regarding the atmospheric, land surface, and ground targets.

The Gaussian filter is often used to smooth full-waveform data. The Gaussian filter has two parameters: weight and width. The weight of filtering can be obtained from the waveform position relative to the height of a Gaussian waveform. The width of the Gaussian pulse, σ, can be obtained from the full width at half maximum (FWHM), using the following equation:

$$FWHM = 2\sqrt{2\ln 2} \times \sigma \tag{9.1}$$

After selecting an appropriate FWHM to filter the waveform smoothly, the position and amplitude width of the waveform after smoothing are estimated. First, an appropriate moving window size is selected to estimate the peak position by moving the window from the beginning to the end of the waveform. The peak position is located where the waveform value in the center of the window is larger than in the other four locations, and the values around the center are larger than those outside the window. Then, the amplitude width of the waveform can be calculated according to the peak position. Finally, the waveform data are decomposed into different Gaussian components, and the smoothed waveform curve can be obtained using the following fitting equations (Hofton et al., 2000):

$$w(t) = \sum_{m-1}^{N} w_m(t)$$

$$w_m(t) = A_m e^{-\frac{(t-t_m)^2}{2\delta_m^2}} \tag{9.2}$$

where $w(t)$ represents the value of a fitted waveform at time t; $w_m(t)$ is the contribution of the mth Gaussian component at time t; N represents the total number of Gaussian components in the fitted waveform; and A_m, t_m and δ_m represent the amplitude, mean, and standard variance of the mth Gaussian component. A_m, t_m, and δ_m are determined by the least squares method, which minimizes the difference between the fitting and actual waveforms.

9.2.2 Waveform Parameter Extraction

After smoothing and decomposing the waveform data, various parameters for retrieving canopy structure attributes can be extracted according to the fitted Gaussian components of the waveform. The vital parameters are several key positions, including pBeg, peak0, pCentroid, peakG, and pEnd (Fig. 9.4, the

Table 9.4 Explanation of key position parameters and height index of a waveform.

Symbol	Meaning
pBeg	Beginning position of the effective signal where the voltage value representing the waveform data first appears greater than the preset threshold value. The threshold is determined by the background noise plus a specific multiplier times the standard deviation
peak0	Position of the highest Gaussian peak after Gaussian decomposition
pCentroid	Position of the center of the waveform
peakG	Position of the lowest Gaussian peak (Gaussian ground peak) after Gaussian decomposition
pEnd	Position where the effective signal ends. It represents the position where the waveform data is greater than the preset threshold for the first time from back to front. It can also be understood as the waveform data located at the preset threshold for the last time.
l	$l = pBeg - peak0$, the length of waveform front, which can indicate the effect of vegetation canopy and topography on the first echo
HOME	$HOME = pCentroid - peakG$. The high half-wave energy is often used to indicate information about canopy closure, vertical structure, and biomass.
dPeak	$dPeak = peak0 - peakG$, peak length, indicating vegetation height
dEcho	$dEcho = pBeg - peakG$, waveform length, indicating canopy height
t	$t = peakG - pEnd$, the width of waveform trailing edge, which can indicate terrain and ground characteristics
wEcho	$wEcho = pBeg - pEnd$, the full waveform height, considered the range of valid signals

specific meaning of each parameter is shown in Table 9.4). For example, peakG is the location of the lowest Gaussian component that is often regarded as the location of the ground and can be used to determine ground elevation. pBeg is the beginning position of the effective signal, and the height difference between pBeg and peakG represents the canopy height. Additionally, other important waveform height indices can be extracted using these parameters and the energy information of the waveform, such as the leading edge extent, the height of mean energy, peak length, waveform distance, trailing edge extent (t), and full-waveform height.

The following Python code explains how to use the GLAS 01 and GLAS14 products to extract the length of the leading edge of the waveform, the trailing edge of the waveform, the width of the waveform, and the length of the waveform.

```python
# Extracting waveform parameters from ICESat GLAS data
# -*- coding: UTF-8 -*-
import glob, os, h5py
import numpy as np
#Set up GLAS01 和 GLAS14 file
HDF_list_GLA14="GLAH14_GLAS.H5"
HDF_list_GLA01="GLAH01_GLAS.H5"
#Read GLAS14 file
f_GLA14 = h5py.File(HDF_list_GLA14,'r')
#Read the laser emission record label in GLAS14 to generate a unique identification ID
recNdx=f_GLA14['Data_40HZ']['Time']['i_rec_ndx']
shotNum=f_GLA14['Data_40HZ']['Time']['i_shot_count']
GLA14_ID=recNdx[:]*100.0+shotNum
#Read the latitude and longitude information recorded in GLAS14
lat_dataset = f_GLA14['Data_40HZ']['Geolocation']['d_lat']
lon_dataset = f_GLA14['Data_40HZ']['Geolocation']['d_lon']
#Calculate RH100 in GLAS waveform parameters
d_SigBegOff = f_GLA14['Data_40HZ']['Elevation_Offsets']['d_SigBegOff']
d_gpCntRngOff = f_GLA14['Data_40HZ']['Elevation_Offsets']['d_gpCntRngOff']
RH100_all=d_gpCntRngOff[:,0]-d_SigBegOff
#Read GLAS01 file
GLA01 = h5py.File(HDF_list_GLA01,'r')
#Read the laser emission record label in GLAS01 and generate a unique ID
f_01_recNdx=GLA01['Data_40HZ']['Time']['i_rec_ndx']
f_01_shotNum=GLA01['Data_40HZ']['Time']['i_shot_count']
GLA01_ID=f_01_recNdx[:]*100.0+f_01_shotNum
#Match the numbers in GLAS14 and GLAS01 and generate mask data
mask_14 = np.in1d(GLA14_ID, GLA01_ID)
if np.sum(mask_14)==0:
#     print "no match..."
continue
mask_01 = np.in1d(GLA01_ID, GLA14_ID)
#Match the numbers in GLAS14 and GLAS01 and generate mask data
f_01_rec_wf=GLA01['Data_40HZ']['Waveform']['RecWaveform']['r_rng_wf']
```

```
f_01_rec_wf=np.asarray(f_01_rec_wf)
#Read the waveform type, the start position of the waveform signal, and the end position of the waveform
signal in GLAS01
waveformType = GLA01['Data_40HZ']['Waveform']['Characteristics']['i_waveformType']
#Read the end position of the waveform signal in GLAS01
trailingEdge = GLA01['Data_40HZ']['Waveform']['Characteristics']['i_LastThrXingT']
#Read the start position of the waveform signal in GLAS01
leadingEdge = GLA01['Data_40HZ']['Waveform']['Characteristics']['i_NextThrXing']
#Calculate the width of the waveform
waveformExtent = (trailingEdge[:]-leadingEdge[:])*299792458.0/1000000000.0
#Read the waveform data in GLAS01
waveformRec = GLA01['Data_40HZ']['Waveform']['RecWaveform']['r_rng_wf']
#Read the recording position, energy, and time of each laser in GLAS01
recWfLocationIndex = GLA01['Data_40HZ']['Waveform']['RecWaveform']['i_rec_wf_location_index']
recWfLocationTable = GLA01['ANCILLARY_DATA'].attrs['rec_wf_sample_location_table']
RespEndTime = GLA01['Data_40HZ']['Waveform']['RecWaveform']['i_RespEndTime']
#According to the waveform energy to determine the first and last half of the highest energy position
countShot = 0
firstHalfAmp = np.empty_like(waveformType, dtype=np.float32)
lastHalfAmp = np.empty_like(waveformType, dtype=np.float32)
# waveformType[:]==0
firstHalfAmp[:] = -9999
lastHalfAmp[:] = -9999
for waveformRow in waveformRec:
    waveformRow=waveformRec[1000,]
    countShot=1000
    if waveformType[countShot] == 0:
        firstHalfAmp[countShot] = -9999
        lastHalfAmp[countShot] = -9999
        continue
    elif waveformType[countShot] == 1:
        index = np.where(waveformRow[:] >= (np.max(waveformRow[:])/2))
        if np.size(index[0]) > 0:
            firstHalfAmp[countShot] = RespEndTime[countShot] +
```

```python
recWfLocationTable[recWfLocationIndex[countShot]-1, index[0][-1]]
            lastHalfAmp[countShot] = RespEndTime[countShot] +
recWfLocationTable[recWfLocationIndex[countShot]-1,index[0][0]]
        else:
                firstHalfAmp[countShot] = -9999
                lastHalfAmp[countShot] = -9999
    else:
            index = np.where(waveformRow[0:200] >= (np.max(waveformRow[0:200])/2))
            if np.size(index[0]) > 0:
            firstHalfAmp[countShot] = RespEndTime[countShot] +
recWfLocationTable[recWfLocationIndex[countShot]-1, index[0][-1]]
            lastHalfAmp[countShot] = RespEndTime[countShot] +
recWfLocationTable[recWfLocationIndex[countShot]-1. index[0][0]]
        else:
                firstHalfAmp[countShot] = -9999
                lastHalfAmp[countShot] = -9999
countShot += 1
#Calculate the leading edge length of the waveform
leadingEdgeExtent = (firstHalfAmp[:] - leadingEdge[:])*299792458.0/1000000000.0
#Calculate the length of the trailing edge of the waveform
trailingEdgeExtent = (trailingEdge[:] - lastHalfAmp[:])*299792458.0/1000000000.0
#Remove abnormal data
index_mask = np.where(firstHalfAmp == -9999)
leadingEdgeExtent[index_mask] = -9999
trailingEdgeExtent[index_mask] = -9999
#Use mask data to filter the results to remove mismatched points
we = waveformExtent[mask_01]/2.0
lee =leadingEdgeExtent[mask_01]/2.0
tee =trailingEdgeExtent[mask_01]/2.0
#Close the opened GLAS file
GLA01.close
f_GLA14.close
```

Besides canopy height, quantile heights can also be extracted from the waveform using its energy occupation. A waveform quantile height refers to a location from which, to the end position of the effective signal (pEnd), the accumulated energy occupies a certain percentage of the corresponding energy of the whole waveform. Waveform quantile heights of 0%, 25%, 50%, 75%, and 100% are expressed as RH0, RH25, RH50, RH75, and RH100, respectively. Different waveform quantiles have different ecological meanings. For example, RH50 is often closely related to forest biomass and thus can be used for aboveground biomass (AGB) estimation.

Total waveform energy, E_{total}, is a main waveform energy parameter, which represents the area enclosed by a waveform curve and the vertical line from pBeg to pEnd in Fig. 9.5. E_{total} consists of two parts: canopy echo energy (E_{canopy}), which represents the cumulative value of canopy part in a waveform, and ground echo energy (E_{ground}). Based on the three waveform energy parameters, we can define two parameters (Equation (9.3)), the canopy return energy ratio (R_{canopy}) and ground return energy ratio (R_{ground}). These two parameters are closely related to canopy cover, with smaller R_{ground} values and larger R_{canopy} values indicating higher canopy cover within a laser footprint.

$$R_{canopy} = \frac{E_{canopy}}{E_{total}}$$

$$R_{ground} = \frac{E_{ground}}{E_{total}}$$

(9.3)

9.3 Ice, Cloud, and Land Elevation Satellite 2 Spaceborne LiDAR Data

9.3.1 Introduction to Ice, Cloud, and Land Elevation Satellite 2

ICESat-2 is the successor of ICESat and was launched by NASA on September 15, 2018. Its primary mission is to quantify ice sheet contributions to sea level change, assess change in ice sheet signatures, particularly at outlet glaciers, and estimate sea ice thickness (Neuenschwander & Pitts, 2019). Although ICESat-2 mainly aims to monitor changes in the cryosphere, it also measures the Earth's terrestrial surfaces globally, which enables its application in terrestrial ecosystem studies.

Unlike the ICESat GLAS and GEDI, the Advanced Topographic Laser Altimeter System (ATLAS) aboard ICESat-2 employs a photon counting technology. It can detect the reflected laser pulse at the single-photon level and record multiple range measurements for each laser footprint. The ICESat-2 ATLAS has three pairs of laser emitters, and each pair emits a strong laser beam and a weak laser beam, both at 532 nm. Thus, the ICESat-2 ATLAS has six ground tracks for each orbit. Along each ground track, the ICESat-2 ATLAS continuously measures the Earth's surface using footprints with a diameter of 17 m at an interval of 70 cm. For each

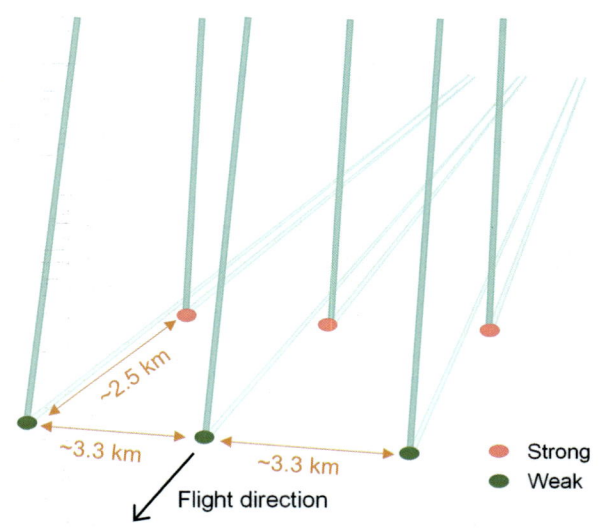

Figure 9.5 Illustration of laser beams and ground tracks of Advanced Topographic Laser Altimeter System aboard Ice, Cloud, and land Elevation Satellite 2. *(Revised from Neuenschwander, A., & Pitts, K. (2019). The ATL08 land and vegetation product for the ICESat-2 Mission. Remote Sensing of Environment, 221, 247–259.)*

footprint, an average of 10 signal photons are expected to be detected and recorded by the ATLAS instrument over highly reflective surfaces such as snow or ice sheets under clear skies, and between 0 and 44 photons would be detected over vegetated land for the strong beam. Ground tracks of ICESat-2 ATLAS within each beam pair are approximately 90 m apart, and ground tracks among neighbor pairs are separated by 3.3 km (Fig. 9.5).

9.3.2 Introduction to Ice, Cloud, and Land Elevation Satellite 2 Data Products

ICESat-2 ATLAS contains 21 products on four levels, including Levels 1, 2, 3A, and 3B in the fifth released version (Table 9.5). Level 1 products (i.e., ATL01 and ATL02) are the reformatted and converted telemetry data, respectively. Level 2 products include ATL03, which reports the latitude, longitude, and elevation of each photon, and ATL04 provides along-track normalized relative backscatter profiles. Level 3A and 3B products comprise surface-specific data sets, among which the ATL08 and ATL18 products are related to forest studies. ATL08 contains terrain height, canopy height, and canopy cover data, and ATL18 provides gridded products for these metrics. However, ATL18 products had not been published when this chapter was written in June 2022.

Table 9.5 Summary of Ice, Cloud, and land Elevation Satellite 2 (ICESat-2) standard data products.

Product level		Product name	Description
Level 1	ATL01	Reformatted telemetry	Raw ICESat-2 ATLAS data are decompressed, ordered in time, and reformatted to HDF5.
	ATL02	Science unit converted Telemetry	Corrected ATL01, which may be achieved by removing the biases in timing and pointing measurements and correcting the influences of temperature and voltage variations on ATLAS electronics
Level 2	ATL03	Global geolocated photon data	Precise latitude, longitude, and elevation for every received photon arranged by beam in the along-track direction. Photons are classified by signal versus background, as well as by surface type (land ice, sea ice, land, ocean), including all geophysical corrections (e.g., Earth tides and atmospheric delay). They are segmented into several minute granules.
	ATL04	Normalized backscatter profiles	Along-track, normalized relative backscatter profiles used to identify atmospheric layers and their boundaries, including clouds, aerosols, and blowing snow
Level 3A	ATL06	Land ice elevation	Geolocated estimates of land ice surface heights and ancillary parameters used to interpret estimates and assess their quality
	ATL07	Sea ice height	Along-track surface height and type (e.g., snow-covered ice, open water) for the ice-covered seas of the northern and southern hemispheres
	ATL08	Land and vegetation height	Terrain height, canopy height, and canopy cover at 100 m, fixed-length steps along the ground track.
	ATL09	Atmospheric layer characteristics	Along-track cloud and other significant atmosphere layer heights, blowing snow, integrated backscatter, and optical depth
	ATL10	Sea ice freeboard	Estimates of along-track, total freeboard for the ice-covered seas of the northern and southern hemispheres along segments generated in the ATL07 product

Table 9.5 Summary of Ice, Cloud, and land Elevation Satellite 2 (ICESat-2) standard data products.—cont'd

Product level		Product name	Description
	ATL12	Ocean surface height	Estimates of sea surface height at a given point on the open ocean surface at a given time, plus parameters needed to assess the quality of the height estimates and interpret and aggregate the data over greater distances
	ATL13	Inland water height	Along-track water surface height, slope, and roughness for each ATLAS beam, plus aspect and maximum slope (or gradient) of the planar surface between adjacent strong beams
Level 3B	ATL11	Annual land ice height	Time series of surface heights
	ATL14	Gridded annual ice sheet height	DEMs for Antarctica and Greenland posted at 125 m
	ATL15	Ice sheet height change	Maps of height change for different time windows
	ATL16	Gridded weekly atmosphere	Polar cloud fraction, blowing snow frequency, ground detection frequency
	ATL17	Gridded monthly atmosphere	Polar cloud fraction, blowing snow frequency, ground detection frequency
	ATL18	Gridded land and vegetation height	Gridded ground surface height, canopy height, and canopy cover estimates
	ATL19	Gridded monthly sea surface height: Open ocean	Gridded monthly estimates of sea surface height using all ICESat-2 tracks from the beginning to the end of each month
	ATL20	Gridded monthly sea ice freeboard	Sea ice freeboard from all ICESat-2 tracks between the beginning and the end of each month
	ATL21	Sea surface height anomaly: Ice-covered ocean	Daily and monthly averages of sea surface height anomaly from all ICESat-2 tracks between the beginning and the end of each month
	ATL22	Mean inland surface water data	Mean inland surface water data derived from continuous along-track inland surface water data (ATL13)

These ICESat-2 products are available at NSIDC. Like ICESat and GEDI, ICESat-2 provides HDF5 format files, which support flexible and efficient reading and writing of complex data types at high volumes. In addition, users can read, process, analyze, and export ICESat-2 ATL03 and

ATL08 data as figures and as .las, .csv, and .kml files by Photon Research and Engineering Analysis Library.

9.3.3 Ice, Cloud, and Land Elevation Satellite 2 Data Processing and Feature Extraction

As mentioned above, the ICESat-2 ATLAS employs a photon–counting LiDAR system that transmits low–power laser pulses and uses detectors with high sensitivity to return echoes at the single–photon level. The XYZ location of each detected photon is determined by combining satellite attitude and posture information and is stored in the ATL03 product (as shown in Fig. 9.6). Because the detectors used in ICESat ATLAS are highly sensitive, the raw data contain a lot of solar background noise. The green laser (532 nm) used in ICESat-2 ATLAS also increases the magnitude of noise over vegetation regions. Thus, fetching useful signals from raw ICESat-2 data is one of the essential data processing steps before applying ICESat-2 data in forest ecosystem studies. In the production of the ICESat-2 ATL08 data, the solar background noise is removed using the differential, regressive, and Gaussian adaptive nearest neighbor method (DRAGANN) to obtain more precise land and vegetation height. DRAGANN assumes signal photons will be closer in space to each other than to random noise photons (Neuenschwander & Pitts, 2019). Thus, for every photon, if the number of photons within its neighbor is greater than a threshold, it is regarded as a signal photon; otherwise, it is noise. An example of identifying signal photons using DRAGANN is shown in Fig. 9.6.

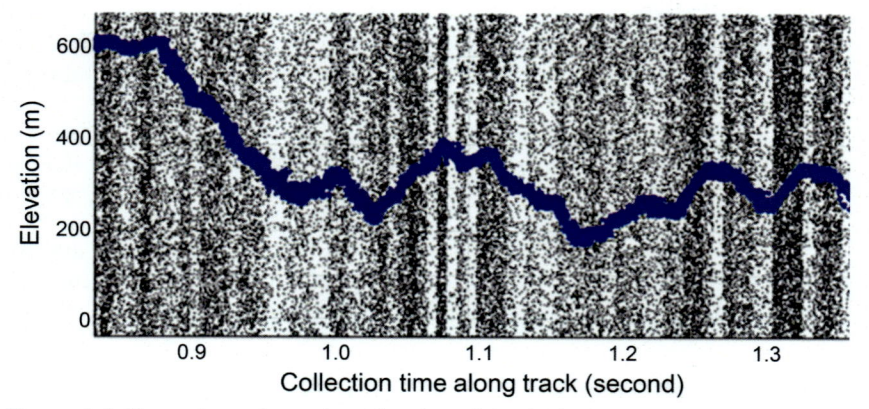

Figure 9.6 Illustration of raw Ice, Cloud, and land Elevation Satellite 2 Advanced Topographic Laser Altimeter System data and identified signal photons (blue points) and noise (black points) using the differential, regressive, and Gaussian adaptive nearest neighbor method. *(Revised from Neuenschwander, A., & Pitts, K. (2019). The ATL08 land and vegetation product for the ICESat-2 Mission. Remote Sensing of Environment, 221, 247–259.)*

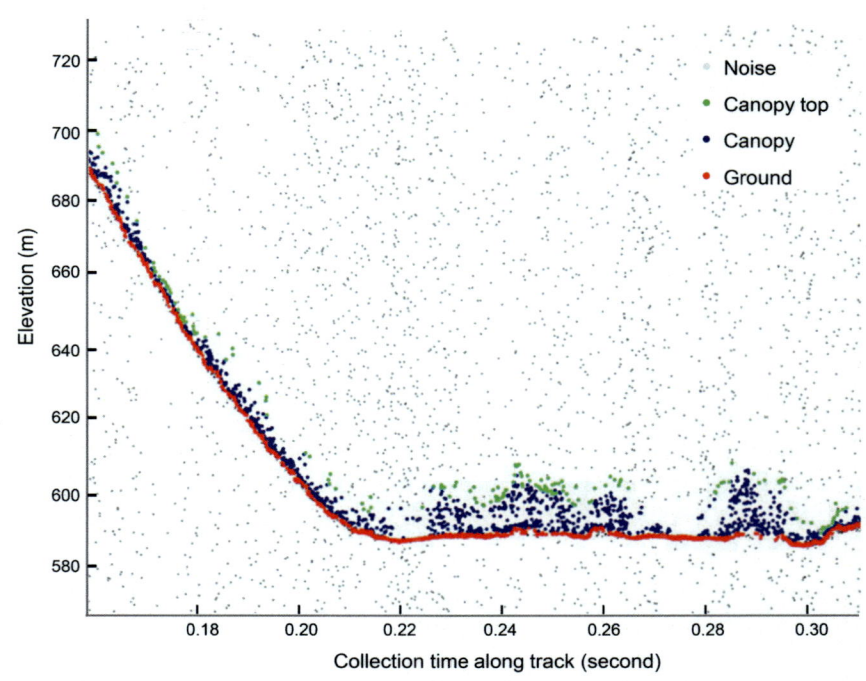

Figure 9.7 An example of the classified photons of the Ice, Cloud, and land Elevation Satellite 2 Advanced Topographic Laser Altimeter System photons. *(From Neuensch-wander, A., & Pitts, K. (2019). The ATL08 land and vegetation product for the ICESat-2 Mission. Remote Sensing of Environment, 221, 247–259.)*

After identifying signal photons, signal photons can be classified into the ground, canopy, and canopy top photons (Fig. 9.7), and canopy height metrics within a segment can then be calculated by analyzing the height differences between canopy photons and ground photons. ATL08 used a fixed segment with a step size of 100 m to provide continuous data and improve the ease of use of the final products for the user community. For each segment, there are approximately 140 signal photons on average. If the 100 m segment has fewer than 50 labeled signal photons, no calculations will be made, and invalid values will be reported. Otherwise, the RH of every canopy photon is calculated using the height difference between its location and the interpolated ground line beneath that photon. RH25, RH50, RH60, RH70, RH75, RH80, RH85, RH90, RH95, and RH98 are then derived based on all photons within the segment. Additionally, canopy closure can be defined as the proportion of canopy photons to the total number of signal (ground + canopy) photons within each 100 m segment.

9.3.4 Comparison of Ice, Cloud, and Land Elevation Satellite 2 and Global Ecosystem Dynamics Investigation

The two recent spaceborne LiDAR missions, ICESat-2 and GEDI, were launched in 2018 by NASA to observe the elevation of the Earth's surface (Table 9.6). The primary mission of ICESat-2 is to measure the height of the Earth's changing surface (such as the elevation of ice sheets, glaciers, sea ice, and forests), whereas GEDI's mission is to obtain the structural and functional attributes of terrestrial ecosystems. Therefore, ICESat-2 provides near-global spatial coverage, whereas GEDI only collects data between 51.6° N and 51.6° S, where most forests are located. Nonetheless, both ICESat-2 and GEDI provide datasets for measuring ground and canopy height and facilitating forest biomass mapping.

In addition, GEDI emits near-infrared laser beams (1064 nm), which are more sensitive to vegetation than the green laser beams (532 nm) used by ICESat-2 ATLAS. GEDI's footprint diameter is approximately 25 m, which is larger than that of ICESat-2 at 17 m. GEDI has eight ground tracks from three lasers, which include four power tracks and four coverage tracks, whereas ICESat-2 contains three pairs of lasers that provide six

Table 9.6 Comparison of Ice, Cloud, and land Elevation Satellite 2 (ICESat-2) and Global Ecosystem Dynamics Investigation (GEDI) in mission specification and sensor configurations.

	ICEsat-2	GEDI
Launch date	September 15, 2018	December 5, 2018
Wavelength	532 nm (green)	1064 nm (near-infrared)
Geographic coverage	Global	51.6° N to 51.6° S
Measurement strategy	Six tracks (three strong beams, three weak beams)	Eight tracks from three lasers (four power tracks, four coverage tracks)
Native measurement resolution	~17 m footprints returning 0—4 photons	~25 m footprints returning full waveforms
Biomass product resolution(s)	NA	25 m footprint estimates; 1 km gridded estimates
Footprint size	17 m circular footprint	25 m circular footprint
Positioning system	GNSS	GNSS, IMU, and three-star tracker information

GNSS, global navigation satellite system; *IMU*, inertial measurement unit.

ground tracks from three weak beams and three strong beams with a 4:1 energy ratio. AGB estimated from strong beams is reported to be more accurate than that from the weak ones, with a higher coefficient of determination (R^2 increased by 0.23) and lower error (root-mean-square error (RMSE) decreased by 30.52%) (Narine et al., 2020).

The full-waveform large-footprint systems used by ICESat GLAS produced a large diameter footprint (~ 70 m), which limited its estimation of terrain and canopy height owing to the potential terrain relief within each footprint. Particularly in areas of high relief, vegetation signals convolve with terrain signals; thus, terrain surfaces cannot be easily disentangled for surface interpretation (Bolton et al., 2013; Nelson et al., 2009). Fortunately, ICESat-2 ATLAS has a much smaller footprint, and thus, both terrain and canopy surface estimation are feasible even in areas with high relief (Kwok et al., 2020). Although GEDI's footprints are larger than that of ICESat-2, GEDI can record full-waveform information rather than a few photons, and the footprint density of GEDI is denser than that of ICESat-2, which leads to the fact that GEDI can directly provide a gridded biomass product at 1-km resolution (Table 9.2).

9.4 Upscaling Forest Attribute Estimations Using Spaceborne LiDAR Data

9.4.1 Upscaling Forest Height Estimates

Spatially continuous estimates of forests height at national to global scales are critical for quantifying forest carbon storage, understanding forest ecosystem processes, and developing forest management and restoration policies to mitigate global climate change (Asner, 2009; Berigan et al., 2012; Li et al., 2015). Although near-surface (e.g., backpack, unmanned aerial vehicle (UAV)) or aerial LiDAR systems can accurately measure forest height from individual tree to forest stand levels, it is impractical to use these LiDAR systems to collect large-scale forest canopy height observations considering their limited spatial coverage (Hudak et al., 2002; Li et al., 2020). Spaceborne LiDAR provides an alternative means for collecting global forest canopy height observations. However, these spaceborne LiDAR systems share a common problem of having discrete ground sampling footprints. For example, footprints of ICESat GLAS are separated by approximately 170 m along a single track and tens of kilometers across multiple tracks (Schutz et al., 2005). Although the footprint density of the recently launched GEDI and ICESat-2 ATLAS is much higher than that of

ICESat GLAS, they still cannot directly provide wall-to-wall forest height observations.

To overcome the spatial discontinuity problem of spaceborne LiDAR data, researchers have proposed two major upscaling methods, including regression-based and interpolation-based methods. The regression-based methods tried to fuse spaceborne LiDAR data with passive optical images to map wall-to-wall forest canopy height. In the regression-based methods, forest height measurements at spaceborne LiDAR footprints are treated as ground truth to build a regression model that uses optical images and environmental variables as predictors (Lefsky, 2010; Simard et al., 2011; Su et al., 2017). Taking one of our studies on Sierra Nevada forests in California (Su et al., 2017) as an example, the regression-based method using ICESat GLAS data will be introduced in detail. The method proposed by Su et al. (2017) has three steps: (1) estimating forest height within the airborne LiDAR footprint using the stepwise regression method; (2) building a regression model to estimate forest heights at ICESat GLAS footprints; and (3) extrapolating the forest height estimations from ICESat GLAS footprints to the whole Sierra Nevada forests using a random forest (RF) regression method (Fig. 9.8). In the first two steps, airborne LiDAR data was used as a bridge to build a relationship between plot-level in situ measurements and ICESat GLAS data because the spatial distributions of field plots and GLAS footprints did not perfectly match. Then a model for estimating forest height from ICESat GLAS data is derived as

$$LH_{GLAS} = -0.163WE + 0.404LE - 10.82\tan(slope) + 14.808 \qquad (9.4)$$

where LH_{GLAS} represents the forest height within the GLAS laser footprint, WE represents the waveform length, LE represents the waveform leading edge length, and *slope* represents the terrain slope within the footprint.

Finally, the continuous forest height product of Sierra Nevada is generated by using the RF regression model built on these forest heights from ICESat GLAS footprints and optical images, and environmental variables. The results are shown in Fig. 9.9. The average height of the forest in the whole Sierra Nevada area was 14.86 m, and the standard deviation was 11.11 m. To evaluate its accuracy, the final forest height product was compared with manually measured data, airborne LiDAR forest height data, and GLAS tree height data which were not used in the model-building process. The evaluation showed that the height product correlates well with measured data in the Sierra Nevada mountains ($R^2 = 0.60$, RMSE = 5.45 m, Fig. 9.10a). However, there was substantial overestimation for trees much shorter than

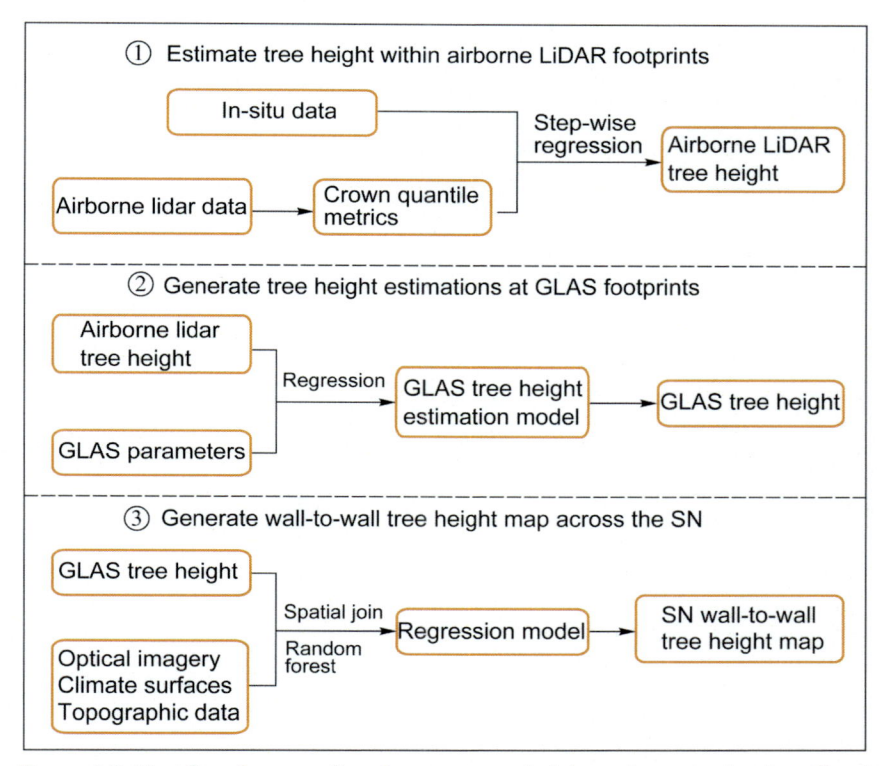

Figure 9.8 Workflow for upscaling forest canopy height estimates using Ice, Cloud, and land Elevation Satellite Geoscience Laser Altimeter System (GLAS) data across the Sierra Nevada (SN) mountains, California, USA.

33 m and underestimation for much taller ones. Additionally, differences between the estimated values and the forest heights derived from airborne LiDAR data obey a normal distribution, with an average of − 5.17 m and a standard deviation of 10.24 m (Fig. 9.10b). Most (72%) of the tree height estimates from the GLAS data were lower than those from airborne LiDAR data.

In Su et al. (2017), the regression-based method for mapping large-scale forest height was only evaluated in the Sierra Nevada mountains. Its transferability to different locations, vegetation types, and spatial scales is still unclear. Therefore, Jin et al. (2018) selected 16 study sites (100 km^2 each) with full airborne LiDAR coverage across the United States and used the LiDAR-derived canopy height along with optical imagery, topographic data, and climate surfaces to evaluate the transferability of the RF regression

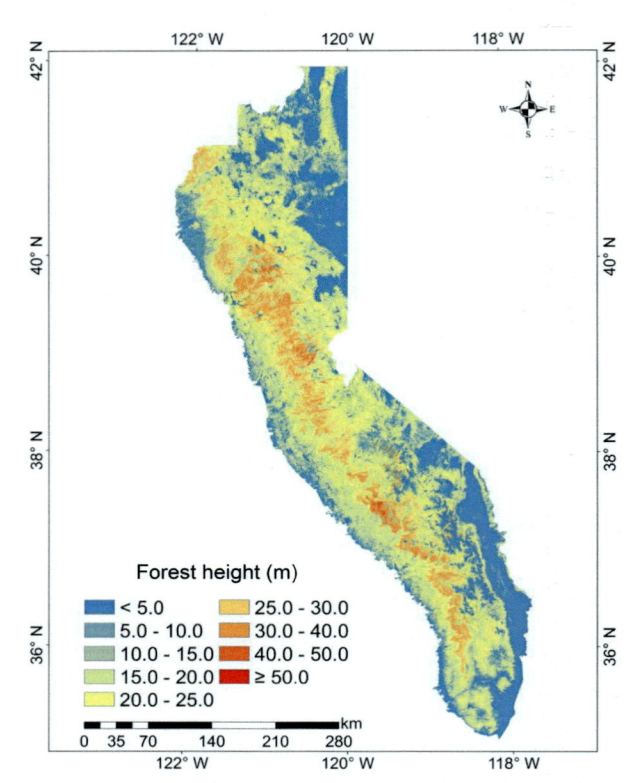

Figure 9.9 Estimation of forest height distribution in the Sierra Nevada mountains, California, USA, using Ice, Cloud, and land Elevation Satellite Geoscience Laser Altimeter System data.

method for mapping large-scale forest height (Fig. 9.11). The results show that the RF regression method trained at a certain location or vegetation type cannot be transferred to other locations or vegetation types (Fig. 9.12). However, by training the RF model using samples from all sites with various vegetation types, a more robust model can be achieved for predicting canopy height at different locations and vegetation types with R^2 higher than 0.6 and RMSE lower than 6 m. Moreover, the influence of spatial scales on the RF prediction accuracy is noticeable when the spatial extent of the study site is less than 50 km^2 or the spatial resolution of the training pixel is finer than 500 m. The canopy height prediction accuracy increases with the spatial extent and the targeted spatial resolution (Figs. 9.13 and 9.14).

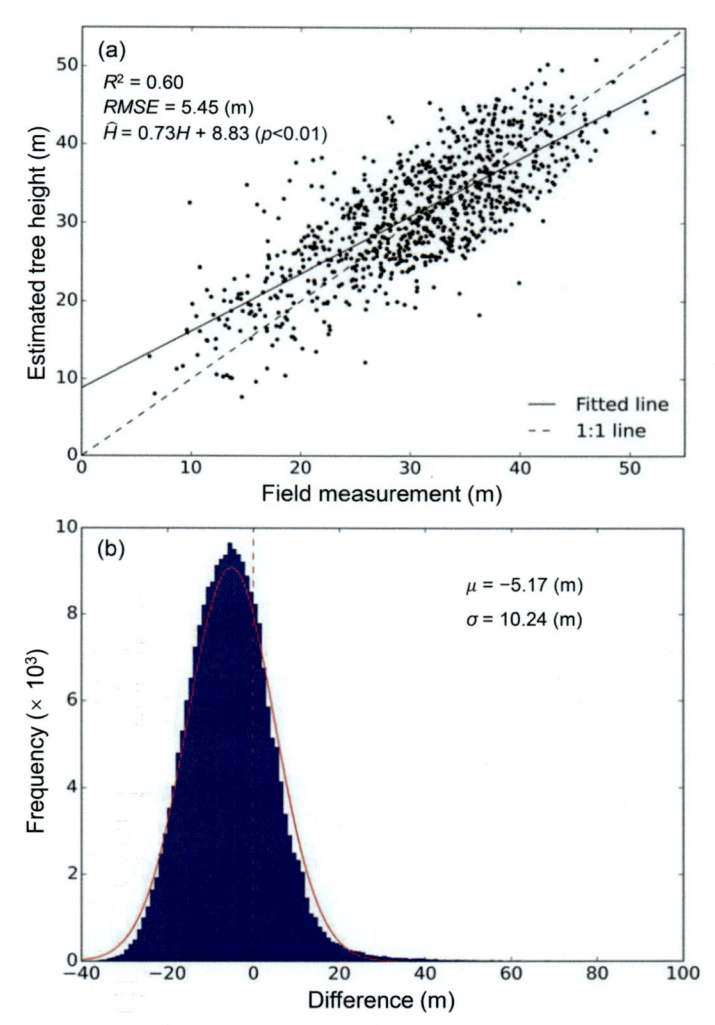

Figure 9.10 Precision analysis of the forest height products of Sierra Nevada mountains, California, USA: (a) Scatter plot of estimated forest height $\hat{H}$ and measured values H and (b) histogram of the difference between the predicted forest height and the height derived from airborne LiDAR data. R^2 and RMSE are the coefficient of determination and root-mean-square error, respectively. μ and σ represent the mean and standard deviation of the differences between the predicted forest height and the height derived from airborne LiDAR data.

As mentioned above, regression-based methods underestimate the forest height in tall forests, mainly because the optical image, a vital dataset used in regression-based methods, is saturated in dense forest regions (McCombs

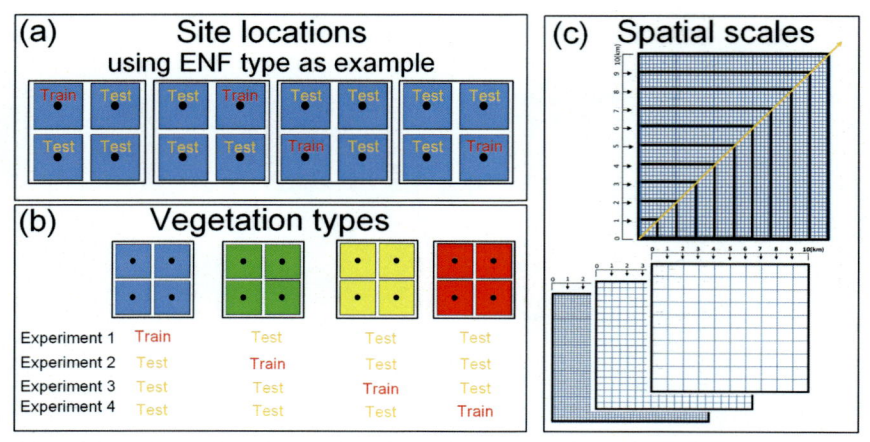

Figure 9.11 Illustration of random forest (RF) regression model transferability evaluation for various geolocations, vegetation types, and spatial scales. (a) RF model transferability evaluation among different study sites within the same vegetation type; (b) RF model transferability evaluation for different vegetation types; and (c) RF model transferability evaluation for different spatial extents and spatial resolutions. *ENF*, evergreen needleleaf forest.

et al., 2003; Su et al., 2017). Additionally, the methods developed for ICESat GLAS data are primarily regression-based and unable to take advantage of the denser footprints of ICESat-2 ATLAS and GEDI data in mapping wall-to-wall forest height.

A spatial interpolation-based method was developed to take full advantage of dense spaceborne LiDAR observations of ICESat-2 ATLAS and GEDI data in mapping large-scale forest heights (Liu et al., 2022). The interpolation-based method uses observations from known locations to predict values at unknown locations. Two major difficulties should be addressed when using spatial interpolation to map large-scale forest heights. First, spaceborne LiDAR footprints are generally evenly distributed along ground tracks instead of randomly distributed, which is a fundamental requirement of spatial interpolation. As a result, forest height interpolated from a single spaceborne LiDAR platform may display a strong strip effect across ground tracks. Fusing spaceborne LiDAR data from different platforms (e.g., GEDI and ICESat-2 ATLAS) can increase the randomness of footprint distribution, alleviating the strip effect. Second, spatial interpolation is based on Tobler's first law of geography that assuming "everything is related to everything else, but near things are more related than distant things" (Tobler, 1970). However, this might be violated for forest height interpolation because forest height can also be influenced by local environmental factors (e.g., slope, aspect, water availability, solar radiation) (Simard et al., 2011; Tao et al., 2016). Thus, Liu et al. (2022) hypothesized

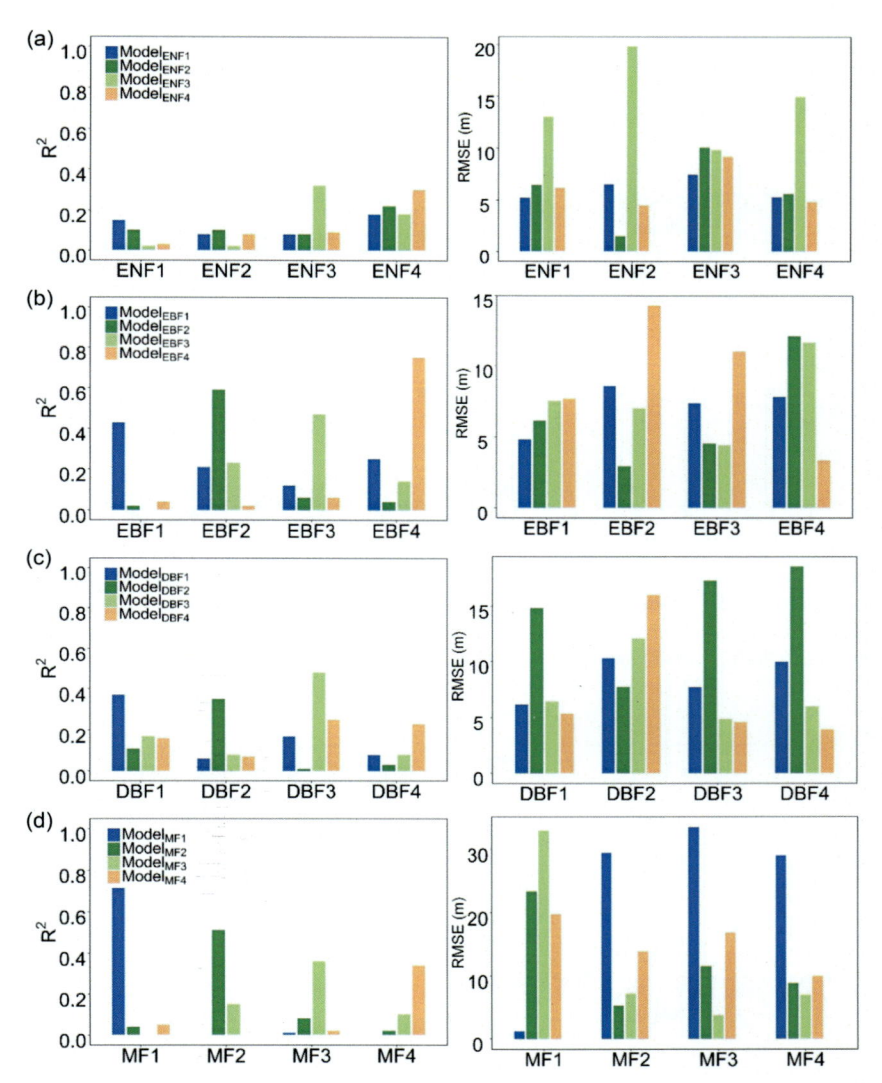

Figure 9.12 Random forest model transferability among different study sites of each vegetation type. The *x*-axis represents different study sites of a certain vegetation type. The *y*-axis is R^2 or RMSE calculated by directly comparing the prediction results and the independent LiDAR-derived validation pixels within the corresponding study site. (a–d) represent the evaluation results among study sites with different vegetation types, including evergreen needleleaf forest (ENF), evergreen broadleaf forest (EBF), deciduous broadleaf forest (DBF), and mixed forest (MF).

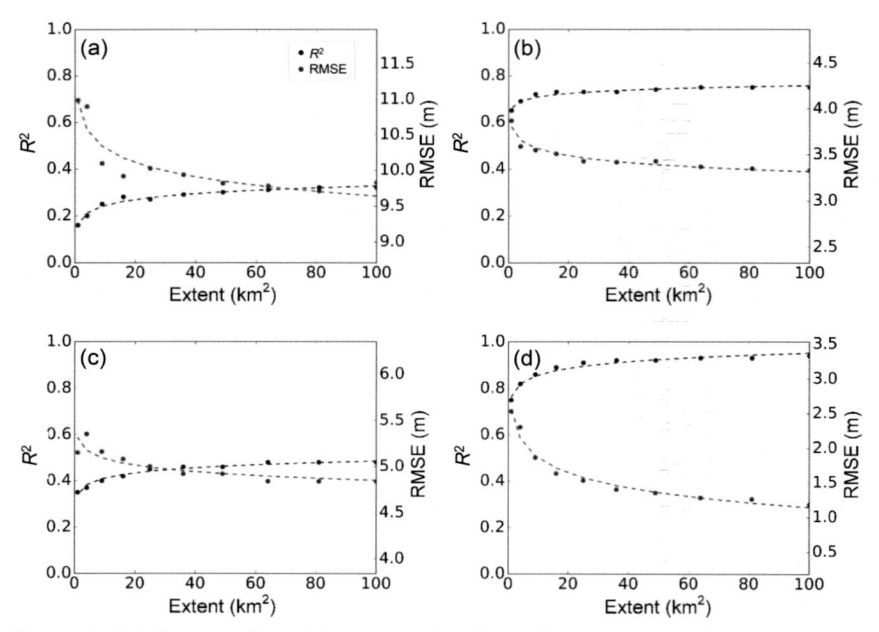

Figure 9.13 Influence of spatial extent on height prediction accuracy from the random forest regression model. (a–d) represent the evaluation results at sites evergreen needleleaf forest 3 (ENF3), evergreen broadleaf forest 4 (EBF4), deciduous broadleaf forest 3 (DBF3), and mixed forest 1 (MF1), respectively.

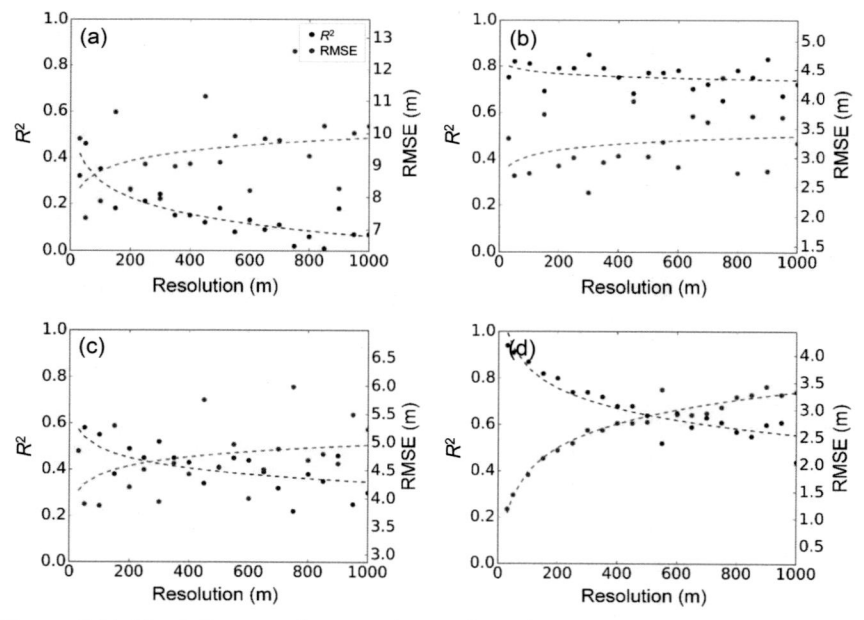

Figure 9.14 The influence of targeted spatial resolution on tree height prediction accuracy. (a–d) represent the evaluation results at sites evergreen needleleaf forest 3 (ENF3), evergreen broadleaf forest 4 (EBF4), deciduous broadleaf forest 3 (DBF3), and mixed forest 1 (MF1), respectively.

that forest height differences between two locations could be partly reflected by differences in environmental factors and spectral values observed from satellite remote sensing images, and feature distance should be used in forest height interpolation from spaceborne LiDAR data instead of spatial distance. Nevertheless, adjusting interpolation weights of different features to predict reasonable forest canopy heights remains a challenge.

To overcome these challenges in spatial interpolation for mapping forest height, Liu et al. (2022) proposed a neural network–guided interpolation (NNGI) method (Fig. 9.15). The NNGI method was inspired by a Kriging interpolation algorithm introduced by (Chiles & Delfiner, 2009):

$$Z(l_0) = \sum_{l_i \in \mathcal{N}(l_0)} \lambda_i(l_0) Z(l_i) + \lambda_0(l_0) \tag{9.5}$$

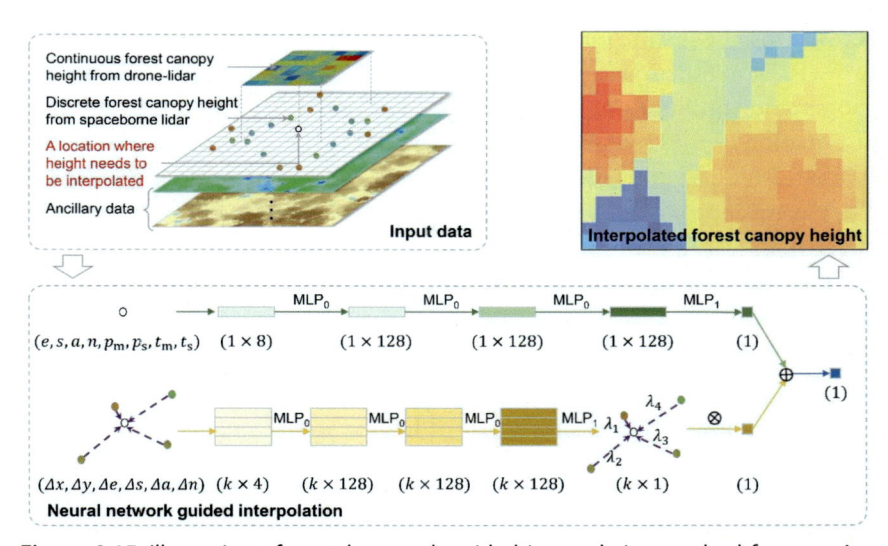

Figure 9.15 Illustration of neural network-guided interpolation method for mapping forest canopy height. MLP represents multilayer perceptron, with MLP_0 comprising a linear transformation function, a layer normalization function, and a nonlinear activation function, and MLP_1 comprises a dropout function and a linear transformation function. Numbers in ( ) represent the dimension of data or features; k represents the number of neighbors at a location to be interpolated; λ represents the spatial interpolation weight learned from MLP; x, y, e, s, a, n, p_m, p_s, t_m, and t_s, represents longitude, latitude, elevation, slope, aspect, normalized difference vegetation index, annual mean precipitation, annual precipitation seasonality, annual mean temperature, and annual temperature seasonality; Δ represents the difference of the corresponding feature between a location to be interpolated and its neighbors.

where $Z(l_0)$ is an unknown value to be interpolated at location l_0; $Z(l_i)$ is a known value at location l_i, which is the ith element in the neighbor set $\mathcal{N}(l_0)$ of l_0; $\lambda_i(l_0)$ is a weight placed on $Z(l_i)$; and λ_0 is a constant depending on the location of l_0. The weight $\lambda_i(l_0)$ is calculated from the variogram and covariogram. Forest height values at the two random locations $Z(l_0)$ and $Z(l_i)$ are not solely dependent on their spatial distance and are also influenced by environmental conditions. The third law of geography, hypothesizing that items with similar geographic conditions are more related (Zhu et al., 2018), may provide a possible solution.

We revised the Kriging interpolation Eq. (9.5) based on the third law of geography as follows:

$$Z(l_0) = \sum_{l_i \in \mathcal{N}(l_0)} \psi(\boldsymbol{d}_{l_0,l_i}) Z(l_i) + \varphi(\boldsymbol{v}_{l_0}) \tag{9.6}$$

where $\psi: \mathrm{R}^n \to \mathrm{R}$ is a function for modeling the weight placed on $Z(l_i)$ considering the feature distance $\boldsymbol{d}_{l_0,l_i}$ between the location pair (l_0, l_i), and $\varphi: \mathrm{R}^n \to \mathrm{R}$ is a function for modeling the constant $\lambda_0(l_0)$ considering the geographic conditions $\boldsymbol{v}_{l_0}$ at location l_0. To interpolate forest height, $Z(l_i)$ in Equation (9.6) is specified as spaceborne LiDAR forest height $H(l_i)$ from ICESat-2 ATLAS and GEDI, and ψ and φ are the two functions to be solved. They were approximated using multilayer perceptron (MLP), which is very effective in learning arbitrary functions (Hornik et al., 1989).

The MLP designed to approximate function ψ was composed of four layers, of which the first three layers had the same architecture:

$$\begin{cases} \boldsymbol{u}_1 = \boldsymbol{d}_{l_0,l_i} \\ \boldsymbol{u}_{j+1} = \mathrm{MLP}_0\left(\boldsymbol{w}_j^{\mathrm{T}} \boldsymbol{u}_i\right) = \rho\left(\mathrm{LN}\left(\boldsymbol{w}_j^{\mathrm{T}} \boldsymbol{u}_j\right)\right), j = 1, 2, 3 \end{cases} \tag{9.7}$$

where $\boldsymbol{w}_j$ is the weight to be learned in the jth layer during the training process; $\boldsymbol{u}_{j+1}$ is the output of the jth layer and the input to the $(j+1)$th layer, LN is a layer normalization function to make MLP invariant to per training-case feature shifting and scaling, and ρ is the nonlinear activation function. Before feeding the output $\boldsymbol{u}_4$ into the fourth layer, we used dropout to set 20% of the $\boldsymbol{u}_4$ units to 0 at each training step to eliminate overfitting. Finally, the fourth layer generated the interpolation weight λ_i for neighbor l_i using linear transformation:

$$\lambda_i = \psi(\boldsymbol{d}_{l_0,l_i}) = \boldsymbol{w}_4 \boldsymbol{u}_4' \tag{9.8}$$

where $\boldsymbol{u}_4'$ was the result of dropout.

The MLP designed to approximate the function φ also contained four layers, of which the weight at the jth layer was w'_j (Fig. 9.15) and served as a regression model to estimate the constant $\lambda_0(l_0)$ like the regression-Kriging method (Hengl et al., 2007). The input to the designed MLP to approximate the function φ was a geographic condition vector $(e, s, a, n, p_m, p_s, t_m, t_s)$, where $e, s, a, n, p_m, p_s, t_m$, and t_s represent elevation, slope, aspect, normalized difference vegetation index (NDVI), mean annual temperature, temperature seasonality, mean annual precipitation, and precipitation seasonality.

Following Eq. (9.5), the final interpolated forest height $\widehat{h}_0$ at l_0 was

$$\widehat{h}_0 = \sum_{l_i \in \mathcal{N}(l_0)} \lambda_i H(l_i) + \varphi(v_{l_0}) \tag{9.9}$$

In these MLP training processes, our objective was to make the error ε between the estimated $\widehat{h}_0$ and ground truth h_0 ($\varepsilon = \left| h_0 - \widehat{h}_0 \right|$) as small as possible. However, forest height may follow a normal distribution rather than a uniform distribution, and using this criterion as the loss function may lead to interpolated results with a narrower distribution than ground truth. To solve this problem, we constructed a loss function based on mean absolute percentage error as follows:

$$\text{loss} = \frac{1}{n} \sum_{i=1}^{n} \left| \frac{h_i - \widehat{h}_i}{h_i} \right| + \tau ||w||_2 \tag{9.10}$$

where n is the number of training samples, and τ was a tradeoff, set as 0.00001. The first term on the right-hand side of Eq. (9.10) represents the mean absolute percentage error, and the second term is L_2 regularization of all weights w (including w_j and w'_j). This loss function was minimized by updating w using an adaptive moment estimation optimizer during the training process (Kingma & Ba, 2014).

To evaluate the NNGI method for mapping forest height at large scales, we collected more than 140 km^2 UAV LiDAR data across China (Fig. 9.16b) to train the proposed NNGI method and mapped the forest canopy height distribution of China at 30 m resolution for the year 2019 (Fig. 9.17). The results showed that the average forest canopy height in China is 15.90 m with a standard deviation of 5.77 m. We evaluated the interpolated forest height product of China by over 1,100,000 GEDI validation footprints ($R^2 = 0.55$, RMSE $= 5.32$ m, Fig. 9.18a), about 33 km^2 UAV LiDAR validation data ($R^2 = 0.58$, RMSE $= 4.93$ m, Fig. 9.18b), and over 59,000 field plot

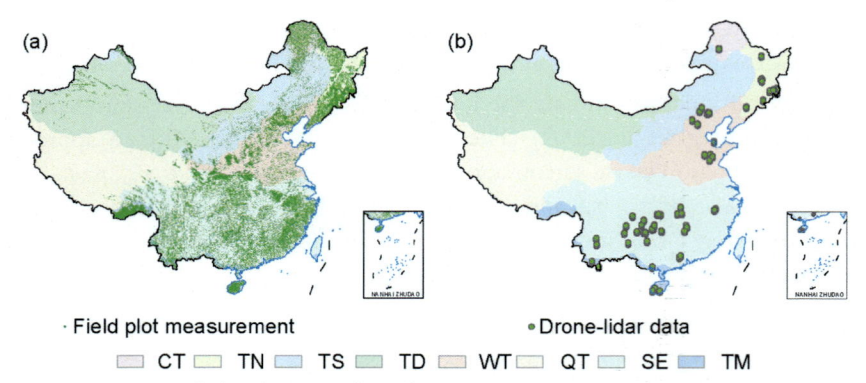

Field plot measurement · Drone-lidar data •

CT   TN   TS   TD   WT   QT   SE   TM

Figure 9.16 Spatial distribution of (a) field plot measurements and (b) unmanned aerial vehicle LiDAR data used to evaluate the neural network-guided interpolation method. CT, TN, TS, TD, WT, QT, SE, and TM represent the vegetation division of cold temperate needleleaf forest, temperate needleleaf-broadleaf mixed forest, temperate steppe, temperate desert, warm temperate deciduous-broadleaf forest, Qinghai—Tibet Plateau alpine vegetation, subtropical evergreen broadleaf forest, and tropical monsoon forest-rainforest, respectively.

measurements ($R^2 = 0.60$, RMSE $= 4.88$ m, Figs. 9.16a and 9.18a). The resulting product benefited from the interpolation–based mapping strategy, with almost no saturation effect in areas with tall forest canopies (Fig. 9.19). The high mapping accuracy demonstrates the feasibility of the proposed NNGI method for monitoring spatially continuous forest height at national to global scales by integrating multi–platform spaceborne LiDAR data and optical images, enabling opportunities to provide more accurate quantification of terrestrial carbon storage and a better understanding of forest ecosystem processes. The generated wall–to–wall forest height product for China is publicly available through the internet (www.3decology.org/dataset-software, accessed on June 15th, 2022).

9.4.2 Upscaling Leaf Area Index Estimates

Currently available spaceborne LiDAR data is mainly obtained by ICESat GLAS, ICESat-2 ATLAS, and GEDI. Compared with ground–based and airborne LiDAR data, spaceborne LiDAR data have the advantage of providing LAI information at larger scales. Many studies have focused on estimates of LAI from spaceborne LiDAR waveform data (Drake et al., 2002; Farid et al., 2008; Tang et al., 2012). Drake et al. (2002) attempted to estimate LAI using the ground return ratio. Solberg et al. (2007) used the $\ln(N_a/N_b)$ to calculate the penetration of spaceborne LiDAR. They studied

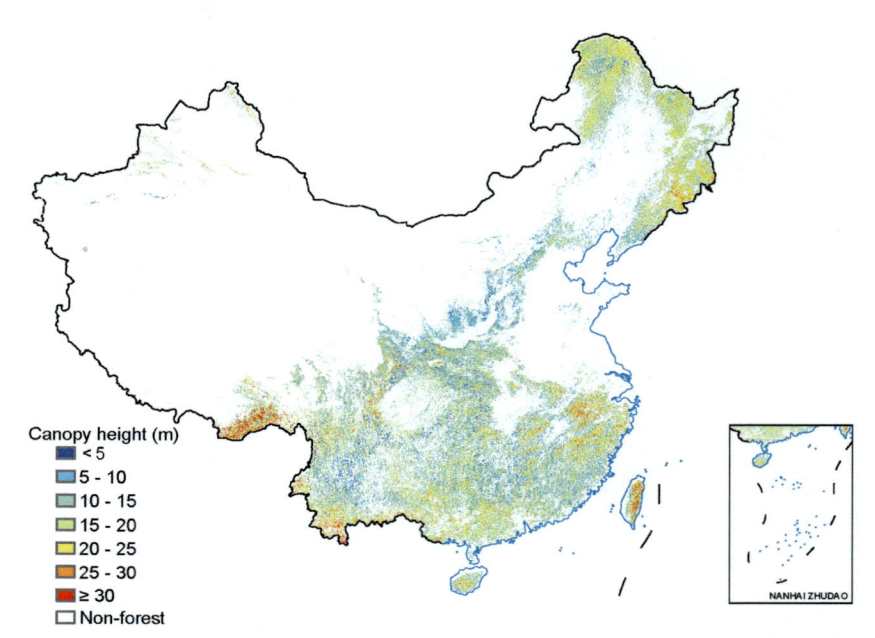

Canopy height (m)
- ■ < 5
- ■ 5 - 10
- □ 10 - 15
- □ 15 - 20
- □ 20 - 25
- ■ 25 - 30
- ■ ≥ 30
- □ Non-forest

NANHAI ZHUDAO

Figure 9.17 Neural network-guided interpolation-derived forest height of China at 30 m resolution for 2019.

its relationship with LAI, where N_a is the integral energy of the whole waveform, and N_b is the integral of the waveform on the ground below 1 m. Thus, the major method for estimating LAI from spaceborne waveform LiDAR data uses the Gaussian decomposition algorithm to decompose acquired waveform data into two parts—ground and vegetation—and the penetration index of spaceborne LiDAR can then be obtained by calculating the ratio of ground energy E_g to total energy E_t (Eq. 9.11), which can be combined with the field-measured LAI to establish an estimation model of spaceborne LiDAR LAI (Fig. 9.19). Using a similar method, GEDI provides LAI products in L2B data. The following is a Python script for extracting LAI from ICESat GLAS waveform data.

$$E_g = \sum_{GroBeg}^{GroEnd} E_i$$

$$E_t = \sum_{SigBeg}^{SigEnd} E_i$$

(9.11)

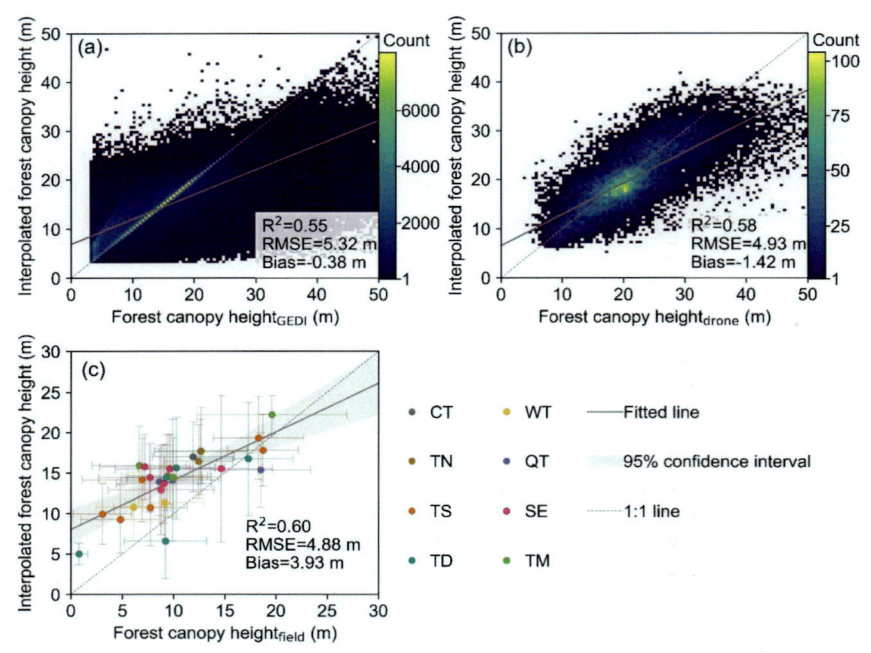

Figure 9.18 Accuracy assessment of the neural network-guided interpolation-derived forest height compared with (a) Global Ecosystem Dynamics Investigation validation footprints, (b) unmanned aerial vehicle LiDAR validation data, and (c) field measurements. CT, TN, TS, TD, WT, QT, SE, and TM represent the vegetation division of cold temperate needleleaf forest, temperate needleleaf-broadleaf mixed forest, temperate steppe, temperate desert, warm temperate deciduous broadleaf forest, Qinghai—Tibet Plateau alpine vegetation, subtropical evergreen broadleaf forest, and tropical monsoon forest—rainforest, respectively.

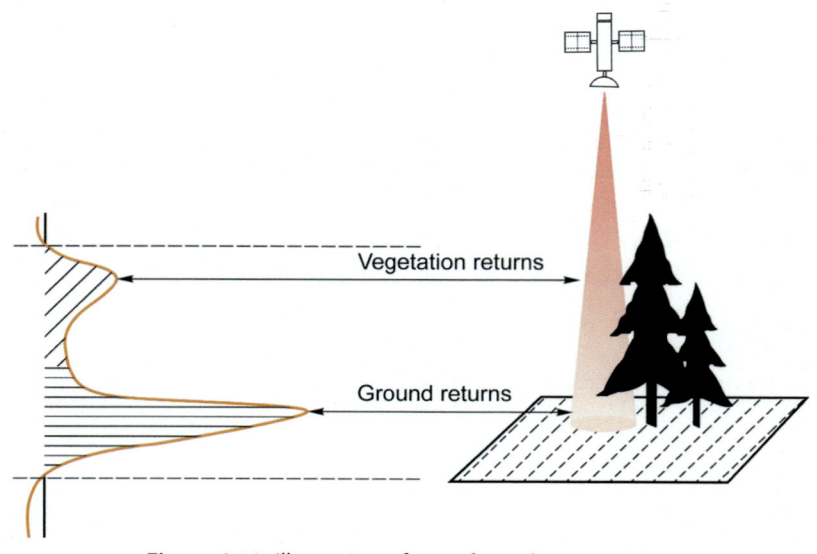

Figure 9.19 Illustration of waveform decomposition.

```python
# The following code is for LAI extraction from ICESat GLAS data
# -*- coding: UTF-8 -*-
import glob, os, h5py
import numpy as np
# Open GLAS14 data
HDF_list_GLA14="GLAH14_LAI.H5"
f_GLA14 = h5py.File(HDF_list_GLA14,'r')
# Calculate LAI
# Read the energy of each Gaussian wave
GLA14_Garea = f_GLA14['Data_40HZ']['Waveform']['d_Garea']
# Read the number of Gaussian waves in each spot
GLA14_numPk = f_GLA14['Data_40HZ']['Waveform']['i_numPk']
a=GLA14_numPk[:]
b=np.array(GLA14_Garea)
# Extract the ground echo energy of each spot
E_g=np.zeros((len(a),1),dtype='f')
for i in range(1,7): # 1,2,3,4,5,6
    index=np.where(GLA14_numPk[:]==i)
    if len(index[0]) != 0 :
        temp_dat=GLA14_Garea[:,i-1]
        E_g[index] = temp_dat[index][0]
        index = np.where(GLA14_Garea[:,i-1]>1000)
        b[index,i-1]=0
# Calculate the total echo energy of each spot
Et= np.sum(b,axis=1)
# Calculate leaf area index
LPI= E_g.reshape(len(a),)[:]/E_t
lai =-1.1*np.log(LPI)
# Filter should be executed by using the cloud, saturated signal, and signal-to-noise ratio when we used LAI
# Close GLAS14 data
f_GLA14.close
```

9.4.3 Upscaling Biomass Estimates

Forest ecosystems account for approximately 30% of global land cover, providing approximately 75% of primary productivity and 80% of global vegetation biomass (Beer et al., 2010; Bonan, 2008). Global forest ecosystems are the largest terrestrial carbon sink and play an important role in maintaining the global carbon cycle and carbon balance (Pan et al., 2011). Therefore, in the context of global climate change, it is of great significance to accurately estimate the spatial and temporal distributions of forest carbon sinks in the terrestrial biomass (Galbraith et al., 2010; Keith et al., 2009).

The biomass mentioned in this chapter refers to AGB, including the biomass of the stems, branches, and leaves. Because of the complexity of the forest vertical structure, passive remote sensing (optical remote sensing and radar) often encounters a saturation problem (Baccini et al., 2008; Luckman et al., 1997), which limits its capacity to accurately estimate forest biomass. LiDAR can accurately estimate forest structure parameters closely related to biomass, such as tree height and volume (Boudreau et al., 2008; Clark et al., 2004; Popescu et al., 2011). It has obvious advantages in forest biomass estimation (Clark et al., 2011; Nelson et al., 2009; Swatantran et al., 2011). Owing to the characteristics of the discrete distribution of spaceborne LiDAR footprints and their low point density (the average density in China is approximately 0.14 pt/km^2) (Yang et al., 2015), multi-source data fusion has become a focus of current research to obtain accurate and continuous forest canopy height and biomass estimates at large scales (Baccini et al., 2008; Boudreau et al., 2008; Simard et al., 2011).

By fusing ICESat GLAS data, airborne LiDAR data, and optical remote sensing data, Su et al. (2016) mapped the forest AGB distribution in China at a spatial resolution of 1 km. The specific estimation method and process are shown in Fig. 9.20.

GLAS-derived parameters, e.g., waveform extent, leading-edge extent, and trailing-edge extent, do not typically have direct biological meanings (Lefsky et al., 2007). Additional forest attributes derived from airborne LiDAR data or field plot measurements are usually required to convert

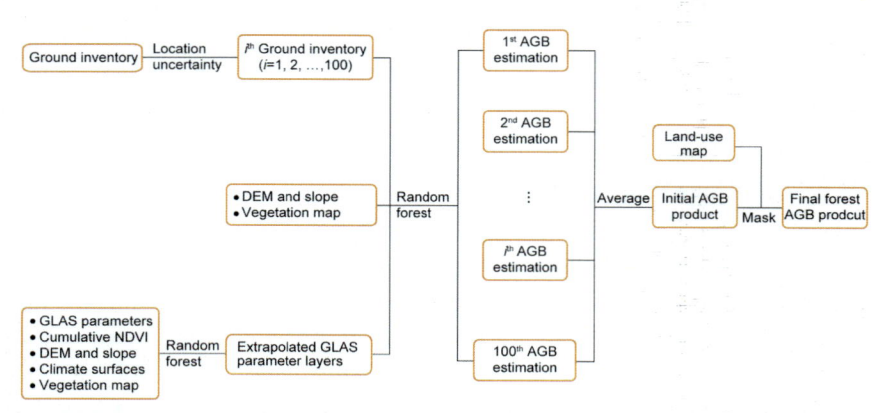

Figure 9.20 Estimation of forest aboveground biomass (AGBs) in China by fusing Ice, Cloud, and land Elevation Satellite Geoscience Laser Altimeter System (GLAS), airborne, and optical remote sensing data. *DEM*, digital elevation model; *NDVI*, normalized difference vegetation index.

GLAS parameters to biologically meaningful parameters (e.g., canopy height and AGB) (Boudreau et al., 2008; Lefsky, 2010; Saatchi et al., 2011). However, airborne LiDAR data are only available for a few small areas because of flight mission cost and duration limitations. Moreover, in our cases, the probability of the collected plot measurements overlapping with GLAS footprints was too small because of the huge gap between two adjacent GLAS tracks, and the considerable uncertainty within the plot location further increased the difficulty of matching the GLAS footprints with plot measurements. To solve this issue, we extrapolated the three GLAS parameters (i.e., waveform extent, leading-edge extent, and trailing-edge extent) into continuous spatial layers using the RF regression method. GLAS footprints that did not fall in the closed-forested area group or the open-forested area group from the land-use map were excluded to minimize the influence of nonforested areas on the extrapolation results (Liu et al., 2002). Moreover, seven ancillary predictors were used in the RF regression model, namely cumulative NDVI, elevation, slope, and four climate surfaces (i.e., annual mean temperature, annual temperature seasonality, annual total precipitation, and annual precipitation seasonality).

In building the RF regression model, the plot location uncertainties were too large to be neglected. Without considering the surveying accuracy, most of the latitudes and longitudes provided by the inventory data were accurate to $0.01°$ (corresponding to ~ 1 km), with some only accurate to $0.1°$ (corresponding to ~ 10 km). These huge plot location uncertainties could result in significant differences in the corresponding values of predictor variables and therefore influence the forest AGB estimation result. To minimize the influence of plot location uncertainty, Su et al. (2016) introduced the uncertainty field model of Guo et al. (2008) into the AGB mapping procedure (Fig. 9.20). This method hypothesized that the real plot center was randomly located within a circular buffer zone of the provided plot location, and the radius of the buffer was determined by the corresponding plot location uncertainty. Su et al. (2016) assumed that the plot location could not be more than 11 or 10 km (determined by the accuracy of the given plot location) from a given location in any direction. A 11 or 10 km buffer was created around each plot, and then 100 sets of ground inventory data were randomly generated within the buffers.

The AGB mapping procedure based on the RF regression method was performed for each set of ground inventory data with a location uncertainty (Fig. 9.20). Then, 100 predicted forest AGB layers were generated, and the average of these forest AGB layers was taken as the initial forest AGB estimation result. The final forest AGB mapping result was computed from the initial forest AGB estimation result by setting the nonforested pixels at 0 Mg/ha. In this study, if a 1 km pixel was not covered by either of the two forested groups from the land-use map (i.e., the closed-forested area group and the open-forested area group), it was treated as a nonforested pixel, and vice versa. It should be noted that the RF prediction process for each run was operated separately on each vegetation zone using the same RF regression tree model obtained from the corresponding run, considering the computation efficiency. The Qinghai—Tibet Plateau (QT) alpine vegetation and the temperate steppe vegetation zones were merged and treated as one zone in this process, which were both dominated by grassland and shrubs and had relatively small forested areas.

The produced wall-to-wall forest AGB product of China is shown in Fig. 9.21. The average forest AGB in China is approximately 120 Mg/ha with a standard deviation of 61 Mg/ha. The forest AGB in the south is significantly higher than in the north. The accuracy of the wall-to-wall forest AGB estimation in China was evaluated using 266 independent validation plot measurements. As shown in Fig. 9.22, the R^2 between the predicted and field-measured AGB is 0.75, and the RMSE is 42.39 Mg/ha. Although the slope and intercept for the correlation between estimated and

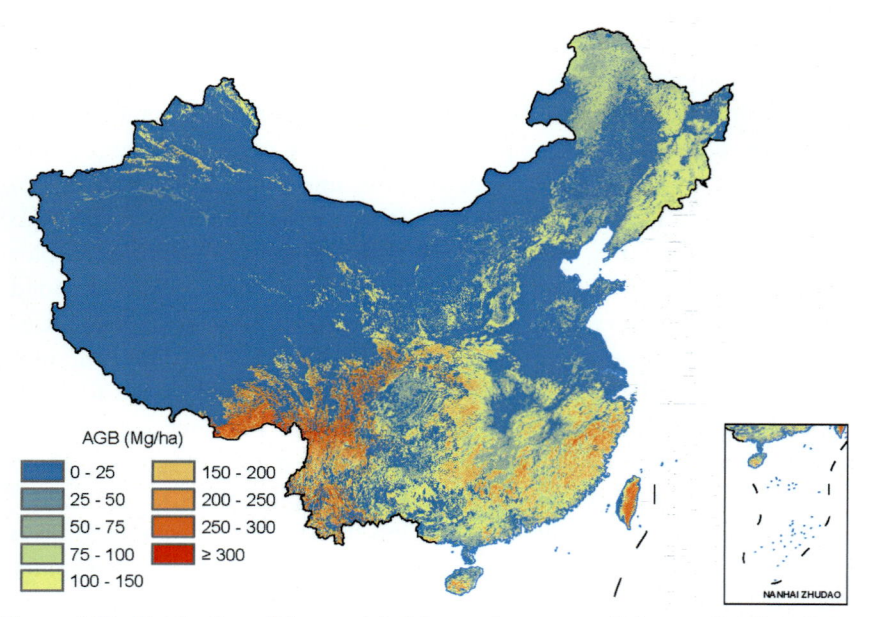

Figure 9.21 Distribution of the modeled forest aboveground biomass (AGB) in China.

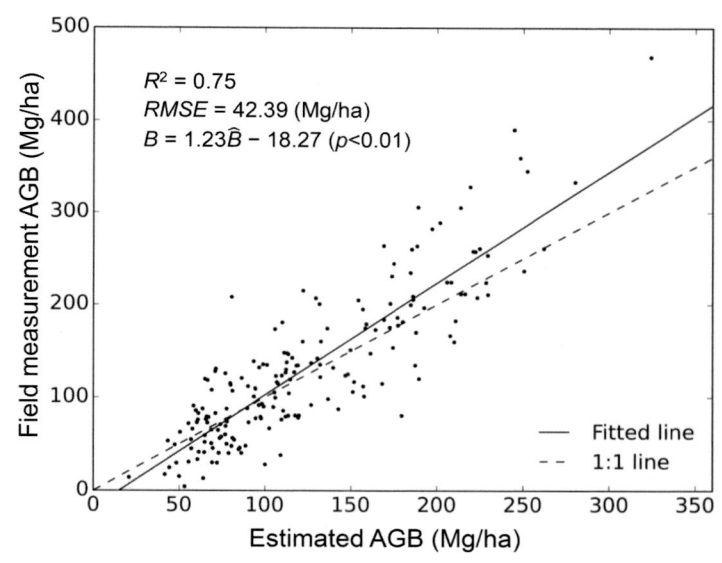

Figure 9.22 Evaluation of the modeled forest aboveground biomass (AGB) using validation ground inventory dataset at the plot level.

field–measured AGB are larger than one and smaller than zero, the fitted line is still very close to the 1:1 line (Fig. 9.22). The negative intercept suggests that the proposed AGB estimation method tends to slightly overestimate AGB densities in low-value areas (<87 Mg/ha). After forest AGB density reaches 87 Mg/ha, the proposed AGB estimation method tends to underestimate it, and this underestimation effect becomes more pronounced with further increases in forest AGB density.

The estimated mean AGB density for each vegetation zone was further compared with that calculated from all measurements in each vegetation zone. As shown in Table 9.7, the differences between the predicted mean AGB density and plot-measured mean AGB density for most vegetation zones are smaller than 10 Mg/ha. The temperate desert has the largest difference, i.e., ∼20 Mg/ha. The RMSE for each vegetation zone, calculated from the validation plot measurements, is shown in Table 9.7. The RMSEs for most vegetation zones range from 23 to 33 Mg/ha. The subtropical evergreen broadleaf forest and QT alpine vegetation have relatively large RMSEs (more than 45 Mg/ha). The high RMSE for the QT alpine vegetation may be caused by the small number of validation

Table 9.7 Comparison of predicted aboveground biomass (AGB) values with all measured points in each vegetation area.

Vegetation area	Forecast AGB average	Measuring AGB average	Root-mean-square error
Boreal forest	72.75	74.48	32.81
Temperate mixed forest	104.02	92.79	26.08
Temperate deciduous broadleaf forest	72.29	82.12	32.56
Subtropical evergreen broadleaf forest	133.22	139.43	45.86
Tropical monsoon rainforest	191.43	180.90	23.30
Temperate grassland	56.11	64.70	26.20
Temperate desert	81.45	101.63	13.86
Qinghai—Tibet Plateau alpine vegetation	168.20	160.42	50.12

Note: Root-mean-square error is only calculated from the verification samples; unit: Mg/ha.

plots (only three) within this vegetation zone. The temperate desert has the smallest RMSE ($\sim$ 14 Mg/ha) among all vegetation zones.

Moreover, because plot location uncertainty was introduced into the forest AGB modeling procedure, the R^2 between the predicted AGB and plot validation dataset increased from 0.64 to 0.75, and the RMSE decreased from 50 to 42 Mg/ha. At the pixel level, the absolute value of the uncertainty introduced by the plot location increased with forest AGB density and contributed to about 10% of the final AGB estimation (Fig. 9.23). In the tropical monsoon forest—rain forest in the most southeasterly part of QT, the uncertainty brought on by the plot location can reach nearly 40 Mg/ha.

Overall, this section has explained how to use in situ measurement data, ICESat GLAS data, optical image data, and topographic and climatic data to produce continuous forest AGB products on a large scale. The feasibility and accuracy of the methods fusing multi-source remote sensing data to estimate large-scale forest AGB have been well verified. Currently, we can use ICESat-2 ATLAS and GEDI data to map forest AGB at higher resolutions, helping to deepen our understanding of the spatial and temporal dynamics of forest carbon sinks.

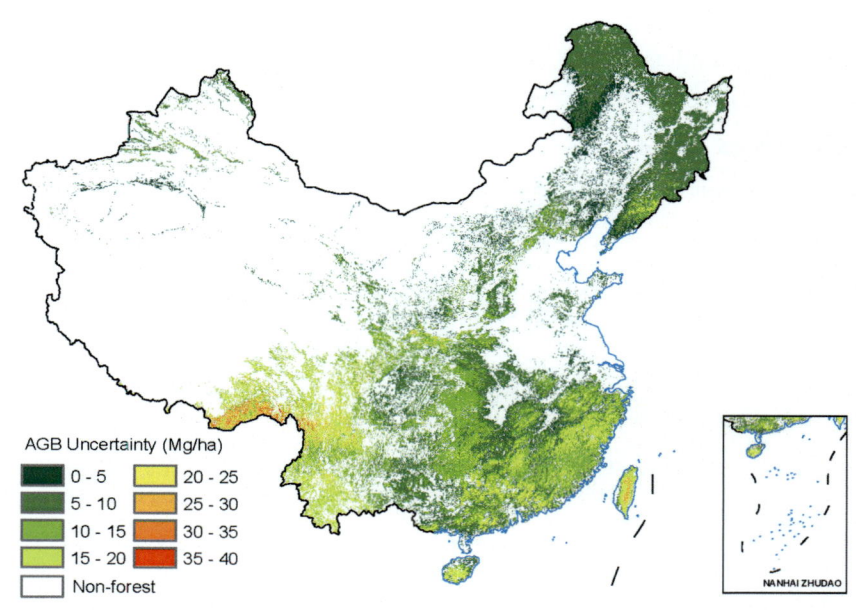

Figure 9.23 Uncertainty of the modeled forest aboveground biomass (AGB) of China at the pixel level.

9.5 Chapter Summary

Unlike the near-surface or airborne LiDAR systems introduced in previous chapters, spaceborne LiDAR systems, including ICESat GLAS, ICESat-2 ATLAS, and GEDI, measure the Earth's surface in different modes. Among the three spaceborne LiDAR systems, ICESat GLAS and GEDI use waveform sensors, which can record more echo information than the point cloud and must be processed using different methods, while ICESat-2 ATLAS uses photon counting technology. This chapter introduced these spaceborne LiDAR systems and their data products. We then used GLAS data as an example to introduce the processing methods for waveform data, including waveform smoothing, waveform decomposition, and canopy attribute extraction. For ICESat-2 ATLAS, which uses a photon counting technology with high-sensitive detectors, we also discussed how to extract canopy structure attributes from signal photons after removing solar background noise. Finally, taking the works of our team as examples, we described regression-based and interpolation-based methods for producing continuous products of forest height and biomass, which provide opportunities for more accurate quantification of terrestrial carbon storage and a better understanding of forest ecosystem processes.

References

Abshire, J. B., Sun, X., Riris, H., Sirota, J. M., McGarry, J. F., Palm, S., Yi, D., & Liiva, P. (2005). Geoscience laser altimeter system (GLAS) on the ICESat mission: On-orbit measurement performance. *Geophysical Research Letters, 32*(21), L21S02.

Asner, G. P. (2009). Tropical forest carbon assessment: Integrating satellite and airborne mapping approaches. *Environmental Research Letters, 4*(3), 034009.

Baccini, A., Laporte, N., Goetz, S., Sun, M., & Dong, H. (2008). A first map of tropical Africa's above-ground biomass derived from satellite imagery. *Environmental Research Letters, 3*(4), 045011.

Beer, C., Reichstein, M., Tomelleri, E., Ciais, P., Jung, M., Carvalhais, N., Rödenbeck, C., Arain, M. A., Baldocchi, D., & Bonan, G. B. (2010). Terrestrial gross carbon dioxide uptake: Global distribution and covariation with climate. *Science, 329*(5993), 834–838.

Berigan, W. J., Gutierrez, R. J., & Tempel, D. J. (2012). Evaluating the efficacy of protected habitat areas for the California spotted owl using long-term monitoring data. *Journal of Forestry, 110*(6), 299–303.

Bolton, D. K., Coops, N. C., & Wulder, M. A. (2013). Investigating the agreement between global canopy height maps and airborne Lidar derived height estimates over Canada. *Canadian Journal of Remote Sensing, 39*(Suppl. 1), S139–S151.

Bonan, G. B. (2008). Forests and climate change: Forcings, feedbacks, and the climate benefits of forests. *Science, 320*(5882), 1444–1449.

Boudreau, J., Nelson, R. F., Margolis, H. A., Beaudoin, A., Guindon, L., & Kimes, D. S. (2008). Regional aboveground forest biomass using airborne and spaceborne LiDAR in Québec. *Remote Sensing of Environment, 112*(10), 3876–3890.

Chiles, J.-P., & Delfiner, P. (2009). *Geostatistics: modeling spatial uncertainty.* John Wiley & Sons.

Clark, M. L., Clark, D. B., & Roberts, D. A. (2004). Small-footprint lidar estimation of sub-canopy elevation and tree height in a tropical rain forest landscape. *Remote Sensing of Environment, 91*(1), 68–89.

Clark, M. L., Roberts, D. A., Ewel, J. J., & Clark, D. B. (2011). Estimation of tropical rain forest aboveground biomass with small-footprint lidar and hyperspectral sensors. *Remote Sensing of Environment, 115*(11), 2931–2942.

Drake, J. B., Dubayah, R. O., Clark, D. B., Knox, R. G., Blair, J. B., Hofton, M. A., Chazdon, R. L., Weishampel, J. F., & Prince, S. (2002). Estimation of tropical forest structural characteristics using large-footprint lidar. *Remote Sensing of Environment, 79*(2–3), 305–319.

Dubayah, R., Blair, J. B., Goetz, S., Fatoyinbo, L., Hansen, M., Healey, S., Hofton, M., Hurtt, G., Kellner, J., & Luthcke, S. (2020). The global ecosystem dynamics investigation: High-resolution laser ranging of the Earth's forests and topography. *Science of Remote Sensing, 1*, 100002.

Farid, A., Goodrich, D., Bryant, R., & Sorooshian, S. (2008). Using airborne lidar to predict Leaf Area Index in cottonwood trees and refine riparian water-use estimates. *Journal of Arid Environments, 72*(1), 1–15.

Galbraith, D., Levy, P. E., Sitch, S., Huntingford, C., Cox, P., Williams, M., & Meir, P. (2010). Multiple mechanisms of Amazonian forest biomass losses in three dynamic global vegetation models under climate change. *New Phytologist, 187*(3), 647–665.

Guo, Q., Liu, Y., & Wieczorek, J. (2008). Georeferencing locality descriptions and computing associated uncertainty using a probabilistic approach. *International Journal of Geographical Information Science, 22*(10), 1067—1090.

Hengl, T., Heuvelink, G. B., & Rossiter, D. G. (2007). About regression-kriging: From equations to case studies. *Computers & Geosciences, 33*(10), 1301—1315.

Hofton, M. A., Minster, J.-B., & Blair, J. B. (2000). Decomposition of laser altimeter waveforms. *Ieee Transactions on Geoscience and Remote Sensing, 38*(4), 1989—1996.

Hornik, K., Stinchcombe, M., & White, H. (1989). Multilayer feedforward networks are universal approximators. *Neural Networks, 2*(5), 359—366.

Hudak, A. T., Lefsky, M. A., Cohen, W. B., & Berterretche, M. (2002). Integration of lidar and Landsat ETM+ data for estimating and mapping forest canopy height. *Remote Sensing of Environment, 82*(2—3), 397—416.

Jin, S., Su, Y., Gao, S., Hu, T., Liu, J., & Guo, Q. (2018). The transferability of Random Forest in canopy height estimation from multi-source remote sensing data. *Remote Sensing, 10*(8), 1183.

Keith, H., Mackey, B. G., & Lindenmayer, D. B. (2009). Re-evaluation of forest biomass carbon stocks and lessons from the world's most carbon-dense forests. *Proceedings of the National Academy of Sciences, 106*(28), 11635—11640.

Kingma, D. P., & Ba, J. (2014). Adam: A method for stochastic optimization. *arXiv preprint arXiv:1412.6980*.

Kwok, R., Cunningham, G., Markus, T., Hancock, D., Morison, J., Palm, S., Farrell, S., & Ivanoff, A. (2020). ATLAS/ICESat-2 L3A sea ice freeboard, version 3. *Boulder, Colorado USA: NSIDC: National Snow and Ice Data Center, 10*, 5067.

Lefsky, M. A. (2010). A global forest canopy height map from the moderate resolution imaging spectroradiometer and the Geoscience Laser Altimeter System. *Geophysical Research Letters, 37*(15).

Lefsky, M. A., Keller, M., Pang, Y., De Camargo, P. B., & Hunter, M. O. (2007). Revised method for forest canopy height estimation from Geoscience Laser Altimeter System waveforms. *Journal of Applied Remote Sensing, 1*(1), 013537.

Li, L., Guo, Q., Tao, S., Kelly, M., & Xu, G. (2015). Lidar with multi-temporal MODIS provide a means to upscale predictions of forest biomass. *ISPRS Journal of Photogrammetry and Remote Sensing, 102*, 198—208.

Li, W., Niu, Z., Shang, R., Qin, Y., Wang, L., & Chen, H. (2020). High-resolution mapping of forest canopy height using machine learning by coupling ICESat-2 LiDAR with Sentinel-1, Sentinel-2 and Landsat-8 data. *International Journal of Applied Earth Observation and Geoinformation, 92*, 102163.

Liu, J., Liu, M., Deng, X., Zhuang, D., Zhang, Z., & Luo, D. (2002). The land use and land cover change database and its relative studies in China. *Journal of Geographical Sciences, 12*, 275—282.

Liu, X., Su, Y., Hu, T., Yang, Q., Liu, B., Deng, Y., Tang, H., Tang, Z., Fang, J., & Guo, Q. (2022). Neural network guided interpolation for mapping canopy height of China's forests by integrating GEDI and ICESat-2 data. *Remote Sensing of Environment, 269*, 112844.

Luckman, A., Baker, J., Kuplich, T. M., Yanasse, C. D. C. F., & Frery, A. C. (1997). A study of the relationship between radar backscatter and regenerating tropical forest biomass for spaceborne SAR instruments. *Remote Sensing of Environment, 60*(1), 1—13.

Luthcke, S., Rebold, T., Thomas, T., & Pennington, T. (2020). In *Algorithm theoretical basis document (ATBD) for GEDI waveform geolocation for L1 and L2 products*.

McCombs, J. W., Roberts, S. D., & Evans, D. L. (2003). Influence of fusing lidar and multispectral imagery on remotely sensed estimates of stand density and mean tree height in a managed loblolly pine plantation. *Forest Science, 49*(3), 457—466.

Narine, L. L., Popescu, S. C., & Malambo, L. (2020). Using ICESat-2 to estimate and map forest aboveground biomass: A first example. *Remote Sensing, 12*(11), 1824.

Nelson, R., Ranson, K., Sun, G., Kimes, D., Kharuk, V., & Montesano, P. (2009). Estimating Siberian timber volume using MODIS and ICESat/GLAS. *Remote Sensing of Environment, 113*(3), 691—701.

Neuenschwander, A., & Pitts, K. (2019). The ATL08 land and vegetation product for the ICESat-2 Mission. *Remote Sensing of Environment, 221*, 247—259.

Pan, Y., Birdsey, R. A., Fang, J., Houghton, R., Kauppi, P. E., Kurz, W. A., Phillips, O. L., Shvidenko, A., Lewis, S. L., & Canadell, J. G. (2011). A large and persistent carbon sink in the world's forests. *Science, 333*(6045), 988—993.

Patterson, P. L., Healey, S. P., Ståhl, G., Saarela, S., Holm, S., Andersen, H.-E., Dubayah, R. O., Duncanson, L., Hancock, S., & Armston, J. (2019). Statistical properties of hybrid estimators proposed for GEDI—NASA's global ecosystem dynamics investigation. *Environmental Research Letters, 14*(6), 065007.

Popescu, S. C., Zhao, K., Neuenschwander, A., & Lin, C. (2011). Satellite lidar vs. small footprint airborne lidar: Comparing the accuracy of aboveground biomass estimates and forest structure metrics at footprint level. *Remote Sensing of Environment, 115*(11), 2786—2797.

Saatchi, S. S., Harris, N. L., Brown, S., Lefsky, M., Mitchard, E. T., Salas, W., Zutta, B. R., Buermann, W., Lewis, S. L., & Hagen, S. (2011). Benchmark map of forest carbon stocks in tropical regions across three continents. *Proceedings of the National Academy of Sciences, 108*(24), 9899—9904.

Schutz, B. E., Zwally, H. J., Shuman, C. A., Hancock, D., & DiMarzio, J. P. (2005). Overview of the ICESat mission. *Geophysical Research Letters, 32*(21), L21S01.

Simard, M., Pinto, N., Fisher, J. B., & Baccini, A. (2011). Mapping forest canopy height globally with spaceborne lidar. *Journal of Geophysical Research: Biogeosciences, 116*(G4), G04021.

Solberg, S., Weydahl, D. J., & Næsset, E. (2007). SAR forest canopy penetration depth as an indicator for forest health monitoring based on leaf area index (LAI). In *Proceedings of the 5th international symposium on retrieval of bio-and geophysical parameters from SAR data for land applications* (Bari, Italy).

Su, Y., Guo, Q., Xue, B., Hu, T., Alvarez, O., Tao, S., & Fang, J. (2016). Spatial distribution of forest aboveground biomass in China: Estimation through combination of spaceborne lidar, optical imagery, and forest inventory data. *Remote Sensing of Environment, 173*, 187—199.

Su, Y., Ma, Q., & Guo, Q. (2017). Fine-resolution forest tree height estimation across the Sierra Nevada through the integration of spaceborne LiDAR, airborne LiDAR, and optical imagery. *International Journal of Digital Earth, 10*(3), 307—323.

Swatantran, A., Dubayah, R., Roberts, D., Hofton, M., & Blair, J. B. (2011). Mapping biomass and stress in the Sierra Nevada using lidar and hyperspectral data fusion. *Remote Sensing of Environment, 115*(11), 2917—2930.

Tang, H., Dubayah, R., Swatantran, A., Hofton, M., Sheldon, S., Clark, D. B., & Blair, B. (2012). Retrieval of vertical LAI profiles over tropical rain forests using waveform lidar at La Selva, Costa Rica. *Remote Sensing of Environment, 124*, 242—250.

Tao, S., Guo, Q., Li, C., Wang, Z., & Fang, J. (2016). Global patterns and determinants of forest canopy height. *Ecology, 97*(12), 3265—3270.

Tobler, W. R. (1970). A computer movie simulating urban growth in the Detroit region. *Economic Geography, 46*(Suppl. 1), 234—240.

Yang, T., Wang, C., Li, G., Luo, S., Xi, X., Gao, S., & Zeng, H. (2015). Forest canopy height mapping over China using GLAS and MODIS data. *Science China Earth Sciences, 58*(1), 96—105.

Zhu, A. X., Lu, G., Liu, J., Qin, C. Z., & Zhou, C. (2018). Spatial prediction based on Third law of geography. *Annals of GIS, 24*(4), 225—240.

CHAPTER 10

LiDAR-Based Three-Dimensional Radiative Transfer Models and Applications

Contents

Modeling radiative transfer (RT) in heterogeneous forests is essential to understanding their biophysical processes and retrieving information from remote sensing data. Although a range of three-dimensional (3-D) RT models have been reported, it is still challenging to obtain realistic 3-D scenes and input parameters using traditional field surveys or optical imagery. Light detection and ranging (LiDAR) can capture highly detailed point clouds of forests to provide 3-D scenes and canopy structural information for parameterizing 3-D RT models. This chapter introduces the most common 3-D RT models. We then focus on the voxel-based RT (VBRT) model, which uses octree and high-resolution voxels rather than coarse-resolution turbid medium voxels to represent forest scenes. First, we describe 3-D scene construction and parameterization using LiDAR point cloud data in detail. We then examine and compare simulated bidirectional reflectance factor and orthogonal photographs using VBRT and two additional 3-D RT models. Finally, we summarize the applications of 3-D RT models combined with other RT or process models to simulate the directional reflectance spectra of forests and ecological processes of forest ecosystems, such as gas exchange and solar-induced chlorophyll fluorescence.

LiDAR Principles, Processing and Applications in Forest Ecology
ISBN 978-0-12-823894-3
https://doi.org/10.1016/B978-0-12-823894-3.00010-4

10.1 Principles of Three-Dimensional Radiative Transfer Models

Modeling the interactions between light and land surfaces is useful for understanding many biophysical processes, such as exchanges of energy, carbon, and water between the biosphere and atmosphere (Chen et al., 2008; Magney et al., 2016; Ni-Meister & Gao, 2011; Widlowski et al., 2006). Three-dimensional (3-D) radiative transfer (RT) models have been widely applied in such research (Disney et al., 2000; van Leeuwen et al., 2015; Widlowski et al., 2011). In addition, understanding the nature of the interaction between light and the land surface is necessary to effectively retrieve information from remotely sensed data for research activities in various domains (Govaerts & Verstraete, 1998; Kimes et al., 2000; Kimes & Kirchner, 1982; Koetz et al., 2006; North, 1996; Woodcock et al., 1997). Highly detailed realistic 3-D RT models can serve as a virtual laboratory to simulate the bidirectional reflectance distribution function, radiative fluxes, and remotely sensed data under controlled experimental conditions (e.g., illumination, viewing, and spectral properties), which can then be analyzed to test hypotheses, model assumptions, and quantitative retrieval algorithms (Disney et al., 2011; Goodwin et al., 2007; Morton et al., 2014; Widlowski et al., 2014, 2015; Woodcock et al., 1997). For example, a spatially explicit 3-D shortwave and longwave RT model was coupled with soil-and-canopy energy balance and a canopy physiology model to simulate the energy and carbon fluxes in a heterogeneous oak woodland (Kobayashi et al., 2012). Using an extended version of physically-based ray tracer (PBRT) (Pharr et al., 2004), Stuckens et al. (2009) investigated the impact of common assumptions on canopy RT models such as neglecting leaf asymmetry. Morton et al. (2014) used the three-dimensional forest light interaction model (FLIGHT) model (North, 1996) to simulate changes in light detection and ranging (LiDAR) and optical remote sensing metrics to test whether changes in leaf area or leaf reflectance drive the appearance of a seasonal green up of forests in southern Amazonia.

Numerous methods have been developed to model RT in terrestrial environments (Disney et al., 2006; Gastellu-Etchegorry et al., 1996; Govaerts & Verstraete, 1998; Kimes & Kirchner, 1982; Kobayashi & Iwabuchi, 2008; Lewis, 1999; Li et al., 1995; North, 1996; Widlowski et al., 2006). Whereas one-dimensional (1-D) models can successfully capture the propagation of radiation in spatially homogenous media, 3-D models are more suitable for spatially heterogeneous and complex scenes (Govaerts & Verstraete, 1998; Kimes & Kirchner, 1982). Examples of 3-D RT models include CanSPART (Haverd et al., 2012), Raytran (Govaerts & Verstraete, 1998), Rayspread (Widlowski et al., 2006), the invertible forest reflectance model (Schlerf & Atzberger, 2006), price (Essery et al., 2008), a radiosity-graphics combined method (RGM) (Qin & Gerstl, 2000), radiosity applicable to porous individual objects (Huang et al., 2013), large-scale remote sensing data and image simulation (LESS) (Qi et al., 2019), discrete anisotropic radiative transfer (DART) (Gastellu-Etchegorry et al., 1996, 2017), PBRT (Pharr et al., 2004), voxel-based radiative transfer (VBRT) (Li et al., 2018), forest light environmental simulator (FLiES) (Kobayashi & Iwabuchi, 2008), and librat (Lewis, 1999; Widlowski et al., 2015).

Normally, 3-D scenes are described by a set of simple geometric primitives such as triangles, discs, cones, spheres, cylinders, and ellipsoids (Chen et al., 2000; Kobayashi & Iwabuchi, 2008; Widlowski et al., 2015). Using different combinations of these geometric primitives with defined location, size, shape, orientation, and scattering properties, we can generate 3-D scenes of different complexities (Chen et al., 2000; Widlowski et al., 2006). For convenience, we use triangle-based representations to refer to explicit 3-D representations hereafter. However, it should be noted that explicit 3-D representations of scenes are not limited to triangle-based representations. Alternatively, a scene can be subdivided into a set of 3-D rectangular cells, known as voxels (Gastellu-Etchegorry et al., 1996; Kimes & Kirchner, 1982; Widlowski et al., 2014). Each voxel is associated with specific information such as the element, location, and scattering properties (Gastellu-Etchegorry et al., 1996). The side length of a voxel is known as the voxel size or resolution. Usually, a voxel is relatively large (e.g., 0.5 m) and is assumed to be a turbid medium. Fig. 10.1 illustrates three typical examples of different ways of scene characterization.

(a) (b) (c)

Figure 10.1 Three typical scene characterizations using different models: (a) simple geometric primitives at a coarse scale, (b) voxels, and (c) triangles at a fine scale.

Both analytical and numerical methods have been developed to solve the RT equation (Chen et al., 2000; Disney et al., 2000; Gastellu-Etchegorry et al., 1996; Govaerts & Verstraete, 1998; Ni et al., 1999). Analytical solutions are fast but rely on rigorous model assumptions and are therefore limited to relatively simple scenes (Disney et al., 2000). By contrast, numerical solutions can deal with highly complex scenes with minimal assumptions but are slow to converge (Disney et al., 2000; Widlowski et al., 2014). Various numerical methods have been developed and applied to 3-D RT simulations, such as the discrete ordinate method (Gastellu-Etchegorry et al., 2004) and the RGM (Qin & Gerstl, 2000). With the availability of increasing computing power, additional numerical methods such as Monte Carlo path tracing have been widely used to simulate the light regime and directional reflectance characteristics of realistic 3-D scenes (Govaerts & Verstraete, 1998; Pharr et al., 2004; Qi et al., 2019; Widlowski et al., 2006).

10.2 Voxel-Based Radiative Transfer Model

10.2.1 Voxel-Based Scene Model

Geometries can be represented by triangles or voxels, both of which have advantages and disadvantages (Laine & Karras, 2010). In this section, we focus on the voxel-based method for scene representation because it is now frequently used to process LiDAR point cloud data (Cifuentes et al., 2014; Hosoi & Omasa, 2007). Suppose we have a scene with the dimensions of 10 m × 10 m × 10 m. A single tree approximated by a cylinder and sphere is located at the center (Fig. 10.2a). The scene is then subdivided into regularly spaced voxels of the same size. Each voxel has a unique element type (e.g., air, foliage, or wood) associated with defined scattering properties. The distribution of different types of voxels thus describes the architectures of the scene (Fig. 10.2b). As we increase the resolution of the voxels, the scene is represented more accurately (Fig. 10.2c). In this section, we use very fine-resolution voxels (e.g., 0.01 m), each of which is a solid prototypical scatterer; thus, we do not have to make further statistical simplifications about the architecture of elements within voxels.

It is important to consider memory usage during voxelization. If we decrease the voxel size by N, memory usage increases by N^3. For a large scene voxelized at high resolution, the memory requirement can be unaffordable. To solve this issue, we can use an octree data structure to store voxels (Laine & Karras, 2010; Rodrigues et al., 2000). For example, by scanning the tree in Fig. 10.2a, we generate a point cloud, as shown in

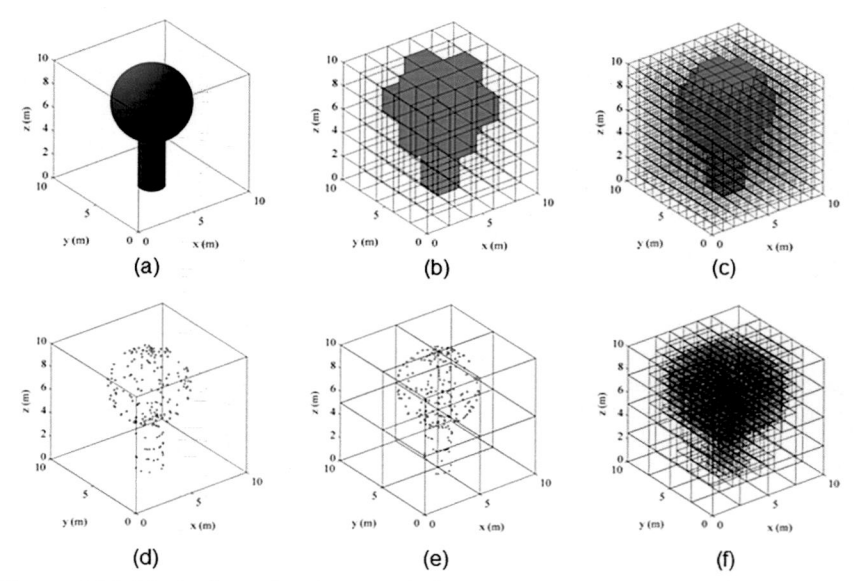

Figure 10.2 Three-dimensional scene (10 m × 10 m × 10 m) with a single tree approximated by a sphere and cylinder (a). Voxelization of the scene with different resolutions: 2 m (b) and 1 m (c). Generation of voxel octree: the bounding box (root node) containing point cloud (d); subdivision of the root node into eight octants (e); and the final voxel octree (f). The colors represent different elements: Foliage (*dark green*) and wood (*dark yellow*).

Fig. 10.2d. The bounding box that contains the point cloud is the first and largest voxel, which is the root node of the octree. If a node is not empty (i.e., it contains points inside) and the voxel size is larger than a defined threshold, we subdivide it into exactly eight children (octants) recursively; otherwise, we stop the recursion. Consequently, the root node is subdivided into eight voxels, as shown in Fig. 10.2e. By defining a threshold of 1 m, we finally obtain the voxel octree in Fig. 10.2f. Compared with using fixed–size (1 m) voxels, as in Fig. 10.2c, we can reduce memory usage dramatically with a voxel octree. Small voxels are only required at locations where the point density is high. For areas with sparse points, we can use large and sparse voxels to save memory. Fig. 10.3 shows the voxelization of real point cloud data. A camphor tree (*Cinnamomum camphora*) was scanned with a RIEGL VZ–400 (RIEGL Laser Measurement System GmbH, Horn, Austria) 3-D laser scanner. A leaf—wood separation algorithm was used to classify the point cloud into two classes, leaf and wood (Tao et al., 2015). By voxelizing the classified point cloud with 0.01 m resolution, the architectures of the tree can be accurately described.

Figure 10.3 Illustration of the voxelization of a camphor tree (*Cinnamomum camphora*): original photograph (a); terrestrial light detection and ranging (LiDAR) point cloud acquired by RIEGL VZ-400 3-D laser scanner (b); and voxelization with a resolution of 0.01 m (c). Tree height: 8.8 m scanning azimuth angles: 0, 90, 180, and 270 degrees. Scanning distance: 5 m away from the trunk. Scanner height: 1.5 m above the ground. Missing data near the apex can be observed in the point cloud.

10.2.2 Radiative Transfer Modeling

Once there is access to a 3-D scene with defined geometry and scattering properties, the next step is to simulate the interaction between light and the scene elements. In this section, we used an unbiased algorithm, Monte Carlo path tracing, to trace the light path from the light source to the camera (Fig. 10.4) or vice versa (Kajiya, 1986; Pharr et al., 2004). The

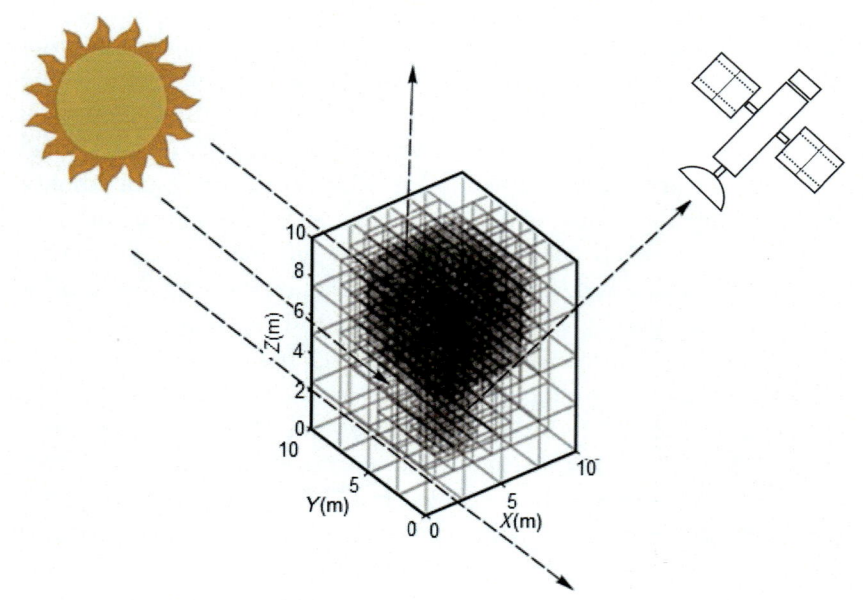

Figure 10.4 An illustration of Monte Carlo path tracing.

behavior of light transport in a scene can be described with the rendering equation (Kajiya, 1986):

$$L_o(p, \omega_o) = L_e(p, \omega_o) + \int_\Omega L_i(p, \omega_i) f_r(p, \omega_i, \omega_o) \cos \theta_i d\omega_i \qquad (10.1)$$

where $L_o(P, \omega_o)$, $L_e(P, \omega_o)$, and $L_i(P, \omega_i)$ refer to exitant, emitted, and incident radiance at point P in outgoing direction ω_o and incoming direction ω_i, respectively; $f_r(P, \omega_i, \omega_o)$ refers to the bidirectional scattering distribution function (BSDF) at point P; and θ_i is the angle between the surface normal and incident directions. For Lambertian surfaces, the BSDF is a constant that is independent of the incoming and outgoing directions (Schaepman–Strub et al., 2006). The integration domain Ω is a hemisphere over the surface for opaque materials (e.g., wood) and a full sphere for transparent materials (e.g., foliage) (Lafortune, 1995). The integral quantifies the fraction of incident radiance scattered by the surface. According to the rendering equation, the emitted plus the scattered radiances equal the total exitant radiance so that the energy is in balance. Notably, radiance refers to radiant flux emitted, reflected, transmitted, or received by a surface per unit projected area per unit solid angle ($W/m^2/sr$) (Schaepman–Strub et al., 2006). Furthermore, we assume that the radiance distribution in the scene is in equilibrium (Pharr et al., 2004).

To solve the rendering equation, we need to estimate the integral that considers incoming light from all directions. Because we aim to simulate RT in highly complex scenes and an analytical solution may be difficult, we use a Monte Carlo integration. The integral can be estimated using N directions randomly sampled over Ω. Hence, Eq. (10.1) can be rewritten as:

$$L_o(p, \omega_o) = L_e(p, \omega_o) + \frac{1}{N} \sum_{i=1}^{N} \frac{L_i(p, \omega_i) f_r(p, \omega_i, \omega_o) \cos \theta_i}{P_r(\omega_i)} \qquad (10.2)$$

where $P_r(\omega_i)$ refers to the sampling probability density function (Lafortune, 1995). If we sample uniformly over a hemisphere, $P_r(\omega_i)$ equals $1/2\pi$. Importance sampling can be used to improve efficiency; that is, $P_r(\omega_i)$ is higher for directions with larger BSDF values (Talbot et al., 2005).

We can use path tracing to solve Eq. (10.2). Let us set up a camera looking in a certain direction. The aim is to estimate the radiance reaching each pixel. From each pixel, we shoot a ray k_1 into the scene. If it hits a surface at point P_1, we evaluate $L_o(P_1, \omega_{o1})$ for ray k_1 using Eq. (10.2), where ω_{o1} is the direction from P_1 to the pixel. The incoming radiance

$L_i(P_1, \omega_{i1})$ at point P_1 is estimated by shooting another ray k_2 in a random direction according to $P_r(\omega_i)$. If ray k_2 hits another surface at point P_2, $L_o(P_2, \omega_{o2})$ is evaluated again, contributing to $L_i(P_1, \omega_{i1})$, where ω_{o2} is the direction from P_2 to P_1. Obviously, we also need to shoot a ray k_3 randomly to estimate $L_o(P_2, \omega_{o2})$. Thus, the radiance is evaluated recursively until the ray exits the scene or hits light sources. We can obtain the radiance $L_o(P_1, \omega_{o1})$ for the first shot ray coming from a pixel when the recursion terminates. However, we may find endless recursion because of multiple scattering. To address this issue and improve efficiency, a Russian roulette approach can be used to terminate recursion while keeping the results unbiased (Lafortune & Willems, 1994). At each intersection, we generate a uniform random number r between [0, 1]: the recursion is terminated if $r < q$ (termination probability) or continues otherwise. In this section, Russian roulette was applied when the scattering order was larger than five, and q was empirically set as 0.5. According to our test, the simulation results were not sensitive to q. To reduce aliasing and variance, we can shoot multiple rays from random positions within each pixel and average their radiances (Pharr et al., 2004).

During path tracing, a scene can be easily traversed using the voxel octree data structure. We implemented the VBRT model using C programming language and compiled it using GNU compiler collection on a Linux system. To speed up the model, parallel computing techniques, including multithreading via POSIX Threads and multiprocessing via the message passing interface, were implemented to run the model on a workstation or high-performance computing platform.

10.3 Parameterizations of Three-Dimensional Radiative Transfer Models Using LiDAR

Fine-scale and accurate characterizations of 3-D scenes are important to parameterize 3-D RT models. Point cloud data in high point density can accurately capture the features of heterogeneous 3-D scenes; therefore, they can be used as input to drive RT models. With rapid advancements in sensor technology, system development, and algorithms, LiDAR and photogrammetry are becoming more popular in forest inventories, and it is increasingly affordable and convenient to obtain detailed 3-D point clouds of forests with high spatial resolution and high accuracy (Bauwens et al., 2016; Liang et al., 2015, 2016; Westoby et al., 2012).

Many methods are available for collecting point cloud data to characterize fine-scale 3-D forest structures, including terrestrial LiDAR (TLS), mobile LiDAR (MLS), unmanned aerial vehicle (UAV) LiDAR (ULS), and photogrammetry (Liang et al., 2016; Wallace et al., 2012). For small-scale forest plots, we can use TLS or handheld MLS to acquire point cloud data from different positions to reduce occlusions. Especially at the top of canopies, the missing points in TLS and MLS caused by occlusions can be filled by point clouds derived from ULS or UAV photogrammetry (Flener et al., 2013; Harwin & Lucieer, 2012; Wallace et al., 2012). Airborne LiDAR (ALS) is an efficient way to cover large areas, but the point density is too low to derive the fine-scale structures of individual trees. However, structural attributes at the stand level, such as tree location, tree height, diameter at breast height, leaf area index, and crown size, can be estimated from ALS data. By contrast, TLS can provide high-density point cloud data to derive the fine-scale structures of individual trees, but it is not efficient or practical to use TLS for covering large areas. Therefore, we can use TLS data to construct detailed 3-D models of representative sample trees of each species and then duplicate, shift, scale, and rotate them according to the stand-level attributes derived from ALS data to reconstruct large-scale 3-D forest scenes.

After the preprocessing of point cloud data (e.g., registration, alignment, upscaling, and classification), the point cloud is voxelized using an octree data structure. Each voxel has a unique label, such as foliage, wood, or ground, and it is assumed to be solid with Lambertian surface scattering properties. The spectral characteristics of different elements can be derived from field measurements or taken from the literature. We used high-resolution voxels at the scale of centimeters (e.g., 1 cm or 2 cm) so that it was not necessary to make statistical simplifications such as the turbid medium assumption in DART or to use a large number of primitives or facets such as in PBRT or other rendering models. Other user-defined parameters include direct and diffuse light sources, camera position and viewing angle, number of samples per pixel, number of bands, and image resolution (Fig. 10.5).

Once the 3-D scene and simulation configurations are established, rays are emitted from each pixel and traced backward to solve the RT equation and estimate the radiance for each pixel. Orthogonal, perspective, and hemispherical photographs of the scene can be simulated, and the bidirectional reflectance factor (BRF) can be calculated by aggregating the radiance values of all pixels and addressing the incident solar irradiance.

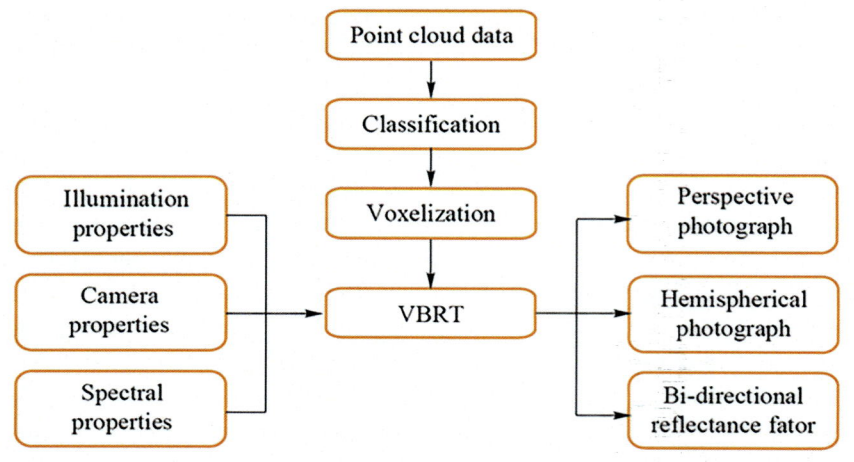

Figure 10.5 Workflow of three-dimensional simulations based on voxel-based radiative transfer (VBRT) model and LiDAR point clouds.

Comparisons of the simulated photographs and BRF patterns using PBRT and DART are presented in Figs. 10.6 and 10.7.

We found that all three models produced good representations of both artificial and realistic scenes reconstructed from LiDAR point clouds, with

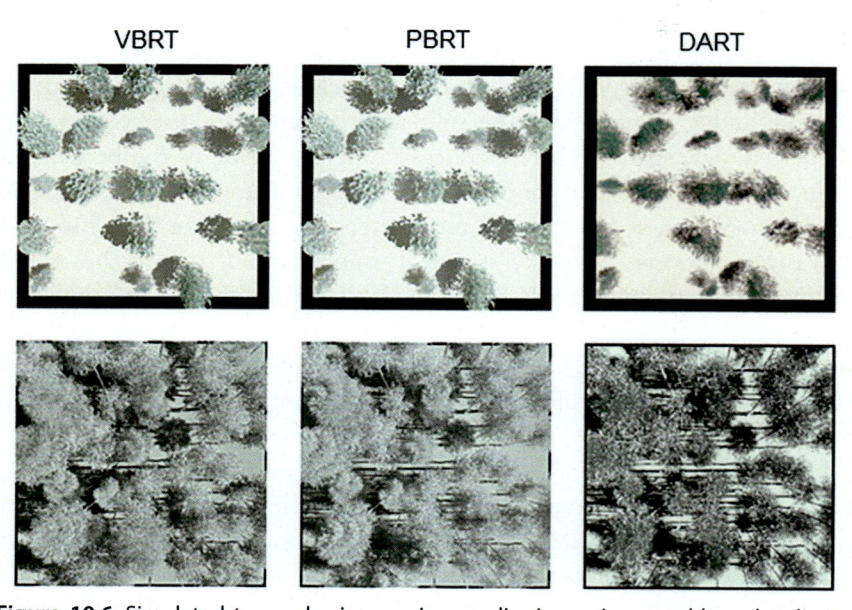

Figure 10.6 Simulated true color images in a nadir view using voxel-based radiative transfer (VBRT), physically-based ray tracer (PBRT), and discrete anisotropic radiative transfer (DART) for an artificial scene with rectangular patches (*top row*) and a realistic scene reconstructed from point clouds (*bottom row*). The voxel size is 0.01 and 0.3 m in VBRT and DART, respectively.

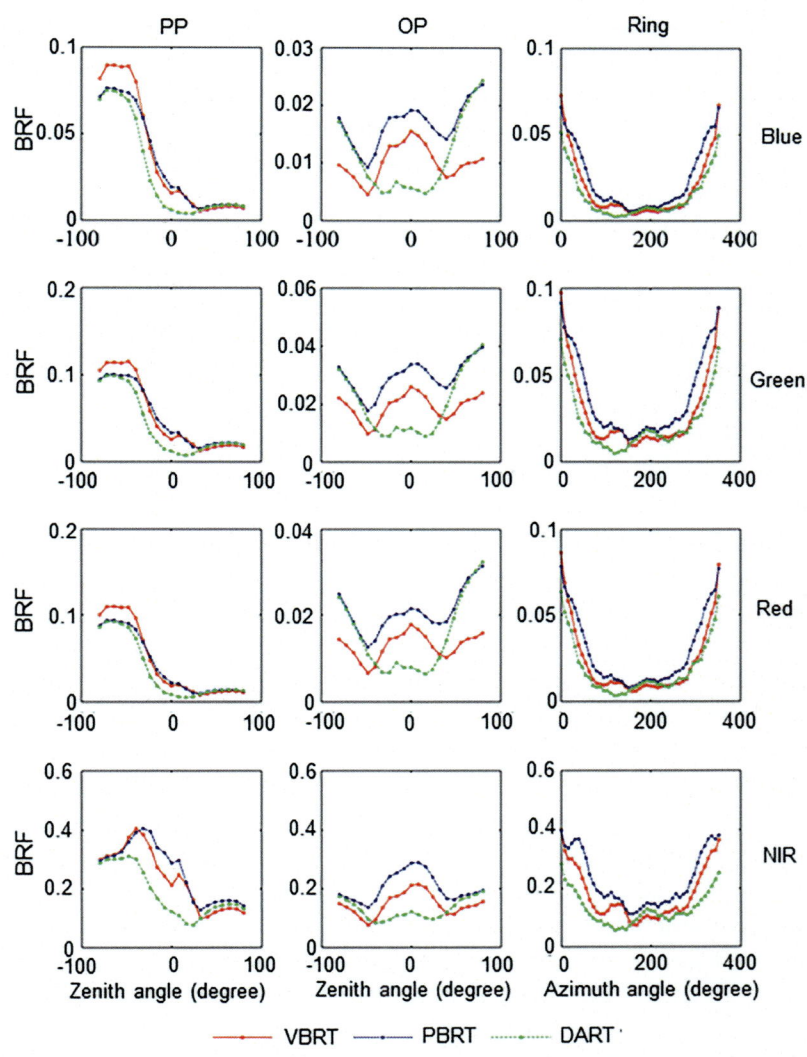

Figure 10.7 Simulated bidirectional reflectance factor (BRF) in the principal plane (PP), orthogonal plane (OP), and a view zenith ring of 37 degrees (ring) for the realistic scene with a solar zenith angle of 45 degrees. *DART*, discrete anisotropic radiative transfer; *PBRT*, physically-based ray tracer; *VBRT*, voxel-based radiative transfer.

similar patterns of highlights and shadows. The canopy in the simulated photographs using DART had a slightly darker color, likely because of its lower nadir reflectance values in the green band (Fig. 10.7). BRF values simulated by the three models along the 37–degree view zenith ring generally agreed well. In contrast, nadir BRF values simulated by DART diverged from the others. Furthermore, BRF values in the principal plane simulated by DART were also lower. These may be affected by the different mechanisms in the models: the voxels in DART are filled with turbid media that absorb and scatter the incident radiation, whereas PBRT and VBRT mainly address the scattering at the surfaces of patches or voxels. Notably, the octree of voxels in VBRT is easier to obtain than reconstructing a 3-D tree model using triangles or patches for leaves and branches, particularly for large trees or forests, and sufficient details are maintained in the octrees of voxels. This highlights an advantage of the VBRT model, indicating that quantitative comparisons of model simulation results are essential.

As VBRT uses an octree of solid voxels, estimation of the biophysical properties of individual voxels is not required. For other voxel-based RT models, particularly those using voxels filled by turbid media like DART, an essential step is to estimate the input parameters in RT simulations, such as leaf area density (LAD) and average leaf inclination angles. Various methods have been reported to estimate voxel LAD from TLS or ALS point clouds. A widely applied approach is based on the gap fraction theory that the ratio of transmitted energy through a voxel and the energy entering it can be used to estimate its LAD (Béland et al., 2014; Grau et al., 2017; Hosoi et al., 2010; Hosoi & Omasa, 2006). The inclination angle is usually calculated as the normal of a fitted surface for neighboring points (Zheng & Moskal, 2012). However, estimating the voxel level average leaf inclination angle is challenging, and a spherical or ellipsoidal leaf orientation distribution is often adopted. For shrubs with large and distinguishable leaves, a mesh reconstruction method can extract the leaf area and angle simultaneously (Bailey & Mahaffee, 2017). However, applying this method to forests with many trees may be difficult. After these essential voxel biophysical properties are estimated, 3-D simulations can be performed. Schneider et al. (2014) demonstrated that simulated at-sensor radiance with 3-D scene reconstruction using voxel grids showed better agreement with the observed radiance than individual tree-based reconstruction. Two factors constraining 3-D RT model parameterization are the high spectral variability of leaves between individual species and small-scale structural properties such as needle clumping and leaf angle distribution. Both have a strong influence on model output.

10.4 Integrating Three-Dimensional Radiative Transfer Models with Other Models

In addition to VBRT models, voxel-based 3-D reconstruction and RT simulations have been adopted in other studies and broader applications. Free software such as AMAPVox (Vincent et al., 2017) is now available and has been widely applied to reconstruct 3-D forest scenes and estimate voxel biophysical attributes from ALS point clouds (Kükenbrink et al., 2021; Weiser et al., 2021; Winiwarter et al., 2022). The reconstructed 3-D forest scenes are incorporated into 3-D RT models such as DART to simulate and analyze the light transmission within forest canopies (Kükenbrink et al., 2021).

Various studies have combined 3-D RT canopy reflectance models with other models to simulate the reflectance spectra of forests. The reflectance and transmittance spectra of individual leaves are usually essential input parameters of canopy reflectance models. Hence, RT models of individual leaves, such as the PROSPECT model (Féret et al., 2017, 2021; Jacquemoud & Baret, 1990), are often combined with canopy reflectance models. For example, simulated top-of-canopy (TOC) reflectance spectra using DART and PROSPECT were used to explore the dissimilarity in reflectance caused by different tree species (Ebengo et al., 2021). Variations in leaf optical properties and the spatial distribution of woody material can be addressed in a 3-D canopy reflectance model, which is challenging for 1-D models such as the scattering by arbitrarily inclined leaves (SAIL) model (Verhoef, 1984). These changes can result in an improved match between measured and simulated canopy reflectance (Ebengo et al., 2021). Furthermore, 3-D and 1-D RT models can be combined to simulate the canopy reflectance of mixed and structurally complex vegetation ecosystems to achieve a good balance of accuracy and efficiency. For example, Melendo-Vega et al. (2018) combined FLIGHT and PROSAIL (PROSPECT + SAIL) to simulate the seasonal dynamics of the reflectance characteristics of a multilayered tree—grass ecosystem, where trees had a fractional cover of approximately 20% and the grass understory dominated the reflectance dynamics owing to its strong seasonal variability. In this case, a 1-D homogeneous RT model was insufficient to address the impact of trees on reflectance characteristics. Still, it was challenging and unnecessary to create 3-D models for all grass individuals. The results demonstrated that the performance of PROSAIL decreased if the tree and grass layers had distinct optical or biophysical properties or the illumination and observation angles were high (Melendo-Vega et al., 2018).

Simulated TOC reflectance spectra can further be used as the bottom boundary condition for an atmosphere RT model to obtain the top-of-atmosphere reflectance spectra; these are essential for simulating the realistic light conditions within forests under various atmospheric conditions as well as for analyzing the impact of the atmosphere on satellite imagery (Kobayashi & Iwabuchi, 2008). The second simulation of the satellite signal in the solar spectrum (6S) model (Vermote et al., 1997) and the moderate resolution atmospheric transmission model (MODTRAN) (Berk et al., 1999) are among the most widely used 1-D atmosphere RT models. These models usually divide the atmosphere into parallel layers and use different predefined or custom profiles of molecule and aerosol properties to depict atmospheric absorption and scattering. Such coupled simulations can help evaluate the impact of aerosol optical depth on downward direct and diffuse radiation and thus the radiation distribution in forest canopies (Kobayashi & Iwabuchi, 2008).

Three-dimensional RT simulations also play a critical role in simulating and analyzing various ecological processes in forest ecosystems. The distribution of photosynthetic active radiation and gas exchanges such as CO_2 and H_2O within forests can be precisely simulated using DART and the ecohydrological soil—plant—atmosphere continuum model (García-Tejera et al., 2017). This 3-D simulation demonstrates the substantial 3-D heterogeneity of forest gas exchange, highlighting that simplifications in canopy structure such as a big leaf approximation lead to larger uncertainties—up to a factor of 10 for estimates of net CO_2 assimilation and transpiration (Damm et al., 2020). Solar-induced chlorophyll fluorescence (SIF) is another important indicator of forest ecosystem functions, such as gross primary production (Zhang et al., 2020). One-dimensional and 3-D RT models have been extensively combined with SIF models, such as Fluspect (Vilfan et al., 2016) and the soil canopy observation of photochemistry and energy fluxes (SCOPE) model (Van der Tol et al., 2009), to simulate canopy SIF radiance under various canopy structures. The commonly used 3-D RT and SIF combinations include DART and SCOPE (Malenovský et al., 2021), DART and Fluspect (Regaieg et al., 2021), FLiES—SIF (Sakai et al., 2020), and FluorFLIGHT (Hernández-Clemente et al., 2017). Compared with 1-D RT and SIF simulations, 3-D RT and SIF simulations can better present the spatial distributions of sun- and shade-adapted leaves and woody material and therefore highlight the impact of clumping and wood shadowing on canopy SIF emissions (Malenovský et al., 2021). Vertical and horizontal heterogeneity has also been addressed in 3-D simulations; the relative errors in simulated nadir SIF radiance based on simplified canopies of homogeneous layers can amount to 55% for oblique sun directions (Regaieg et al., 2021).

These studies demonstrate the advantages of incorporating 3-D scenes and RT models to simulate interactions between radiation and heterogeneous forests, which is essential for simulating and analyzing the physiological and ecological processes of forest ecosystems. Nonetheless, large-scale modeling is still challenging owing to the lack of suitable datasets, 3-D models, and high computational resource requirements. For a 100 m heterogeneous forest plot, the memory usage of DART can be 236 GB or 93 GB when using triangles or turbid voxels at a resolution of 0.2 m, with a running time of 14 and 3 h, respectively (Qi et al., 2019). In contrast, the LESS model can significantly reduce the resource requirement to 5 GB and 0.6 h (Qi et al., 2019), which is a more feasible approach for large-scale simulations of 3-D heterogeneous forest scenes. However, new algorithms and techniques, such as GPU computing, are emerging (Bailey et al., 2014), and computer infrastructure is also evolving. All of these lay a foundation for more practical and efficient 3-D simulations of realistic, large-scale forest scenes in the future.

10.5 Chapter Summary

This chapter briefly introduced the principles of 3-D radiative transfer modeling. Three-dimensional point cloud data from LiDAR or photogrammetry are important inputs that drive 3-D RT models. We described a voxel-based RT model because it is designed for RT simulations with point cloud data as the model input. The RT simulation results can be validated by field measurements or benchmark datasets, such as the radiation transfer model intercomparison initiative (Widlowski et al., 2015). We also introduced a range of remote sensing and ecological applications that combine 3-D RT and other models, such as SIF and gas exchange models. These studies highlight the superiority of 3-D RT over 1-D RT in handling horizontal and vertical heterogeneous forest canopies and the necessity of using 3-D RT to reduce the errors in model simulation results. Although challenges exist in large-scale 3-D simulations, a promising future is expected with advances in algorithms and techniques.

References

Bailey, B. N., & Mahaffee, W. F. (2017). Rapid measurement of the three-dimensional distribution of leaf orientation and the leaf angle probability density function using terrestrial LiDAR scanning. *Remote Sensing of Environment, 194*, 63−76.

Bailey, B., Overby, M., Willemsen, P., Pardyjak, E., Mahaffee, W., & Stoll, R. (2014). A scalable plant-resolving radiative transfer model based on optimized GPU ray tracing. *Agricultural and Forest Meteorology, 198*, 192−208.

Bauwens, S., Bartholomeus, H., Calders, K., & Lejeune, P. (2016). Forest inventory with terrestrial LiDAR: A comparison of static and hand-held mobile laser scanning. *Forests, 7*(6). Article 127.

Béland, M., Baldocchi, D. D., Widlowski, J.-L., Fournier, R. A., & Verstraete, M. M. (2014). On seeing the wood from the leaves and the role of voxel size in determining leaf area distribution of forests with terrestrial LiDAR. *Agricultural and Forest Meteorology, 184*, 82−97.

Berk, A., Anderson, G. P., Bernstein, L. S., Acharya, P. K., Dothe, H., Matthew, M. W., Adler-Golden, S. M., Chetwynd, J. H., Jr., Richtsmeier, S. C., & Pukall, B. (1999). MODTRAN4 radiative transfer modeling for atmospheric correction. In *Optical spectroscopic techniques and instrumentation for atmospheric and space research III*.

Chen, Q., Baldocchi, D., Gong, P., & Dawson, T. (2008). Modeling radiation and photosynthesis of a heterogeneous savanna woodland landscape with a hierarchy of model complexities. *Agricultural and Forest Meteorology, 148*(6), 1005−1020.

Chen, J. M., Li, X., Nilson, T., & Strahler, A. (2000). Recent advances in geometrical optical modelling and its applications. *Remote Sensing Reviews, 18*(2−4), 227−262.

Cifuentes, R., Van der Zande, D., Farifteh, J., Salas, C., & Coppin, P. (2014). Effects of voxel size and sampling setup on the estimation of forest canopy gap fraction from terrestrial laser scanning data. *Agricultural and Forest Meteorology, 194*, 230−240.

Damm, A., Paul-Limoges, E., Kukenbrink, D., Bachofen, C., & Morsdorf, F. (2020). Remote sensing of forest gas exchange: Considerations derived from a tomographic perspective. *Global Change Biology, 26*, 2717−2727.

Disney, M. I., Lewis, P., Gomez-Dans, J., Roy, D., Wooster, M. J., & Lajas, D. (2011). 3D radiative transfer modelling of fire impacts on a two-layer savanna system. *Remote Sensing of Environment, 115*(8), 1866−1881.

Disney, M., Lewis, P., & North, P. (2000). Monte Carlo ray tracing in optical canopy reflectance modelling. *Remote Sensing Reviews, 18*(2−4), 163−196.

Disney, M., Lewis, P., & Saich, P. (2006). 3D modelling of forest canopy structure for remote sensing simulations in the optical and microwave domains. *Remote Sensing of Environment, 100*(1), 114−132.

Ebengo, D. M., de Boissieu, F., Vincent, G., Weber, C., & Féret, J.-B. (2021). Simulating imaging spectroscopy in tropical forest with 3D radiative transfer modeling. *Remote Sensing, 13*(11).

Essery, R., Bunting, P., Rowlands, A., Rutter, N., Hardy, J., Melloh, R., Link, T., Marks, D., & Pomeroy, J. (2008). Radiative transfer modeling of a coniferous canopy characterized by airborne remote sensing. *Journal of Hydrometeorology, 9*(2), 228−241.

Féret, J.-B., Berger, K., de Boissieu, F., & Malenovský, Z. (2021). PROSPECT-PRO for estimating content of nitrogen-containing leaf proteins and other carbon-based constituents. *Remote Sensing of Environment, 252*.

Féret, J.-B., Gitelson, A., Noble, S., & Jacquemoud, S. (2017). PROSPECT-D: Towards modeling leaf optical properties through a complete lifecycle. *Remote Sensing of Environment, 193*, 204−215.

Flener, C., Vaaja, M., Jaakkola, A., Krooks, A., Kaartinen, H., Kukko, A., Kasvi, E., Hyyppä, H., Hyyppä, J., & Alho, P. (2013). Seamless mapping of river channels at high resolution using mobile liDAR and UAV-photography. *Remote Sensing, 5*(12), 6382−6407.

García-Tejera, O., López-Bernal, Á., Testi, L., & Villalobos, F. J. (2017). A soil-plant-at-mosphere continuum (SPAC) model for simulating tree transpiration with a soil multi-compartment solution. *Plant and Soil, 412*(1−2), 215−233.

Gastellu-Etchegorry, J.-P., Demarez, V., Pinel, V., & Zagolski, F. (1996). Modeling radi-ative transfer in heterogeneous 3-D vegetation canopies. *Remote Sensing of Environment, 58*(2), 131−156.

Gastellu-Etchegorry, J.-P., Lauret, N., Yin, T., Landier, L., Kallel, A., Malenovský, Z., Al Bitar, A., Aval, J., Benhmida, S., & Qi, J. (2017). Dart: Recent advances in remote sensing data modeling with atmosphere, polarization, and chlorophyll fluorescence. *IEEE Journal of Selected Topics in Applied Earth Observations and Remote Sensing,* (10), 6.

Gastellu-Etchegorry, J. P., Martin, E., & Gascon, F. (2004). Dart: A 3D model for simulating satellite images and studying surface radiation budget. *International Journal of Remote Sensing, 25*(1), 73−96.

Goodwin, N. R., Coops, N. C., & Culvenor, D. S. (2007). Development of a simulation model to predict LiDAR interception in forested environments. *Remote Sensing of Environment, 111*(4), 481−492.

Govaerts, Y. M., & Verstraete, M. M. (1998). Raytran: A Monte Carlo ray-tracing model to compute light scattering in three-dimensional heterogeneous media. *IEEE Transactions on Geoscience and Remote Sensing, 36*(2), 493−505.

Grau, E., Durrieu, S., Fournier, R., Gastellu-Etchegorry, J.-P., & Yin, T. (2017). Estimation of 3D vegetation density with terrestrial laser scanning data using voxels. A sensitivity analysis of influencing parameters. *Remote Sensing of Environment, 191*, 373−388.

Harwin, S., & Lucieer, A. (2012). Assessing the accuracy of georeferenced point clouds produced via multi-view stereopsis from unmanned aerial vehicle (UAV) imagery. *Remote Sensing, 4*(6), 1573−1599.

Haverd, V., Lovell, J., Cuntz, M., Jupp, D., Newnham, G., & Sea, W. (2012). The canopy semi-analytic P gap and radiative transfer (CanSPART) model: Formulation and application. *Agricultural and Forest Meteorology, 160*, 14−35.

Hernández-Clemente, R., North, P. R. J., Hornero, A., & Zarco-Tejada, P. J. (2017). Assessing the effects of forest health on sun-induced chlorophyll fluorescence using the FluorFLIGHT 3-D radiative transfer model to account for forest structure. *Remote Sensing of Environment, 193*, 165−179.

Hosoi, F., Nakai, Y., & Omasa, K. (2010). Estimation and error analysis of woody canopy leaf area density profiles using 3-D airborne and ground-based scanning lidar remote-sensing techniques. *IEEE Transactions on Geoscience and Remote Sensing, 48*(5), 2215−2223.

Hosoi, F., & Omasa, K. (2006). Voxel-based 3-D modeling of individual trees for estimating leaf area density using high-resolution portable scanning lidar. *IEEE Transactions on Geoscience and Remote Sensing, 44*(12), 3610−3618.

Hosoi, F., & Omasa, K. (2007). Factors contributing to accuracy in the estimation of the woody canopy leaf area density profile using 3D portable lidar imaging. *Journal of Experimental Botany, 58*(12), 3463−3473.

Huang, H., Qin, W., & Liu, Q. (2013). Rapid: A radiosity applicable to porous individual objects for directional reflectance over complex vegetated scenes. *Remote Sensing of Environment, 132*, 221−237.

Jacquemoud, S., & Baret, F. (1990). Prospect: A model of leaf optical properties spectra. *Remote Sensing of Environment, 34*(2), 75−91.

Kajiya, J. T. (1986). *The rendering equation.* ACM Siggraph Computer Graphics.

Kimes, D. S., & Kirchner, J. A. (1982). Radiative transfer model for heterogeneous 3-D scenes. *Applied Optics, 21*(22), 4119−4129.

Kimes, D. S., Knyazikhin, Y., Privette, J. L., Abuelgasim, A. A., & Gao, F. (2000). Inversion methods for physically-based models. *Remote Sensing Reviews, 18*(2), 381—439.

Kobayashi, H., Baldocchi, D. D., Ryu, Y., Chen, Q., Ma, S., Osuna, J. L., & Ustin, S. L. (2012). Modeling energy and carbon fluxes in a heterogeneous oak woodland: A three-dimensional approach. *Agricultural and Forest Meteorology, 152*(1), 83—100.

Kobayashi, H., & Iwabuchi, H. (2008). A coupled 1-D atmosphere and 3-D canopy radiative transfer model for canopy reflectance, light environment, and photosynthesis simulation in a heterogeneous landscape. *Remote Sensing of Environment, 112*(1), 173—185.

Koetz, B., Morsdorf, F., Sun, G., Ranson, K. J., Itten, K., & Allgöwer, B. (2006). Inversion of a lidar waveform model for forest biophysical parameter estimation. *IEEE Geoscience and Remote Sensing Letters, 3*(1), 49—53.

Kükenbrink, D., Schneider, F. D., Schmid, B., Gastellu-Etchegorry, J.-P., Schaepman, M. E., & Morsdorf, F. (2021). Modelling of three-dimensional, diurnal light extinction in two contrasting forests. *Agricultural and Forest Meteorology, 296*.

Lafortune, E. (1995). *Mathematical models and Monte Carlo algorithms for physically based rendering*. Katholieke Universiteit Leuven.

Lafortune, E. P., & Willems, Y. D. (1994). *Using the modified phong reflectance model for physically based rendering* (cw reports).

Laine, S., & Karras, T. (2010). Efficient sparse voxel octrees. *IEEE Transactions on Visualization and Computer Graphics, 17*(8), 1048—1059.

van Leeuwen, M., Coops, N., & Black, T. (2015). Using stochastic ray tracing to simulate a dense time series of gross primary productivity. *Remote Sensing, 7*(12), 17272—17290.

Lewis, P. (1999). Three-dimensional plant modelling for remote sensing simulation studies using the botanical plant modelling system. *Agronomie, 19*(3—4), 185—210.

Liang, X., Kankare, V., Hyyppä, J., Wang, Y., Kukko, A., Haggrén, H., Yu, X., Kaartinen, H., Jaakkola, A., & Guan, F. (2016). Terrestrial laser scanning in forest inventories. *ISPRS Journal of Photogrammetry and Remote Sensing, 115*, 63—77.

Liang, X., Wang, Y., Jaakkola, A., Kukko, A., Kaartinen, H., Hyyppä, J., Honkavaara, E., & Liu, J. (2015). Forest data collection using terrestrial image-based point clouds from a handheld camera compared to terrestrial and personal laser scanning. *IEEE Transactions on Geoscience and Remote Sensing, 53*(9), 5117—5132. Article 7109840.

Li, W., Guo, Q., Tao, S., & Su, Y. (2018). Vbrt: A novel voxel-based radiative transfer model for heterogeneous three-dimensional forest scenes. *Remote Sensing of Environment, 206*, 318—335.

Li, X., Strahler, A. H., & Woodcock, C. E. (1995). A hybrid geometric optical-radiative transfer approach for modeling albedo and directional reflectance of discontinuous canopies. *IEEE Transactions on Geoscience and Remote Sensing, 33*(2), 466—480.

Magney, T. S., Eitel, J. U., Griffin, K. L., Boelman, N. T., Greaves, H. E., Prager, C. M., Logan, B. A., Zheng, G., Ma, L., & Fortin, E. A. (2016). LiDAR canopy radiation model reveals patterns of photosynthetic partitioning in an Arctic shrub. *Agricultural and Forest Meteorology, 221*, 78—93.

Malenovský, Z., Regaieg, O., Yin, T., Lauret, N., Guilleux, J., Chavanon, E., Duran, N., Janoutová, R., Delavois, A., Meynier, J., Medjdoub, G., Yang, P., van der Tol, C., Morton, D., Cook, B. D., & Gastellu-Etchegorry, J.-P. (2021). Discrete anisotropic radiative transfer modelling of solar-induced chlorophyll fluorescence: Structural impacts in geometrically explicit vegetation canopies. *Remote Sensing of Environment, 263*, 112564.

Melendo-Vega, J., Martín, M., Pacheco-Labrador, J., González-Cascón, R., Moreno, G., Pérez, F., Migliavacca, M., García, M., North, P., & Riaño, D. (2018). Improving the performance of 3-D radiative transfer model FLIGHT to simulate optical properties of a tree-grass ecosystem. *Remote Sensing, 10*(12), 2061.

Morton, D. C., Nagol, J., Carabajal, C. C., Rosette, J., Palace, M., Cook, B. D., Vermote, E. F., Harding, D. J., & North, P. R. J. (2014). Amazon forests maintain consistent canopy structure and greenness during the dry season. *Nature, 506*(7487), 221–224.

Ni-Meister, W., & Gao, H. (2011). Assessing the impacts of vegetation heterogeneity on energy fluxes and snowmelt in boreal forests. *Journal of Plant Ecology, 4*(1–2), 37–47.

Ni, W., Li, X., Woodcock, C. E., Caetano, M. R., & Strahler, A. H. (1999). An analytical hybrid GORT model for bidirectional reflectance over discontinuous plant canopies. *IEEE Transactions on Geoscience and Remote Sensing, 37*(2), 987–999.

North, P. R. J. (1996). Three-dimensional forest light interaction model using a Monte Carlo method. *IEEE Transactions on Geoscience and Remote Sensing, 34*(4), 946–956.

Pharr, M., Jakob, W., & Humphreys, G. (2004). *Physically based rendering: From theory to implementation.* Morgan Kaufmann.

Qin, W., & Gerstl, S. A. W. (2000). 3-D scene modeling of semidesert vegetation cover and its radiation regime. *Remote Sensing of Environment, 74*(1), 145–162.

Qi, J., Xie, D., Yin, T., Yan, G., Gastellu-Etchegorry, J.-P., Li, L., Zhang, W., Mu, X., & Norford, L. K. (2019). Less: LargE-scale remote sensing data and image simulation framework over heterogeneous 3D scenes. *Remote Sensing of Environment, 221,* 695–706.

Regaieg, O., Yin, T., Malenovský, Z., Cook, B. D., Morton, D. C., & Gastellu-Etchegorry, J.-P. (2021). Assessing impacts of canopy 3D structure on chlorophyll fluorescence radiance and radiative budget of deciduous forest stands using DART. *Remote Sensing of Environment, 265,* 112673–112695.

Rodrigues, J., Loke, R., & du Buf, J. (2000). Fast segmentation of 3D data using an octree. In *Proceedings of the 11th portuguese conference on pattern recognition RECPAD, Porto, May 11–12* (p. 185e189).

Sakai, Y., Kobayashi, H., & Kato, T. (2020). FLiES-SIF version 1.0: Three-dimensional radiative transfer model for estimating solar induced fluorescence. *Geoscientific Model Development, 13*(9), 4041–4066.

Schaepman-Strub, G., Schaepman, M., Painter, T., Dangel, S., & Martonchik, J. (2006). Reflectance quantities in optical remote sensing—definitions and case studies. *Remote Sensing of Environment, 103*(1), 27–42.

Schlerf, M., & Atzberger, C. (2006). Inversion of a forest reflectance model to estimate structural canopy variables from hyperspectral remote sensing data. *Remote Sensing of Environment, 100*(3), 281–294.

Schneider, F. D., Leiterer, R., Morsdorf, F., Gastellu-Etchegorry, J.-P., Lauret, N., Pfeifer, N., & Schaepman, M. E. (2014). Simulating imaging spectrometer data: 3D forest modeling based on LiDAR and in situ data. *Remote Sensing of Environment, 152,* 235–250.

Stuckens, J., Somers, B., Delalieux, S., Verstraeten, W., & Coppin, P. (2009). The impact of common assumptions on canopy radiative transfer simulations: A case study in citrus orchards. *Journal of Quantitative Spectroscopy and Radiative Transfer, 110*(1), 1–21.

Talbot, J. F., Cline, D., & Egbert, P. K. (2005). Importance resampling for global illumination. *Proceedings of Eurographics Symposium on Rendering,* 139–146.

Tao, S., Guo, Q., Su, Y., Xu, S., Li, Y., & Wu, F. (2015). A geometric method for wood-leaf separation using terrestrial and simulated lidar data. *Photogrammetric Engineering & Remote Sensing, 81*(10), 767–776.

Van der Tol, C., Verhoef, W., Timmermans, J., Verhoef, A., & Su, Z. (2009). An integrated model of soil-canopy spectral radiances, photosynthesis, fluorescence, temperature and energy balance. *Biogeosciences, 6*(12), 3109–3129.

Verhoef, W. (1984). Light scattering by leaf layers with application to canopy reflectance modeling: The SAIL model. *Remote Sensing of Environment, 16*(2), 125–141.

Vermote, E. F., Tanré, D., Deuze, J. L., Herman, M., & Morcette, J. J. (1997). Second simulation of the satellite signal in the solar spectrum, 6S: An overview. *IEEE Transactions on Geoscience and Remote Sensing, 35*(3), 675−686.

Vilfan, N., van der Tol, C., Muller, O., Rascher, U., & Verhoef, W. (2016). Fluspect-B: A model for leaf fluorescence, reflectance and transmittance spectra. *Remote Sensing of Environment, 186*, 596−615.

Vincent, G., Antin, C., Laurans, M., Heurtebize, J., Durrieu, S., Lavalley, C., & Dauzat, J. (2017). Mapping plant area index of tropical evergreen forest by airborne laser scanning. A cross-validation study using LAI2200 optical sensor. *Remote Sensing of Environment, 198*, 254−266.

Wallace, L., Lucieer, A., Watson, C., & Turner, D. (2012). Development of a UAV-LiDAR system with application to forest inventory. *Remote Sensing, 4*(6), 1519−1543.

Weiser, H., Winiwarter, L., Anders, K., Fassnacht, F. E., & Höfle, B. (2021). Opaque voxel-based tree models for virtual laser scanning in forestry applications. *Remote Sensing of Environment, 265*, 112641.

Westoby, M., Brasington, J., Glasser, N., Hambrey, M., & Reynolds, J. (2012). 'Structure-from-Motion' photogrammetry: A low-cost, effective tool for geoscience applications. *Geomorphology, 179*, 300−314.

Widlowski, J.-L., Côté, J.-F., & Béland, M. (2014). Abstract tree crowns in 3D radiative transfer models: Impact on simulated open-canopy reflectances. *Remote Sensing of Environment, 142*, 155−175.

Widlowski, J.-L., Lavergne, T., Pinty, B., Verstraete, M., & Gobron, N. (2006). Rayspread: A virtual laboratory for rapid BRF simulations over 3-D plant canopies. In *Computational methods in transport* (pp. 211−231). Springer.

Widlowski, J.-L., Mio, C., Disney, M., Adams, J., Andredakis, I., Atzberger, C., Brennan, J., Busetto, L., Chelle, M., & Ceccherini, G. (2015). The fourth phase of the radiative transfer model intercomparison (RAMI) exercise: Actual canopy scenarios and conformity testing. *Remote Sensing of Environment, 169*, 418−437.

Widlowski, J. L., Pinty, B., Clerici, M., Dai, Y., De Kauwe, M., De Ridder, K., Kallel, A., Kobayashi, H., Lavergne, T., & Ni-Meister, W. (2011). RAMI4PILPS: An intercomparison of formulations for the partitioning of solar radiation in land surface models. *Journal of Geophysical Research: Biogeosciences, 116*(G2).

Winiwarter, L., Esmorís Pena, A. M., Weiser, H., Anders, K., Martínez Sánchez, J., Searle, M., & Höfle, B. (2022). Virtual laser scanning with HELIOS++: A novel take on ray tracing-based simulation of topographic full-waveform 3D laser scanning. *Remote Sensing of Environment, 269*, 112772.

Woodcock, C. E., Collins, J. B., Jakabhazy, V. D., Li, X., Macomber, S. A., Wu, Y., & Company, A. S. (1997). Inversion of the li-strahler canopy reflectance model for mapping forest structure. *IEEE Transactions on Geoscience and Remote Sensing, 35*(2), 405−414.

Zhang, Z., Zhang, Y., Porcar-Castell, A., Joiner, J., Guanter, L., Yang, X., Migliavacca, M., Ju, W., Sun, Z., Chen, S., Martini, D., Zhang, Q., Li, Z., Cleverly, J., Wang, H., & Goulas, Y. (2020). Reduction of structural impacts and distinction of photosynthetic pathways in a global estimation of GPP from space-borne solar-induced chlorophyll fluorescence. *Remote Sensing of Environment, 240*, 111722.

Zheng, G., & Moskal, L. M. (2012). Leaf orientation retrieval from terrestrial laser scanning (TLS) data. *IEEE Transactions on Geoscience and Remote Sensing, 50*(10), 3970−3979.

CHAPTER 11

Visualization and Reconstruction of Forest Ecosystems

Contents

In recent years, studies have increasingly been conducted on tree growth mechanisms and the simulation and prediction of changes in forest ecosystems. To achieve the goals of these studies, three-dimensional (3-D) visualization and reconstruction of forest landscapes have become a focus in many fields, including physical geography, ecology, botany, and forest management. Visualizing forest ecosystems provides an intuitive and convenient method to investigate the processes of tree growth and the interaction between trees and environmental factors. Three-dimensional visualization of simulated tree dynamics enables researchers to observe the structural change and organ distribution of trees during their life cycles more intuitively and effectively, which assists in tree growth analysis. Moreover, establishing and simulating the geometric structure of trees facilitates the calculation and prediction of important biophysical attributes of forests, such as the light regime inside the canopy and the leaf area index.

LiDAR Principles, Processing and Applications in Forest Ecology
ISBN 978-0-12-823894-3
https://doi.org/10.1016/B978-0-12-823894-3.00011-6

The 3-D reconstruction of vegetation is generally divided into two scalesdmicroscopic and macroscopic. Microscopic reconstruction aims to establish a fine-scale 3-D model of an individual tree by detailed modeling of leaves, branches, and trunks, whereas macroscopic reconstruction aims to reconstruct a landscape containing many trees and related terrain features.

11.1 Three-Dimensional Reconstruction of Individual Trees

Individual tree modeling is the foundation of three-dimensional (3-D) landscape modeling. Trees can be accurately described by solid geometric primitives. The methods for individual trees can be divided into three categories: rule-based, sketch-based, and image-based. These methods are used in various fields, with advantages and disadvantages for each. Rule-based methods are the most commonly used in the 3-D reconstruction of individual trees, but they require user expertise in biology and mechanics. Furthermore, before users can generate an ideal tree model, many parameters must be determined, such as tree height and diameter at breast height (DBH) (Holton, 1994; Lin et al., 2011; Lintermann & Deussen, 1999). Sketch-based methods are used to construct a tree based on a model hypothesis (such as a catheter model) (Chen, Neubert, et al., 2008; Cornea et al., 2007; Okabe et al., 2006). Using the intuitive human interface offered by this method, users can construct a complex tree model through a series of steps. However, the algorithm used in sketch-based methods is complicated, and it is challenging to reconstruct various tree types (Lin et al., 2011). In contrast to the two modeling methods above, image-based methods usually do not require users to participate in the generation of tree models. Image-based methods obtain images of a tree from different perspectives and generate point clouds of the tree using photogrammetry (Dai et al., 2010; Neubert et al., 2007; Shlyakhter et al., 2001). The generated point clouds can accurately represent the external morphology of a tree. Still, they cannot capture internal canopy structures and small organisms in sufficient detail owing to mutual occlusion between branches and leaves.

In recent years, the cost of data acquisition has decreased with the development of light detection and ranging (LiDAR) technology, which is widely used in forestry, ecology, botany, and related areas to capture detailed structural information about trees. Using the point clouds provided by LiDAR scanning to build fine-scale 3-D models of trees has become a research focus in recent years (Huang, 2013). Many studies have been carried out on the 3-D reconstruction of individual trees using point cloud data. For example, Gorte and Winterhalder (2004) projected the LiDAR point cloud data of trees to a 3-D space using coordinate transformation, constructed the topology of branches using various views of point cloud data in the 3-D space, and analyzed the connections between branches according to their topology, which resulted in a good performance in a 3-D model reconstruction. Pfeifer and Winterhalder (2004) and Thies et al. (2004) adopted cylinders to represent tree branches, extracted the related parameters from LiDAR point clouds, and reconstructed 3-D tree models using these parameters, proving that the method applied to LiDAR point clouds. Hosoi and Omasa (2012) constructed 3-D models of individual trees and forests using a primitive method of discrete point clouds. Côté et al. (2012) combined terrestrial LiDAR data with airborne LiDAR data. They used the L-Architect model to construct a 3-D model of a small forest on southeast Vancouver Island, Canada. The tree species included *Pseudotsuga menziesii*, *Thuja plicata*, *Tsuga chinensis*, and *Pinus thunbergii*.

Compared with spaceborne and airborne LiDAR, terrestrial LiDAR—or terrestrial laser scanning (TLS)—provides more reliable data for the 3-D modeling of individual trees owing to its high point density and positioning accuracy. After essential preprocessing, such as point cloud registration and denoising, 3-D modeling is achieved by extracting the topological structure between branches and leaves and estimating the corresponding structural parameters from acquired fine point clouds. The steps include (1) separating branch points, i.e., dividing the initial point cloud into branch points and leaf points, (2) extracting the skeleton, (3) adding twigs, (4) surface reconstruction of the truck and branches, and (5) adding leaves (Fig. 11.1).

11.1.1 Branch Point Separation

Pulses emitted by a laser scanner are intercepted by branches or leaves, and hence, the returned beams contain spatial information such as their orientation and distance from the scanner. This forms the basis of the 3-D modeling of individual trees. The first step in the 3-D reconstruction of individual trees is to separate the branch points from leaf points, i.e., branch point separation. Two methods are usually used to separate branches and leaves: manual separation and automatic separation using specific algorithms. Hosoi and Omasa (2006) separated the branch points and leaf points

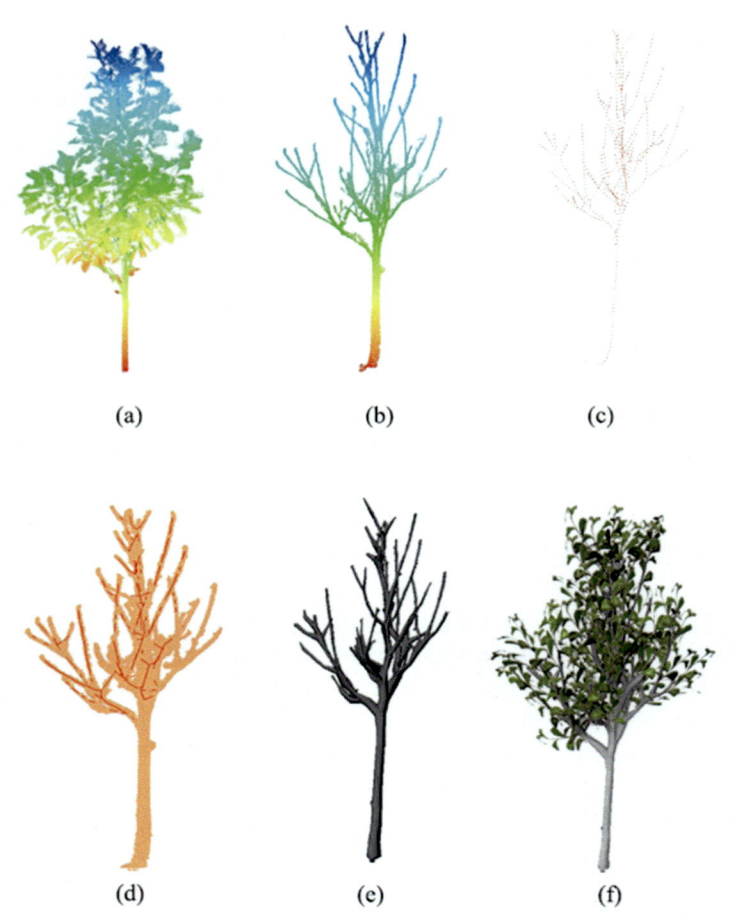

(a) (b) (c)

(d) (e) (f)

Figure 11.1 The process for 3-D reconstruction of individual trees. (a) Point cloud data of an individual tree, (b) separated branch points, (c) skeleton points, (d) skeleton lines, (e) surface reconstruction, and (f) leaf addition.

of TLS point clouds using a manual method, which was relatively accurate but time-consuming. Automatic separation algorithms can be divided into two main groups: one uses the attributes recorded by LiDAR point clouds (including intensity, reflectivity, and waveform information), and the other uses the morphological geometry of branches and leaves (Luo & Sohn, 2010). Côté et al. (2009, 2011) used the reflectance difference between branches and leaves in the infrared band to separate branch points and leaf points, as branches have higher reflectance than leaves. Yao et al. (2011) and Yang et al. (2013) separated branches and leaves based on differences in the widths and heights of their echo signals. Xu et al. (2007) used geometric information about the tree to separate the branch points. They regarded points with the shortest path to the tree root node below a certain empirical threshold as branch points and the others as leaf points. Tao et al. (2015) used the Hough transform to fit circles to separate branch points from the original point cloud and then separated branch points from leaf points according to their geometries. The Hough transform algorithm is quite accurate but requires high-quality point cloud data. In summary, separation algorithms solely relying on certain attribute information from the tree point clouds are rare. It is more common to combine a variety of information to separate branch and leaf points; this approach accounts for a large proportion of current studies on branch and leaf classification algorithms (Huang, 2013).

11.1.2 Extraction of the Skeleton

The skeleton of a tree, also known as the central axis, is an ideal representation of the original connectivity and topological structure of trees using objects such as thin curves. The essence of the skeleton is an abstract form, which is an intuitive and effective way to describe the topological characteristics of a single tree. In recent years, there have been many studies on skeleton extraction algorithms using point clouds, with a special focus on enhancing the efficiency and robustness of the algorithms. The relevant algorithms for extracting the skeleton from branch points can be divided into four categories: clustering-based, level set-based, graph theory-based, and Laplacian operator-based (Huang, 2013). In clustering-based methods, the Euclidean distance between each point of a branch and its neighboring points is calculated to find the nearest neighboring point and construct a neighborhood graph. Then the neighbor relationship is used to group all the points into clusters, and the centers of neighboring clusters are connected to build the skeleton, given that each point has the shortest distance to the root node (Xu et al., 2007; Yan et al., 2009). For methods based on the level set, a neighborhood graph is first constructed to obtain a mesh for all the trunk points; then, the distance between each point and a user-defined root node is calculated, and a new connectivity network of the trunk is built according to the shortest distance. All points within a specific distance range of a root node are grouped into a level set, and finally, the centers of all level sets are connected to constitute the skeleton (Côté et al., 2009, 2011; Verroust & Lazarus, 1999). Graph theory-based methods use an algorithm that first voxelizes the point cloud and builds an octree according to the spatial connectivity of the voxels and then merges or deletes the nodes using rules such as erosion and dilation to extract the skeleton (Bucksch et al., 2010; Bucksch & Lindenbergh, 2008). The Laplacian operator-based algorithm has been one of the most widely applied algorithms in recent years. The basic calculation process first constructs a Laplacian weighting matrix for neighboring points, then iteratively solves the equation and updates the weights to shrink the point cloud data to a series of skeleton points, and finally uses the minimum spanning tree algorithm to connect all the skeleton points (Cao et al., 2010).

11.1.3 Addition of Twigs

In point clouds acquired using LiDAR scanning, parts of the twigs may be missing or incomplete owing to occlusion by leaves or branches. In addition, twig points may be misclassified as leaf points during leaf and wood separation because there are only subtle spectral and geometric differences. Therefore, to make the final 3-D model more realistic and complete, it is necessary to add twigs in the model artificially according to the growth rules of trees. The addition of twigs is generally based on the skeleton from step (2) and the leaf points from step (1). The basic process of adding twigs starts from the edge nodes of the existing skeleton and searches for points that meet certain requirements of distance and angle in the neighboring leaf points; the requirements are based on the growth rules of the specific tree. These points are regarded as the twig points of the trunk and are added to the skeleton (Côté et al., 2009, 2011; Runions et al., 2007; Xu et al., 2007).

11.1.4 Surface Reconstruction of Trunks and Branches

After constructing the complete skeleton, the trunk and branches are segmented into sections, and the radius of each section is determined to reconstruct the surface before reconstructing the whole tree. The process is as follows: (1) the point cloud of the trunk is segmented into sections, and their radius values are estimated by fitting a cylinder (more generally a truncated cone) or B-spline function; (2) after the reconstruction of each section of the trunk, the connection between sections of the trunk is smoothed following the corresponding trunk growth rules (Livny et al., 2010; Xu et al., 2007); and (3) textures are added to trunks and branches to enhance the visual reality of the model. Various 3-D reconstruction software packages provide trunk textures for trees, including the option to apply textures to reconstructed trunks.

11.1.5 Addition of Leaves

As the final and most important step for the 3-D reconstruction of individual trees, the addition of leaves is performed after the reconstruction of the trunk and branches, which greatly increases the integrity and authenticity of the tree model. Because leaves in point clouds are often incomplete owing to occlusion and movement caused by wind, it is difficult to reconstruct all the leaves directly from point clouds. At present, adding leaves is mainly achieved manually following the same rules for adding twigs in step (3). Côté et al. (2011) also suggested that layered leaf area index values and simulated canopy reflectance values using radiative transfer modeling should be used to verify the 3-D model of trees. This can enhance the integrity and authenticity of the model and give it ecological value.

11.2 Three-Dimensional Reconstruction of Forest Landscapes

The concepts and technology of landscape visualization have been proposed for many years. Visualizations are widely used in various industrial and civil applications, such as urban planning and construction, garden landscape design, and commercial exhibitions. The 3-D visualization of forest landscapes is significant for forest management. Because forests usually occupy a large area and are composed of diverse vegetation types and complex terrain, the visual simulation of forests at the local landscape level can intuitively display the structural pattern of forests and predict their dynamics when combined with dynamic vegetation models. This is important for providing practical guidelines for ecosystem management and forestry production (Dongli, 2005). Early studies on forest visualization usually displayed the forest landscape in two-dimensional (2-D) plots, which showed the overall distribution and condition of forests but could not fully meet the requirements of forest management. Taking logging planning as an example, forestry enterprises and management departments require 3-D perspectives that can show the cutting structure and layout, which in turn can be adopted to identify changes in the logging area and layout and serve as a logging plan or in public evaluation for decision-makers (Song & Ling, 1997).

In recent years, owing to improvements in 3-D data acquisition capabilities and modeling techniques (Chen, Zhang, et al., 2008), landscape visualization technology has developed rapidly and has shifted from 2-D to 3-D visualization (Ding & Zhang, 2005; Yu et al., 2008). The 3-D visualization of forest landscapes allows the virtual construction and simulation of the growth and change processes of forests using computer modeling and visualization techniques. After 3-D landscapes have been artificially reconstructed by configuring the relevant influencing factors, the structure, function, and dynamic processes of forests can be further understood. The interaction and influence of related factors can be explored, providing a reference for forest management planning and related decision-making (Gao, 2008).

The 3-D visualization of forest landscapes includes the reconstruction of terrain and vegetation. Previous studies have usually achieved this with satellite remote sensing images, field measurements, and forest resource inventory data. The underlying topography was generated from publicly available topographic data, and the distribution and specific biophysical parameters of vegetation were obtained through forest surveys. However, owing to the limitations of data accuracy, the accuracy and fidelity of overall landscape topography may be greatly reduced. For example, the commonly used Shuttle Radar Topography Mission data products are at a spatial resolution of 30 m, and the terrain information contains errors; thus, the data cannot accurately represent the terrain under forests. The emergence of LiDAR has provided more effective data for the 3-D visualization of forest landscapes.

The terrain and tree distribution information required for forest landscape reconstruction can be obtained from LiDAR point clouds, and they can be visualized using 3-D modeling software. For simulating large scenes, VUE performs better than other visualization software in terms of maintaining more detail and consuming fewer hardware resources. VUE can randomly generate ground elevation and single tree models and reconstruct the forest landscape according to real terrain and single tree information. At present, the basic steps for 3-D forest landscape reconstruction using airborne LiDAR include (1) processing the raw point cloud, (2) processing the digital elevation model (DEM) data, (3) establishing the tree models, and (4) reconstructing the 3-D forest landscape using VUE (Fig. 11.2).

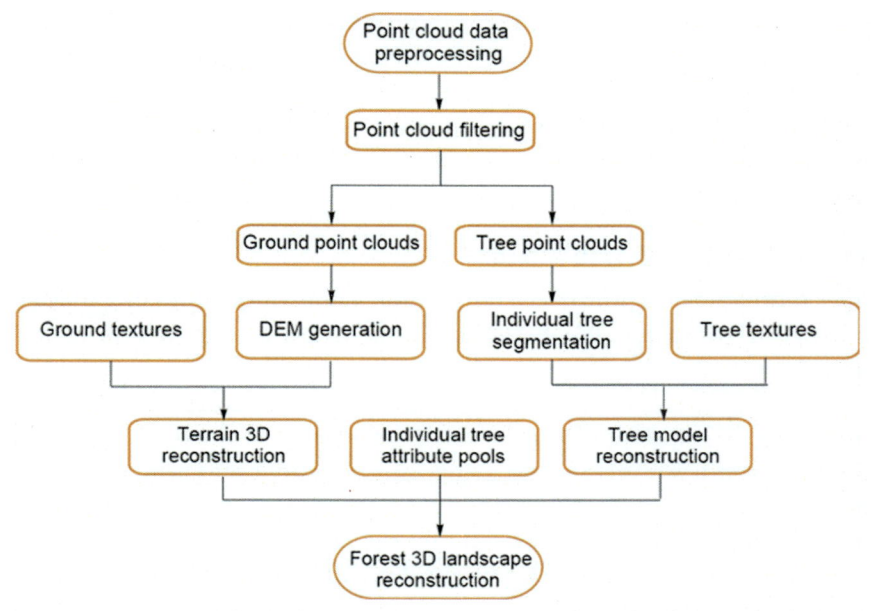

Figure 11.2 The workflow of a three-dimensional (3-D) forest landscape visualization. *DEM*, digital elevation model.

11.2.1 Processing Raw Point Clouds

Raw airborne LiDAR data are filtered to segregate the forest points from the ground points. A highly accurate DEM is generated by spatially interpolating the obtained terrain point cloud data. The point cloud data of the forest are segmented into individual trees. Then, morphological attributes, such as tree height, crown width, DBH, and tree location, are extracted according to the obtained point cloud data.

11.2.2 Processing the Digital Elevation Model

The workflow to generate a high–precision DEM from point clouds was described above. The data format of the DEM is then converted into a format that the software can recognize. The coordinates of the DEM data also should be converted from a geographic coordinate system or projected coordinate system into the relative coordinate system of the reconstructed 3-D scene. This process only involves the translation and scaling of coordinates. The origin of the DEM is usually translated to the origin of the 3-D scene to facilitate this process. A scaling factor is applied according to the actual scene to achieve the best display. If the display needs further improvement, the terrain texture, ambient lighting, and other settings can be used. The terrain texture can be generated from field photographs or satellite imagery.

11.2.3 Creation of Tree Models

The morphological attributes and location information of trees extracted from point cloud data can be used to assist with the construction of tree models. In a virtual 3-D scene, individual trees are the basic unit, but they have a highly complex structure. The number of trees and level of detail for individual trees directly affect the fidelity of the reconstruction. However, limitations in hardware are also a concern when trying to achieve a more realistic virtual scene because modeling a large number of trees will require much greater computational power and resources.

There are many efficient algorithms for building 3-D tree models, such as L-system, fractal model, particle system, bifurcation matrix model and texture mapping, and iterative function system. In addition, many 3-D tree modeling software platforms, such as Maya and SpeedTree, use these algorithms. SpeedTree is reputable in 3-D virtual tree modeling and is widely used in the film, television, and gaming industries, as well as in 3-D scene simulations. SpeedTree consists of three main modules—SpeedTree Modeler, SpeedTree Compiler, and SpeedTree SDK—for creating tree models, managing materials and texture maps, and rendering trees or forests in combination with other programs. Forest or tree models can be customized using SpeedTree SDK, and the attributes of individual trees extracted from point cloud data can be applied to create a large number of trees in a forest scene. The 3-D tree models created in SpeedTree can be exported to commonly used data formats such as .obj and .stl, which are useable in various software programs.

11.2.4 Three-Dimensional Reconstruction of Forest Landscapes in VUE

VUE is a powerful 3-D landscape-generating software that can render large landscape scenes, including terrain, vegetation, light sources, atmosphere, and other objects. It can simulate the refraction and reflection of light and mutual shadowing between objects under realistic illumination conditions. The OpenGL engine in VUE can maximize the power of professional

GPUs. VUE also offers interfaces for Python scripts, with which users can implement customized functions, such as extracting the location and morphological information of individual trees to reconstruct a forest landscape (Fig. 11.3).

After constructing terrain with a high–fidelity texture, the next step is to reconstruct trees on top of the terrain. First, customized tree models based on attributes extracted from point clouds are imported into the model library of VUE, and these can be "planted" in the reconstructed forest scene. Then, a virtual ecosystem is created in VUE using high-precision DEM data. The VUE ecosystem module is used to randomly scatter the imported tree models over the whole terrain. Since the location and attributes of each tree are extracted from LiDAR point cloud data, we can use Python scripts to adjust the location and proportion of randomly "planted" trees in the virtual ecosystem according to the location and height information from real trees. Finally, a virtual duplicate of the real forest scene with high fidelity can be realized (Fig. 11.4).

After constructing a 3-D model of the terrain and the forest landscape, the main light source, ambient light, and atmosphere must be configured to render an image. In addition, its saturation and other attributes must be adjusted to obtain a realistic image. The sun is usually the main light source. By adjusting the positions of the light source and the camera, various illumination effects can be achieved for different times of the day. In

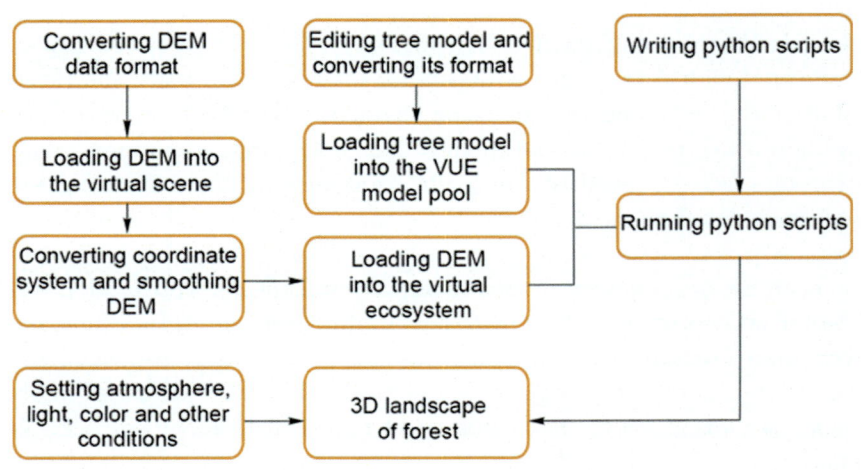

Figure 11.3 A flowchart for three-dimensional (3-D) forest landscape reconstruction using VUE. *DEM*, digital elevation model.

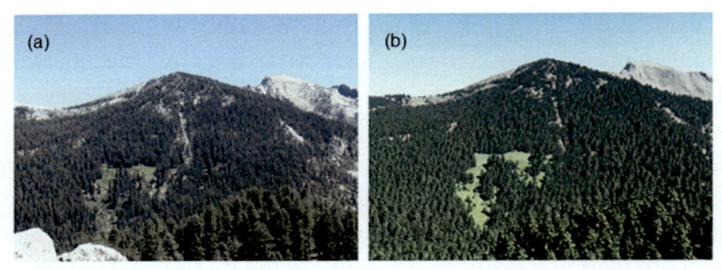

Figure 11.4 (a) A photograph of a real forest landscape and (b) a rendered image based on the reconstructed landscape.

addition, a careful choice of ambient light can further enhance the fidelity of the landscape, which is also affected by attributes of the atmosphere, such as its transparency and the distribution of clouds.

11.3 Applications of Three-Dimensional Forest Reconstruction

Vegetation visualization at the landscape level is widely used in fields such as forest management, demonstrations and simulations of forest fires, and forest tourism using virtual reality. A few examples are provided in the following sections.

11.3.1 Applications in Forest Management

It is challenging to intuitively evaluate the impact of management on a forest landscape using traditional field surveys and mapping techniques. Compared with 2-D images, 3-D data for forests contain more and richer information, including elevation and textures, that reflects more natural conditions that people can use to view and check forest conditions. Long-term forecasts are essential for forest management, and visualizing the projected dynamics of the forest landscape is valuable and beneficial to decision-makers.

With the development of technology, forest landscape visualization will become an important tool for managers. In the virtual forest landscape, the vegetation structure, patterns, and landscape factors can be simulated and evaluated, including logging and pest control. These provide useful information for managers to make and adjust their management practices. In addition, tree growth models can be incorporated into the visualization to demonstrate the dynamics of forests and estimate future timber production.

Three-dimensional visualization of the forest landscape is expected to be an essential component of stand data for forestry planning and management in the future.

11.3.2 Applications in Forest Fire Research

Forest fires are unplanned, unwanted, and uncontrolled. They can cause substantial damage to forests and threaten the life and property of local communities. Because forest fires may be caused by chance at random locations, it is challenging to locate the ignition point and send an alarm immediately. The dense vegetation, various types of fuels, and the complex terrain in forests make early detection, monitoring, and evaluation of forest fires very difficult. Therefore, it is critical to establish a comprehensive response plan to forest fires in addition to actively preventing them.

Constructing virtual forests using 3-D reconstruction, virtual visualization, and a geographic information system (GIS) to simulate and visualize forest fires and firefighting is a new approach. Visualization of the virtual forest landscape is important for studies of forest fires. When combined with forest fire models, the disturbance pattern caused by forest fires can be simulated, such as the scale, frequency, coverage, and landscape changes. Current studies include fire spread modeling, visualization of simulated forest fires, and forest fire simulation systems. The spread of forest fires is affected by many factors, such as wind speed, wind direction, terrain, and fuel distribution. Studies on the spread of forest fires that combine GIS, remote sensing, and visualization are the basis for decision-making and management actions. Current applications are still dominated by 2-D fire spread models—that is, actual fire spread conditions are projected onto a 2-D surface for display. Therefore, simulating and displaying 3-D forest fire spread is greatly needed and has a promising future.

11.3.3 Application in Forest Virtual Tourism

With the rapid development of augmented reality and virtual reality technologies, the 3-D reconstruction of forest landscapes has become a critical component of the virtual tourism industry, and it plays an important role in improving the tourism experience and operation mechanism. The combination of high-tech and traditional tourism will be a growth industry in the future. The 3-D virtual reconstruction of forest tourist sites using 3-D remote sensing mapping and GIS will provide tourists with a new

Figure 11.5 Three-dimensional (3-D) reconstruction of forest landscape for virtual tourism (the picture shows the authors' group members wearing 3-D glasses to browse a forest landscape reconstructed from light detection and ranging point clouds).

immersive experience that integrates multidimensional information and feeling into the virtual landscape (Fig. 11.5).

11.4 Chapter Summary

Realistic 3-D reconstruction of vegetation is an important approach for studies in many disciplines, including ecology, botany, and physical geography. LiDAR provides new data for the 3-D reconstruction of vegetation, such as individual trees and forest landscapes. At the individual tree level, LiDAR can provide realistic and complete information on the branch structure. At the landscape level, it can provide accurate information on topography and vegetation. Based on these valuable data, 3-D reconstruction of real forest scenes can be achieved by combining fractal theory and computer visualization. Three-dimensional reconstructions have broad applications in exploring the ecological mechanisms of forest ecosystems, forest management, burn severity evaluation, and virtual tourism.

References

Bucksch, A., & Lindenbergh, R. (2008). CAMPINO—a skeletonization method for point cloud processing. *ISPRS Journal of Photogrammetry and Remote Sensing, 63*(1), 115—127.

Bucksch, A., Lindenbergh, R., & Menenti, M. (2010). SkelTre. *The Visual Computer, 26*(10), 1283—1300.

Cao, J., Tagliasacchi, A., Olson, M., Zhang, H., & Su, Z. (2010). Point cloud skeletons via laplacian based contraction. In *2010 shape modeling international conference, aix-en-provence, France.*

Chen, N. W., Zhang, X. Y., & Lu, X. M. (2008). Mapping direct use value of ecosystem services: A GIS-based approach. *China Environmental Science, 28*(7), 661—666 (in Chinese).

Chen, X., Neubert, B., Xu, Y.-Q., Deussen, O., & Kang, S. B. (2008). Sketch-based tree modeling using markov random field. In *ACM SIGGRAPH Asia 2008 papers* (pp. 1—9) (Singapore).

Cornea, N. D., Silver, D., & Min, P. (2007). Curve-skeleton properties, applications, and algorithms. *IEEE Transactions on Visualization and Computer Graphics, 13*(3), 530.

Côté, J.-F., Fournier, R. A., & Egli, R. (2011). An architectural model of trees to estimate forest structural attributes using terrestrial LiDAR. *Environmental Modelling & Software, 26*(6), 761—777.

Côté, J.-F., Fournier, R. A., Frazer, G. W., & Niemann, K. O. (2012). A fine-scale architectural model of trees to enhance LiDAR-derived measurements of forest canopy structure. *Agricultural and Forest Meteorology, 166*, 72—85.

Côté, J.-F., Widlowski, J.-L., Fournier, R. A., & Verstraete, M. M. (2009). The structural and radiative consistency of three-dimensional tree reconstructions from terrestrial lidar. *Remote Sensing of Environment, 113*(5), 1067—1081.

Dai, M., Zhang, X., & Hongjun, L. I. (2010). Tree reconstruction based on range image with strong noise. *Chinese Journal of Stereology & Image Analysis, 19*(10), 1468—1474 (in Chinese).

Ding, S. Y., & Zhang, M. L. (2005). Urban landscape dynamics of kaifeng city from 1988 to 2002. *Geographical Research, 24*(1), 28—37 (in Chinese).

Dongli, L. B. C. (2005). Research and application of forest landscape visualization system in North America. *World Forestry Research, 18*(5), 57—64 (in Chinese).

Gao, G. (2008). Progress in the study of 3D modeling technology of forestry landscape. *World Forestry Research, 21*(2), 22—25 (in Chinese).

Gorte, B., & Winterhalder, D. (2004). Reconstruction of laser-scanned trees using filter operations in the 3D raster domain. *International Archives of Photogrammetry, Remote Sensing and Spatial Information Sciences, 36*(Part 8), W2.

Holton, M. (1994). Strands, gravity and botanical tree imagery. *Computer Graphics Forum, 13*(1), 57—67.

Hosoi, F., & Omasa, K. (2006). Voxel-based 3-D modeling of individual trees for estimating leaf area density using high-resolution portable scanning lidar. *IEEE Transactions on Geoscience and Remote Sensing, 44*(12), 3610—3618.

Hosoi, F., & Omasa, K. (2012). Estimation of vertical plant area density profiles in a rice canopy at different growth stages by high-resolution portable scanning lidar with a lightweight mirror. *ISPRS Journal of Photogrammetry and Remote Sensing, 74*, 11—19.

Huang, H. (2013). Tree geometrical 3D modeling from terrestrial laser scanned point clouds: A review. *Scientia Silvae Sinicae, 49*(4), 123—130 (in Chinese).

Lin, D., Chen, C. C., Tang, L. Y., Zou, J., & Wang, Q. M. (2011). 3D tree modeling based on the parametric curve and its integral. *Journal of Fuzhou University(Natural Science Edition), 39*(03) (in Chinese).

Lintermann, B., & Deussen, O. (1999). Interactive modeling of plants. *IEEE Computer Graphics and Applications, 19*(1), 56—65.

Livny, Y., Yan, F., Olson, M., Chen, B., Zhang, H., & El-Sana, J. (2010). Automatic reconstruction of tree skeletal structures from point clouds. In *ACM SIGGRAPH Asia 2010 papers* (pp. 1—8).

Luo, C., & Sohn, G. (2010). A knowledge based hierarchical classification tree for 3D facade modeling using terrestrial laser scanning data. Geomatics—shaping Canada's Competitive Landscape. In *Proceedings of the Canadian geomatics conference 2010 and ISPRS com I symposium, Calgary, AB, Canada.*

Neubert, B., Franken, T., & Deussen, O. (2007). Approximate image-based tree-modeling using particle flows. In *ACM SIGGRAPH 2007 papers* (pp. 88(-es)).

Okabe, M., Owada, S., & Igarashi, T. (2006). Interactive design of botanical trees using freehand sketches and example-based editing. In *ACM SIGGRAPH 2006 courses* (pp. 18(-es), Boston, USA).

Pfeifer, N., & Winterhalder, D. (2004). Modelling of tree cross sections from terrestrial laser scanning data with free-form curves. *International Archives of Photogrammetry, Remote Sensing and Spatial Information Sciences, 36*(Part 8), W2.

Runions, A., Lane, B., & Prusinkiewicz, P. (2007). Modeling trees with a space colonization algorithm. *NPH, 7*(63—70), 6.

Shlyakhter, I., Rozenoer, M., Dorsey, J., & Teller, S. (2001). Reconstructing 3D tree models from instrumented photographs. *IEEE Computer Graphics and Applications, 21*(3), 53—61.

Song, T., & Ling, W. (1997). *Statistics and simulation of forest spatial data.* Journal of Beijing Forestry University (in Chinese).

Tao, S., Guo, Q., Xu, S., Su, Y., Li, Y., & Wu, F. (2015). A geometric method for wood-leaf separation using terrestrial and simulated lidar data. *Photogrammetric Engineering & Remote Sensing, 81*(10), 767—776.

Thies, M., Pfeifer, N., Winterhalder, D., & Gorte, B. G. (2004). Three-dimensional reconstruction of stems for assessment of taper, sweep and lean based on laser scanning of standing trees. *Scandinavian Journal of Forest Research, 19*(6), 571—581.

Verroust, A., & Lazarus, F. (1999). Extracting skeletal curves from 3D scattered data. In *Proceedings shape modeling International'99. International conference on shape modeling and applications* (pp. 194—201). Aizu-Wakamatsu, Japan: IEEE.

Xu, H., Gossett, N., & Chen, B. (2007). Knowledge and heuristic-based modeling of laser-scanned trees. *ACM Transactions on Graphics (TOG), 26*(4), 19—es.

Yang, X., Strahler, A. H., Schaaf, C. B., Jupp, D. L., Yao, T., Zhao, F., Wang, Z., Culvenor, D. S., Newnham, G. J., & Lovell, J. L. (2013). Three-dimensional forest reconstruction and structural parameter retrievals using a terrestrial full-waveform lidar instrument (Echidna®). *Remote Sensing of Environment, 135*, 36—51.

Yan, D.-M., Wintz, J., Mourrain, B., Wang, W., Boudon, F., & Godin, C. (2009). Efficient and robust reconstruction of botanical branching structure from laser scanned points. In *2009 11th IEEE international conference on computer-aided design and computer graphics* (pp. 572—575). Yellow Mountain City, China: IEEE.

Yao, T., Yang, X., Zhao, F., Wang, Z., Zhang, Q., Jupp, D., Lovell, J., Culvenor, D., Newnham, G., & Ni-Meister, W. (2011). Measuring forest structure and biomass in New England forest stands using Echidna ground-based lidar. *Remote Sensing of Environment, 115*(11), 2965—2974.

Yu, C., Yuanman, H. U., Rencang, B. U., Zhitao, M., Qianggen, D. U., & Jiaming, Z. (2008). Landscape visualization and its applications. *Chinese Journal of Ecology, 27*(8), 1422—1429 (in Chinese).

CHAPTER 12

Forest Dynamics Monitoring

Contents

Forests are an important part of terrestrial ecosystems, not only providing timber but also conserving water and soil, regulating the climate, and protecting species diversity. Accurate real-time monitoring of forest ecosystems is important for protecting the environment and achieving sustainable development. The two categories of forest dynamics monitoring are growth monitoring and disturbance monitoring. Growth monitoring measures changes in tree height, density, trunk size, volume, and other traits to investigate tree growth and provide a detailed and reliable data source for forest resource management. Forest disturbances can be divided into natural and human disturbances according to their causes. Natural disturbances refer to changes resulting from natural environmental conditions, such as wildfires, storms, volcanic eruptions, crustal movements, flooding, and pests and diseases. Human disturbances are transformations in nature or ecological construction resulting from human activities, such as deforestation, grazing, and burning wasteland for cultivation. Monitoring these forest dynamics requires high spatiotemporal resolution and accuracy because forest ecosystems have a long growth cycle, are susceptible to disturbance, and show substantial heterogeneity in their spatial structure and function. This chapter introduces the principles, methods, and challenges of monitoring forest ecosystem dynamics, using light detection and ranging as the main data source and other remote sensing data with high spatiotemporal resolution as auxiliary data.

LiDAR Principles, Processing and Applications in Forest Ecology
ISBN 978-0-12-823894-3
https://doi.org/10.1016/B978-0-12-823894-3.00012-8

12.1 Forest Growth Dynamics Monitoring

Accurate real-time monitoring of changes in forest structural and functional traits is the key to understanding ecological health, quantifying resource reserves, and formulating management measures for forest ecosystems. Traditional monitoring methods rely mostly on long-term forest inventories, such as continuous forest inventory and forest management inventory. These inventories estimate forest growth through long-term surveys of various parameters, such as tree height, diameter at breast height (DBH), and volume. However, in situ forest inventories are always time-consuming and labor-intensive. Long-term, repeated forest surveys are particularly difficult in remote mountain forests that are hard to access. Light detection and ranging (LiDAR) data have many advantages for estimating forest structural traits such as tree height, canopy size and cover, biomass, and volume and provide a good data source for monitoring forest dynamics. In recent years, with the development of LiDAR technology and the accumulation of LiDAR data in forests, monitoring forest ecosystems by multitemporal LiDAR has become a research focus. This section introduces the application of LiDAR technology to monitor forest dynamics and discusses research progress and future developments from various perspectives, including forest trait observations, monitoring methods, and potential applications.

12.1.1 Forest Growth Monitoring

Although LiDAR has the unique characteristic of measuring the three-dimensional structure of forests and inferring structural and functional traits, it is difficult to directly compare multitemporal point cloud datasets owing to different point densities, incidence angles, and other parameters. As a result, using LiDAR to monitor forest dynamics is still challenging. Current LiDAR-based forest dynamics monitoring methods are area-based or individual tree-based, depending on the spatial scale (Maltamo et al., 2014). Area-based methods are typically used to analyze changes in multiple trees, such as changes in biomass and plant volume per unit area, because their basic units are always larger than the canopy cover of an individual tree (~ 100 m^2). Individual tree-based methods monitor changes in traits at the individual tree scale after separating individual trees from point cloud data, enabling the monitoring of structural and functional changes for every tree. However, their reliability is greatly affected by the accuracy of individual tree segmentation, which depends not only on the quality of the LiDAR data but also on forest canopy closure and terrain conditions. To some extent, incorrect segmentation is inevitable in individual tree segmentation. In addition, monitoring forest dynamics at the individual tree scale usually requires multiphase segmentation results, which may accumulate segmentation errors in the repeated procedure. Thus, these two monitoring methods have advantages and disadvantages in the monitoring of forest dynamics. Besides, direct and indirect methods can be used to estimate the changes in forest traits by using LiDAR data. Direct methods directly simulate changes in vegetation from changes in multitemporal LiDAR data, whereas indirect methods compare the differences between the estimated forest traits before and after the change.

Initially, LiDAR-based monitoring of changes in forest dynamics was applied primarily to monitor changes in tree height. Hyyppä et al. (2003) first estimated changes in tree height in Iceland through bitemporal LiDAR data (1998 and 2000). They segmented individual trees from bi-temporal LiDAR data and used the change in the maximum height within individual tree point clouds to measure the height change of a single tree. They then calculated the average change in tree height and compared it with the measured change in tree height. The error in the tree height change estimated from LiDAR data was within 15 and 5 cm at the individual tree and forest stand scales, respectively. This study proved that LiDAR could monitor tree height changes. However, the maximum height estimated from LiDAR was always an underestimate owing to incomplete treetop detection. This phenomenon may occur in both area-based and individual tree-based monitoring methods. According to Hopkinson et al. (2008), the error can reach 10% when calculating the change in tree height after 3 years.

Based on Hyyppä et al. (2003), the individual tree-based monitoring method was further developed by Yu et al. (2004) and extended to monitoring crown sizes and volume dynamics. They found that changes in the maximum tree height derived from LiDAR data effectively reflected tree growth and performed better than the average and median changes in tree height. For the estimation of tree crown volume, the method based on changes in the digital surface model (DSM) produced better performance. Vastaranta et al. (2012) also utilized individual tree segmentation to detect dynamic changes in forest gaps. They classified tree gap changes into five types: gap generation, decreasing, increasing, disappearing, and unchanged. The above studies indicate that the methods based on individual tree segmentation have unique potential for monitoring forest structural traits.

LiDAR-based monitoring of forest dynamics still has some shortcomings. First, large errors can exist in data registration and individual tree segmentation. Furthermore, the rapid updating of LiDAR sensors and different weather conditions during LiDAR scanning also hinder multi-temporal LiDAR registration. Improving the matching accuracy of multi-temporal individual tree segmentation results has become a bottleneck in LiDAR-based forest dynamics monitoring. To break this bottleneck, we proposed a rule-based individual tree segmentation method for bi-temporal LiDAR data (Ma et al., 2018). In this study, we firstly determined segmented individual trees that can be accurately matched from bi-temporal LiDAR data using two criteria (1) the treetop from later-phase LiDAR data was the only treetop located in the tree crown from LiDAR data in the earlier phase; and (2) the distance between the treetops from the bi-temporal LiDAR data was less than 2 m (Fig. 12.1). If segmented individual trees from bi-temporal LiDAR data are mismatched, they need to be segmented and matched again according to a series of rules (Fig. 12.2). The rule-based individual tree segmentation method can reduce the matching error, thus improving the accuracy of forest dynamics monitoring.

Existing individual tree segmentation algorithms cannot precisely extract information about understory forest structure, and underestimation is a common problem. Hamraz et al. (2017) proposed a hierarchical segmentation method for airborne LiDAR data in which they segmented the data into canopy layers. For each layer, they segmented individual trees independently. The segmentation results showed that using the stratification procedure strongly improved the detection accuracy of understory trees (from 46% to 68%), with useful application prospects in forests with complex forest structures. However, this method is still in the early stage of development and has not been effectively applied in forest dynamics monitoring. The development of methods to improve the accuracy of tree segmentation and comparison from multiphase LiDAR data is the future direction of forest dynamics monitoring at the individual tree scale.

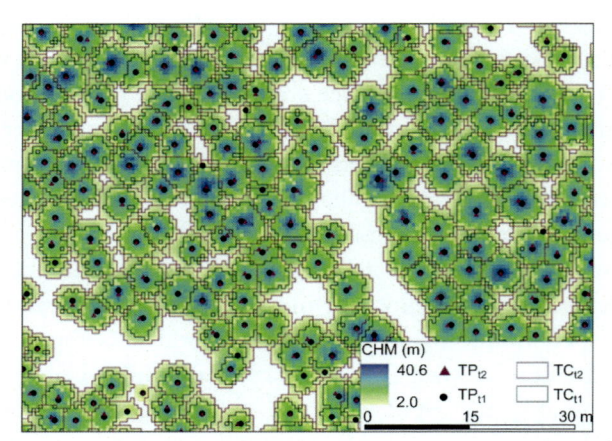

Figure 12.1 Example of tree crown segmentation results and corresponding treetop locations. TP_{t1} and TP_{t2} are treetops and TC_{t1} and TC_{t2} are the tree crowns detected from light detection and ranging (LiDAR) data in earlier and later periods, respectively. *CHM*, canopy height model.

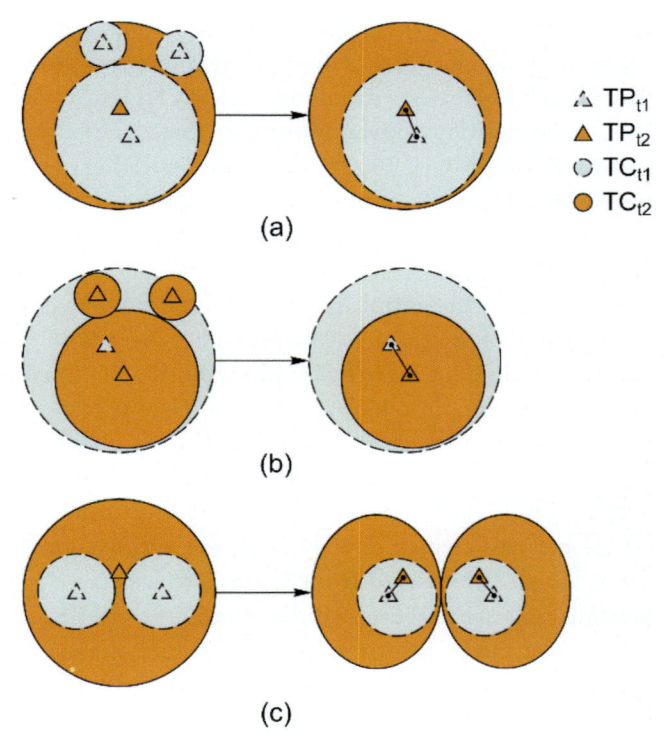

Figure 12.2 Schematic shows how to rematch mismatched tree crowns under three scenarios. (a) When the earlier main canopy includes two or more canopies compared with the later period, this indicates that some small trees in the earlier canopy did not survive. Therefore, only the main crown before and after the change is matched, and the smaller crowns before the change are manually deleted. (b) When the later canopy includes two or more canopies compared with the earlier period, this indicates that by the later period, new small trees have grown. As with (a), only the main crown before and after the change is matched, and the smaller crowns are deleted. (c) When two tree canopies of similar size in the earlier period appear in a later canopy, this indicates a missegmentation in the later period. In this scenario, it is necessary to resegment the later canopy, calculate the highest point value of the two divided canopies as the top of the tree, and then match the two trees according to the principle of proximity. TC_{t1} and TC_{t2}, tree crowns detected in earlier and later periods; TP_{t1} and TP_{t2}, treetops detected in earlier and later periods.

Area-based methods are mainly used to estimate forest structural changes at the plot and stand scales, including average tree height, basal area, aboveground biomass, and timber volume. Næsset (2004) monitored changes in mean tree height, basal area, and timber volume using bi-temporal LiDAR data (1999 and 2001). This research reconstructed the variation in 56 forest stands, with a 350 m^2 plot as a statistical unit, using the percentile of different height layers calculated from LiDAR data. Solberg et al. (2006) monitored seasonal changes in the leaf area index using LiDAR data from different months in the same year at a scale of 100 m^2. These area-based methods typically rely on forest attributes obtained through regression models. However, forest attributes obtained through these models usually have large uncertainty, decreasing the accuracy of forest monitoring. In addition, area-based methods often rely on bi-temporal monitoring data with a short time interval, which reduces the generality of the results predicted by the model.

To assess different methods, Bollandsås et al. (2013) compared biomass changes in the forests of southeastern Norway between 2005 and 2008 simulated from three different LiDAR-based methods. The three methods were (1) indirect estimation, in which the biomass change was estimated as the difference between predicted biomass based on height quantiles extracted from the two LiDAR datasets (2) modeling the biomass change directly through changes in different variables extracted from LiDAR data; (3) modeling the biomass change directly using the average annual change rate of different variables extracted from LiDAR data between 2005 and 2008. Their results indicate that the two direct methods are better than the indirect method. The direct estimation of the absolute change in biomass (second method) is slightly better than direct estimation of the change rate.

In addition to monitoring tree growth dynamics, LiDAR is also applicable for research on forest disturbance, such as pests, diseases, and snowstorms. Vastaranta et al. (2012) successfully monitored the effect of snow and insect damage on tree canopies using multitemporal LiDAR data. They found that the area-based method performs better than the individual tree-based method, especially when the point density is low or bi-temporal data are difficult to register.

12.1.2 Construction of Forest Growth Models

Forest growth modeling is a quantitative simulation of the process of forest ecosystem change. Model construction relies on the understanding of the mechanisms of forest dynamic change, and it uses quantitative methods such as mathematical models to simulate and predict future forest changes (Weiskittel et al., 2011). Forest growth models have a long development history. They are widely used to estimate yield, simulate carbon stocks, and explore the role of forest structure in climate change. According to their simulation scales, traditional growth models can be categorized as individual tree, stand, and landscape-scale simulations (Weiskittel et al., 2011). In the book *Forest Growth and Yield Modeling*, Weiskittel et al. (2011) concluded that forest growth simulations at different scales are useful for managers. For example, simulations at the stand scale can be used to establish a series of management measures regarding planting density, tree species choices, fertilization, and thinning strategies. Simulations at regional and national scales can provide data support for important policy decisions such as forest carbon stock assessments, ecological aesthetics value assessments, and biodiversity conservation (Table 12.1).

Forest growth simulation can be grouped into three broad categories: (1) process-based, (2) statistical, and (3) hybrid models. Statistical models rely on data analysis to quantify the relationship between two related variables of forest growth and establish regression models to simulate and predict forest growth. Therefore, statistical models require a mass of detailed forest survey data, including different forest management methods, many parameters, and long-term monitoring results. Statistical methods are widely used to develop forest management policies because of their easy accessibility and high simulation accuracy. To simulate the forest as a whole, traditional models usually focus on forest dynamics at individual tree and forest stand scales. In contrast, process-based models monitor forest changes from a microcosmic perspective, such as stems, leaves, and roots. Hence, process-based models are usually used to explore the processes and principles of forest system change, as well as responses to different environments. In this section, we focus on statistical models and introduce the application of LiDAR in forest growth model construction.

In recent decades, statistical models have been developed to simulate forest dynamics at various scales, such as individual trees or forest stands. Detailed vegetation structural traits, such as tree volume, crown size, and stem structure, are used for model construction. Because model construction requires long-term continuous observations of typical plots in different regions, the depth and breadth of forest monitoring data become a major limiting factor. The growing demand for fast, large-area, and accurate forest growth simulations, as well as the demand for substantial amounts of data, has led to the increasing importance of LiDAR data in constructing forest growth models.

Table 12.1 The role of forest growth monitoring at various scales for forest management.

Monitoring scale	Application	References
Even-aged stand scale	Planting density; fertilization strategy; species selection; thinning strategy	Hann (1980)
Uneven-aged stand scale	Thinning strategy; ecological aesthetics value assessment; biodiversity conservation	Bettinger et al. (2016)
Regional or national scale	Forest carbon stock assessment; allowable harvest; biodiversity conservation	Bettinger et al. (2005)

LiDAR data can provide detailed and accurate three-dimensional structural information for forest growth simulation, including tree height, canopy size, trunk thickness, volume, and biomass. These structural traits are important inputs for forest growth simulation and can be used to derive forest age, productivity, and competitiveness (Jakubowski et al., 2013; Korhonen et al., 2011; Su et al., 2016; Tompalski et al., 2016). As shown in Fig. 12.3, changes in tree height and diameter vary with initial tree age, tree height, and DBH, and rates of change vary by stand types. The emergence of LiDAR data compensates for the shortcomings of traditional models that rely solely on forest survey data, such as few samples, single data sources, and insufficient updates (Londo, 2010; Song et al., 2016; Vepakomma et al., 2011). Ma et al. (2018) analyzed the growth of tree height, canopy area, and volume using bi-temporal LiDAR data over 5 years, finding that the original tree height and crown size were positively correlated with the change in crown size (Pearson's correlation coefficient $r > 0.4$) but had no significant relationship with changes in tree height. The monitoring of large

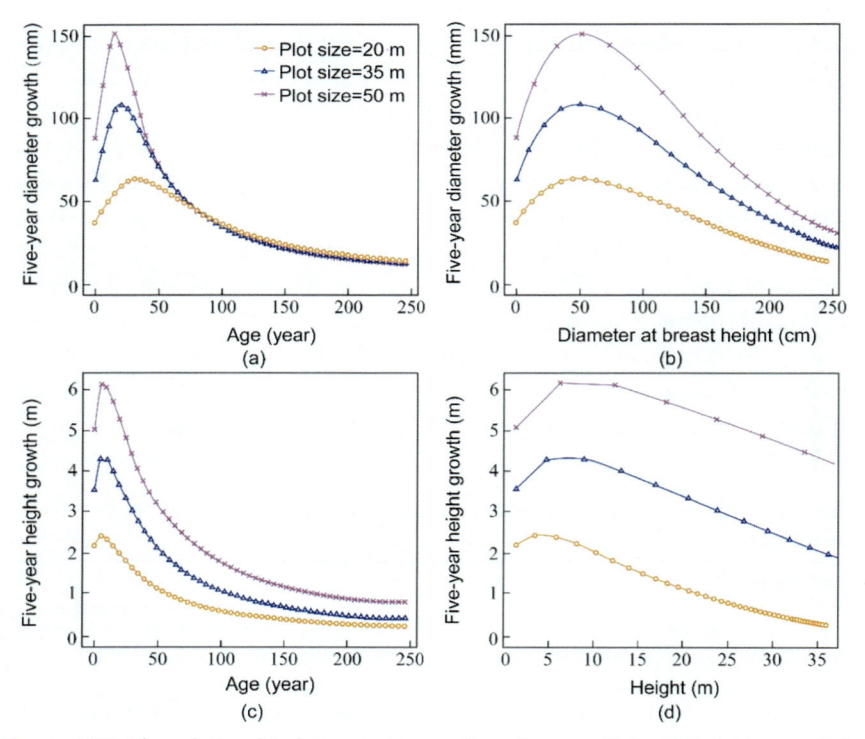

Figure 12.3 The relationship between 5-year diameter growth (a, b), height growth (c, d), and forest age, diameter at breast height, and tree height. *(Revised from Weiskittel, A. R., Hann, D. W., Kershaw Jr, J. A., & Vanclay, J. K. (2011). Forest growth and yield modeling. John Wiley & Sons.)*

areas, multiforest stands, and long-term forest structural changes based on LiDAR data provides a powerful data foundation for the construction and updating of statistical growth models.

LiDAR also has advantages for topographic detection under forest canopies, which is difficult in traditional forest inventories (Adams et al., 2014). The terrain under forest canopies can characterize the hydrological environment (e.g., topographic wetness index) and light conditions (e.g., potential solar insulation) and provide data support for simulating forest growth potential (Cook, 1942). Ma et al. (2018) found a strong positive correlation between the change in tree height and the topographic humidity index at stand scale ($r = 0.82$) in the Sierra Nevada mountains, California, USA, concurring with the understanding that local forest growth is limited by water resources. The terrain features extracted by LiDAR are of great significance for constructing forest growth models.

The development of LiDAR also brings new possibilities for the simulation of forest competition indices. Competition indices are crucial traits in forest growth models and are often used to quantify the degree of tree competitiveness during growth (Biging & Dobbertin, 1995; Twery & Weiskittel, 2013; Wensel et al., 1987). Depending on whether the distance between trees is considered, competition indices can be divided into three categories: distance-dependent, distance-independent, and semi—distance-independent. In practical calculations, distance-dependent indices assign different weights to trees at different distances within the estimation range, whereas distance-independent indices are calculated by a unified standard without concern for distance. Theoretically, distance-dependent indices depict the spatial heterogeneity of forest competition in more detail, while distance-independent indices are easier to calculate when comparing the two methods. Semi—distance-independent indices have the advantages of both. First, the neighborhood used to compute competition indices (usually a circle with a specific radius centered on the target tree) is selected. Competitive indices are then computed using distance-independent methods in the selected neighborhood (Fig. 12.4). Semi—distance-independent indices have been adopted by an increasing number of forest growth models because of their high accuracy and computational efficiency (Contreras et al., 2011; Ledermann, 2010).

Regardless of the competition indices, the most critical step in the calculation process is obtaining accurate tree positions and structural properties. In recent years, with the development of individual tree segmentation and structure extraction algorithms, the advantages of LiDAR

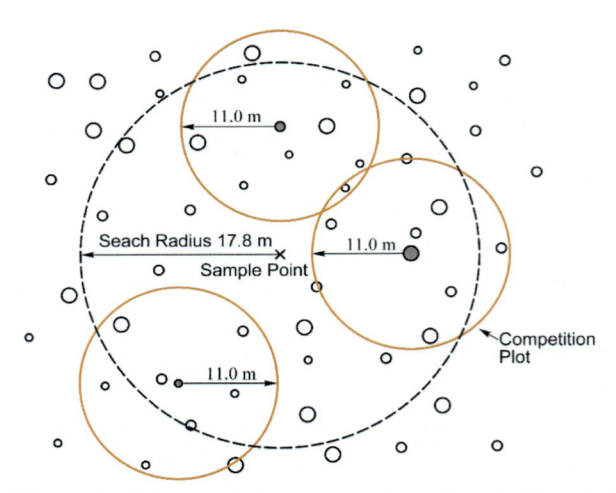

Figure 12.4 Schematic diagram of the computation of semi−distance-independent competition indices. The gray solid circles represent the three cored trees, where the open circles within a radius of 11 m are competing trees in the neighborhood. The size of each small circle represents the size of the crown. *(Revised from Contreras, M. A., Affleck, D., & Chung, W. (2011). Evaluating tree competition indices as predictors of basal area increment in western Montana forests.* Forest Ecology and Management, 262(11), *1939−1949.)*

data in calculating tree competition indices at large scales have become increasingly prominent. Ma et al. (2018) computed the competition indices for 114,000 individual trees in two conifer-dominant Sierra Nevada forests using airborne LiDAR data. The method used semi−distance-independent competition indices, including tree number, crown competition of all tree crowns, crown competition of 66% tree height (CC66), and crown competition of higher trees (CCHT). The competition indices were calculated as shown in Eqs. (12.1)−(12.3). The results showed that CC66 and CCHT better described the competition of tree growth than the other indices and had a significant negative correlation with the growth of canopy area and volume.

$$CCT = \frac{A_{N(\text{CHM}>2)} - A_{S(\text{CHM}>2)}}{A_N} \qquad (12.1)$$

$$CC66 = \frac{A_{N(\text{CHM}>0.66H_S)} - A_{S(\text{CHM}>0.66H_S)}}{A_N} \qquad (12.2)$$

$$CCHT = \frac{A_{N(\text{CHM}>H_S)}}{A_N} \qquad (12.3)$$

where A is the area of the defined neighborhood, which is a circle with a 15 m radius centered on the target treetop, A_S is the crown area of the target tree, and H_S is the height of the target tree.

Overall, the application of LiDAR in forest growth monitoring and simulation is increasingly important. It is worth noting that with the accumulation of LiDAR data, there are some shortcomings in its application. For example, multitemporal data matching and individual tree segmentation of complex forests are the keys to improving the efficiency of forest growth monitoring. In addition, the cost of large-scale, long-term forest dynamics monitoring using LiDAR is still very high. The future development of forest dynamics monitoring will likely need to combine LiDAR data, forest inventory, and optical remote sensing data.

12.2 Forest Fire Monitoring and Fire Severity Assessment

Forests are one of the most important ecosystems, and protecting existing forests is of great significance for sustainable development. Forest fires are one of the biggest threats to forests. For example, from 1950 to 2005, there were more than 13,000 forest fires per year, on average, in China. Fire areas accounted for 5% of the total forest area, resulting in substantial economic and ecological losses (Shu et al., 2003). Identifying how to monitor the occurrence of forest fires, reduce and accurately assess losses, and determine appropriate forest restoration strategies is the key to current forest management.

12.2.1 Real-Time Monitoring of Forest Fires

Timely discovery and early warning of forest fires are critical tasks to stop fires from developing, reduce fire damage, and extinguish fires as soon as possible. Forest fire monitoring can be divided into four levels based on the height of the monitoring platform: satellite monitoring, aircraft patrol, near-surface observations, and ground patrol (Wu et al., 2010). Satellite monitoring and near-surface observations are the main fire monitoring methods.

Satellite monitoring detects abnormal surface temperatures using thermal infrared data. Generally, the shorter the wavelength, the higher the sensitivity to high temperatures, such as middle infrared bands. Fire spots have unique spectral characteristics and radiance. Differences between fire spots and background can be used to detect fire spots. Currently, satellite images used in monitoring fires are mainly moderate and low-resolution images, such as those from Earth Observing System/Moderate Resolution Imaging Spectroradiometer (EOS/MODIS) and Fengyun meteorological satellites. Although it is frequently used, satellite monitoring has the disadvantage of low spatial resolution, saturated signals for high temperature, strong reflector interference, and inflexibility, which makes it only feasible for monitoring large-scale, well-developed fires. This risks missing the best firefighting opportunities.

Aircraft patrols, using forest protection aircraft and unmanned aerial vehicles, are one of the most effective means of forest fire monitoring. Near-surface observations include lookout towers and video monitoring. Lookout towers are fixed observation facilities in forest areas, and their distribution should be properly designed to ensure high monitoring coverage with fewer lookout towers. Video monitoring is an intelligent monitoring technique that uses computer technology, video processing, pattern recognition, and artificial intelligence to automatically analyze the image sequences acquired by cameras. By continuously detecting, tracking, and identifying moving targets in the monitored scene, the monitoring system can react in time when an abnormal phenomenon occurs. Through near-surface observation methods, fires can be discovered quickly, which is essential for more effective fire control. However, lookout towers have an extremely limited monitoring range and require considerable manpower. In addition, the theoretical monitoring radius of video monitoring systems is approximately 10−15 km and greatly affected by the terrain.

Ground patrol is done by forest rangers who patrol the forest and report fires to the administration department in time to conduct the firefighting. This method is still used to monitor forest fires in many regions.

LiDAR-based forest fire monitoring is part of near-surface observation (Fig. 12.5), and its use is still in the exploratory stage (Utkin et al., 2003). Owing to its high sensitivity and spatial resolution, LiDAR is a promising tool for forest fire monitoring. LiDAR can work continuously throughout the whole day as an active detection technology. Previous studies showed that small fires with a burning rate of approximately 0.03 kg of wood per second could be promptly detected at a distance of 6.5 km (Utkin et al., 2002).

The principle of monitoring forest fires using LiDAR is that LiDAR continuously emits laser pulses in the monitoring direction within a certain time range, and the theoretical intensity P_r of the echo received by a LiDAR scanner at each moment can be obtained through the transport Eq. (12.4).

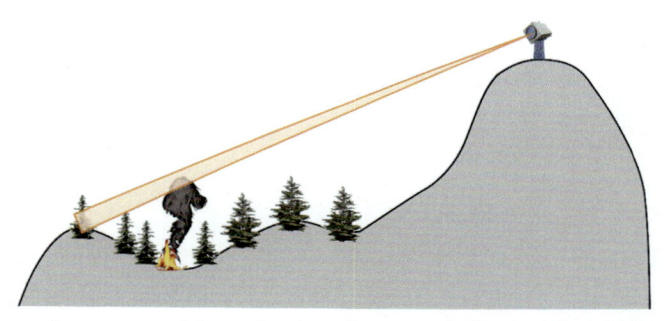

Figure 12.5 Observation of forest fire based on LiDAR. *(Revised from Utkin, A. B., Fernandes, A., Simões, F., Lavrov, A., & Vilar, R. (2003). Feasibility of forest fire smoke detection using lidar.* International Journal of Wildland Fire, 12(2), 159–166.)

$$P_r(R) = E_p \frac{c\langle\beta(R)\rangle}{2} \frac{A_r}{R^2} \tau_t \tau_r \exp\left(-2\int_0^R \alpha(R')\mathrm{d}R'\right) \qquad (12.4)$$

where R is the current distance, E_p is the output laser pulse energy, c is the speed of light, α is the extinction coefficient, $\langle\beta(R)\rangle$ is the mean backscattering coefficient, A_r is the effective receiver area, and τ_t and τ_r are the transmitter and receiver efficiencies.

When smoke exists, there is a peak in the echo energy received by LiDAR (Fig. 12.6). The location of smoke can be calculated using the theoretical intensity, actual intensity, and the corresponding time of the echo. Then, the fire location can be determined by combining the wind direction. It can be determined more accurately through multiple detections. Real-time forest fire monitoring based on LiDAR can determine fire locations very quickly, which greatly helps with firefighting.

12.2.2 Forest Fire Severity Assessment

Fire severity refers to the degree of consumption of combustible forest biomass and surface soil organic matter after a fire, reflecting the impact of forest fires on forest ecosystems (Keeley, 2009). Fire severity directly determines the survival rate of plants and the consumption of surface roots and seed banks and influences ecological processes such as the carbon–nitrogen cycle and litter decomposition. The assessment of fire severity can help to reveal the impact of fire disturbance on various ecological processes and has great significance for guiding ecological restoration and fire management.

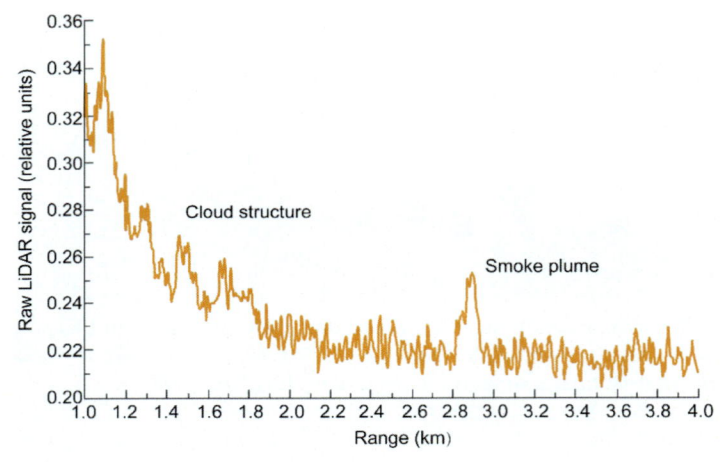

Figure 12.6 The LiDAR signal distribution when detecting smoke. *(Revised from Utkin, A. B., Fernandes, A., Simões, F., Lavrov, A., & Vilar, R. (2003). Feasibility of forest fire smoke detection using lidar.* International Journal of Wildland Fire, 12(2), 159–166.)*

Forest fire severity assessment is mainly conducted using two methods, field investigation and remote sensing evaluation. Field investigation has gone through a long development process from single-factor to comprehensive multifactor assessment. Single-factor assessment of fire severity is based on the statistics of tree mortality (Greene et al., 2004), crown and trunk damage (Keyser et al., 2006), scorch height and burning area (Knapp & Keeley, 2006), color change of solid combustion products (Smith et al., 2005), consumption of soil organic matter (Miyanishi & Johnson, 2002), and other factors. Among multifactor assessments, the composite burn index (CBI) established by Lutes et al. (2006, p. 164) is the most representative and common way to verify the results of remote sensing evaluation. It has become the standard for fire severity assessment in the US Forest Service and has been widely used in North America and Europe. The process of estimating CBI is that first, plots of 20 m × 20 m or 30 m × 30 m in the burned forest are divided into five layers: (1) surface combustible and soil, (2) herb, low shrub, and tree layer (<1 m), (3) tall shrubs and tree layers (1–5 m), (4) intermediate tree layers (5–20 m), and (5) large tree layers (>20 m). Four or five evaluation variables for each layer are then estimated visually and further integrated to obtain the fire severity of the whole plot. The CBI has some shortcomings. Kasischke et al. (2008) pointed out that the CBI overemphasizes the aboveground vegetation and does not pay enough attention to the surface and soil. In addition, the CBI has the same weight for each layer and may not be consistent with spectral reflectance from remote sensing imagery. De Santis and Chuvieco (2009) proposed the geometrically structured composite burn index, which considers the hierarchical coverage and leaf area index.

Remote sensing evaluation has also developed from supervised classification and regression models and finally to spectral indices with clear physical meaning. For supervised classification methods, fire severity is mainly assessed by partitioning regions of interest to visually determine the degree of fire severity (Hall et al., 1980). The regression method is usually based on Landsat as the main data source and establishes the correlation between spectral indices and fire severity data from field investigations.

The normalized burn ratio (NBR) is the most representative spectral index (Key & Benson, 1999). It addresses the change in surface spectral reflectance after the fire, e.g., near-infrared reflection (band 4 of Landsat Thematic Mapper, TM4) is reduced, and mid-infrared reflection (band 7 of Landsat Thematic Mapper, TM7) is increased (Van Wagtendonk et al., 2004). The index (TM4−TM7)/(TM4+TM7) was found to be a good indicator of fire severity by García and Caselles (1991) and was used for mapping Spanish forest fires. It was named the NBR by Key and Benson (1999) and then was developed into the differenced normalized burn ratio (dNBR) (Lutes et al., 2006, p. 164). The calculation Eq. (12.5) is as follows:

$$dNBR = NBR_{Prefire} - NBR_{Postfire} \qquad (12.5)$$

Considering forest fire severity in different times and spaces, Miller and Thode (2007) proposed the relative differential normalized burn ratio (rdNBR). The calculation Eq. (12.6) is as follows:

$$rdNBR = dNBR / \sqrt{|NBR_{Prefire}|} \qquad (12.6)$$

At present, NBR and its derived indices, dNBR and rdNBR, are widely used in the quantitative assessment of forest fire severity (Escuin et al., 2008; French et al., 2008; Soverel et al., 2010) and research on the impact of forest management on fire severity (Wimberly et al., 2009). However, optical remote sensing can only reflect the horizontal distribution and structural changes of vegetation, which do not fully reflect the losses caused by forest fires. In addition, NBR is sensitive to moisture of soil and vegetation, as well as chlorophyll content. Whether NBR and its derivative indices reflect the real postfire situation of forest ecosystems under various fire severities needs further verification.

In the last few years, the advantages of LiDAR in fire severity assessment have been examined (Kane et al., 2014; Montealegre et al., 2014; Wang & Glenn, 2009). At present, research on forest fire severity assessment based on LiDAR is divided into two main categories:

(1) Field data and postfire traits extracted by LiDAR are used to establish a regression equation to accomplish fire severity assessment. For example, Montealegre et al. (2014) combined ALS-derived variables (first returns, all returns above 1 m, and canopy relief ratio) and field-assessed CBI through multivariate statistical analysis to estimate the severity of wildland fires in Zuera, Jaulin, Los Olmos, and Aliaga in Spain.

(2) Changes in vegetation structural traits derived from prefire and postfire LiDAR data are used to assess fire severity. As shown in Fig. 12.7, the digital elevation model and DSM can be extracted from the prefire data for computing the canopy height model (CHM). Based on CHM data and the watershed segmentation algorithm, each tree's location and crown size before the forest fire in the target area are extracted. According to the spatial coverage of each tree, the prefire and postfire LiDAR data can be extracted, as well as the canopy cover, tree height, and average laser intensity. Then, fire severity is obtained by analyzing the changes in the above indices. The results show that changes in canopy cover and laser intensity are in good agreement with forest fire severity (Fig. 12.8). Additionally, tree height showed a good correlation with fire severity in farmland and shrub forest (Wang & Glenn, 2009). The changes in structural traits derived from prefire and postfire LiDAR data can accurately reflect forest fire severity, assess the degree of damage to each tree, and provide the number of damaged trees.

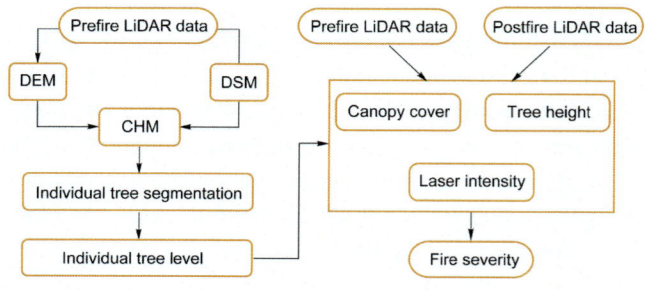

Figure 12.7 Workflow for forest fire severity assessment based on prefire and postfire LiDAR data. *CHM*, canopy height model; *DEM*, digital elevation model; *DSM*, digital surface model.

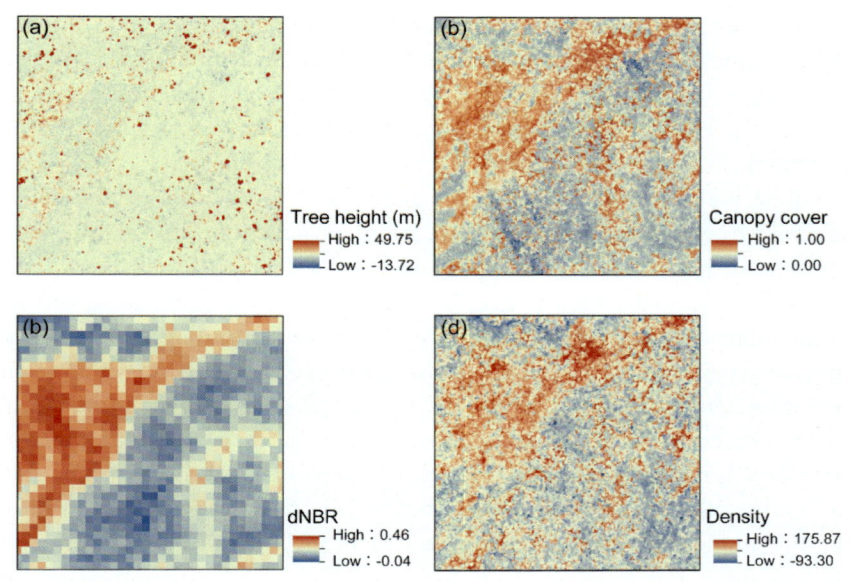

Figure 12.8 Severity assessment of a forest fire in 2012 in the Sierra Nevada, California, based on tree height, canopy cover, and laser intensity extracted from LiDAR data and the differenced normalized burn ratio (dNBR) extracted from optical imagery.

These measures are difficult to obtain with traditional remote sensing methods.

In general, using regression methods based on optical remote sensing to assess forest fire severity cannot reflect the vertical vegetation information, and they show a saturation effect, which limits their applications. LiDAR data lack optical information but are rich in information on vertical vegetation structure, compensating for the deficiencies in traditional remote sensing. Therefore, combining traditional optical remote sensing, high-resolution remote sensing (hyperspectral and high spatial resolution), and LiDAR to comprehensively reflect the severity of forest fires is an important direction in future forest fire research (Chang et al., 2012).

12.2.3 Forest Fuel Load Estimation

Forest fires are one of the most significant natural disturbance factors for forest ecosystems. Moderate fires play an important role in maintaining ecosystem structure, function, and biodiversity. However, uncontrolled forest fires often cause a large loss of forest resources and even threaten the safety of life and property of forest residents. The occurrence of forest fires usually has three necessary conditions: fire sources, fuel, and suitable meteorological conditions. The load of forest fuels determines the degree of fire damage. Therefore, understanding forest fuel loads and their distribution are of great significance for forest management, such as fire prevention, prescribed fires, and the removal of combustible materials.

Forest fuels refer to all organic matter that can be burned in forests, including trees, shrubs, herbs, lichens, moss, surface litter, and humus and peat below the surface. From the perspective of forest management, the classification of forest fuels is mainly to facilitate an understanding of the flammability of various fuels in forests so that corresponding measures for different types of fuels can be taken for the prevention and suppression of forest fires. For example, Zheng et al. (1962) classed forest fuels into dead objects, lichens, mosses, herbs, shrubs, trees, and forest clutter. Different forest fuel types may have distinct combustion characteristics. From the perspective of fire simulation and prediction, the classification of forest fires mainly considers fire behaviors, moisture changes, and the characteristics of different fuels. For example, Deeming and Brown (1975) classified forest fuels into live and dead fuels. Dead fuels can be further divided into fuels with 1 h, 10 h, 100 h, and 1000 h time delays according to the rate of water content dissipation, which corresponds to surface litter of different diameters. In addition, there are classification methods based on the location of forest fuels, such as underground, surface, and airborne fuels. Because forest fuel classification that considers moisture change and characteristics can be combined with forest fire prediction and fire spread models, these classification methods, such as Deeming and Brown (1975), are widely used.

Traditional methods for estimating forest fuel loads generally rely on field investigations and can be divided into direct and indirect methods. Direct methods include the quadrat method and line intercept method. The quadrat method involves selecting surface fuels from a variety of representative sample plots and classifying, drying, and weighing them to obtain forest fuel loads. Generally, the quadrat size for fuels with a 1 h time delay is 1 m × 1 m, for fuels with 10 h or 100 h time delay are 5 m × 5 m or 10 m × 10 m, and for fuels with 1000 h time delay require a larger quadrat. In the line intercept method, many parallel sample lines are set in a sample area, and fuel loads of different types are calculated by counting the number of intersections between different diameter branches and sample lines. Indirect methods generally estimate forest fuel loads by standard photography of surface fuels.

Despite their accuracy, traditional methods are time-consuming and labor-intensive if a wide range of fuel loads are to be obtained to meet the needs of forest management. A combination of methods is required to obtain a wide range of data. Typically, researchers set up plots and then obtain fuel loads for different vegetation types, which are used in predictive models and forest management, such as the US fire rating system. Brandis and Jacobson (2003) used vegetation types from Landsat TM data, litter accumulation and decomposition processes, forest fire history, and vegetation communities to map the distribution of forest fuels in New South Wales, Australia. High-resolution satellite images can help to produce a more detailed fuel distribution map. Jin and Chen (2012) compared the accuracy of the fuel loads derived from QuickBird and Landsat TM images in Northeast China using 70 plots. Their results showed that QuickBird was more accurate than Landsat TM—the coefficient of determination (R^2 of the total amount of dead fuels was 0.354 and 0.173, respectively).

Although many studies of forest fuels have used optical remote sensing methods, the accuracy of estimating fuel loads is relatively low owing to the two-dimensional information acquired by optical remote sensing. LiDAR provides a new technical means for accurately estimating large-scale forest fuel loads. The LiDAR method generally establishes a regression model using the height percentile, point density of different height layers, echo intensity, maximum height, and ground truth. Jakubowksi et al. (2013) used airborne LiDAR to estimate surface fuel loads with an estimated R^2 of 0.40 in the Sierra Nevada Mountains, California, USA. In addition to forest fuel loads, LiDAR data can also help to obtain more accurate forest fire prediction or spread models, such as average tree height (R^2 reached 0.60), maximum tree height (R^2 reached 0.87), and coverage (R^2 reached 0.78). Jakubowksi et al. (2013) combined point cloud data with aerial images to improve the estimation accuracy and achieved an R^2 of 0.48 for the estimated fuel loads. Kelly et al. (2017) evaluated the influence of errors associated with two LiDAR data products, canopy height (CH) and canopy base height (CBH), on simulated fire behavior, and the results showed that errors in CH and CBH did not greatly influence the modeled conditional burn probability, fire size, or fire size distribution.

In general, LiDAR has certain advantages in estimating forest surface fuels compared with optical remote sensing, but it is not outstanding. This is mainly because surface fuels are generally distributed in the understory, with limited LiDAR access. When dealing with larger fuels, such as fuels with 100 h and 1000 h time lags, LiDAR performs better. However,

LiDAR has a great advantage in estimating canopy fuels. For example, Andersen et al. (2005) used LiDAR to estimate canopy fuels (canopy fuel loads, under branch height, CH, and canopy bulk density) in Washington State, USA, and their R^2 was above 0.77.

12.3 Forest Fuel Treatment Detection

Forest fuels provide the material for fires, and fuel treatment can fundamentally solve fire safety problems and improve forest structure and health. Forest fuel treatment usually uses both prescribed fires and mechanical removal (mainly thinning) to reduce fuel loads, reduce the probability of forest fires, and minimize losses.

Intermediate cutting, also known as tending thinning, refers to the regularly repeated cutting of some trees in immature forests to create good environmental conditions for the retained trees and promote their growth and development (Zeng, 1984). As a common method of intermediate cutting and fuel management, thinning simulates the natural effect of changing forest structure. Intermediate cutting can increase the growth space and nutrient area of the retained wood, change the light, temperature, and humidity conditions under the forest canopy, and improve the physical and chemical properties of the soil, thereby promoting the rapid growth of retained wood and shortening the development period. For fuel management, thinning can reduce forest fuel loads and fuel accumulation, reducing the probability of forest fires and the losses of forest fires.

Although most studies have shown that thinning is beneficial for forest ecosystems, some people may expand the scope and intensity of thinning operations to obtain short-term timber profits, which causes irreparable damage to forest ecosystems. It is important for current forest management to detect the actual scope and intensity of thinning operations in the area.

The detection of thinning scope based on LiDAR mainly uses multi-temporal airborne LiDAR data. The data before and after processing can be intuitively judged by visual interpretation (Fig. 12.9), e.g., the number of points in the thinning area is significantly reduced. The change in LiDAR data before and after thinning can also be automatically extracted using CHM and canopy cover combined with pixel-wise thresholding and object of interest (OBI) segmentation (Fig. 12.10). The pixel-wise thresholding method assumes that value changes in forest attributes (tree height or canopy coverage) should conform to the normal distribution and that a variation value within the 95% confidence interval is in a reasonable range.

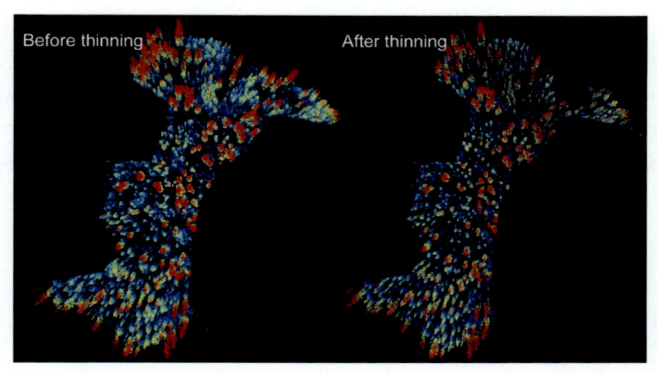

Figure 12.9 Airborne LiDAR data for a forest in Sierra Nevada, California, USA, before thinning (2007) and after thinning (2012).

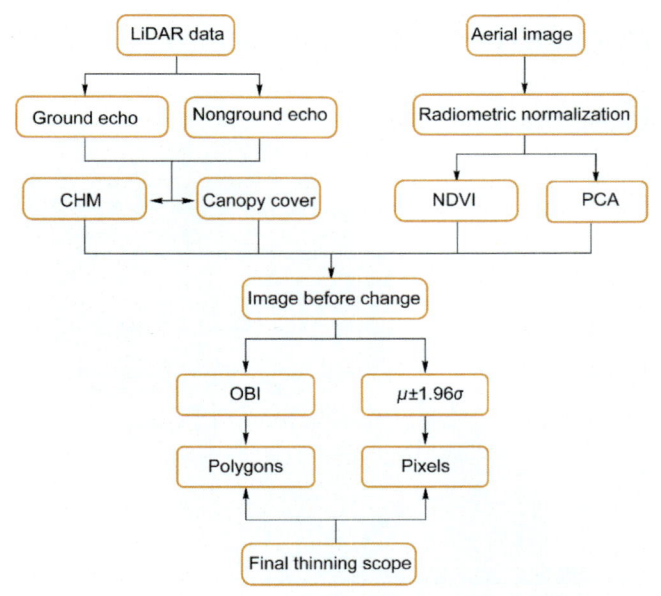

Figure 12.10 Flowchart of forest fuel treatment area detection based on pixel-wise thresholding and object of interest (OBI) segmentation. μ and σ represent the average and standard deviation of the change in forest attributes before and after thinning, respectively. *CHM*, canopy height model; *NDVI*, normalized difference vegetation index; *PCA*, principal component analysis.

Hence, when the change value exceeds $\mu \pm 1.96\sigma$ (μ is the average, σ is the standard deviation), the pixel is regarded as being thinned. However, the method is subject to noise interference during the extraction process, which generates many false-positive points. The OBI method segments images into polygons using remote sensing image processing software such as ENVI, and then potential thinning in the polygons is assessed according to the average degree of change. The object-oriented segmentation method is used to obtain the processing region to filter the results obtained by the pixel variation threshold method to better extract the range of thinning (Fig. 12.11).

12.4 Tree Mortality Analysis Under Environmental Stress

Forest death is the phenomenon of forest decline manifested by the loss of tree crowns during the growing season, the death of whole trees, and the death of widespread forests. Forest death causes substantial ecological impacts, which may lead to changes in the forest ecosystem and have an

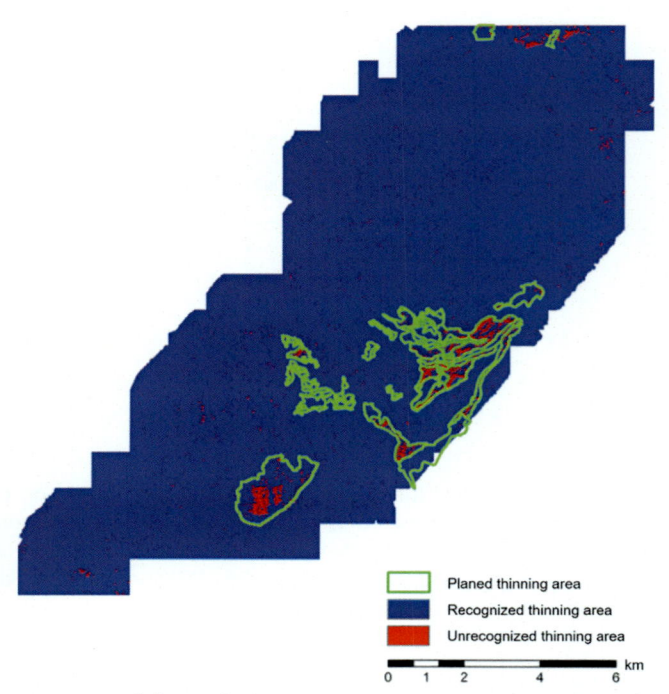

Planed thinning area
Recognized thinning area
Unrecognized thinning area

km
0 1 2   4   6

Figure 12.11 Detected forest fuel treatment areas in the Sierra Nevada, California, USA, using LiDAR data.

important effect on the climate, hydrological processes, and biogeochemical cycles of the Earth system.

In recent years, global warming has led to frequent occurrences of climate extremes worldwide, especially in semi-arid regions, causing large areas of forest death and having important impacts on plant communities, ecological functions, and ecosystem services. The relationship between forest death and climate change has gradually become the focus of many ecologists. With global climate change, the temperature rises, and the possibility of droughts increases. The occurrence of tree deaths caused by droughts is becoming increasingly common. Drought will likely become more widespread, prolonged, and extreme over the next century. Severe drought conditions strongly increase the ambient temperature and decrease precipitation which, in concert, push trees beyond their physiological limits by increasing the vapor pressure deficit. Stovall et al. (2019) combined high-resolution (sub-meter level) airborne three-dimensional LiDAR data and optical data over 40,000 ha within the southern Sierra Nevada forest to locate individual trees and assess mortality across the landscape. The study results showed that tree height is the most important predictor of tree death during drought. As environmental stressors increase, the influences on large trees are nonlinear, suggesting that more frequent and extreme droughts may be most detrimental to the largest trees on Earth.

Giant Sequoia (*Sequoiadendron giganteum*, SEGI) trees, the sole living species in the genus *Sequoiadendron*, are among the largest trees (both in height and volume) and the oldest living trees on Earth (Cook, 1942). A warmer temperature can extend the growing season and thus enhance SEGI growth. However, wildfire frequency within SEGI groves is also higher as the temperature warms (Mutch, 1994). To assess the relative vulnerability of SEGI groves in climate warming and multiyear dry periods, vegetation indices from Landsat imagery for the period 1985–2015 were used to investigate the relative greenness and wetness of giant Sequoia groves and the surrounding forests in the Sierra Nevada, California (Su et al., 2017). The results indicate that the wetness and greenness of SEGI groves show a larger response to the warming climate and drought than in non-grove areas (Fig. 12.12). The influence of droughts on the wetness of SEGI groves reflects the effects of both the multidecadal increase in forest biomass and of warmer drought-year temperatures on the evaporative demand of current grove vegetation, as well as affecting the regolith water storage of rain and snowmelt that sustain the vegetation through seasonal and multiyear dry periods.

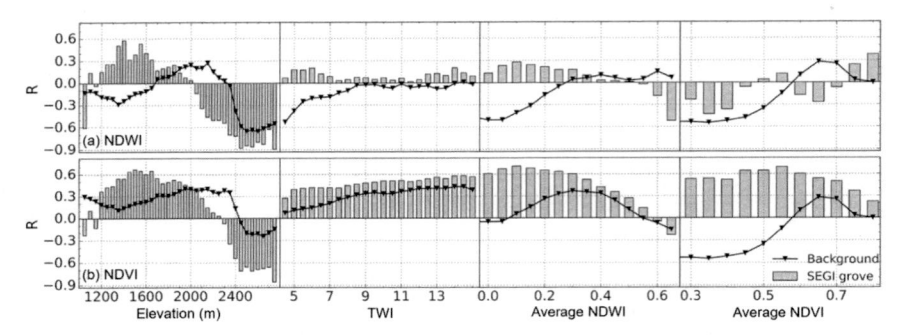

Figure 12.12 Distribution of SEGI groves with a decreasing, increasing, or insignificant trend in (a) normalized difference wetness index (NDWI) and (b) normalized difference vegetation index (NDVI) over the last 3 decades. *TWI*, topographic wetness index; *SEGI, Sequoiadendron giganteum.*

12.5 Chapter Summary

LiDAR provides a powerful data source for forest ecosystem dynamics monitoring (including forest growth monitoring and interference monitoring) because of its advantages in extracting three-dimensional structural traits. Using long time-series LiDAR data for forest dynamics monitoring is also important for forest ecology. Adding the three-dimensional structural information derived from LiDAR to the spectral information derived from traditional optical remote sensing makes it an important data source for future forest dynamics monitoring. Combining long time-series LiDAR data and other remote sensing data for real-time forest dynamics monitoring is of great significance for understanding forest health and establishing management measures.

References

Adams, H. R., Barnard, H. R., & Loomis, A. K. (2014). Topography alters tree growth—climate relationships in a semi-arid forested catchment. *Ecosphere, 5*(11), 1—16.

Andersen, H.-E., McGaughey, R. J., & Reutebuch, S. E. (2005). Estimating forest canopy fuel parameters using LIDAR data. *Remote Sensing of Environment, 94*(4), 441—449.

Bettinger, P., Boston, K., Siry, J., & Grebner, D. L. (2016). *Forest management and planning.* Academic press.

Bettinger, P., Lennette, M., Johnson, K. N., & Spies, T. A. (2005). A hierarchical spatial framework for forest landscape planning. *Ecological Modelling, 182*(1), 25—48.

Biging, G. S., & Dobbertin, M. (1995). Evaluation of competition indices in individual tree growth models. *Forest Science, 41*(2), 360—377.

Bollandsås, O. M., Gregoire, T. G., Næsset, E., & Øyen, B.-H. (2013). Detection of biomass change in a Norwegian mountain forest area using small footprint airborne laser scanner data. *Statistical Methods & Applications, 22*(1), 113—129.

Brandis, K., & Jacobson, C. (2003). Estimation of vegetative fuel loads using Landsat TM imagery in New South Wales, Australia. *International Journal of Wildland Fire, 12*(2), 185–194.

Chang, Y., Chen, H., Hu, Y., Feng, Y., & Li, Y. (2012). Advances in the assessment of forest fire severity and its spatial heterogeneity. *Ziran Zaihai Xuebao, 21,* 28–34.

Contreras, M. A., Affleck, D., & Chung, W. (2011). Evaluating tree competition indices as predictors of basal area increment in western Montana forests. *Forest Ecology and Management, 262*(11), 1939–1949.

Cook, L. F. (1942). *Giant sequoias of California.*

De Santis, A., & Chuvieco, E. (2009). GeoCBI: A modified version of the composite burn index for the initial assessment of the short-term burn severity from remotely sensed data. *Remote Sensing of Environment, 113*(3), 554–562.

Deeming, J. E., & Brown, J. K. (1975). Fuel models in the national fire-danger rating system. *Journal of Forestry, 73*(6), 347–350.

Escuin, S., Navarro, R., & Fernández, P. (2008). Fire severity assessment by using NBR (normalized burn ratio) and NDVI (normalized difference vegetation index) derived from LANDSAT TM/ETM images. *International Journal of Remote Sensing, 29*(4), 1053–1073.

French, N. H., Kasischke, E. S., Hall, R. J., Murphy, K. A., Verbyla, D. L., Hoy, E. E., & Allen, J. L. (2008). Using Landsat data to assess fire and burn severity in the North American boreal forest region: An overview and summary of results. *International Journal of Wildland Fire, 17*(4), 443–462.

García, M. L., & Caselles, V. (1991). Mapping burns and natural reforestation using thematic mapper data. *Geocarto International, 6*(1), 31–37.

Greene, D. F., Noël, J., Bergeron, Y., Rousseau, M., & Gauthier, S. (2004). Recruitment of *Picea mariana, Pinus banksiana,* and *Populus tremuloides* across a burn severity gradient following wildfire in the southern boreal forest of Quebec. *Canadian Journal of Forest Research, 34*(9), 1845–1857.

Hall, D. K., Ormsby, J. P., Johnson, L., & Brown, J. (1980). Landsat digital analysis of the initial recovery of burned tundra at Kokolik River, Alaska. *Remote Sensing of Environment, 10*(4), 263–272.

Hamraz, H., Contreras, M. A., & Zhang, J. (2017). Vertical stratification of forest canopy for segmentation of understory trees within small-footprint airborne LiDAR point clouds. *ISPRS Journal of Photogrammetry and Remote Sensing, 130,* 385–392.

Hann, D. W. (1980). Even-aged management: Basic managerial questions and available or potential techniques for answering them. In *Intermountain forest and range experiment station* (Vol 83). US Department of Agriculture Forest service.

Hopkinson, C., Chasmer, L., & Hall, R. (2008). The uncertainty in conifer plantation growth prediction from multi-temporal lidar datasets. *Remote Sensing of Environment, 112*(3), 1168–1180.

Hyyppä, J., Yu, X., Rönnholm, P., Kaartinen, H., & Hyyppä, H. (2003). Factors affecting laser-derived object-oriented forest height growth estimation. *The Photogrammetric Journal of Finland, 18*(2), 16–31.

Jakubowksi, M. K., Guo, Q., Collins, B., Stephens, S., & Kelly, M. (2013). Predicting surface fuel models and fuel metrics using Lidar and CIR imagery in a dense, mountainous forest. *Photogrammetric Engineering & Remote Sensing, 79*(1), 37–49.

Jakubowski, M. K., Li, W., Guo, Q., & Kelly, M. (2013). Delineating individual trees from LiDAR data: A comparison of vector-and raster-based segmentation approaches. *Remote Sensing, 5*(9), 4163–4186.

Jin, S., & Chen, S.-C. (2012). Application of QuickBird imagery in fuel load estimation in the Daxinganling region, China. *International Journal of Wildland Fire, 21*(5), 583–590.

Kane, V. R., North, M. P., Lutz, J. A., Churchill, D. J., Roberts, S. L., Smith, D. F., McGaughey, R. J., Kane, J. T., & Brooks, M. L. (2014). Assessing fire effects on forest spatial structure using a fusion of Landsat and airborne LiDAR data in Yosemite National Park. *Remote Sensing of Environment, 151*, 89−101.

Kasischke, E. S., Turetsky, M. R., Ottmar, R. D., French, N. H., Hoy, E. E., & Kane, E. S. (2008). Evaluation of the composite burn index for assessing fire severity in Alaskan black spruce forests. *International Journal of Wildland Fire, 17*(4), 515−526.

Keeley, J. E. (2009). Fire intensity, fire severity and burn severity: A brief review and suggested usage. *International Journal of Wildland Fire, 18*(1), 116−126.

Kelly, M., Su, Y., Di Tommaso, S., Fry, D. L., Collins, B. M., Stephens, S. L., & Guo, Q. (2017). Impact of error in lidar-derived canopy height and canopy base height on modeled wildfire behavior in the Sierra Nevada, California, USA. *Remote Sensing, 10*(1), 10.

Key, C. H., & Benson, N. C. (1999). Measuring and remote sensing of burn severity. In *Proceedings joint fire science conference and workshop.*

Keyser, T. L., Smith, F. W., Lentile, L. B., & Shepperd, W. D. (2006). Modeling postfire mortality of ponderosa pine following a mixed-severity wildfire in the black hills: The role of tree morphology and direct fire effects. *Forest Science, 52*(5), 530−539.

Knapp, E. E., & Keeley, J. E. (2006). Heterogeneity in fire severity within early season and late season prescribed burns in a mixed-conifer forest. *International Journal of Wildland Fire, 15*(1), 37−45.

Korhonen, L., Korpela, I., Heiskanen, J., & Maltamo, M. (2011). Airborne discrete-return LIDAR data in the estimation of vertical canopy cover, angular canopy closure and leaf area index. *Remote Sensing of Environment, 115*(4), 1065−1080.

Ledermann, T. (2010). Evaluating the performance of semi-distance-independent competition indices in predicting the basal area growth of individual trees. *Canadian Journal of Forest Research, 40*(4), 796−805.

Londo, H. A. (2010). *The suitability of LiDAR-derived forest attributes for use in individual-tree distance-dependent growth-and-yield modeling.* Mississippi State University.

Lutes, D. C., Keane, R. E., Caratti, J. F., Key, C. H., Benson, N. C., Sutherland, S., & Gangi, L. J. (2006). *FIREMON: Fire effects monitoring and inventory system. Gen. Tech. Rep. RMRS-GTR-164.* Fort Collins, CO: US Department of Agriculture, Forest Service, Rocky Mountain Research Station. 1 CD.

Maltamo, M., Næsset, E., & Vauhkonen, J. (2014). Forestry applications of airborne laser scanning. *Concepts and Case Studies. Manag For Ecosys, 27*, 460.

Ma, Q., Su, Y., Tao, S., & Guo, Q. (2018). Quantifying individual tree growth and tree competition using bi-temporal airborne laser scanning data: A case study in the Sierra Nevada mountains, California. *International Journal of Digital Earth, 11*(5), 485−503.

Miller, J. D., & Thode, A. E. (2007). Quantifying burn severity in a heterogeneous landscape with a relative version of the delta Normalized Burn Ratio (dNBR). *Remote Sensing of Environment, 109*(1), 66−80.

Miyanishi, K., & Johnson, E. (2002). Process and patterns of duff consumption in the mixedwood boreal forest. *Canadian Journal of Forest Research, 32*(7), 1285−1295.

Montealegre, A. L., Lamelas, M. T., Tanase, M. A., & De la Riva, J. (2014). Forest fire severity assessment using ALS data in a Mediterranean environment. *Remote Sensing, 6*(5), 4240−4265.

Mutch, L. (1994). *Growth responses of giant sequoia to fire and climate in sequoia and Kings Canyon National Parks* (master thesis). California, University of Arizona.

Næsset, E. (2004). Practical large-scale forest stand inventory using a small-footprint airborne scanning laser. *Scandinavian Journal of Forest Research, 19*(2), 164−179.

Shu, L., Tian, X., & Kou, X. (2003). The focus and progress on forest fire research. *World Forestry Research, 16*(3), 37−40.

Smith, A. M., Wooster, M. J., Drake, N. A., Dipotso, F. M., Falkowski, M. J., & Hudak, A. T. (2005). Testing the potential of multi-spectral remote sensing for retrospectively estimating fire severity in African Savannahs. *Remote Sensing of Environment, 97*(1), 92−115.

Solberg, S., Næsset, E., Hanssen, K. H., & Christiansen, E. (2006). Mapping defoliation during a severe insect attack on Scots pine using airborne laser scanning. *Remote Sensing of Environment, 102*(3−4), 364−376.

Song, Y., Imanishi, J., Sasaki, T., Ioki, K., & Morimoto, Y. (2016). Estimation of broadleaved canopy growth in the urban forested area using multi-temporal airborne LiDAR datasets. *Urban Forestry & Urban Greening, 16*, 142−149.

Soverel, N. O., Perrakis, D. D., & Coops, N. C. (2010). Estimating burn severity from Landsat dNBR and RdNBR indices across western Canada. *Remote Sensing of Environment, 114*(9), 1896−1909.

Stovall, A. E., Shugart, H., & Yang, X. (2019). Tree height explains mortality risk during an intense drought. *Nature Communications, 10*(1), 1−6.

Su, Y., Bales, R. C., Ma, Q., Nydick, K., Ray, R. L., Li, W., & Guo, Q. (2017). Emerging stress and relative resiliency of giant sequoia groves experiencing multiyear dry periods in a warming climate. *Journal of Geophysical Research: Biogeosciences, 122*(11), 3063−3075.

Su, Y., Guo, Q., Collins, B. M., Fry, D. L., Hu, T., & Kelly, M. (2016). Forest fuel treatment detection using multi-temporal airborne lidar data and high-resolution aerial imagery: A case study in the Sierra Nevada mountains, California. *International Journal of Remote Sensing, 37*(14), 3322−3345.

Tompalski, P., Coops, N. C., White, J. C., & Wulder, M. A. (2016). Enhancing forest growth and yield predictions with airborne laser scanning data: Increasing spatial detail and optimizing yield curve selection through template matching. *Forests, 7*(11), 255.

Twery, M. J., & Weiskittel, A. R. (2013). Forest-management modelling. *Environmental Modelling: Finding Simplicity in Complexity*, 379−398.

Utkin, A. B., Fernandes, A., Simões, F., Lavrov, A., & Vilar, R. (2003). Feasibility of forest-fire smoke detection using lidar. *International Journal of Wildland Fire, 12*(2), 159−166.

Utkin, A., Lavrov, A., Costa, A., Simões, F., & Vilar, R. (2002). Detection of small forest fires by lidar. *Applied Physics B, 74*(1), 77−83.

Van Wagtendonk, J. W., Root, R. R., & Key, C. H. (2004). Comparison of AVIRIS and Landsat ETM+ detection capabilities for burn severity. *Remote Sensing of Environment, 92*(3), 397−408.

Vastaranta, M., Korpela, I., Uotila, A., Hovi, A., & Holopainen, M. (2012). Mapping of snow-damaged trees based on bitemporal airborne LiDAR data. *European Journal of Forest Research, 131*(4), 1217−1228.

Vepakomma, U., St-Onge, B., & Kneeshaw, D. (2011). Response of a boreal forest to canopy opening: Assessing vertical and lateral tree growth with multi-temporal lidar data. *Ecological Applications, 21*(1), 99−121.

Wang, C., & Glenn, N. F. (2009). Estimation of fire severity using pre-and post-fire LiDAR data in sagebrush steppe rangelands. *International Journal of Wildland Fire, 18*(7), 848−856.

Weiskittel, A. R., Hann, D. W., Kershaw, J. A., Jr., & Vanclay, J. K. (2011). *Forest growth and yield modeling*. John Wiley & Sons.

Wensel, L., Meerschaert, W., & Biging, G. (1987). Tree height and diameter growth models for northern California conifers. *Hilgardia, 55*(8), 1−20.

Wimberly, M. C., Cochrane, M. A., Baer, A. D., & Pabst, K. (2009). Assessing fuel treatment effectiveness using satellite imagery and spatial statistics. *Ecological Applications, 19*(6), 1377−1384.

Wu, X., Qin, X., Li, C., Tian, Z., Xiong, Y., Yang, D., & Zhang, R. (2010). Analysis of current forest fire monitoring system in China. *Inner Mongolia Forestry Investigation and Design, 3*, 69–72 (in Chinese).

Yu, X. W., Hyyppa, J., Kaartinen, H., & Maltamo, M. (2004). Automatic detection of harvested trees and determination of forest growth using airborne laser scanning. *Remote Sensing of Environment, 90*(4), 451–462 (in Chinese).

Zeng, X. (1984). What is forest tending and thinning. *Shaanxi Forestry Science and Technology, 2*, 78 (in Chinese).

Zheng, H., Wang, Y., Guo, K., & Xi, X. (1962). *Forest fire prevention.* Agricultural Press (in Chinese).

CHAPTER 13

Applications of LiDAR in Biodiversity Conservation, Ecohydrology, and Ecological Process Modeling of Forest Ecosystems

Contents

In addition to extracting topographical information, structural traits, and functional traits from forest ecosystems, light detection and ranging (LiDAR) can be used in forest biodiversity and ecohydrology studies and ecological modeling with the extracted structural and functional traits. In this chapter, we first introduce the application of remote sensing in biodiversity studies and explore the role and advantages of LiDAR in biodiversity monitoring in forest ecosystems. Then, we examine the application of LiDAR in ecohydrology and discuss how to assimilate the ecological traits obtained using remote sensing, such as tree height and leaf area index, into ecosystem models to improve the ecosystem simulation accuracy. Finally, we introduce LiDAR applications for forest ecological modeling.

LiDAR Principles, Processing and Applications in Forest Ecology
ISBN 978-0-12-823894-3
https://doi.org/10.1016/B978-0-12-823894-3.00013-X

13.1 Forest Biodiversity Studies

Human activities and global climate change have exerted increasing pressure on ecosystems, leading to ecosystem degradation and biodiversity losses. The conservation of biodiversity is an issue of global concern. Accurate monitoring, a focal scientific issue in biodiversity research, is essential for measuring the effectiveness of biodiversity conservation efforts. For a long time, monitoring and assessing biodiversity required field investigation, which is time-consuming and labor-intensive and calls for experienced investigators to avoid human error. These challenges restrict the application of fieldwork to large-scale monitoring and assessment of biodiversity. The rapid development of remote sensing technology offers a new method to address these challenges. This section introduces the progress and advantages of remote sensing in biodiversity research with specific case studies.

13.1.1 Research Progress on Remote Sensing for Biodiversity

Since the launch of Landsat-1 by the National Aeronautics and Space Administration (NASA) in 1972, remote sensing technology has been widely used in environmental exploration, ecological cartography, and understanding and predicting habitat changes. By the early 1990s, researchers had found that satellite imagery provided instantaneous, systematic, and repeatable data, with many advantages in monitoring and assessing biodiversity at large scales (Noss, 1990; Roughgarden et al., 1991; Soulé & Kohm, 1989). The current methods for monitoring biodiversity based on remote sensing technology can be categorized as either direct or indirect (Turner et al., 2003). Direct methods identify species or community types, together with their distribution and abundance. They require remote sensing data with high spatial and spectral resolution and are an important future development direction for biodiversity monitoring. Indirect methods aim to derive indicators or variables through remote sensing data that are considered or confirmed to be closely related to biodiversity. Then, these data are combined with field survey data to construct models to predict species distribution and diversity patterns. Indirect methods are currently the mainstream application of remote sensing in biodiversity research.

Early biodiversity monitoring studies based on direct methods first classified satellite images to obtain land cover data. From those data, vegetation types were distinguished, and a series of landscape indices were calculated, such as the number and area of patches, boundary density, and the Shannon diversity index. Then, based on the assumption that reduced species richness was linked to deforestation and habitat fragmentation, those studies predicted species distribution and assessed the ability of species to cope with disturbances and risks (Stoms & Estes, 1993; Westman et al., 1989). Biodiversity monitoring based on landscape index analysis is suitable for large-scale applications and was widely used from 1980 to the 1990s (Hu et al., 2012). However, since 2000, researchers have pointed out that these methods cannot ensure the accuracy of derived biodiversity information from the perspectives of environmental factors (Griffiths et al., 2000), the resolution of remote sensing data (Saura, 2004), the accuracy of land cover classification (Langford et al., 2006), and internal data from patches (Gillespie, 2005).

Owing to continuous improvements in the spatial and spectral resolution of remote sensing imagery, the relationship between spectral radiance from imagery and species distribution patterns obtained from field surveys can be directly established in small areas to analyze and monitor biodiversity (Nagendra, 2001). This method classifies the particular spectral features of target objects to avoid difficulties in recognizing individual species from remote sensing imagery (Wulder et al., 2004). The theoretical basis of this method is the correlation between spectral variability, species richness, and species diversity, also known as the spectral variation hypothesis (Palmer et al., 2002). The theory predicates spectral variability as an indicator of habitat heterogeneity, and habitats with greater heterogeneity theoretically contain more species. Asner and Martin (2009) used regression analyses of both vegetation chemical properties and spectral reflectance to determine the distribution of various tropical vegetation types. Carlson et al. (2007) applied hyperspectral remote sensing data to analyze the relationship between woody plant diversity and spectral variability to explore the canopy diversity of tropical forests in Hawaii, USA. They further analyzed the relationship between changes in the physiological and biochemical indices of leaves (e.g., water, pigments, nitrogen content) and the spectral variability of hyperspectral data, finding interspecific differences in the physiochemical properties of leaves that were responsible for their spectral variation. Their study identified the link between the physiochemical properties of organisms and spectral variability in remote sensing data, establishing a critical milestone in research on spectral variation theory (Hu et al., 2012). The availability of information about the physiochemical properties of leaves can improve direct tree species classification and biodiversity assessment from remote sensing data.

Compared with direct methods for biodiversity monitoring, indirect approaches mainly aim to acquire four environmental variables using remote sensing data for the estimation and simulation of biodiversity: climate and topography, productivity, habitat conditions, and interference. Under various environmental scenarios, the importance of these variables is different. Still, a consensus has been reached that climate is the determining factor for the patterns of biodiversity at regional and global scales (Hawkins et al., 2003). Climate variables are mainly based on temperature and precipitation—the most frequently used traits are mean annual temperature, annual precipitation, potential evaporation, and actual evaporation. Without considering topographic variables, the accuracy of biodiversity prediction based solely on climate factors can reach approximately 70%—88% (Duro et al., 2007). Topography has also been recognized as a key variable that explains biodiversity differences at regional and landscape scales (Irl et al., 2015; Rosenzweig, 1995). Elevation, topographic complexity, and potential solar radiation intensity are the most common topographic traits used in indirect biodiversity monitoring.

Net primary productivity (NPP) and gross primary productivity (GPP) are other important drivers of large-scale biodiversity patterns because of the positive correlation between productivity and species richness. This means that higher-productivity regions produce more abundant resources for more competitive coexisting species than lower-productivity regions—thus, they support larger biological groups with larger population sizes. For example, Kooistra et al. (2008) simulated the biodiversity of a floodplain located on the lower Rhine River by combining estimated NPP using remote sensing technology with dynamic vegetation models. The results demonstrated that NPP products could validate and initialize dynamic vegetation models to predict plant biodiversity and biomass development. Because vegetation indices are closely related to biomass and productivity and can be easily computed, normalized differential vegetation index (NDVI), enhanced vegetation index (EVI), fraction of photosynthetically active radiation (fPAR), and many other indices representing the photosynthetic rate of vegetation are often used as indicators in biodiversity study. These indices can be used to construct relationships between the field survey data of species richness and the estimated vegetation production function based on remote sensing technology. Krishnaswamy et al. (2009) used multitemporal NDVI calculated from the Mahalanobis distance as a single measure to replace the original reflectance data. They were able to describe changes in forest types consistently and quantitatively. A simple relationship between NDVI and animal diversity has been found in related studies. For example, Seto et al. (2004) discussed the relationship between NDVI calculated from Landsat TM data and the diversity of birds and butterflies in North America. However, EVI is more suitable than NDVI in biodiversity prediction in arid regions. This is because the blue band is included in EVI, which makes it less sensitive to soil and atmosphere than NDVI. EVI is also useful as the stand density increases. NDVI tends to become saturated and thus insensitive to dense vegetation, whereas EVI maintains its sensitivity to dense vegetation (Wei et al., 2008).

Subsequent studies have found that the relationship between biodiversity and productivity varies with different spatial scales (Belote et al., 2011). At local and landscape scales, such as old-growth forests with stable land cover, species distribution, richness, and population size are most affected by habitat availability, distribution, and quality. Information describing the vertical characteristics of habitats has become a key factor for explaining and predicting biodiversity (Davies & Asner, 2014; MacArthur, 1960). Monitoring biodiversity at these small or fine scales requires remote sensing data with high accuracy. A deficiency of traditional optical remote sensing technology and aerial photography is that they cannot depict the three-dimensional structure of habitats (Levick & Rogers, 2008), which is important for biodiversity research. Quantitative descriptions are lacking for the three-dimensional structure and quality of habitats: canopy height, canopy biomass, and especially information on vertical biomass profiles, crown base height, true leaf area index, and timber volume. Thus, the need to depict habitat number and quality cannot be met solely by relying on traditional optical remote sensing technology.

Light detection and ranging (LiDAR) point cloud data can extract three-dimensional structural information on habitats, such as topography underneath the canopy, tree height, crown base height, and canopy area, thus complementing traditional optical remote sensing technology and enhancing the application of remote sensing technology in direct biodiversity quantification. Some studies have shown that LiDAR point cloud data, in combination with hyperspectral data, can successfully identify canopy species, estimate aboveground biomass, extract the height of target species, analyze canopy gaps, and remove shadows (Asner & Martin, 2009; Lucas et al., 2008). Voss and Sugumaran (2008) classified four deciduous and three evergreen species using LiDAR data and Airborne Imaging Spectrometer for Applications hyperspectral data. The effect of shadows was successfully removed with the help of LiDAR data. It was possible to distinguish tall species from short species. Ultimately, the classification accuracy for the data acquired from both fall and summer field campaigns was improved by 19% when LiDAR elevation data were available. Sugumaran and Voss (2007) and Johansen et al. (2010) used object-oriented classification and regression analysis to combine LiDAR intensity and elevation information and spectral bands for image segmentation, which improved classification accuracy by approximately 12%—24%.

Vegetation types are another way to represent biodiversity. Accurate identification of vegetation types is important for estimating and protecting biodiversity. The three-dimensional structural information from LiDAR data can work well with optical remote sensing imagery. The two data sources are often combined to produce better, finer-scale vegetation maps. The authors' research team mapped the vegetation types in the Nevada mountains in the United States based on an airborne LiDAR system and aerial photographs, using a newly developed unsupervised classification method with Bayesian information criteria and k-means automatic clustering (Su et al., 2016), which showed that incorporating LiDAR data could improve classification accuracy and the recognition of more vegetation types than traditional methods that were based solely on optical imagery (Fig. 13.1). With crowdsourced data, highly accurate national-scale vegetation maps can also be produced. The authors' research team successfully updated the Vegetation Map of China (1:1,000,000) using a "crowdsourcing-change detection-classification-expert knowledge" vegetation mapping strategy (Su, Guo, et al., 2020). It contains 12 vegetation type groups, 55 vegetation types/subtypes, and 866 vegetation formation/sub-formation types (Fig. 13.2).

Accurate positional and structural information acquired from LiDAR technology can also be used as data for indirect biodiversity estimation (Mucher et al., 2013; Simonson et al., 2014). Goetz et al. (2010) analyzed the distribution of a tropical bird species using vertical canopy structure

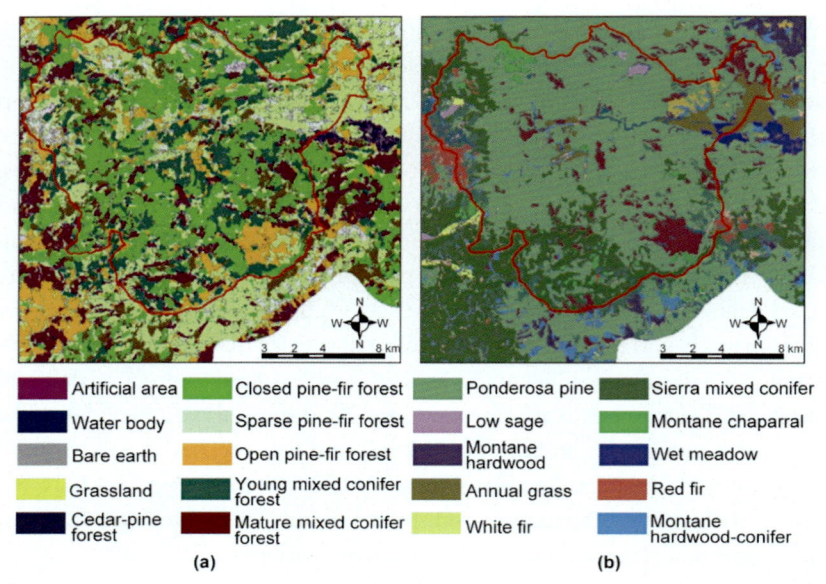

Figure 13.1 Comparison of vegetation classification results. (a) Based on light detection and ranging (LiDAR) data and aerial photos, a vegetation map was produced using a new unsupervised method with Bayesian information criteria and k-means automatic clustering. (b) The original vegetation map adopted by United States Forest Service.

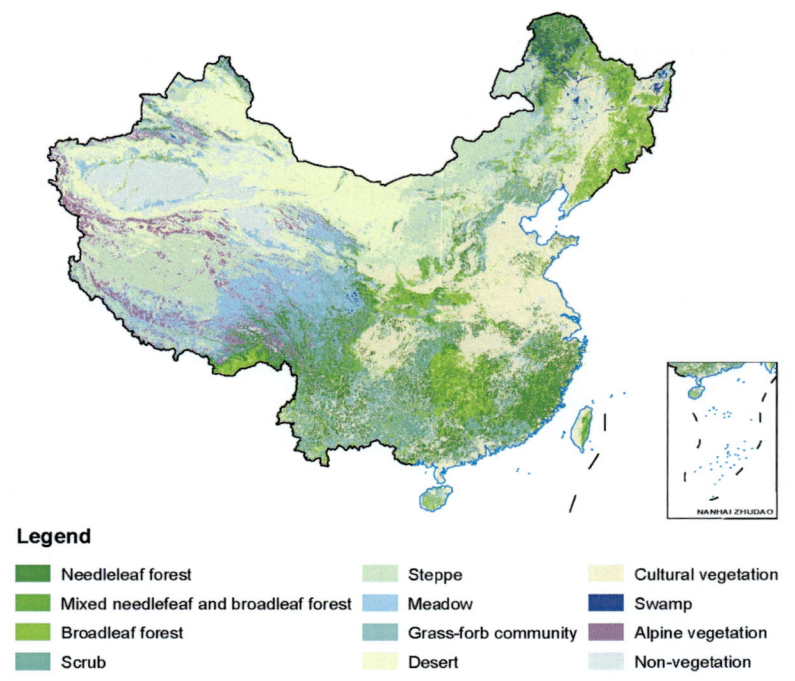

Legend

■	Needleleaf forest	■	Steppe	■	Cultural vegetation
■	Mixed needlefeaf and broadleaf forest	■	Meadow	■	Swamp
■	Broadleaf forest	■	Grass-forb community	■	Alpine vegetation
■	Scrub	■	Desert	■	Non-vegetation

Figure 13.2 Updated Vegetation Map of China (1:1,000,000) presented at the vegetation group type level.

information calculated from LiDAR data; they found that the forest structural data acquired from LiDAR included traits that are important in determining bird diversity. From this conclusion, Goetz et al. (2010) modeled the relationship between bird richness and forest structure and successfully predicted the distribution of tropical bird species in the United States. Muller and Brandl (2009) used LiDAR data to predict forest beetle populations in mountainous regions. They found that environmental variables obtained from LiDAR data were consistent with the results acquired from field survey data. This indicates that LiDAR technology can effectively depict the forest habitats in complex or mountainous regions and establish relationships between information extracted from remote sensing data and information acquired from field surveys at a large scale. Vihervaara et al. (2015) combined the bird distribution data obtained from field surveys with forest vegetation data acquired from airborne LiDAR scanning to evaluate biodiversity at the landscape scale. Many structural traits extracted from aerial LiDAR scanning data were significant for quantifying essential biodiversity variables (Pereira et al., 2013).

13.1.2 Case Studies of LiDAR Applications in Biodiversity Research

LiDAR technology has been proven to play an important role in biodiversity studies. This section will focus on applying LiDAR technology in biodiversity studies, using two projects sponsored by the Sierra Nevada Adaptive Management Project (SNAMP) as examples. The two SNAMP projects selected two forests in the Nevada mountains, one in the south and one in the north. The research objectives focused on management practices for preventing forest fires (e.g., thinning or removing fuels on the surface) and how these practices might affect wildlife, forest health, and watersheds in forest ecosystems. The Pacific fisher and California spotted owl were the two focal wildlife species in the SNAMP projects.

The research on the Pacific fisher took place at the southern station of the SNAMP project, named Sugar Pine (Fig. 13.3). The study area is 36.1 km^2, with the elevation ranging between 758 m (the lowest point) and

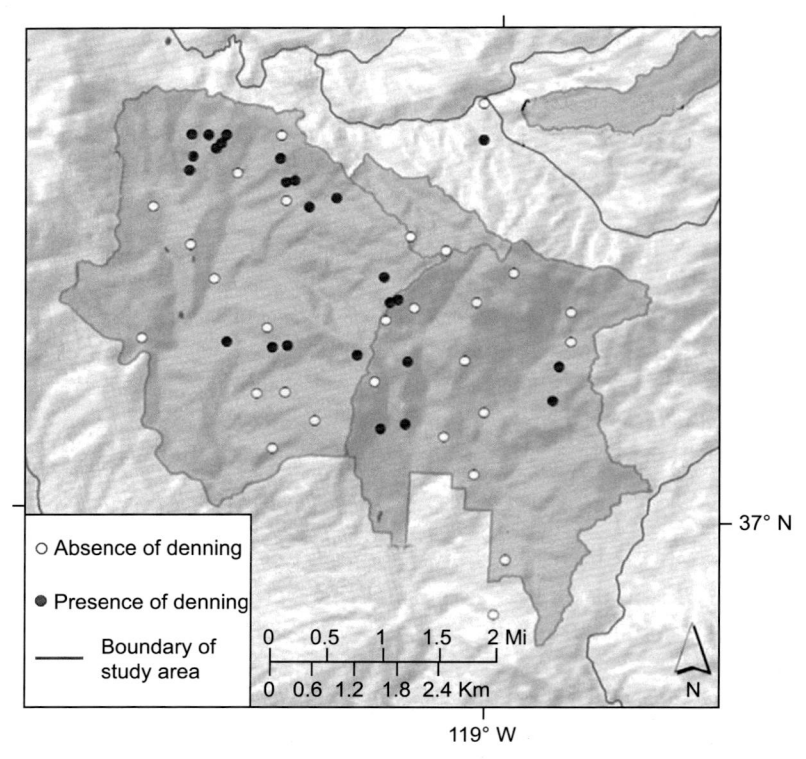

Figure 13.3 The distribution of field quadrats for surveying pacific fisher habitats.

2652 m (the highest point), and is mostly covered by mixed coniferous and broadleaf forests. Pacific fishers were captured in traps, and electronic collars were placed on each captured animal to monitor its daily routine. The authors' research team scanned the entire Sugar Pine area using airborne LiDAR. The point density was 12 points per square meter, and up to four returns could be recorded from each laser pulse. According to the data processing and parameter extraction methods introduced in previous chapters, terrain and vegetation structural parameters were extracted from the LiDAR point cloud data, such as slope, aspect, elevation quantiles, coefficient of variation of elevation, maximum tree height, and canopy cover. The diameter at breast height (DBH) for each quadrat was also calculated by combining field survey data, and a coefficient of determination (R^2) of 0.71 was achieved. The research on Pacific fishers started by focusing on their preference for nesting places, mainly based on the circular field quadrats shown in Fig. 13.3. By analyzing the terrain, vegetation structural parameters, and information on whether Pacific fishers chose to nest inside the quadrats, the habitat preference of Pacific fishers was clarified. Through classification and regression tree analysis, the authors' research team estimated and evaluated the nesting probability of the Pacific fisher. The results showed that the species was most likely to nest in places with high vegetation coverage, large trees, and high vertical structural complexity. Pacific fishers also preferred to nest in regions with steep terrain features. Such information can facilitate better Pacific fisher habitat protection and make it easier for us to monitor the dynamics of the species.

Based on this study and several years of data accumulated from electronic collars, the authors' research team worked with colleagues who had previously focused on Pacific fisher habitats for a more in-depth study of the potential distribution of Pacific fishers. Because the terrain and vegetation structural parameters of the entire study area were acquired, positioning information for Pacific fishers was used in combination with a species distribution model (SDM) to simulate the spatial distribution of Pacific fishers. According to the results in Section 12.3, the authors' research team acquired data on the distribution and intensity of thinning within the study area. Changes in vegetation structural parameters under various levels of thinning were analyzed, and scenarios of 10%, 20%, and 30% thinning were constructed. After importing these scenarios into the SDM, it was concluded that thinning with an intensity of 10%—30% did not significantly influence the spatial distribution of Pacific fishers (Fig. 13.4).

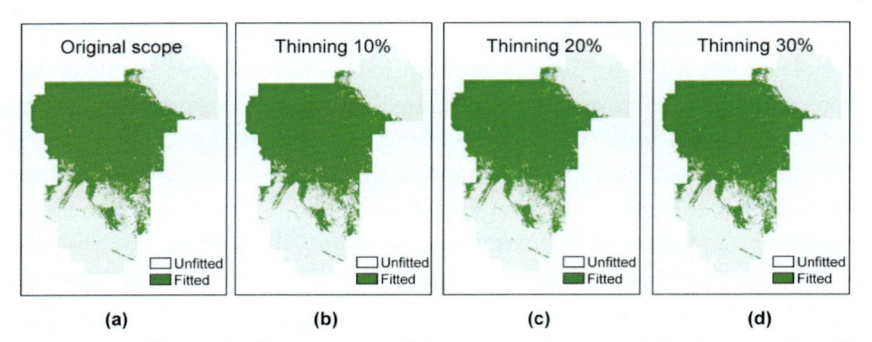

Figure 13.4 Effects of different levels of thinning on the spatial distribution of pacific fishers.

California spotted owls are a symbol of American forests. They are now endangered, with statistics indicating that only approximately 2000 in-dividuals are left in California. Understanding their preferences for habitats and the effects of fire disturbance, forestry management operations, and other human activities is critical for conserving California spotted owls. The research on this species was conducted in Last Chance on the northern site of the SNAMP Project (Fig. 13.5). The total study site area is

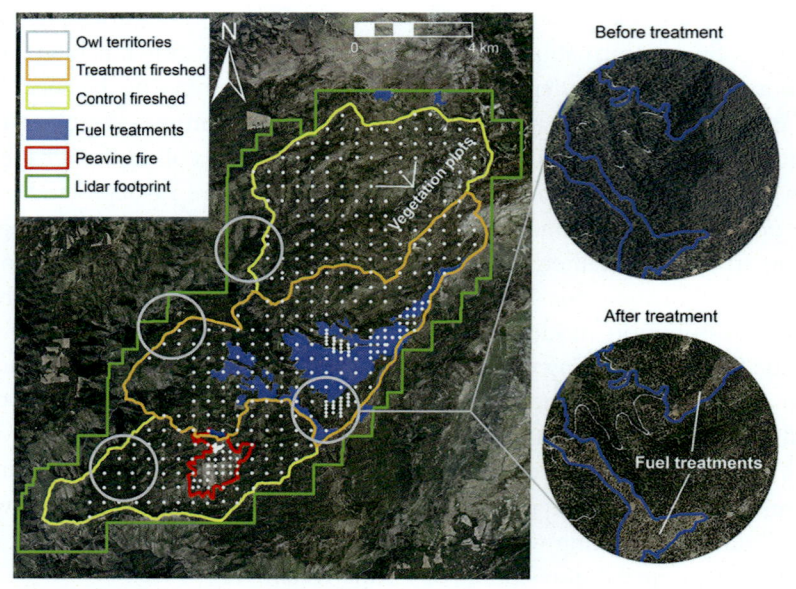

Figure 13.5 The study area for research on the habitats of California spotted owls.

approximately 100 km², with the elevation ranging from 600 m (the lowest point) to 2200 m (the highest point), and is dominated by mixed coniferous and broadleaf forests. The authors' research team acquired LiDAR data for Last Chance. The density of the points was 12 points per square meter, and up to four LiDAR returns could be recorded from each laser pulse. Based on the data processing and parameter extraction methods mentioned in the previous chapters, the terrain features, canopy cover, location of each tree, tree height, DBH, and other traits were extracted.

According to previous studies, two variables, i.e., canopy cover and the density of large trees, were used with the observed distribution of California spotted owl nests to analyze their habitat preferences. A logistic regression method was used to fit their nesting probability, with the results showing that approximately 90% of California spotted owls preferred to build their nests on trees with a DBH larger than 71.3 cm (Fig. 13.6).

To analyze the effects of forest dynamics and fire disturbance on California spotted owls, the authors' research team established the following four scenarios: (1) forest thinning and forest fire disturbance are both present; (2) forest thinning is absent, but forest fire disturbance is present; (3) forest thinning is present, but forest fire disturbance is absent; and (4) both forest thinning and fire disturbance are absent. For scenarios with forest fire disturbance, vegetation information and the distribution of ground fuels acquired from LiDAR data were used as inputs, and the Fire Area Simulator (FARSITE) model was used (Finney, 1998) to simulate the flame length of forest fires within the study area (Fig. 13.7). During the simulation, the FARSITE model adopted the same meteorological conditions that existed when forest fires occurred in this study area in 2001. Based on the four different scenarios, we found that when fire disturbance was absent, forest

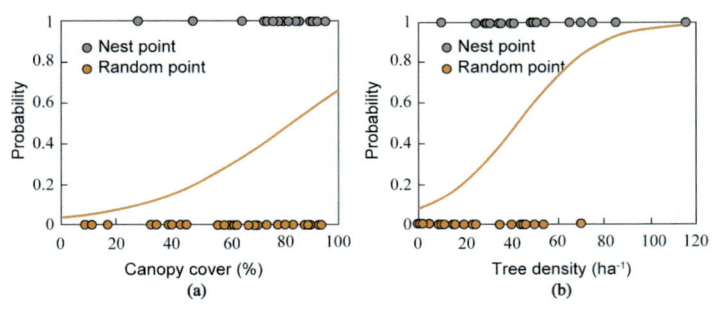

Figure 13.6 Comparison of the predicted habitats of California spotted owls using logistic regression based on (a) canopy cover and (b) the density of large trees.

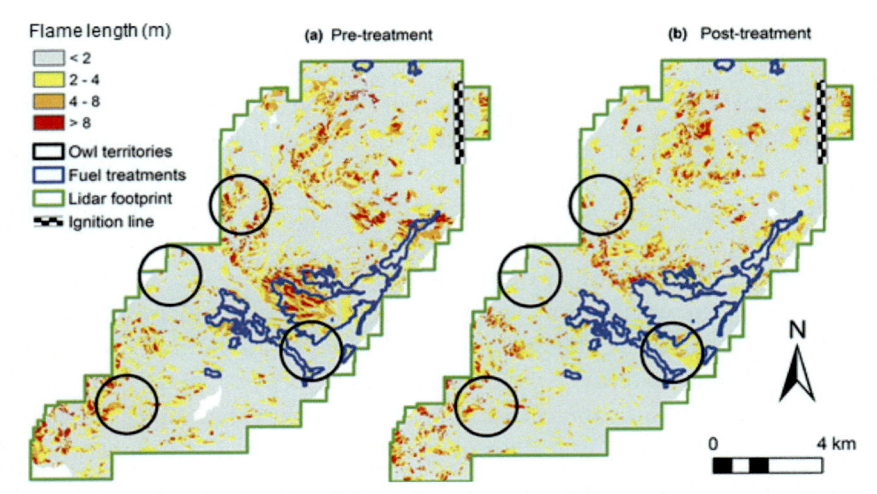

Figure 13.7 The distribution of flame length under different fire scenarios in Last Chancec.

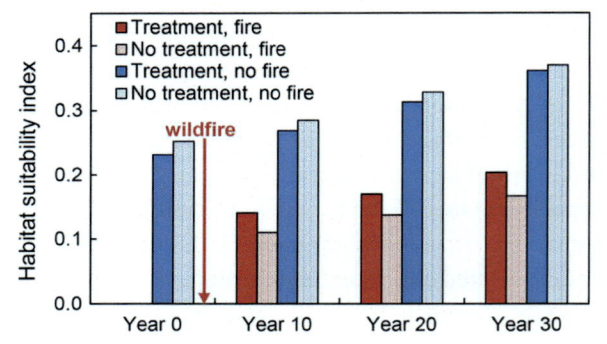

Figure 13.8 Changes in the suitability index of California spotted owl habitats under various scenarios.

thinning had a subtle influence on the distribution of California spotted owls, i.e., slightly reducing their distribution area. When fire disturbance was present, forest thinning significantly affected the distribution of California spotted owls, causing a notable reduction in suitable habitats (Fig. 13.8).

13.1.3 Prospects for Applying LiDAR in Biodiversity Monitoring Networks

Biodiversity monitoring has shifted from single-point observations at fixed stations to integrated multistation observation networks. Several countries, international organizations, and international cooperation projects have established observation, monitoring, and information-sharing networks at regional, national, and global scales, such as the National Ecological Observation Network (NEON), the Committee on Earth Observation Satellites, and the Group on Earth Observations-Biodiversity Observation Network (GEO BON). NEON, an integrated, interdisciplinary, cross-scale biology research network established in United States, combines site-based data, remote sensing data, and current continental-scale datasets to derive a series of data products that can be used to study ecosystem changes along both temporal and spatial dimensions. One of NEON's assignments is to provide detailed aerial photography of regional landscapes and vegetation, and this target is achieved with the help of the airborne observation platform (AOP). The AOP comprises three airplanes that carry an imaging spectrometer acquiring visible to shortwave infrared reflectance spectra, a waveform laser rangefinder, and a high-resolution digital camera. The AOP produces aerial photographs of all plots included in the NEON project under expert instructions or a routine operating mode (Kampe et al., 2010). Sensors integrated with the AOP can provide information on regional land use, vegetation structure, the biochemical and biophysical characteristics of vegetation, and the responses of ecosystems to changes in land use, the climate, and invasive species. NEON was the first long-term project to use LiDAR as the main sensor to conduct observations at fixed stations. The special role of LiDAR in NEON demonstrates its importance in future biodiversity monitoring projects and studies.

There are already complete guidelines for the field observation of biodiversity monitoring networks, and new technologies from different disciplines are continuously incorporated to improve the observations. However, the application of remote sensing technology in biodiversity monitoring networks is still incomplete. GEO BON has established a specialized group (the seventh working group) to integrate remote sensing data into new methods and models to monitor biodiversity, provide information that can indicate changes in biodiversity at large scales, propose priorities for monitoring, regional protection, and biodiversity protection plans, and simulate the trends in biodiversity. To respond to the work of the seventh GEO BON group, several currently existing biodiversity monitoring networks have started relevant assignments, such as establishing relationships among the structural, functional, and biodiversity traits of terrestrial ecosystems and remote sensing data and proposing relevant strategic plans to compensate for the lack of remote sensing technology and data in biodiversity monitoring. The Chinese biodiversity monitoring and research network (Sino BON) has been a leader in incorporating remote sensing technology into biodiversity monitoring. Sino BON has recognized the importance of remote sensing technology in biodiversity research in terms of providing abundant data sources and more comprehensive evaluations of ecosystem status and the diversity of relevant species, quantifying the level of biodiversity degradation, and achieving large-scale monitoring that is challenging for traditional field surveys (Nagendra et al., 2013). LiDAR technology and hyperspectral systems have been adopted to acquire multisource and multiscale remote sensing data of typical Chinese vegetation habitat types in the 13th Five-Year Plan. It is believed that LiDAR will be widely used in all the biodiversity monitoring networks soon and help researchers better understand biodiversity.

13.2 Forest Ecohydrology Studies

13.2.1 Forest Precipitation Distribution

Forest canopies play an important role in the hydrological cycle. The advent of LiDAR makes it possible to obtain the three-dimensional structural traits of forest canopies, which can improve the estimation accuracy of hydrological indices. For example, Yu et al. (2020) used LiDAR data to establish a new structural parameter to quantify rainfall interception (RI). Canopies redistribute rainfall to throughfall and stemflow in forest ecosystems. The canopy RI accounts for 10%–50% of the total rainfall of the forests, which has important ecological and hydrological significance. Research on quantifying RI through structural traits has received widespread attention. The leaf area index (LAI), plant area index (PAI), and tree height are commonly used to quantify RI, but the quantification capabilities of these traits are controversial. Some studies have found that RI increases significantly with increases in LAI (Fathizadeh et al., 2018), whereas others have found that the relationship between the two is not significant (Deguchi et al., 2006) or even that increases in LAI lead to decreased RI (Toba & Ohta, 2005). The cause of this controversy is that LAI does not fully consider the three-dimensional distribution and structural characteristics of tree components, including leaves, branches, and trunks, all of which are related to RI. As a state-of-the-art active remote sensing technology, LiDAR offers an opportunity to rapidly and accurately capture these data at a high resolution. The new structural traits defined by fully considering the structural features related to RI may improve the depiction of RI.

To improve the accuracy of estimating RI, a new structural parameter-
the canopy interception index (CII) was defined based on LiDAR data (Yu
et al., 2020). The study was carried out in the Qingyuan Forest site of the
Chinese Ecosystem Research Network, which is managed by the Chinese
Academy of Sciences and located in Liaoning Province, China (Fig. 13.9).
Four temperate forest types (i.e., Korean pine (*Pinus koraiensis*) plantation
forest (KPF), larch (*Larix* spp.) plantation forest (LPF), mixed broadleaved
forest (MBF), and Mongolian oak (*Quercus mongolica*) forest (MOF)) were
selected. Terrestrial laser scanning was used to collect point cloud data to
calculate CII. To test the ability of CII to quantify RI, it was compared
with other structural traits (i.e., LAI, PAI, and average canopy height) based
on the measured RI of the four forest types.

The results showed that RI increased linearly and significantly with CII
when combining all plots of the four forest types ($R^2 = 0.79$, $P < .001$,
Nash–Sutcliffe efficiency (NSE)=0.79, Fig. 13.10). For a given forest type,
the CII was also significantly related to the RI and explained 58%–63% of
the variation in the RI (Fig. 13.11). The strongest relationship between CII

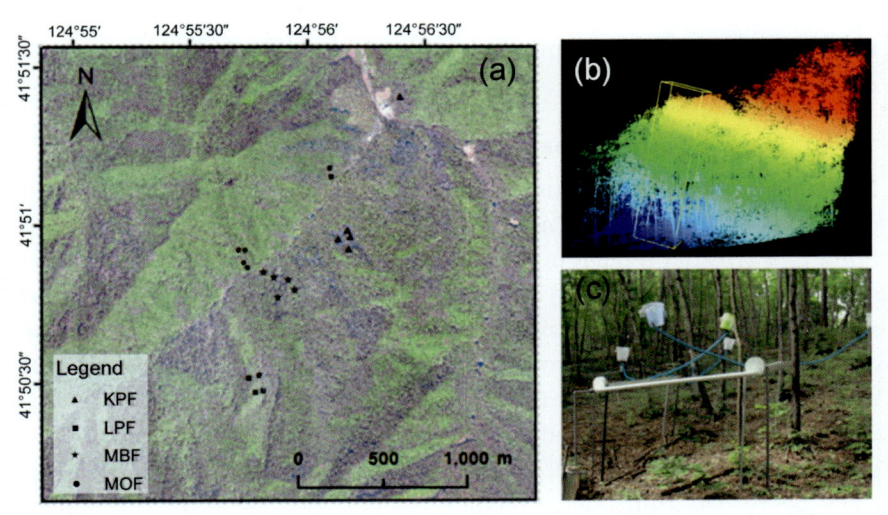

Figure 13.9 Location of the study area and the experimental forest stands: (a) Location of the study sites (b) terrestrial laser scanning point clouds of the study sites
(using the MOF sites as an example, the different colors represent the different height
levels) with a study plot marked by the yellow rectangular box; and (c) photograph of
a study plot (using the MOF study plots as an example) with throughfall collectors. *KPF*,
Korean pine plantation forest; *LPF*, larch plantation forest; *MBF*, mixed broadleaved
forest; *MOF*, Mongolian oak forest.

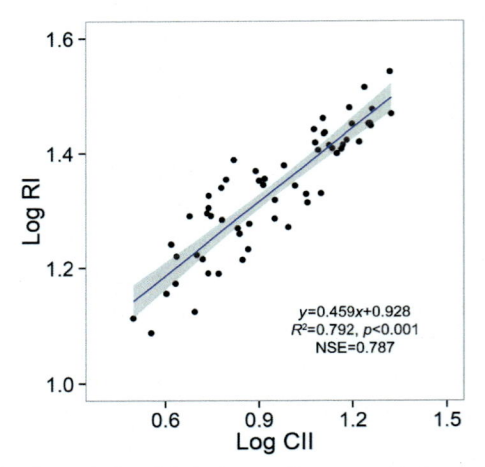

Figure 13.10 The log-log relationship between the canopy interception index (CII) and rainfall interception (RI) for the four forest types combined. The solid blue line is the fit line, whereas the gray band shows the standard error of the regression line. *NSE*, Nash—sutcliffe efficiency.

and RI was found in LPF, followed by KPF MOF and MBF. Therefore, the new CII structural parameter extracted from LiDAR data has a strong capability in quantifying RI.

13.2.2 Snow Depth Estimation

Snow cover is highly sensitive to climate change in the cryosphere. It is closely related to climate change and hydrological cycle processes and is an important research direction in global cryosphere studies. Accurately estimating the spatiotemporal distribution of large-scale snow cover and simulating its features, such as thickness and density, in response to climate change and external meteorological conditions, can contribute to a better understanding of the regional water balance, more effective use of meltwater resources, and knowledge about the hydrological cycle's responses to climate change. Traditional snow cover monitoring systems mainly use snow depth data from meteorological stations. Because there is a limited number of meteorological stations and their locations are not evenly distributed, solely relying on meteorological records to extrapolate and analyze the distribution and thickness of snow cover has several limitations. In regions with complex topographic features, relying on sparse

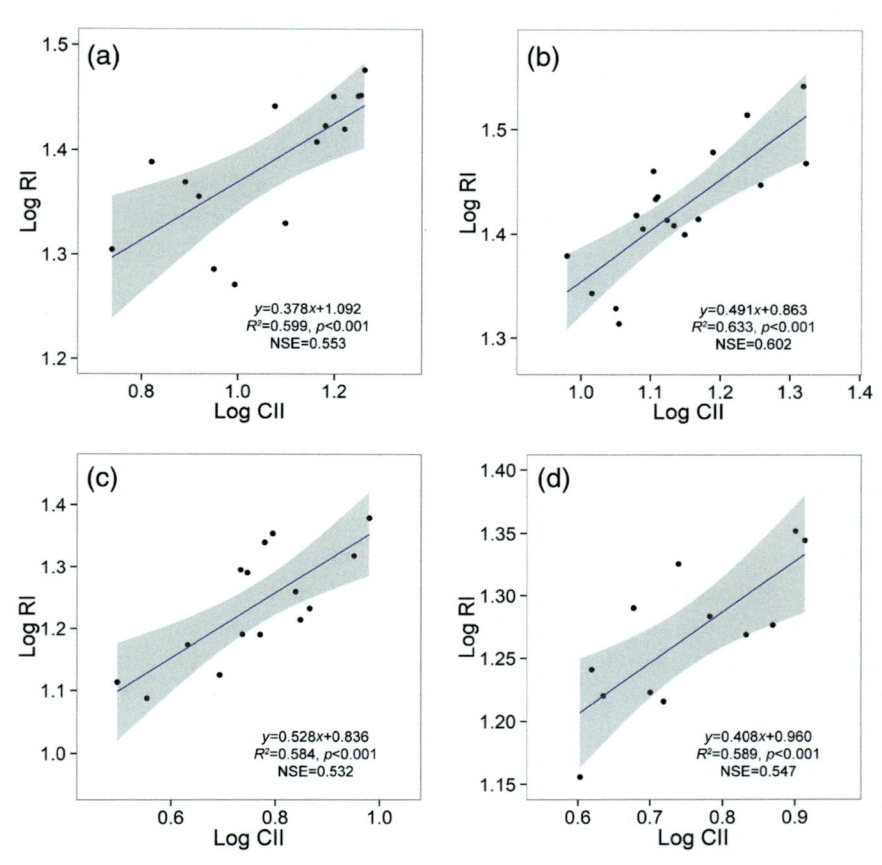

Figure 13.11 The log-log relationships between the canopy interception index (CII) and rainfall interception (RI) for a given forest type: (a) KPF, (b) LPF, (c) MBF, and (d) MOF. The solid blue lines are the fit lines, whereas the gray bands show the standard error of regression lines. *KPF*, Korean pine plantation forest; *LPF*, larch plantation forest; *MBF*; mixed broadleaved forest; *MOF*, Mongolian oak forest; *NSE*, Nash–sutcliffe efficiency.

meteorological station records alone to extrapolate the distribution of snow cover and the potential severity of snow disasters cannot fulfill the needs of snow disaster prevention and rescue. Snow cover products derived from optical remote sensing data, such as the Moderate-Resolution Imaging Spectroradiometer (MODIS), have played an important role in global snow cover monitoring, meltwater runoff modeling, and data assimilation. Studies have shown that MODIS snow cover products have higher accuracy under cloudless conditions than under cloudy conditions because snow cover and clouds have similar reflectance spectra (Arsenault et al., 2014;

Hall & Riggs, 2007; Klein & Barnett, 2003). Therefore, weather conditions can influence the performance of MODIS snow cover products. Although there has been substantial progress in developing snow depth detection methods based on passive microwave remote sensing data, the products still contain major errors in some regions (Kelly et al., 2003). For example, optical and microwave sensors are weak at capturing information below canopies; thus, applying existing inversion methods to extract the snow depth in dense or complex forests can lead to major errors.

LiDAR can directly acquire the three-dimensional structure of ice and snow without penetrating through them. Therefore, they can be used to accurately invert the mass of glaciers, sea ice, and snow cover (Deems et al., 2013). Detailed data on large-scale snow cover acquired from airborne LiDAR have been widely used to analyze the distribution of snow cover thickness and its relationship with environmental factors (Deems et al., 2006; Melvold & Skaugen, 2013; Trujillo et al., 2007). Multitemporal airborne LiDAR data are suitable for inverting changes in glaciers, sea ice, and snow cover to analyze their melting patterns and downstream water supply (Arnold et al., 2006; Harpold et al., 2014; Sturm et al., 2010). In January 2003, NASA launched the Ice, Cloud, and land Elevation Satellite, which provides elevation data on the global surface that have been widely used in studies on glaciers, sea ice, and snow cover (Kwok et al., 2007; Zwally et al., 2008).

Meltwater from snow cover is an important source of water for many regions in spring. Traditional field observations can only obtain snow cover data at scattered locations. Such data cannot reflect the distribution and thickness of large-scale snow cover to allow for an accurate evaluation of the water supply at the beginning of spring. The advent of LiDAR technology has made it possible to collect information on the thickness of large-scale snow cover. This section introduces two studies by the authors' research team. The first was an accuracy assessment of snow cover thickness products derived from LiDAR data for the Critical Zone Observatory (CZO) in the United States. The study areas were seven small drainage basins within four sites of the CZO, with the Jemez River Basin (JRB), Boulder Creek Watershed (BCW), and Kings River Experimental Watershed (KREW) each containing two small drainage basins and the Wolverton Basin (WLVB) site only containing one (Fig. 13.12). The areas of the four sites were different, and their forest types and coverage were also different. They had distinct climate conditions regarding the maximum snow–water equivalent in winter, precipitation, and temperature. The BCW site had the largest scanned area (>400 km^2) and was dominated by

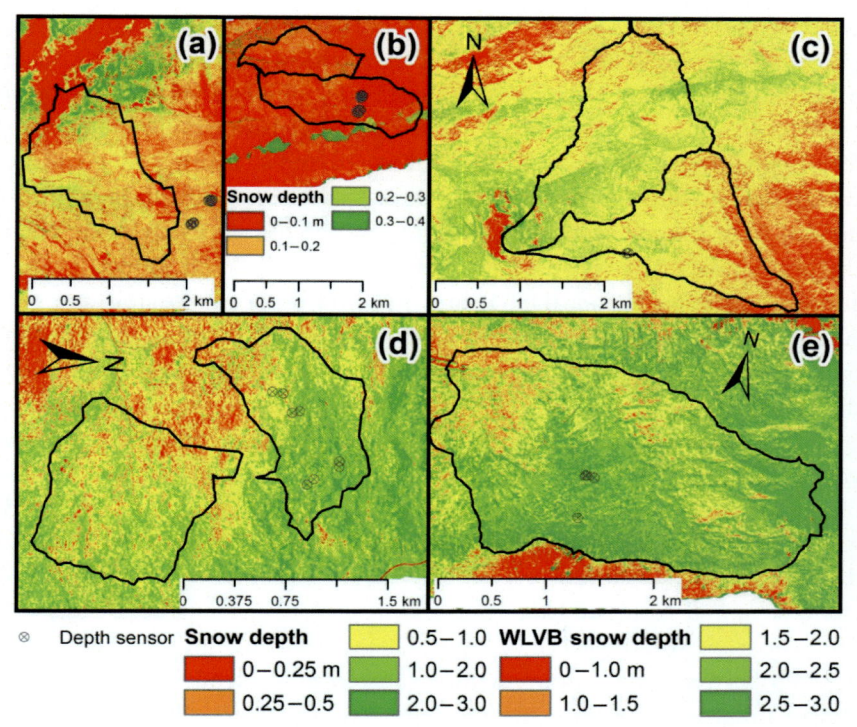

Figure 13.12 The snow depth maps at 1 m spatial resolution derived from LiDAR point cloud data: (a) Boulder Creek Watershed (BCW) Como drainage basin, (b) BCW Gordon drainage basin, (c) Jemez River Basin Jaramillo and History drainage basins, (d) Kings River Experimental Watershed P30 and P303 drainage basins, and (e) Wolverton drainage basin (WLVB).

coniferous forests below 3000 m and rocks and snow above 3000 m. The scanned area for the JRB site was 298 km², and the vegetation types shifted from grasslands at low elevations to coniferous and broadleaf mixed forests at high elevations. The scanned area for the KREW site was 18 km², and the vegetation was mostly fir and broadleaf mixed forests. The scanned area of the WLVB site was 59 km², and the vegetation types shifted from redwood forests at low elevations to alpine meadows at high elevations.

The aerial LiDAR scans were acquired in May 2010 when the snow depth reached its maximum. The footprints of the LiDAR pulses were approximately 15–20 cm; the scanning frequency was approximately 60 Hz; the vertical and horizontal positioning accuracy of the system was approximately 5–10 cm and 20–40 cm, respectively. Each site was scanned on one or two consecutive days to minimize the change in snow cover, except for the BCW site, which was scanned on May 5 and May 20. Field measurements of snow depth were conducted simultaneously with aerial LiDAR scanning, measuring the snow depth on bare ground and under forest canopies. The error in field-measured snow depth was generally within 10 cm but could vary with slope, elevation, and different snow depths. After the filtering process, the ground points were extracted. Then the point cloud data was used to generate a digital terrain model (DTM) and digital surface model (DSM) in a 1 m grid using the kriging interpolation method. The canopy height model (CHM) was obtained using snow-free DTM and DSM differences. The snow depth was derived using the difference in DTMs under snow-covered and snow-free conditions.

At the drainage basin scale, the estimated snow depth from LiDAR data varied between 7 cm and 222 cm (Fig. 13.12), and the range of field-observed snow depth was between 0 cm and 274 cm. The observed values were 19%−28% higher on average than those derived from LiDAR data (except BCW Gordon). The snow depth uncertainty (standard deviation) acquired from LiDAR data was greater than from field-observed values, except for the KREW sites. There was a significant linear relationship between the measured and estimated snow depth values from 60 sample points, with a root mean square error (RMSE) of approximately 23 cm. The verification results for all sites showed that the RMSEs of estimated snow depth varied from approximately 7 cm to 31 cm. Among all the sites, the RMSE corresponding to the P303 sampling point in KREW was the largest, which could be caused by its high elevation. Generally, the comparison between field-observed and LiDAR-based snow depths (Fig. 13.13) demonstrates the high accuracy of the snow depth products derived from LiDAR data, thus confirming the feasibility of using LiDAR technology to estimate snow depth at the drainage basin scale.

The second research project used airborne LiDAR measurements to determine how snow depth varies with canopy structure and the interactions between canopies and terrain from four sites in the southern Sierra Nevada. The study areas—Bull Creek, Shorthair Creek, Providence Creek, and Wolverton Basin—are headwater areas of the Southern Sierra Critical Zone Observatory. The LiDAR data had point densities varying from 0.1 to 10 points per square meter and were divided into 250 m × 250 m tiles using LAStools software. Ground points were extracted from the raw point cloud data and interpolated into a 0.5 m × 0.5 m resolution digital elevation model (DEM) using a simple

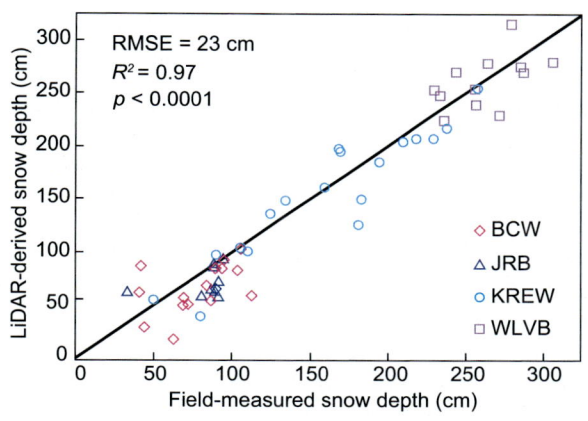

Figure 13.13 Comparison between snow depth values acquired from LiDAR and field observations.

kriging model with a spherical covariance function. All the first returns of the point cloud data were used to generate a DSM. Then, a CHM was produced by subtracting the DEM from the DSM. Using the watershed segmentation algorithm, individual trees were segmented from the CHM. The snow depth pixels beneath each tree were also extracted. The snow depth and tree well (the area next to the tree bole with a shallower snowpack) pattern beneath the segmented trees were evaluated relative to local topographic conditions. The samples were analyzed to determine terrain and vegetation effects on the spatial distribution of snow depth. A machine-learning model, extreme gradient boosting (XGBoost) (Chen & Guestrin, 2016), was applied to predict snow depth at a 0.5 m resolution with topographic and canopy-related predictors, aiming to model the spatial distribution of snow depth at the same resolution as the LiDAR-derived data.

The study found that tree well snow surfaces could be observed from airborne LiDAR with an average ground-point density higher than one point per square meter (Fig. 13.14), and the effect of point density was more significant in dense forests than in sparse forests. As a result of the undersampling of data points under the canopy and oversampling of data points in open and dripping marginal areas, overestimation and underestimation could be offset when forest density was moderate. The dominant direction of a tree well was correlated with the local aspect of the terrain. The gradient of the snow surface in a tree well was correlated with the tree's crown area. Observed from the 0.5 m × 0.5 m resolution raster, the

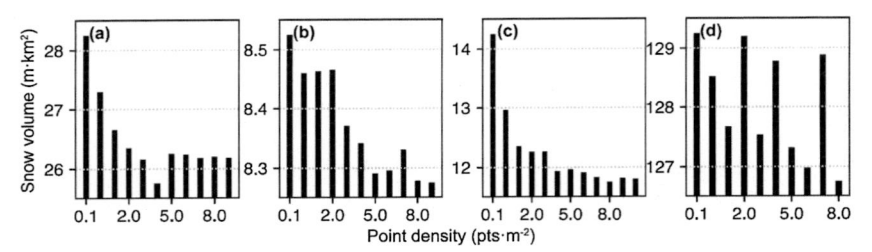

Figure 13.14 Changes in LiDAR-derived total snow volume with LiDAR point density at the southern sierra Critical Zone Observatory (a) Bull site, (b) Shorthair site, (c) Providence site, and (d) Wolverton site.

snow depth was correlated with tree height, surrounding canopy coverage, tree-height standard deviation, and crown volume (Fig. 13.15). Both topographic and vegetation variables were important in spatially predicting the snow depth at different scales using the trained XGBoost model.

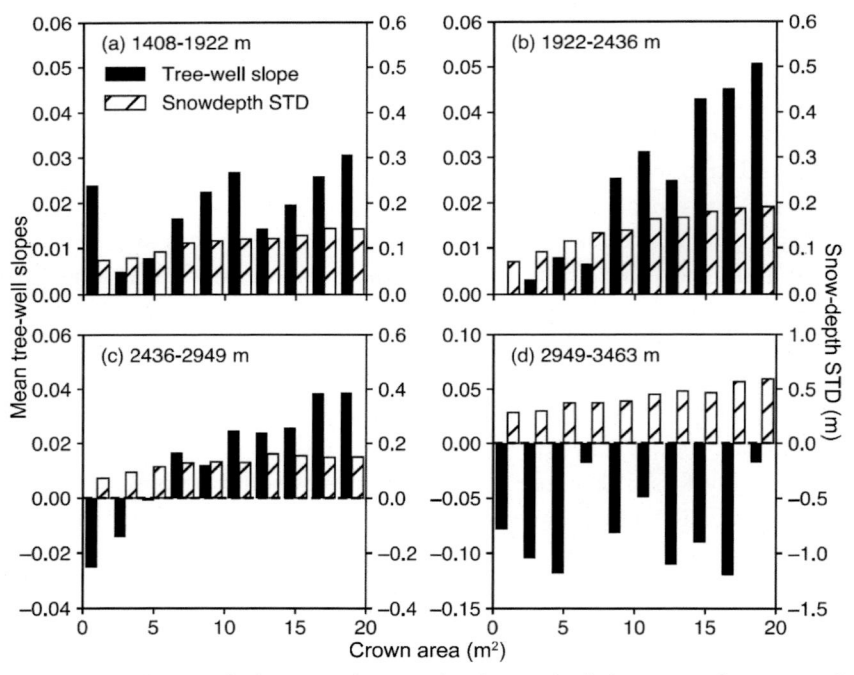

Figure 13.15 Tree well slopes and snow depth standard deviations for tree wells versus crown areas in four elevation bands—(a) 1408—1922 m, (b) 1922—2436 m, (c) 2436—2949 m, and (d) 2949—3463 m—indicating that as the crown area grows larger, the snow depth has a wider distribution, and the tree well gradient becomes steeper except at the highest elevations.

13.3 Applications of LiDAR in Understanding Ecological Theories

The tree crown architecture is composed of traits reflecting the spatial distribution of branches and stems. The combination of crown architecture and leaf and trunk traits defines the spatial representation and characteristics of a tree (Fig. 13.16). Crown architecture can adjust the tree canopy to intercept light and ultimately affect the carbon-water cycle of trees. Its temporal and spatial variation reflects the competition mechanism during the growth of trees and the defense mechanism against changes in climate conditions. Therefore, an in-depth study of the temporal and spatial variability of canopy architecture and its influencing factors is an important prerequisite for further understanding the ecological and physiological responses of trees to global climate change. With the establishment and improvement of the global plant trait database, there are already many studies on the spatial distribution of tree trunk and leaf traits and their response to climatic conditions from local to global scales. Still, these studies have limited numbers of observed traits and a lack of data on three-dimensional canopy architecture traits. Hence, studies on the spatial variation of canopy architecture and its driving factors, especially at the global scale, are still in the early stages and are largely theoretical hypotheses. Further studies will require substantial data on canopy architecture worldwide, particularly three-dimensional datasets, which LiDAR is starting to provide. This section introduces two studies by the authors' research team.

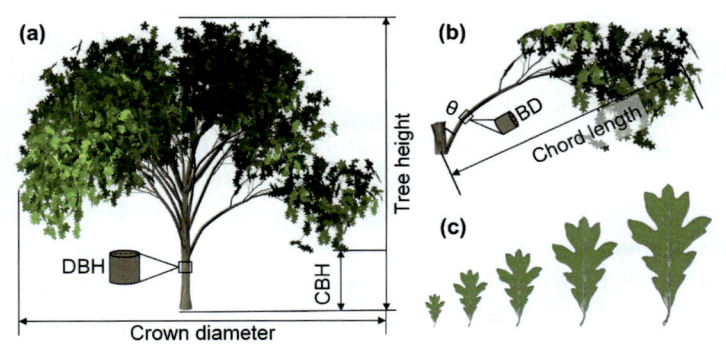

Figure 13.16 A simulated tree demonstrating (a) trunk traits, (b) crown architecture traits, and (c) leaf traits. *BD*, branch diameter; *CBH*, crown base height; *DBH*, diameter at breast height; θ, insertion angle.

The first study examines large-scale geographical variations in crown architecture and the effect of climate variables (Su, Hu, et al., 2020). Taking Q. *mongolica*, a tree species distributed in arid and semiarid areas in Northern China, as an example, ground-based LiDAR technology was used to quantitatively explore the coupling mechanism between trunk and leaf characteristics, the spatial heterogeneity of canopy architecture, and its adaptation mechanism to climatic conditions. This study selected 36 plots from 12 study sites (each has three plots) across Northern China, covering the four common genetic groups of Q. *mongolica* within China. Terrestrial laser scanning was used to obtain point cloud data of the plots. After data registration, abnormal point elimination, filtering, normalization, individual tree segmentation, branch and leaf separation, branch reconstruction, and trait extraction, two trunk traits and nine crown architecture traits were obtained with high accuracy. At the same time, four leaf traits were acquired from field sampling and measurements. By correlating crown architecture traits with trunk and leaf traits, the team used nested analysis of variance, regression analysis, and structural equation modeling to explore the spatial heterogeneity of the canopy architecture of Q. *mongolica* and its influencing factors.

The study results showed that even within the same genetic group, significant spatial variations in the various traits of Q. *mongolica* were present relative to environmental gradients, and these variations were not caused by differences in tree age. The trunk and leaf traits of Q. *mongolica* were mainly controlled by mean annual precipitation (MAP), and some crown architectural traits (related to canopy shape) were affected by both MAP and mean annual temperature (MAT) (Fig. 13.17). These architectural traits related to the canopy shape and the leaf traits show a strong coupling effect, reflecting the role of crown architecture in bridging the main trunk and the leaves to balance the light and water demand of Q. *mongolica*. In a relatively cold and humid environment, Q. *mongolica* looks and behaves more like a tree, grows quickly, and increases the proportion of the top crown as much as possible to compete for light resources. In contrast, in a relatively warm and dry environment, Q. *mongolica* tends to take on a shrub shape to control growth in height and form a multilayered self-shading effect to reduce water loss. The model simulation shows that arid and semiarid regions of Northern China will become drier and warmer; thus, we predict that Q. *mongolica* will gradually transition from a tree to a shrub shape. This phenomenon is not considered in dynamic global vegetation models (DGVMs) and will greatly affect the simulation results of the existing

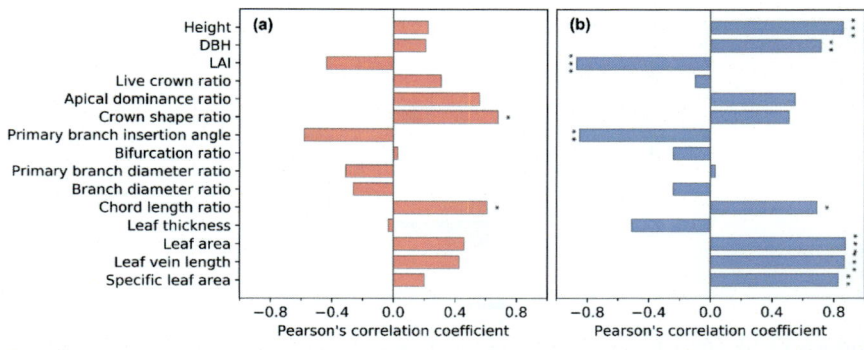

Figure 13.17 Pearson's correlation coefficients between estimated traits and (a) mean annual temperature and (b) mean annual precipitation. If a correlation is significant, the corresponding bar is labeled as ***, **, or *, indicating a confidence level of 99.9%, 99%, or 95%, respectively. *DBH*, diameter at breast height; *LAI*, leaf area index.

carbon cycle model for Northern China. The above results provide a new perspective for understanding tree adaptation and response mechanisms during global climate change and improving the simulation accuracy of global carbon cycle models.

The second project was on the determinants of forest canopy height (Tao et al., 2016). Forest canopy height is an important indicator of forest biomass, species diversity, and other ecosystem functions. However, the climatic determinants that underline its global patterns have not been fully explored. The hydraulic limitation hypothesis proposes that (1) reduced growth in taller trees is caused by decreased photosynthesis resulting from a decrease in hydraulic conductance promoted by a longer root-to-leaf flow path, and (2) this mechanism reduces stand productivity after canopy closure. Therefore, it is important to study the climatic controls of the world's giant trees, especially water availability, quantified by the difference between annual precipitation and annual potential evapotranspiration (P-PET).

Waveform data from the geoscience laser altimeter system (GLAS) were chosen to calculate the canopy top height. GLAS14 (version 34) and GLAS01 data covering May 20 to June 23, 2005, were obtained from the National Snow and Ice Data Center website. Then, RH100 was calculated as the height difference between the signal start and the ground peak, with the ground peak determined as the last Gaussian peak (Fig. 13.18). Data on the field-measured giant trees were collected from a database of the world's giant trees, including the tallest tree on Earth (Fig. 13.18). Climatic indices

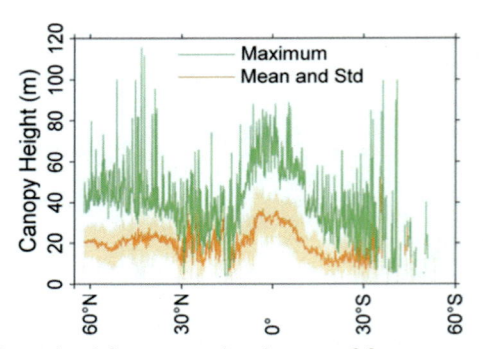

Figure 13.18 The latitudinal frequency distribution of forest canopy heights derived from the geoscience laser altimeter system data and field-measured giant trees.

with significant correlations with forest height were chosen according to previous studies (Klein et al., 2015; Rind et al., 1990), including four temperature indices (annual mean temperature, mean temperature of the wettest quarter, mean temperature of the driest quarter, and mean temperature of the coldest quarter), and four moisture indices (annual precipitation (AP), precipitation of the wettest month, the difference between annual precipitation and annual P–PET, and the difference between annual precipitation and actual annual evapotranspiration (P–AET)). All these indices were obtained from WorldClim at a spatial resolution of 0.00833 degrees (∼1 km) from 1950–2000. The National Forest Resource Inventory Database of the State Forestry Administration, China (FRID), and the Forest Inventory Analysis (FIA) database for the United States were used to validate the results calculated using the GLAS data and field-measured giant trees. Various regression methods, including one polynomial regression (i.e., quadratic) and three peak regressions (i.e., lognormal, Weibull, and Gaussian), were used to explore the relationship.

The results showed that forest canopy means were highest in tropical forests, whereas the maximum forest height had three peaks in tropical and midlatitude forests (approximately 40°N and 40°S). These facts suggested that temperature might not be the primary determinant of global forest canopy heights. Compared with temperature indices, the moisture indices predicted the global forest canopy heights well, with R^2 of up to 0.72. As shown in Fig. 13.19, the canopy height exhibited a hump-shaped trend with increasing moisture levels (P–PET); it initially increased, peaked at approximately 680 mm of P–PET, and then declined, regardless of the moisture index used. This result suggests that an excessive water supply

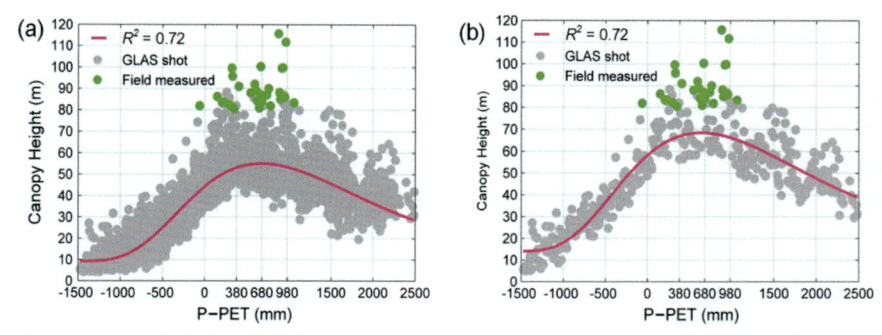

Figure 13.19 Relationships between potential evapotranspiration (P-PET) and global maximum forest canopy heights in every (a) 1 mm increment and (b) 10 mm increment in P-PET. PET was calculated using the Thornthwaite equation. Red lines are the fitted log-normal models. The overall trend is unchanged when Hargreaves PET is used.

negatively affects forest canopy height. Therefore, a close relationship exists between forest canopy heights and water availability in geographic space, which verifies the hydraulic limitation hypothesis.

13.4 Applications of LiDAR in Forest Ecological Modeling

The development of field observation networks (such as FLUXNET and China Flux) and remote sensing technology have accumulated much data for ecological research; however, under many scenarios, models cannot fully use the acquired data. For example, vegetation height can be acquired using LiDAR. Nevertheless, most current ecological and land surface models in meteorology use tree height as an input without considering changes in its value. On the one hand, this is because there are no accurate tree height products for verifying these models, and most of these models are based on populations rather than individuals. On the other hand, current studies about LiDAR data fusion are still based on forcing methods that cannot resolve the problem of the spatiotemporal discontinuity of observational data (in contrast, simulations are often spatially and temporally continuous). A new generation of mechanism models must be developed to improve the fusion of LiDAR data with ecological models. These models can address and simulate vegetation growth and mortality rates. The major structural traits (such as tree height) predicted by the models can be linked with LiDAR data to verify the models. A few such models have been developed, such as the Ecosystem Demography model (ED2), and pre-liminarily applied at regional scales. ED2 is an individual-based ecological model; its most distinctive feature is that the heights of all individual trees are included (Moorcroft et al., 2001). Thus, it has an inherent advantage over other models in terms of linking with LiDAR data. With the development of computer science and LiDAR technology, new-generation ecological models should gradually be able to simulate more vegetation structural traits and tree height and use more LiDAR data to correct and verify the models to improve their simulation accuracy.

The methods for fusing LiDAR data with ecological models have important influences on model improvements. In previous studies, the fusion between remote sensing data and ecological models can be categorized into two types. The first is the driving method, which adopts the status variables or parameters (such as LAI and NDVI) obtained from LiDAR and other remote sensing data as inputs to drive the models directly. The driving method is relatively simple and was widely used in early ecological modeling studies. However, this method is easily influenced by the number of available images, and large errors can be introduced during the parameter inversion of remote sensing data (Fig. 13.20). In contrast, the other method, data assimilation, has developed rapidly in recent years and attracted broad attention. Data assimilation effectively combines models with observations, which means fusing direct or indirect observations from multiple sources and multiple scales within the framework of kinetics. Based on the kinetics models and various weighted observation operators, this method continuously incorporates new

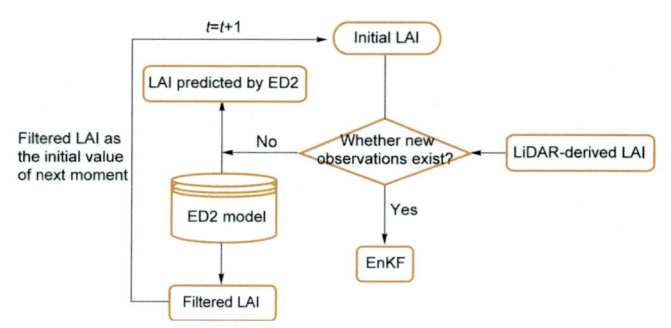

Figure 13.20 Using the Ecosystem Demography model (ED2) and leaf area index (LAI) as an example, the workflow of assimilating LiDAR data using Ensemble Kalman Filter (EnKF).

observations to optimize the model and reduce errors in model predictions (Li, 2014).

Data assimilation was first used in the field of atmosphere and ocean. It has been gradually applied to terrestrial ecosystems and achieves excellent performance. The most commonly used data assimilation algorithms include the sequential Monte Carlo methods, such as the ensemble Kalman filter (EnKF) algorithm (Evensen, 2003). The assimilation of remote sensing data in ecological models falls into two categories. The first directly assimilates the status variables or parameters extracted from remote sensing data into ecological models. Studies in this group date back to the early stage of data assimilation and have inspired many related studies that mainly adopt NDVI, LAI, or other parameters in the assimilation (Migliavacca et al., 2009). The other group assimilates the spectral data directly, which avoids the errors in parameter inversion and theoretically can achieve higher accuracy (Quaife et al., 2008). In recent years, many scholars have tried to improve crop models by assimilating spectral data and have obtained good results (Li et al., 2008). Other researchers have assimilated biomass information acquired from synthetic aperture radar to optimize crop models, achieving a significant increase in the accuracy of biomass simulated by the corrected crop models (Tan et al., 2011). In addition, using multitemporal data as the source of assimilation data also increases the accuracy of the models. An example is shown in Fig. 13.20. The time-series LAI data from remote sensing observations are used as the external assimilation data, and the EnKF algorithm is applied to update the corresponding simulation values. The specific process is as follows. Assuming that the LAI predicted by the ED2 model at time t-1 has been filtered, and a new LAI observation is available at time t, the simulated LAI by ED2 at t is updated using EnKF and the observed LAI at t. The updated or filtered LAI value at t is then used to drive ED2 to simulate LAI at $t + 1$. If a new LAI observation at $t + 1$ is available, the simulated LAI at $t + 1$ can be updated following the same procedure. This progress continues until all LAI observations have been assimilated into the ED2 model, and the updated LAI values at each moment will be the final assimilated LAIs.

With the continuous improvement of ecological models and the development of field observation networks and remote sensing technology, data assimilation methods to fuse the increasingly abundant field observations and remote sensing data into ecological models have become a research focus to improve simulation accuracy. For example, Koffi et al. (2015) attempted to use solar-induced chlorophyll fluorescence (SIF) data derived from remote sensing imagery to constrain the prediction of GPP. Their research identified that by assimilating SIF data and the derived chlorophyll concentration, the GPP estimation accuracy of ecological models could be increased. Viskari et al. (2015) assimilated the phenological changes of forests into ecological models and significantly improved the accuracy of simulated net ecosystem exchange. Another example from the authors' research team is the global pattern of woody residence time and its influence on simulated aboveground biomass (AGB) (Xue et al., 2017), which will be introduced in the following paragraphs.

DGVMs are also useful for mapping global carbon stocks and predicting their future variation. Woody residence time (τ_w) is an important parameter that expresses the balance between mature forest recruitment/growth and mortality. To predict forest growth and biomass accumulation, most DGVMs use the τ_w parameter to represent the time that carbon remains in an ecosystem. Therefore, quantifying τ_w is important for gaining a clearer understanding of the global forest carbon budget balance. The study (Xue et al., 2017) mainly investigated the determinants of forest τ_w based on the field-collected values and its role in improving DGVM simulations, focusing on forest plots without disturbances for at least 100 years. The τ_w values were compiled for 1319 forest sites from published literature covering all forest biomes, including boreal, temperate, and tropical forests. Then, a global, gridded τ_w map was generated using the field-derived τ_w values and other meteorological and physiological variables as predictors using the random forest (RF) method. Based on the generated global τ_w map, a DGVM-integrated biosphere simulator (IBIS) was parameterized. Based on an overview of different model descriptions of τ_w, an assessment was made of using field-derived τ_w data to improve biomass simulations.

The results showed that the absence of AP and GPP significantly increased the mean squared error and node purity and thus lowered the predictive ability of the RF model. Therefore, these variables were the two most important predictors for estimating global τ_w (Fig. 13.21). Compared with using the default τ_w, the predicted AGB based on the improved τ_w had a relatively close relationship with the observed plot AGB values, even though overestimation and underestimation were observed for small and large AGB values (Fig. 13.22). Moreover, τ_w could change the predicted AGB tenfold based on a site-level test using the Monte Carlo method. At the global level, different parameterization schemes of the IBIS using estimated τ_w resulted in a twofold change in the simulated AGB for 2100.

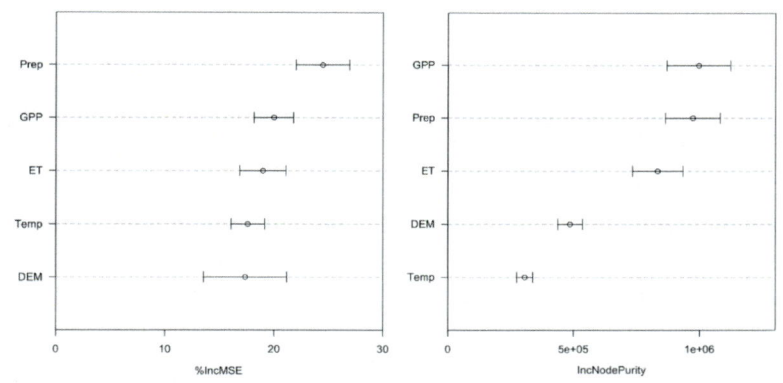

Figure 13.21 The mean importance of variables estimated from 100 runs of the woody residency time random forest model, denoted by (a) the percentage increase of mean squared error (%IncMSE) and (b) the increase in node purity (IncNodePurity). IncNodePurity is a specific parameter for regression trees that is measured by the residual sum of squares. The larger the %IncMSE and IncNodePurity of a variable are, the worse the model performs when the variable is randomized. *DEM*, digital elevation model; *ET*, evapotranspiration; *GPP*, gross primary production; *Prep*, annual precipitation; *Temp*, annual temperature; *%IncMSE*, percentage increase in mean squared error when the variable is randomized.

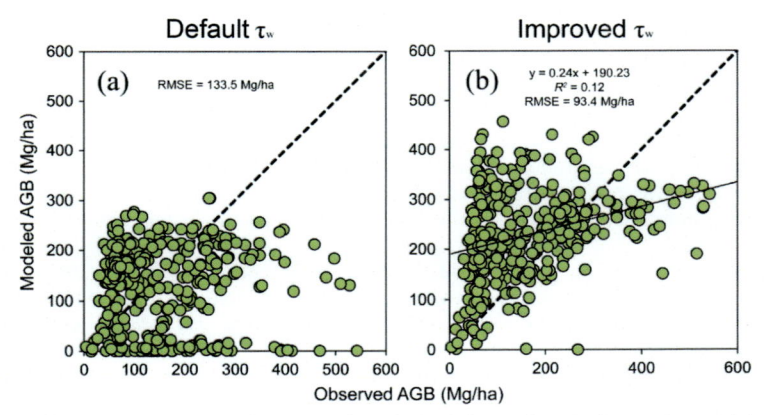

Figure 13.22 Comparison of observed and model-simulated aboveground biomass (AGB) by (a) default and (b) estimated present-day τ_w. Each point in the figure indicates the AGB measured in one or more plots (average for more than one plot) within a $0.5° \times 0.5°$ model grid. The observed AGB was mostly measured from 1990 to 2010, and the model simulations were the averaged values during those corresponding years. The dashed line shows the 1:1 line. Root mean square error (Mg ha^{-1}) is also shown.

The results from the study highlight the influences of various biotic and abiotic variables on forest τ_w.

13.5 Chapter Summary

At present, LiDAR technology has been widely used in forest parameter extraction and has played an important role in biodiversity monitoring, ecohydrology studies, and the simulation and optimization of ecological models. In biodiversity monitoring, LiDAR can provide essential data that can be integrated with optical remote sensing data to resolve the problems of tree species identification and vegetation mapping, providing important input for species distribution simulation and biodiversity monitoring. In ecohydrological studies, LiDAR can be used to accurately extract parameters such as snow coverage and snow depth, which are essential for regional or global hydrological studies. During the simulation and optimization of ecological models, the new three-dimensional parameters of forests provided by LiDAR can substantially increase the simulation accuracy of models. Assimilating LiDAR data into models is an important way of future modeling optimization. This is also one of the development directions of ecological models.

References

Arnold, N. S., Rees, W. G., Devereux, B. J., & Amable, G. S. (2006). Evaluating the potential of high-resolution airborne LiDAR data in glaciology. *International Journal of Remote Sensing, 27*(5–6), 1233–1251.

Arsenault, K. R., Houser, P. R., & De Lannoy, G. J. M. (2014). Evaluation of the MODIS snow cover fraction product. *Hydrological Processes, 28*(3), 980–998.

Asner, G. P., & Martin, R. E. (2009). Airborne spectranomics: Mapping canopy chemical and taxonomic diversity in tropical forests. *Frontiers in Ecology and the Environment, 7*(5), 269–276.

Belote, R. T., Prisley, S., Jones, R. H., Fitzpatrick, M., & de Beurs, K. (2011). Forest productivity and tree diversity relationships depend on ecological context within mid-Atlantic and Appalachian forests (USA). *Forest Ecology and Management, 261*(7), 1315–1324.

Carlson, K. M., Asner, G. P., Hughes, R. F., Ostertag, R., & Martin, R. E. (2007). Hyperspectral remote sensing of canopy biodiversity in Hawaiian lowland rainforests. *Ecosystems, 10*(4), 536–549.

Chen, T., & Guestrin, C. (2016). Xgboost: A scalable tree boosting system. In *Proceedings of the 22nd acm sigkdd international conference on knowledge discovery and data mining*.

Davies, A. B., & Asner, G. P. (2014). Advances in animal ecology from 3D-LiDAR ecosystem mapping. *Trends in Ecology & Evolution, 29*(12), 681–691.

Deems, J. S., Fassnacht, S. R., & Elder, K. J. (2006). Fractal distribution of snow depth from lidar data. *Journal of Hydrometeorology, 7*(2), 285–297.

Deems, J. S., Painter, T. H., & Finnegan, D. C. (2013). Lidar measurement of snow depth: A review. *Journal of Glaciology, 59*(215), 467–479.

Deguchi, A., Hattori, S., & Park, H. T. (2006). The influence of seasonal changes in canopy structure on interception loss: Application of the revised Gash model. *Journal of Hydrology, 318*(1–4), 80–102.

Duro, D., Coops, N. C., Wulder, M. A., & Han, T. (2007). Development of a large area biodiversity monitoring system driven by remote sensing. *Progress in Physical Geography-Earth and Environment, 31*(3), 235–260.

Evensen, G. (2003). The ensemble Kalman filter: Theoretical formulation and practical implementation. *Ocean Dynamics, 53*(4), 343–367.

Fathizadeh, O., Hosseini, S. M., Keim, R. F., & Boloorani, A. D. (2018). A seasonal evaluation of the reformulated Gash interception model for semi-arid deciduous oak forest stands. *Forest Ecology and Management, 409*, 601–613.

Finney, M. A. (1998). FARSITE: Fire area simulator - model development and evaluation. *Usda Forest Service Rocky Mountain Forest and Range Experiment Station Research Paper(RP-4), 1-+.*

Gillespie, T. W. (2005). Predicting woody-plant species richness in tropical dry forests: A case study from south Florida, USA. *Ecological Applications, 15*(1), 27–37.

Goetz, S. J., Steinberg, D., Betts, M. G., Holmes, R. T., Doran, P. J., Dubayah, R., & Hofton, M. (2010). Lidar remote sensing variables predict breeding habitat of a Neotropical migrant bird. *Ecology, 91*(6), 1569–1576.

Griffiths, G. H., Lee, J., & Eversham, B. C. (2000). Landscape pattern and species richness; regional scale analysis from remote sensing. *International Journal of Remote Sensing, 21*(13–14), 2685–2704.

Hall, D. K., & Riggs, G. A. (2007). Accuracy assessment of the MODIS snow products. *Hydrological Processes, 21*(12), 1534–1547.

Harpold, A. A., Guo, Q., Molotch, N., Brooks, P. D., Bales, R., Fernandez-Diaz, J. C., Musselman, K. N., Swetnam, T. L., Kirchner, P., Meadows, M. W., Flanagan, J., & Lucas, R. (2014). LiDAR-derived snowpack data sets from mixed conifer forests across the Western United States. *Water Resources Research, 50*(3), 2749–2755.

Hawkins, B. A., Field, R., Cornell, H. V., Currie, D. J., Guegan, J. F., Kaufman, D. M., Kerr, J. T., Mittelbach, G. G., Oberdorff, T., O'Brien, E. M., Porter, E. E., & Turner, J. R. G. (2003). Energy, water, and broad-scale geographic patterns of species richness. *Ecology, 84*(12), 3105–3117.

Hu, H., Li, X., Du, Y., Zheng, H., Du, B., & He, X. (2012). Research advances in biodiversity remote sensing monitoring. *Chinese Journal of Ecology, 31*(6), 1591–1596 (in Chinese).

Irl, S. D. H., Harter, D. E. V., Steinbauer, M. J., Puyol, D. G., Fernandez-Palacios, J. M., Jentsch, A., & Beierkuhnlein, C. (2015). Climate vs. topography - spatial patterns of plant species diversity and endemism on a high-elevation island. *Journal of Ecology, 103*(6), 1621–1633.

Johansen, K., Phinn, S., & Witte, C. (2010). Mapping of riparian zone attributes using discrete return LiDAR, QuickBird and SPOT-5 imagery: Assessing accuracy and costs. *Remote Sensing of Environment, 114*(11), 2679–2691.

Kampe, T. U., Johnson, B. R., Kuester, M., & Keller, M. (2010). Neon: The first continental-scale ecological observatory with airborne remote sensing of vegetation canopy biochemistry and structure. *Journal of Applied Remote Sensing, 4.* Article 043510.

Kelly, R. E., Chang, A. T., Tsang, L., & Foster, J. L. (2003). A prototype AMSR-E global snow area and snow depth algorithm. *Ieee Transactions on Geoscience and Remote Sensing, 41*(2), 230–242.

Klein, A. G., & Barnett, A. C. (2003). Validation of daily MODIS snow cover maps of the Upper Rio Grande River Basin for the 2000-2001 snow year. *Remote Sensing of Environment, 86*(2), 162–176.

Klein, D. J., McKown, M. W., & Tershy, B. R. (2015). Deep learning for large scale biodiversity monitoring. In *Bloomberg data for good exchange conference*.

Koffi, E. N., Rayner, P. J., Norton, A. J., Frankenberg, C., & Scholze, M. (2015). Investigating the usefulness of satellite-derived fluorescence data in inferring gross primary productivity within the carbon cycle data assimilation system. *Biogeosciences, 12*(13), 4067–4084.

Kooistra, L., Wamelink, W., Schaepman-Strub, G., Schaepman, M., van Dobben, H., Aduaka, U., & Batelaan, O. (2008). Assessing and predicting biodiversity in a floodplain ecosystem: Assimilation of net primary production derived from imaging spectrometer data into a dynamic vegetation model. *Remote Sensing of Environment, 112*(5), 2118–2130.

Krishnaswamy, J., Bawa, K. S., Ganeshaiah, K. N., & Kiran, M. C. (2009). Quantifying and mapping biodiversity and ecosystem services: Utility of a multi-season NDVI based Mahalanobis distance surrogate. *Remote Sensing of Environment, 113*(4), 857–867.

Kwok, R., Cunningham, G. F., Zwally, H. J., & Yi, D. (2007). Ice, cloud, and land elevation satellite (ICESat) over arctic sea ice: Retrieval of freeboard. *Journal of Geophysical Research-Oceans, 112*(C12). Article C12013.

Langford, W. T., Gergel, S. E., Dieterich, T. G., & Cohen, W. (2006). Map misclassification can cause large errors in landscape pattern indices: Examples from habitat fragmentation. *Ecosystems, 9*(3), 474–488.

Levick, S., & Rogers, K. (2008). Patch and species specific responses of savanna woody vegetation to browser exclusion. *Biological Conservation, 141*(2), 489–498.

Li, X. (2014). Characterization, controlling, and reduction of uncertainties in the modeling and observation of land-surface systems. *Science China-Earth Sciences, 57*(1), 80–87.

Li, C., Wang, J., Wang, X., Liu, F., & Li, R. (2008). Methods for integration of remote sensing data and crop model and their prospects in agricultural application. *Transactions of the Chinese Society of Agricultural Engineering, 24*(11), 295–301 (in Chinese).

Lucas, R. M., Lee, A. C., & Bunting, P. J. (2008). Retrieving forest biomass through integration of CASI and LiDAR data. *International Journal of Remote Sensing, 29*(5), 1553–1577.

MacArthur, R. (1960). On the relative abundance of species. *The American Naturalist, 94*(874), 25–36.

Melvold, K., & Skaugen, T. (2013). Multiscale spatial variability of lidar-derived and modeled snow depth on Hardangervidda, Norway. *Annals of Glaciology, 54*(62), 273–281.

Migliavacca, M., Meroni, M., Busetto, L., Colombo, R., Zenone, T., Matteucci, G., Manca, G., & Seufert, G. (2009). Modeling gross primary production of agro-forestry ecosystems by assimilation of satellite-derived information in a process-based model. *Sensors, 9*(2), 922–942.

Moorcroft, P. R., Hurtt, G. C., & Pacala, S. W. (2001). A method for scaling vegetation dynamics: The ecosystem demography model (ED). *Ecological Monographs, 71*(4), 557–585.

Mucher, S., Roupioz, L., Kramer, H., Wolters, M., Bogers, M., Lucas, R., Bunting, P., Petrou, Z., Kosmidou, V., Manakos, I., Padoa-Schioppa, E., Ficetola, G. F., Bonardi, A., Adamo, M., & Blonda, P. (July 02–05, 2013). LiDAR as a valuable information source for habitat mapping. [Gi_forum 2013: Creating the gisociety]. Geoinformatics Forum, Salzburg, AUSTRIA.

Muller, J., & Brandl, R. (2009). Assessing biodiversity by remote sensing in mountainous terrain: The potential of LiDAR to predict forest beetle assemblages. *Journal of Applied Ecology, 46*(4), 897–905.

Nagendra, H. (2001). Using remote sensing to assess biodiversity. *International Journal of Remote Sensing, 22*(12), 2377–2400.

Nagendra, H., Lucas, R., Honrado, J. P., Jongman, R. H. G., Tarantino, C., Adamo, M., & Mairota, P. (2013). Remote sensing for conservation monitoring: Assessing protected areas, habitat extent, habitat condition, species diversity, and threats. *Ecological Indicators, 33*, 45–59.

Noss, R. F. (1990). Indicators for monitoring biodiversity - a hierarchical approach. *Conservation Biology, 4*(4), 355–364.

Palmer, M. W., Earls, P. G., Hoagland, B. W., White, P. S., & Wohlgemuth, T. (2002). Quantitative tools for perfecting species lists. *Environmetrics, 13*(2), 121–137.

Pereira, H. M., Ferrier, S., Walters, M., Geller, G. N., Jongman, R. H. G., Scholes, R. J., Bruford, M. W., Brummitt, N., Butchart, S. H. M., Cardoso, A. C., Coops, N. C., Dulloo, E., Faith, D. P., Freyhof, J., Gregory, R. D., Heip, C., Hoft, R., Hurtt, G., Jetz, W., ... Wegmann, M. (2013). Essential biodiversity variables. *Science, 339*(6117), 277–278.

Quaife, T., Lewis, P., De Kauwe, M., Williams, M., Law, B. E., Disney, M., & Bowyer, P. (2008). Assimilating canopy reflectance data into an ecosystem model with an Ensemble Kalman Filter. *Remote Sensing of Environment, 112*(4), 1347–1364.

Rind, D., Goldberg, R., Hansen, J., Rosenzweig, C., & Ruedy, R. (1990). Potential evapotranspiration and the likelihood of future drought. *Journal of Geophysical Research-Atmospheres, 95*(D7), 9983–10004.

Rosenzweig, M. L. (1995). *Species diversity in space and time.* Cambridge University Press.

Roughgarden, J., Running, S. W., & Matson, P. A. (1991). What does remote-sensing do for ecology. *Ecology, 72*(6), 1918–1922.

Saura, S. (2004). Effects of remote sensor spatial resolution and data aggregation on selected fragmentation indices. *Landscape Ecology, 19*(2), 197–209.

Seto, K. C., Fleishman, E., Fay, J. P., & Betrus, C. J. (2004). Linking spatial patterns of bird and butterfly species richness with Landsat TM derived NDVI. *International Journal of Remote Sensing, 25*(20), 4309–4324.

Simonson, W. D., Allen, H. D., & Coomes, D. A. (2014). Applications of airborne lidar for the assessment of animal species diversity. *Methods in Ecology and Evolution, 5*(8), 719–729.

Soulé, M. E., & Kohm, K. A. (1989). *Research priorities for conservation biology* (Vol. 1). Island Press.

Stoms, D. M., & Estes, J. E. (1993). A remote-sensing research agenda for mapping and monitoring biodiversity. *International Journal of Remote Sensing, 14*(10), 1839–1860.

Sturm, M., Taras, B., Liston, G. E., Derksen, C., Jonas, T., & Lea, J. (2010). Estimating snow water equivalent using snow depth data and climate classes. *Journal of Hydrometeorology, 11*(6), 1380–1394.

Sugumaran, R., & Voss, M. (April 11–13, 2007). Object-oriented classification of LIDAR-fused hyperspectral imagery for tree species identification in an urban environment. 2007 urban remote sensing joint event.

Su, Y. J., Guo, Q. H., Fry, D. L., Collins, B. M., Kelly, M., Flanagan, J. P., & Battles, J. J. (2016). A vegetation mapping strategy for conifer forests by combining airborne LiDAR data and aerial imagery. *Canadian Journal of Remote Sensing, 42*(1), 1–15.

Su, Y., Guo, Q., Hu, T., Guan, H., Jin, S., An, S., Chen, X., Guo, K., Hao, Z., Hu, Y., Huang, Y., Jiang, M., Li, J., Li, Z., Li, X., Li, X., Liang, C., Liu, R., Liu, Q., ... Ma, K. (2020a). An updated vegetation map of China (1:1000000). *Science Bulletin, 65*(13), 1125–1136.

Su, Y., Hu, T., Wang, Y., Li, Y., Dai, J., Liu, H., Jin, S., Ma, Q., Wu, J., & Liu, L. (2020b). Large-scale geographical variations and climatic controls on crown architecture traits. Journal of Geophysical Research: Biogeosciences, 125(2), e2019JG005306.

Tan, Z., Liu, X., Zhang, X., & Wu, L. (2011). Simulation of dynamics of crop biomass by assimulation SAR data into crop growth model. *Chinese Agricultural Science Bulletin, 27*(27), 161−167 (in Chinese).

Tao, S. L., Guo, Q. H., Li, C., Wang, Z. H., & Fang, J. Y. (2016). Global patterns and determinants of forest canopy height. *Ecology, 97*(12), 3265−3270.

Toba, T., & Ohta, T. (2005). An observational study of the factors that influence interception loss in boreal and temperate forests. *Journal of Hydrology, 313*(3−4), 208−220.

Trujillo, E., Ramirez, J. A., & Elder, K. J. (2007). Topographic, meteorologic, and canopy controls on the scaling characteristics of the spatial distribution of snow depth fields. *Water Resources Research, 43*(7). Article W07409.

Turner, W., Spector, S., Gardiner, N., Fladeland, M., Sterling, E., & Steininger, M. (2003). Remote sensing for biodiversity science and conservation. *Trends in Ecology & Evolution, 18*(6), 306−314.

Vihervaara, P., Mononen, L., Auvinen, A. P., Virkkala, R., Lu, Y. H., Pippuri, I., Packalen, P., Valbuena, R., & Valkama, J. (2015). How to integrate remotely sensed data and biodiversity for ecosystem assessments at landscape scale. *Landscape Ecology, 30*(3), 501−516.

Viskari, T., Hardiman, B., Desai, A. R., & Dietze, M. C. (2015). Model-data assimilation of multiple phenological observations to constrain and predict leaf area index. *Ecological Applications, 25*(2), 546−558.

Voss, M., & Sugumaran, R. (2008). Seasonal effect on tree species classification in an urban environment using hyperspectral data, LiDAR, and an object-oriented approach. *Sensors, 8*(5), 3020−3036.

Wei, Y. C., Wu, B. F., Zhang, X. W., & Du, X. (2008). Advances in Remote Sensing Research for Biodiversity Monitoring. *Advances in Earth Science, 23*(9), 924−931 (in Chinese).

Westman, W. E., Strong, L. L., & Wilcox, B. A. (1989). Tropical deforestation and species endangerment: The role of remote sensing. *Landscape Ecology, 3*(2), 97−109.

Wulder, M. A., Hall, R. J., Coops, N. C., & Franklin, S. E. (2004). High spatial resolution remotely sensed data for ecosystem characterization. *Bioscience, 54*(6), 511−521.

Xue, B. L., Guo, Q. H., Hu, T. Y., Xiao, J. F., Yang, Y. H., Wang, G. Q., Tao, S. L., Su, Y. J., Liu, J., & Zhao, X. Q. (2017). Global patterns of woody residence time and its influence on model simulation of aboveground biomass. *Global Biogeochemical Cycles, 31*(5), 821−835.

Yu, Y., Gao, T., Zhu, J. J., Wei, X. H., Guo, Q. H., Su, Y. J., Li, Y. M., Deng, S. Q., & Li, M. C. (2020). Terrestrial laser scanning-derived canopy interception index for predicting rainfall interception. *Ecohydrology, 13*(5). Article e2212.

Zwally, H. J., Yi, D. H., Kwok, R., & Zhao, Y. H. (2008). ICESat measurements of sea ice freeboard and estimates of sea ice thickness in the Weddell Sea. *Journal of Geophysical Research-Oceans, 113*(C2). Article C02s15.

CHAPTER 14

LiDAR Applications in Other Ecosystems

Contents

Besides forest ecosystems, light detection and ranging (LiDAR) has been broadly used for various applications in other areas, such as agricultural, grassland, urban, and wetland ecosystems. The applications of LiDAR in these ecosystems mainly focus on extracting and modeling vegetation structural attributes and estimating vegetation functional attributes. Owing to the uniqueness of each ecosystem, selecting an appropriate LiDAR platform is important, and near-surface LiDAR platforms—such as backpack, mobile vehicle, and unmanned aerial vehicle (UAV)—have shown significant advantages. For example, specially designed field mobile phenotyping platforms with LiDAR as the core component have shown promise at accurately and rapidly extracting crop phenotypic attributes for breeding and precise cultivation management. In urban ecosystems, backpack or mobile LiDAR systems are more flexible and efficient due to UAV flight restrictions. This chapter first introduces the representative applications of LiDAR in these ecosystems, after which the corresponding challenges and future perspectives are highlighted.

14.1 Applications of LiDAR in Agricultural Ecosystems

Agricultural ecosystems are artificial ecosystems that focus on crop production and aim to improve crop quality and yield through seeding, fertilization, irrigation, weeding, and pest control activities, thus providing more benefits for people. Simply pursuing production is the previous intent of agricultural ecosystems. However, as people gain an understanding of the

LiDAR Principles, Processing and Applications in Forest Ecology
ISBN 978-0-12-823894-3
https://doi.org/10.1016/B978-0-12-823894-3.00014-1

value of agricultural ecosystem services, developing scientific and sustainable agricultural ecosystems with high efficiency and high yield has become a key point in this field. Achieving high efficiency and high yield in agricultural ecosystems means that agricultural measures must be very efficient and refined, requiring real-time knowledge about crop growth and stress. Because multisource mobile remote sensing platforms with light detection and ranging (LiDAR) as the core component can precisely acquire the three-dimensional (3-D) structural attributes of crops and greatly improve crop growth estimation, they have attracted increasing attention for use in agricultural ecosystems.

Phenotypic measurement is the key to achieving precision cultivation and breeding, but the slow development of specific technology has become a bottleneck in the breeding field. Current progress in LiDAR-based crop phenotyping and structural modeling mainly focuses on target detection and phenotype extraction. Target detection aims to detect the location of individual plants or organs. Until recently, most target detection studies have focused on detecting panicles (Xiong et al., 2017), blooms (Xu, Li, et al., 2018), and roots (Pound et al., 2013; Yasrab et al., 2019) from images. LiDAR-based target detection is difficult because processing massive, irregular, and unordered point cloud data presents many challenges. Funded by the Strategic Leading Science and Technology Projects of the Chinese Academy of Sciences and the Research Equipment R&D Project of the Chinese Academy of Sciences, the authors' research team independently developed a series of sensor-to-plant high-throughput crop phenotype acquisition systems with a LiDAR sensor as the core component (Crop 3D), which covered molecular breeding from indoor screening to field breeding and large-scale promotion (Guo et al., 2018).

The indoor Crop 3D system combines high-resolution cameras, thermal imagers, imaging spectrometers, and other sensors (Fig. 14.1a) to integrate data acquisition, transfer, storage, and processing. The system can collect

Figure 14.1 (a) Crop 3D indoor high-throughput crop phenotyping monitoring system and (b) Crop 3D field mobile phenotyping monitoring system.

multisource data to identify plant attributes, including morphological structures and physiological and biochemical attributes. The outdoor Crop 3D system monitors crop phenotype dynamics in different environmental conditions to better assess crop growth, rapid selection of high-quality varieties, and precise management. It is a mobile LiDAR system that can rapidly and nondestructively measure the structural and phenotypic attributes of crops in the field (Fig. 14.1b). It includes a sensor integration unit, remote control system, power supply system, and drive system. The width and height of the mobile platform can be adjusted to suit the needs of different crop settings. In addition, the mobile phenotypic data acquisition platform can be used in agricultural ecosystems to quantify the effects of various treatments (water and fertilizer additions) on the fine-scale management of cropland.

Data processing software was also developed for the platform. For example, with LiDAR point clouds, the metrics of plant height, plant width, leaf number, leaf width, and leaf inclination angle can be extracted (Fig. 14.2) to systematically analyze the phenotypic changes in crop dynamics to provide accurate and substantial data support for breeding.

As a supplement and extension of the Crop 3D phenotype monitoring platform, we also developed a larger gantry platform—LiDragon (Fig. 14.3a)—which can move along rails and cover a larger crop field. The platform has been located at the Xiangshan Base of Institute of Botany, Chinese Academy of Sciences, since the end of 2018. Jin et al. (2018) built a Faster R-CNN (region-based convolutional neural network) deep learning network to detect individual maize plants from terrestrial LiDAR data acquired from LiDragon, laying the foundation for monitoring and

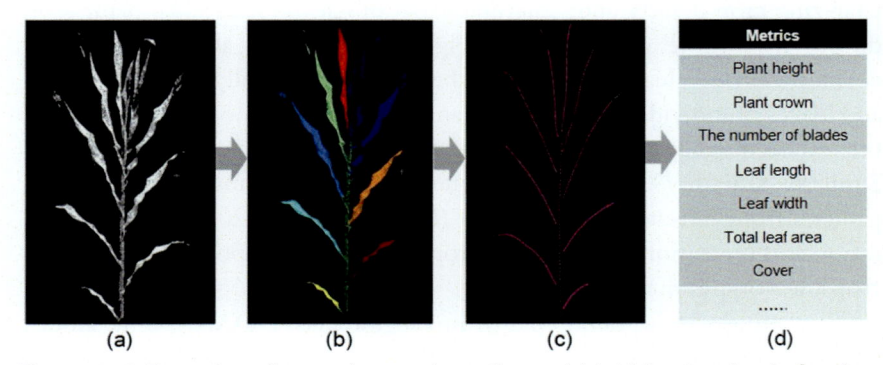

Metrics
Plant height
Plant crown
The number of blades
Leaf length
Leaf width
Total leaf area
Cover
......

(a)   (b)   (c)   (d)

Figure 14.2 Extraction of crop phenotypic attributes: (a) initial point cloud of maize, (b) leaf segmentation, (c) extraction of skeleton, and (d) list of extracted attributes.

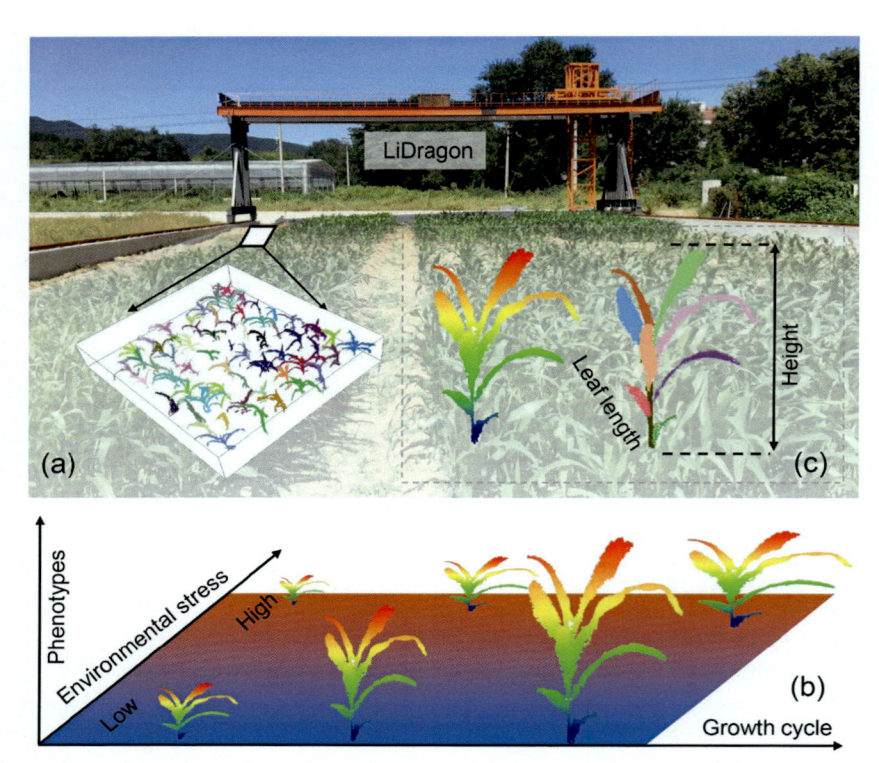

Figure 14.3 Light detection and ranging (LiDAR) enables accurate crop phenotyping and growth monitoring. (a) Individual organ-level crop structural phenotype extraction using the LiDragon field phenotype platform; (b) crop growth monitoring during the whole growth cycle and under different environmental stresses; and (c) stem-leaf separation and phenotypic attribute extraction from LiDAR data.

analyzing crop density and structural variations at the individual plant level during the whole growth cycle (Fig. 14.3b). Jin et al. (2019) further developed a deep learning-based method for automatically separating stems and leaves of individual maize plants and extracting phenotypic attributes at the leaf, stem, and individual levels from terrestrial LiDAR data (Fig. 14.3c). Similarly, a density-based clustering method can detect individual sorghum panicles from terrestrial LiDAR data (Malambo et al., 2019). The classification and separation of individual crop organs make it possible to measure crop phenotypes more precisely.

Extracting crop phenotypes from segmented LiDAR point clouds mainly relies on geometric methods. Compared with traditional manual and image-based methods, LiDAR-based methods have the advantages of being nondestructive, accurate, and robust (Perez-Sanz et al., 2017; Wang et al., 2009; Yang et al., 2013). Currently, terrestrial and mobile LiDAR are the most commonly used systems for extracting fine phenotypes at the individual plant level (Crommelinck & Höfle, 2016; Guo et al., 2018; Madec et al., 2017; Walter et al., 2019), whereas unmanned aerial vehicle (UAV) and airborne LiDAR systems are generally applied in deriving large-scale community-level phenotypes (Su et al., 2016; ten Harkel et al., 2020). For example, wheat canopy height and cover estimated from mobile LiDAR data were consistent with manual measurements and outperformed imagery-based methods (Jimenez-Berni et al., 2018; Madec et al., 2017; Walter et al., 2019). Jin et al. (2019) extracted individual maize heights from terrestrial LiDAR data that were in high agreement with manual measurements—the coefficient of determination (R^2) was greater than 0.91—and systematically evaluated the accuracy of the extracted phenotypes at stem and leaf levels, such as leaf inclination, leaf length, and stem diameter. The accuracy of phenotype extraction shows that LiDAR is a promising and valuable tool for the structural modeling of crops and promotes the development of functional modeling in agriculture.

Furthermore, LiDAR data can be adapted to estimate crop biomass and yield, which are indispensable functional indicators for crop breeding and agricultural management and are closely linked to crop structure (Jimenez-Berni et al., 2018; Walter et al., 2019). For example, Walter et al. (2019) estimated wheat biomass from the projected volume of LiDAR point clouds ($r = 0.86$), while (Jimenez-Berni et al., 2018) estimated wheat biomass from a LiDAR-derived 3-D voxel index and 3-D profile index with R^2 of 0.93 and 0.92, respectively. Jin et al. (2020) estimated maize biomass at the plot, individual, leaf group, and individual organ levels. They demonstrated the potential of terrestrial LiDAR data for accurately estimating crop biomass at fine levels. Biomass estimation from LiDAR-derived vegetation indices has been evaluated for various crop types (ten Harkel et al., 2020). Based on biomass estimation, crop yield can be modeled through the conversion of the harvest index (Mohamad et al., 1994). Moreover, various LiDAR systems, especially field LiDAR phenotyping systems, are increasingly used to monitor temporal (e.g., diurnal, seasonal) changes in crop phenotypes (e.g., height, leaf area, and biomass) under different environmental stresses, which allows timely observations of phenotype dynamics (Fig. 14.3) (Calders et al., 2015; Su et al., 2019).

The successful application of LiDAR in agricultural ecosystems has greatly improved the accuracy of crop growth modeling, as LiDAR can capture the structure of crops efficiently and accurately from the organ to canopy levels during the whole growth period. This indicates that applying LiDAR data to monitor and model crop growth can revolutionize crop phenotyping, monitoring, and functional modeling (Bietresato et al., 2016; Jin et al., 2021; Lin, 2015). In addition, the crop phenotypic information extracted from LiDAR data can be integrated into phenomics and multiomics analysis for breeding and precise cultivation management (Tao et al., 2022). However, challenges still exist when applying LiDAR in agricultural ecosystems. The high cost of field phenotyping systems has been a major obstacle preventing their broad use. The recent development of near-surface LiDAR systems (e.g., backpack and unmanned ground vehicle systems) has reduced the costs and improved the efficiency of LiDAR data collection, which may further boost the application of LiDAR in agricultural ecosystems (Heun et al., 2019). Another issue is that conventional LiDAR data lack spectral information and, thus, LiDAR must be integrated with other remote sensing sensors (e.g., hyperspectral, thermal, and fluorescence sensors) to explore more diverse structural and functional phenotypes (Du et al., 2017; Eitel et al., 2014; Yang & Kang, 2018). The accumulation of multisource LiDAR data in various spatial and temporal resolutions calls for new algorithms and models, such as deep learning, which also brings new opportunities to precise agriculture and modern breeding (Jin et al., 2021; Tao et al., 2022).

14.2 Applications of LiDAR in Grassland Ecosystems

Grassland ecosystems are open terrestrial ecosystems with perennial herbaceous plants as the main producers. They can be divided into artificial grassland ecosystems and natural grassland ecosystems. Taking China as an example, grassland is widely distributed and is the largest terrestrial ecosystem in China, occupying approximately 400 million hectares and accounting for 41.7% of the land area (Kang et al., 2007). In addition to providing traditional ecological services, grassland ecosystems are important for livestock grazing. For a long time, owing to overuse and deregulation, grassland ecosystems have faced serious problems, such as overgrazing, indiscriminate mining, and frequent occurrence of rats and pests, leading to serious degradation, desertification, salinization, soil erosion, and sandstorms. More sustainable use of grassland ecosystems is a national concern in China. To achieve this goal, it is important to understand the growth and yield of grassland ecosystems to effectively control the carrying capacity of livestock.

Grassland is quite dense, with relatively less vegetation than other ecosystems. Traditional grassland resource inventories rely heavily on field sampling, which can be inaccurate and time-consuming (Anderson et al., 2018; Cooper et al., 2017; Xu et al., 2020). Beyond the spectral information provided by optical remote sensing (El Hajj et al., 2016; Jansen et al., 2018; Mermoz et al., 2014), we see increasing applications of LiDAR in grassland ecosystems to extract a range of structural attributes.

Using near-surface LiDAR platforms to obtain 3-D structural information for natural grassland ecosystems has unique advantages. Because the terrain of grassland ecosystems is relatively flat, we can use mobile LiDAR to obtain 3-D point clouds. In addition, as there is no vision occlusion in a grassland ecosystem, it is safer to use UAV LiDAR there than in an urban or forest ecosystem. Applications of LiDAR in grassland ecosystems still focus on extracting structural attributes and modeling functional attributes. However, evaluating the structural attributes of grassland at the individual plant level is difficult even with LiDAR owing to the low height and high density of grass (Fig. 14.4a) (Xu et al., 2020). At present, structural attributes

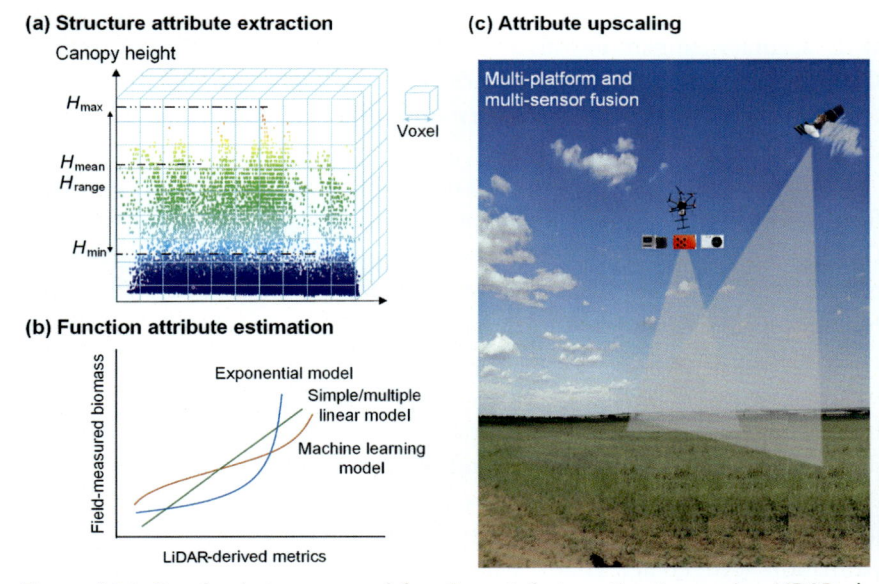

Figure 14.4 Grassland structure and function attribute estimations using LiDAR observations. (a) Community-level structure attribute estimation through the voxel-based method; (b) function attribute (biomass as an example) estimation using the regression-based methods; and (c) structure and function attribute upscaling through the integration of multiplatform and multisensor remote sensing data.

(e.g., canopy height, canopy volume, and canopy cover) at the community level are widely used for monitoring and modeling grassland ecosystems. Canopy height can be extracted using point-based (Mitchell et al., 2011; Streutker & Glenn, 2006), canopy height model (CHM)-based (Wijesingha et al., 2019; Xu et al., 2020), and voxel-based methods (Anderson et al., 2018; Mitchell et al., 2011; Schulze-Brüninghoff et al., 2019). According to previous studies point-based methods often underestimate the grass height for LiDAR data with relatively low point density (Streutker & Glenn, 2006). CHM- and voxel-based methods also can be strongly influenced by the choice of pixel (or voxel) size (Anderson et al., 2018). It is worth noting that compared with terrestrial LiDAR data, UAV LiDAR data may more easily suffer from information loss at canopy tops, which is often ignored in previous studies and can result in underestimated structural attributes of grassland such as canopy height (Zhao et al., 2022). Canopy volume can be calculated from LiDAR data using volume surface differencing approaches (Cooper et al., 2017), voxel counting approaches (Greaves et al., 2015; Schulze-Brüninghoff et al., 2019), and convex hull approaches (Schulze-Brüninghoff et al., 2019). Greaves et al. (2015) showed that a volume surface differencing approach could generate canopy volume estimations with accuracy similar to other approaches but with higher efficiency. Canopy cover is commonly calculated as the ratio of the number of vegetation LiDAR echoes to the total number of LiDAR echoes (Xu et al., 2020).

Using LiDAR-derived structural attributes as predictors, grassland function attributes can also be estimated using regression-based methods (Fig. 14.4b). Aboveground biomass—a functional indicator of grassland productivity—has been extensively studied by relating destructive field biomass measurements to attributes extracted from LiDAR data (e.g., canopy height, canopy volume, canopy cover) to generate wall-to-wall biomass products (Cooper et al., 2017; Greaves, 2017; Greaves et al., 2015). It was suggested that LiDAR-derived canopy height and canopy cover are strongly correlated with aboveground biomass (Schulze-Brüninghoff et al., 2019; Xu et al., 2020); random forest and stepwise linear regression were found to produce more robust regression models for aboveground biomass estimation (Xu et al., 2020). Moreover, it has been reported that the integration of LiDAR with spaceborne optical imagery can be used to extrapolate and upscale aboveground biomass with higher accuracy than using only optical imagery (Fig. 14.4c) (Schaefer & Lamb, 2016).

LiDAR can extract the accurate 3-D structure of vegetation and obtain terrain attributes such as slope and aspect. By combining the structural and topographic attributes of grassland extracted by LiDAR, the restoration of degraded grassland can be monitored. Using full-waveform airborne LiDAR or a combination of hyperspectral images and airborne LiDAR, species in different habitats can be classified and mapped (Fisher et al., 2018; Marcinkowska-Ochtyra et al., 2018; Zlinszky et al., 2014). The precise classification of vegetation species provides the possibility for habitat biodiversity assessment. Moeslund et al. (2019) used airborne LiDAR to successfully assess regional biodiversity using multiple grassland biodiversity variables in multiple habitat types. The spatial heterogeneity of grassland vegetation has a strong correlation with biodiversity. Jansen et al. (2019) used the grassland structure attributes extracted by LiDAR to estimate the biomass and evaluated the effect of different grazing intensities on the spatial heterogeneity of the grassland.

Taking advantage of LiDAR in monitoring the structural attributes and dynamics of grassland, The Chinese Academy of Sciences built the first Test Area of Ecological Grass-Based Livestock Husbandry in Hulun Buir, Inner

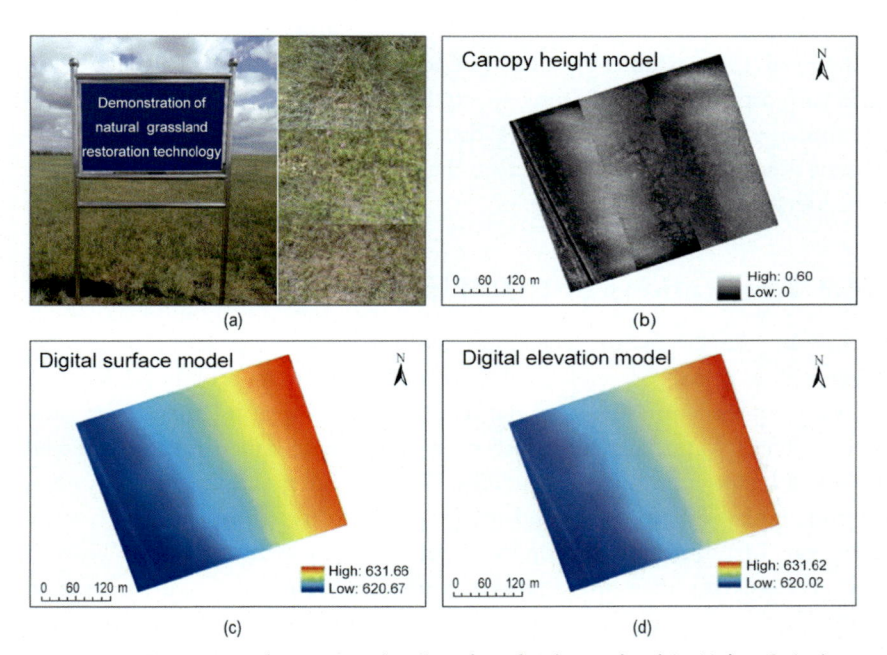

Figure 14.5 Estimating the canopy height of artificial grassland in Hulun Buir, Inner Mongolia, China, based on unmanned aerial vehicle (UAV) LiDAR data.

Mongolia, China, to evaluate the potential of intensive management of artificial grassland as an important way to achieve sustainable use of natural grassland. We used a UAV LiDAR system to obtain the structural attributes of grassland, such as height and coverage, to support the decision-making and management system for husbandry. In August 2016, we carried out UAV LiDAR scanning in the test area and obtained point cloud data covering 1 km^2, in which the flying height was approximately 70 m, the flying speed was approximately 4.5 m/s, and the density of the point cloud was 225 points/m^2. Ground points were obtained using the previously described filtering algorithm, and high-precision digital elevation models, digital surface models, and canopy height models were generated (Fig. 14.5). The results show that, through physiological regulation, the proportion of high-quality pasture *Leymus chinensis* has increased in the test area. Although there was a serious drought in Hulun Buir grassland in Inner Mongolia, China in 2016, the planting of artificial grassland has achieved initial success.

In summary, applying LiDAR in grassland ecosystems has achieved initial success, but many challenges must still be addressed. For example, filtering ground points in grassland, which is a prerequisite for extracting structural attributes, is difficult because of the dense canopy coverage. Structural attributes derived from LiDAR are limited in grassland ecosystems, and more studies are needed to develop ecologically meaningful grassland attributes. Combining spectral data with structural and functional attributes extracted from LiDAR data and exploring the relationship between them is another key direction that needs more attention in the future (Reddersen et al., 2014).

14.3 Applications of LiDAR in Urban Ecosystems

Economic development accelerates the urbanization process, which has triggered a series of ecological and environmental problems, such as excessive population density, traffic congestion, heat island effects, and atmospheric, water, air, and noise pollution. These have seriously affected the health and comfort of residents. The urban green space ecosystem, dominated by urban forests and integrated with trees and flowers, has the functions of refreshing air, adjusting climate, and reducing noise. Thus, it has great significance for ecosystem services. With the concept of an "ecological city" put forward in the 1980s, urban greening has received increasing attention. However, owing to limited space in urban areas, it is particularly important to optimize the green space structure and pursue

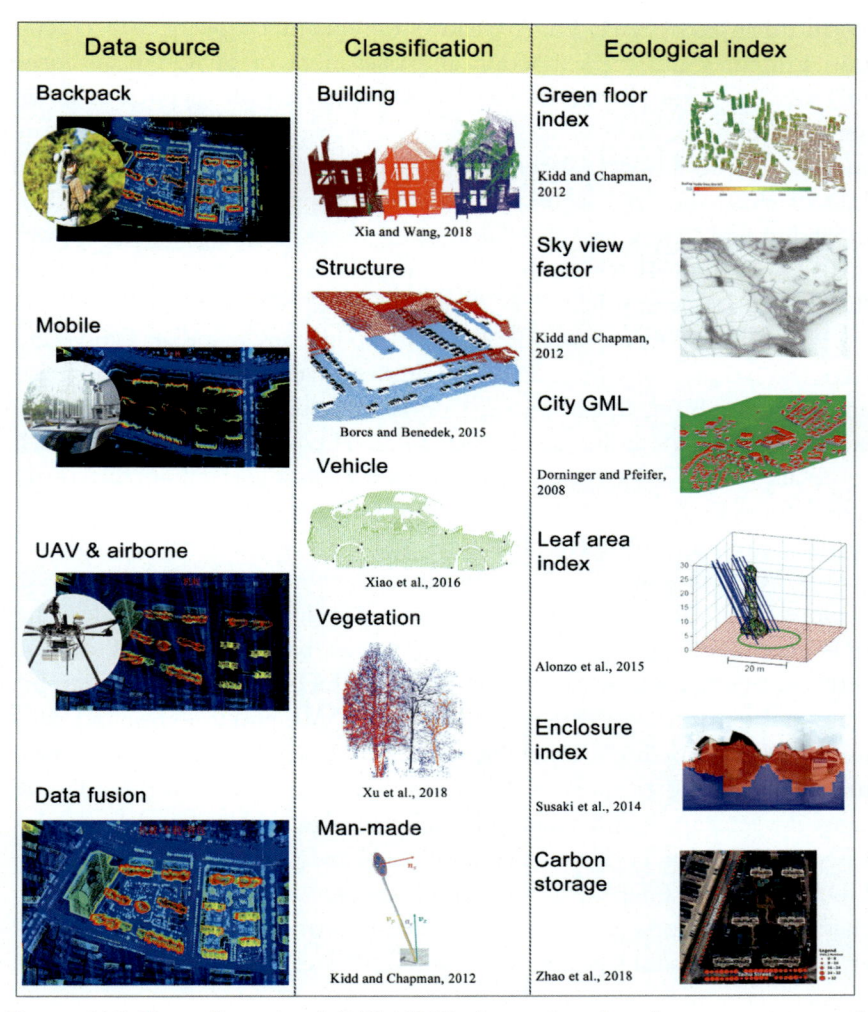

Figure 14.6 Three-dimensional (3-D) LiDAR observations in urban ecosystems and ecological index examples derived from classified LiDAR point clouds. Mobile and backpack LiDAR data can cover areas along and off the road, respectively, while unmanned aerial vehicle (UAV) and airborne LiDAR data can cover a large urban area. The fusion of multiplatform LiDAR data can provide more comprehensive 3-D urban observations. *GML*, geography markup language.

greater ecological benefits. The development of near–surface LiDAR platforms provides a new solution for quantitatively evaluating urban green space in three dimensions.

In urban ecosystems, LiDAR allows the quantification of 3-D vegetation structures and the extraction and assessment of buildings. Airborne LiDAR data were one of the first LiDAR datasets to be applied in urban applications (Goodwin et al., 2009; Yan et al., 2015). Recently, near-surface mobile LiDAR systems such as backpack LiDAR and mobile LiDAR have become dominant in urban applications owing to their flexibility and ability to collect data along roads (Fig. 14.6) (Zhao et al., 2018). UAV LiDAR systems have been used less frequently because of strict flying regulations in urban areas (Zhang et al., 2003).

Unlike in other ecosystems, LiDAR applications in urban ecosystems have the unique prerequisite of classifying vegetation (e.g., trees and grass) and artificial objects (e.g., buildings, powerlines, bridges, vehicles, and railways) from LiDAR point clouds (Fig. 14.6). The purpose of classification is to obtain distinct urban features, including buildings, vegetation, and roads. Then, the 3-D shape of the city can be quantified, and the characteristics and spatial distribution of different features and the ecological relationship between urban landscape structure, climate, and environment can be analyzed. Various classification algorithms have been introduced and are divided into four general categories: pixelwise classification based on LiDAR-derived surfaces, object-based segmentation based on LiDAR-derived surfaces, hierarchical semantic segmentation based on LiDAR point clouds, and deep learning-based methods. Pixelwise classification methods apply traditional machine learning (e.g., artificial neural networks, maximum likelihood, support vector machines, random forests, Gaussian mixture modeling, rule-based classification, conditional random fields, and Markov random fields) to LiDAR-derived surfaces (e.g., digital surface models, and intensity maps) to classify urban objects (Charaniya et al., 2004; Huang et al., 2008; Niemeyer et al., 2011; Samadzadegan et al., 2010; Xu, Xu, et al., 2018). Object-based segmentation methods, which use a user-defined procedure to classify segmented objects from LiDAR-derived surfaces (Alonzo et al., 2014; Borcs & Benedek, 2015), are designed to avoid the salt-and-pepper effect resulting from pixelwise classification (Höfle et al., 2012; Zhang et al., 2015). Numerous studies have reported that object-based methods can achieve an overall accuracy of greater than 80%, which is higher than that derived from pixelwise methods (Hu et al., 2014; MacFaden et al., 2012; O'Neil-Dunne et al., 2013; Soilán et al., 2016; Xiao et al., 2016).

Vegetation structure and function attributes, such as tree height, diameter at breast height, crown diameter, crown base height, leaf area index, tree species, vegetation volume, vegetation coverage, and above-ground biomass, are essential indicators for the green vegetation in urban environments and can be extracted from classified LiDAR point clouds (Fig. 14.6) (Richardson et al., 2009; Susaki et al., 2014; Yan et al., 2015). The accuracy of these extracted attributes is usually higher than that in forested areas owing to the relatively simple tree species composition and regular tree arrangement in urban ecosystems (Alonzo et al., 2015). One attribute that has drawn the attention of researchers is urban green space—public and private urban spaces primarily covered by vegetation (Caynes et al., 2016). LiDAR offers a new way to quantify urban green space from two-dimensional horizontal arrangements to 3-D horizontal and vertical arrangements, which greatly benefits studies of urban functional connectivity, urban planning and management, and biodiversity (Chance et al., 2016; Susaki et al., 2014; Yu et al., 2016). More specifically, the 3-D green volume of urban vegetation, defined as the spatial volume occupied by growing plant stems and leaves, can well reflect the ecological functions of urban vegetation and has become a critical indicator for evaluating urban greening (Takahashi et al., 2010; Zhou & Sun, 1995).

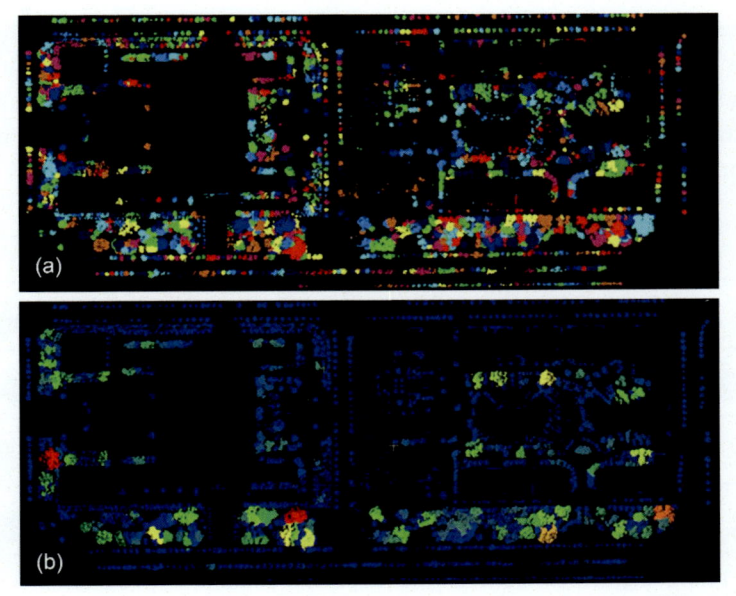

Figure 14.7 Urban three-dimensional green volume estimation using dynamic delta wing LiDAR data. (a) Individual tree segmentation from point clouds and (b) the estimated green volume from low (*blue*) to high (*red*). The data was acquired in Hengyang City, Hunan Province, China, during August 2016.

With the commercial application of LiDAR technology, LiDAR point cloud data have become an effective data source for 3-D green volume estimation. The core of estimating the urban 3-D green volume is accurately estimating the number of trees and the crown volume of each tree. The height and shape information recorded in LiDAR data can be used for classifying different objects based on the height criterion, e.g., tall buildings can be easily and quickly identified. Then, height and canopy morphology information can be combined to extract urban vegetation. After individual tree segmentation, the point cloud of each tree can be marked (Fig. 14.7a). Finally, crown reconstruction (Schöpfer et al., 2005) or the convex hull method (Kato et al., 2009) is used to estimate the green volume of individual trees (Fig. 14.7b). From the estimated 3-D green volume, the distribution of urban vegetation green volume can be seen: the green volume of street trees is low and basically the same, while the green volume of parks and forests at the lower part is significantly higher. Furthermore, urban green spaces such as street trees can provide aesthetic value and ecological services. With the help of mobile LiDAR and questionnaire surveys, several shape indexes, such as the richness and evenness of tree peaks, can be quickly extracted to quantify the fluctuation and continuity of canopy lines of street trees for assessing their aesthetic value (Hu et al., 2022). Different groups of street trees with a similar amount of green volume may have divergent shapes of greenness and aesthetic values (Fig. 14.8). This is an important supplement to the urban green space theory for assessing urban environments and living conditions.

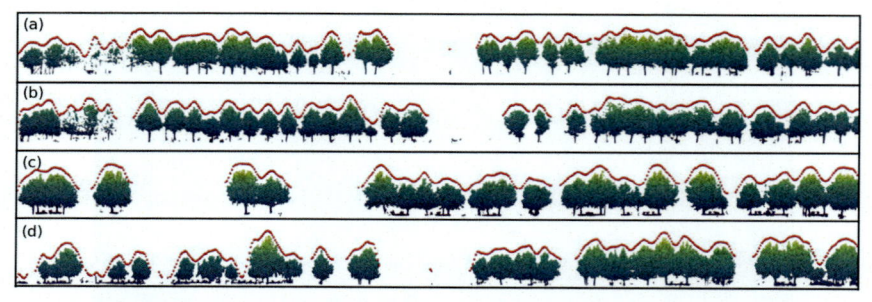

Figure 14.8 Disagreement between the shape and amount of greenness. The street trees in (a) and (b) have similar amounts of greenness (green volume: 30.09 m^3/m and 31.70 m^3/m), but the richness of their peaks is 0.34 and 0.05, respectively. The street trees in (c) and (d) have similar amounts of greenness (green volume: 30.96 m^3/m and 33.17 m^3/m), but the evenness of their peaks is 0.68 and 0.08, respectively.

In addition to vegetation indices, attributes related to buildings can also be obtained from LiDAR points, such as building height, building mass, building volume, building 3-D models, urban envelope index, and sky openness (Fig. 14.6) (Bonczak & Kontokosta, 2019; Dorninger & Pfeifer, 2008; Kidd & Chapman, 2012; Susaki et al., 2014; Xia & Wang, 2018). These 3-D building indices can be used together with urban vegetation attributes to study the microclimate, flora, and fauna in urban ecosystems, as well as for urban planning and management. The 3-D arrangement of buildings and vegetation can alter the spatial and temporal variability of solar radiation (Gage & Cooper, 2017; Lindberg & Grimmond, 2011; Rose & Levinson, 2013; Yu et al., 2009), and assimilating 3-D information on buildings and vegetation into climate models can improve the accuracy of microclimate simulations (Guinot et al., 2006; Tian et al., 2019; Zhao et al., 2015). Moreover, urban ecosystems have become an important habitat for flora and fauna, and anthropogenic activities are important factors influencing the biodiversity of urban ecosystems. Integrating LiDAR-derived information into biodiversity studies can help better describe the spatial distribution of vegetation and animal species in an urban environment (Plowright et al., 2017). Modeled vegetation and animal distributions can further be used to analyze and predict human health risks (e.g., exposure to pollen and mosquitos) (Pecero-Casimiro et al., 2019). In addition, LiDAR data can be used to model the risks related to powerlines, flooding, etc., and thereby provide essential supporting information for management plans that ensure the health of urban ecosystems (Fewtrell et al., 2011; Ortega et al., 2019; Tooke et al., 2014; Yang & Kang, 2018).

Overall, LiDAR applications in urban ecosystems share common characteristics with applications in other ecosystems to extract vegetation information. Still, they involve more effort in recognizing and processing artificial urban components. Currently, the major obstacles lie in handling the huge amount of LiDAR data at a city scale, especially for data collected by mobile LiDAR systems. Moreover, incorporating LiDAR-derived 3-D attributes into urban ecological studies and modeling needs further exploration.

14.4 Applications of LiDAR in Wetland Ecosystems

Wetlands are transitional zones between land and water. They are rich in terrestrial and aquatic flora and fauna resources, forming a natural gene pool and a unique biological environment that no other single ecosystem can match. The unique soil and climate conditions in wetlands can support complex and complete flora and fauna communities, which are vital for maintaining biodiversity, ecological security, and human well-being. For example, the water conservation capacity of wetlands is 5—8 times of that of other terrestrial ecosystems, and wetlands provide habitats to greater than 40% of the world's species (Hu et al., 2017). As wetland ecosystems are usually periodically flooded, it is difficult to conduct field surveys and accurately obtain the information and condition of wetland vegetation, such as tree height and biomass, over a large area.

In the past 30 years, remote sensing has been used in wetland surveys and monitoring, supplemented by geographic information systems and global positioning systems (Cyranoski, 2009; Owers et al., 2016; Szantoi et al., 2013). Through remote sensing image interpretation, information about wetland type, area, distribution, average altitude, and watershed can be obtained. Multispectral remote sensing data with medium spatial resolution, such as the Landsat series, Advanced Land Observing Satellite (ALOS), and Sentinel-2, are widely used for dynamic monitoring of wetland distribution (Mao et al., 2020), while high-resolution multispectral images, such as QuickBird, GeoEye-1, and WorldView-2, are used for detailed investigation of wetland vegetation and hydrological conditions (Zhu et al., 2015). Also, hyperspectral imagery has unique advantages for wetland vegetation species classification (Wan et al., 2020). Microwave remote sensing data, such as SRTM (Shuttle Radar Topography Mission), ALOS-2 PALSAR-2 (Phased Array L-band Synthetic Aperture Radar-2), and Sentinel-1, are widely used in mapping wetland vegetation height and surface biomass (Aslan et al., 2016; Simard et al., 2006). Many passive remote sensing datasets and methods have been applied in wetland surveys and monitoring, but they may fail to obtain fine-scale vertical structure information of vegetation. For example, information on shrubs and individual trees of wetlands is difficult to obtain using traditional remote sensing methods. In addition, wetlands are increasingly fragmented owing to human disturbance, and a detailed inventory method suitable for small- and medium-scale wetlands is also urgently needed.

In the face of these challenges in wetland surveys and monitoring in local areas, LiDAR has a strong advantage in obtaining information on the vertical structure and topography of wetland vegetation (Guo et al., 2017). Spaceborne LiDAR, such as the Geoscience Laser Altimeter System aboard Ice, Cloud, and land Elevation Satellite (ICESat), can be used to obtain wetland vegetation height combined with the SRTM digital elevation model (Simard et al., 2019). Airborne LiDAR can directly obtain wetland topography and can be combined with multispectral data to improve the classification of wetland vegetation species (Chadwick, 2011) and monitor the dynamic changes in wetland vegetation canopy cover (Lymburner et al., 2020). UAV LiDAR is more maneuverable and flexible for obtaining high-precision wetland information, and its cost is gradually decreasing. Therefore, UAV LiDAR has promising prospects in wetland surveys and monitoring. For example, Yin and Wang (2019) verified the feasibility of UAV LiDAR in identifying individual mangrove trees and pointed out that it was more challenging to delineate clumped mangrove crowns. Wang et al. (2020) used UAV LiDAR data as a bridge to link field measurements and satellite imagery for accurately estimating the biomass of mangroves in the northeast of Hainan Island, China. The authors' research team also mapped the mangrove biomass in Zhanjiang City, Guangdong Province, China using a small eight-rotor UAV with a Velodyne VLP-16 sensor (Velodyne Lidar, San Jose, CA, USA). Two flights were conducted with a flying height of 60 m and speed of 4.5 m/s. A mangrove forest of 1 km^2 was scanned—the point density of the obtained data was 293 points/m^2. After preprocessing the point cloud data, a CHM with a spatial resolution of 1 m was generated according to the methods described in the previous section. Then, the watershed algorithm was used to segment individual mangrove trees. The mangrove biomass in the whole region was estimated using allometry equations based on each tree's height and crown width (Fig. 14.9). By comparing the LiDAR and the field data, we found that the

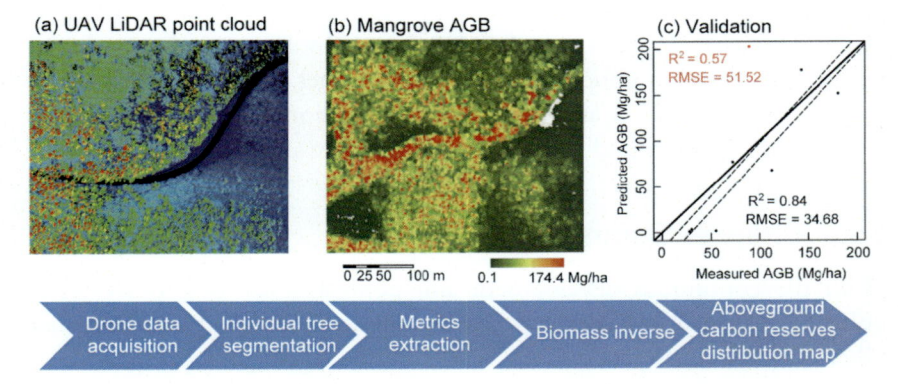

Figure 14.9 Estimating the aboveground biomass (AGBs) of a mangrove forest in Leizhou, Guangdong, using unmanned aerial vehicle (UAV) LiDAR data. *RMSE*, root-mean-square error.

estimated mangrove biomass using UAV LiDAR was very accurate. After removing an abnormal record, the R^2 reached 0.84, and the root-mean-square error was approximately 35 Mg/ha (Guo et al., 2017). Combined with ground samples, spaceborne LiDAR, satellite imagery, and climate data, an upscaling method was constructed to estimate global mangrove aboveground biomass using random forest regression (Hu et al., 2020). In addition, high-resolution 3-D point clouds acquired using terrestrial LiDAR can provide detailed structural information on complex wetland vegetation, which is otherwise difficult to measure accurately. Owers et al. (2018) proved that terrestrial LiDAR was a nondestructive method suitable for estimating the biomass of coastal wetland vegetation with complex structures. Backpack LiDAR systems also have a high potential to obtain 3-D information on wetland ecosystems as it is relatively easy and flexible to walk through the wetland with a backpack. In contrast, mobile LiDAR systems mounted on a vehicle are rarely used for wetland surveys and monitoring because the muddy substrate and complex vegetation hinder the moving of vehicles.

Although LiDAR shows great potential in wetland surveys and monitoring, most LiDAR systems are based on near-surface platforms or aircraft. Thus, they have a limited detection range and thus cannot provide global coverage as Landsat 8 or Sentinel-2 can. In September 2018, the National Aeronautics and Space Administration successfully launched the ICESat-2 satellite with a LiDAR system. Although it has global coverage, the laser footprints are not continuous (Li et al., 2020). In addition, most wetlands have open shallow water that strongly absorbs laser energy. When a LiDAR pulse hits the water, an effective return cannot be obtained (Bandini et al., 2020). Therefore, in the future, LiDAR data will still need to be combined with high-resolution optical remote sensing images and water level gauges to study the vegetation and hydrological conditions of wetlands, such as tree height, aboveground biomass, and stagnant water conditions.

14.5 Chapter Summary

With the continuous development and popularization of LiDAR technology, LiDAR data have been broadly applied in agricultural, grassland, urban, and wetland ecosystems. The divergent characteristics of these ecosystems and the objectives of corresponding studies require different data collection methods and promote the emergence and rapid development of new near-surface LiDAR platforms, such as backpack, mobile vehicle,

UAV, and gantry. These platforms have various advantages such as high mobility, flexibility, and efficiency and low cost and significantly facilitate studies and applications in the abovementioned ecosystems, such as crop growth monitoring and modeling, breeding, grassland restoration, and urban green space evaluation. In general, these applications focus on extracting vegetation structural attributes and the inversion of vegetation functional attributes through modeling. Although excellent results have been achieved, there is still great potential in the future for applying LiDAR in various ecosystems. First, extracting the structural information of various objects more accurately and effectively from massive 3-D point cloud data is a continuing expectation of the research community. Second, different studies have different objectives and thus study areas with various scales and locations. Also, different LiDAR platforms and systems are selected to collect data. This poses a great challenge in comparing different studies and analyzing their potential links. The connection between different scales is also an issue worthy of attention. Finally, the combination of LiDAR and optical remote sensing has become a focus in the fields of remote sensing and ecology in recent years. Combining multisource remote sensing data may provide a new paradigm for the large-scale monitoring and management of ecosystems.

References

Alonzo, M., Bookhagen, B., McFadden, J. P., Sun, A., & Roberts, D. A. (2015). Mapping urban forest leaf area index with airborne lidar using penetration metrics and allometry. *Remote Sensing of Environment, 162*, 141–153.

Alonzo, M., Bookhagen, B., & Roberts, D. A. (2014). Urban tree species mapping using hyperspectral and lidar data fusion. *Remote Sensing of Environment, 148*, 70–83.

Anderson, K. E., Glenn, N. F., Spaete, L. P., Shinneman, D. J., Pilliod, D. S., Arkle, R. S., McIlroy, S. K., & Derryberry, D. R. (2018). Estimating vegetation biomass and cover across large plots in shrub and grass dominated drylands using terrestrial lidar and machine learning. *Ecological Indicators, 84*, 793–802.

Aslan, A., Rahman, A. F., Warren, M. W., & Robeson, S. M. (2016). Mapping spatial distribution and biomass of coastal wetland vegetation in Indonesian Papua by combining active and passive remotely sensed data. *Remote Sensing of Environment, 183*, 65–81.

Bandini, F., Sunding, T. P., Linde, J., Smith, O., Jensen, I. K., Köppl, C. J., Butts, M., & Bauer-Gottwein, P. (2020). Unmanned Aerial System (UAS) observations of water surface elevation in a small stream: Comparison of radar altimetry, LIDAR and photogrammetry techniques. *Remote Sensing of Environment, 237*, 111487.

Bietresato, M., Carabin, G., Vidoni, R., Gasparetto, A., & Mazzetto, F. (2016). Evaluation of a LiDAR-based 3D-stereoscopic vision system for crop-monitoring applications. *Computers and Electronics in Agriculture, 124*, 1–13.

Bonczak, B., & Kontokosta, C. E. (2019). Large-scale parameterization of 3D building morphology in complex urban landscapes using aerial LiDAR and city administrative data. *Computers, Environment and Urban Systems, 73*, 126—142.

Borcs, A., & Benedek, C. (2015). Extraction of vehicle groups in airborne lidar point clouds with two-level point processes. *IEEE Transactions on Geoscience and Remote Sensing, 53*, 1475—1489.

Calders, K., Schenkels, T., Bartholomeus, H., Armston, J., Verbesselt, J., & Herold, M. (2015). Monitoring spring phenology with high temporal resolution terrestrial LiDAR measurements. *Agricultural and Forest Meteorology, 203*, 158—168.

Caynes, R. J. C., Mitchell, M. G. E., Wu, D. S., Johansen, K., & Rhodes, J. R. (2016). Using high-resolution LiDAR data to quantify the three-dimensional structure of vegetation in urban green space. *Urban Ecosystems, 19*(4), 1749—1765.

Chadwick, J. (2011). Integrated LiDAR and IKONOS multispectral imagery for mapping mangrove distribution and physical properties. *International Journal of Remote Sensing, 32*(21), 6765—6781.

Chance, C. M., Coops, N. C., Plowright, A. A., Tooke, T. R., Christen, A., & Aven, N. (2016). Invasive shrub mapping in an urban environment from hyperspectral and LiDAR-derived attributes. *Frontiers in Plant Science, 7*, 1528.

Charaniya, A. P., Manduchi, R., & Lodha, S. K. (2004). Supervised parametric classification of aerial LiDAR data. In *2004 conference on computer vision and pattern recognition workshop*.

Cooper, S. D., Roy, D. P., Schaaf, C. B., & Paynter, I. (2017). Examination of the potential of terrestrial laser scanning and structure-from-motion photogrammetry for rapid nondestructive field measurement of grass biomass. *Remote Sensing, 9*(6), 531.

Crommelinck, S., & Höfle, B. (2016). Simulating an autonomously operating low-cost static terrestrial LiDAR for multitemporal maize crop height measurements. *Remote Sensing, 8*(3).

Cyranoski, D. (2009). Putting China's wetlands on the map. *Nature, 458*(7235), 134—134.

Dorninger, P., & Pfeifer, N. (2008). A comprehensive automated 3D approach for building extraction, reconstruction, and regularization from airborne laser scanning point clouds. *Sensors, 8*(11).

Du, L., Shi, S., Yang, J., Wang, W., Sun, J., Cheng, B., Zhang, Z., & Gong, W. (2017). Potential of spectral ratio indices derived from hyperspectral LiDAR and laser-induced chlorophyll fluorescence spectra on estimating rice leaf nitrogen contents. *Optics Express, 25*(6), 6539—6549.

Eitel, J. U. H., Magney, T. S., Vierling, L. A., Brown, T. T., & Huggins, D. R. (2014). LiDAR based biomass and crop nitrogen estimates for rapid, non-destructive assessment of wheat nitrogen status. *Field Crops Research, 159*, 21—32.

El Hajj, M., Baghdadi, N., Zribi, M., Belaud, G., Cheviron, B., Courault, D., & Charron, F. (2016). Soil moisture retrieval over irrigated grassland using X-band SAR data. *Remote Sensing of Environment, 176*, 202—218.

Fewtrell, T. J., Duncan, A., Sampson, C. C., Neal, J. C., & Bates, P. D. (2011). Benchmarking urban flood models of varying complexity and scale using high resolution terrestrial LiDAR data. *Physics and Chemistry of the Earth, Parts A/B/C, 36*(7), 281—291.

Fisher, R. J., Sawa, B., & Prieto, B. (2018). A novel technique using LiDAR to identify native-dominated and tame-dominated grasslands in Canada. *Remote Sensing of Environment, 218*, 201—206.

Gage, E. A., & Cooper, D. J. (2017). Relationships between landscape pattern metrics, vertical structure and surface urban Heat Island formation in a Colorado suburb. *Urban Ecosystems, 20*(6), 1229—1238.

Goodwin, N. R., Coops, N. C., Tooke, T. R., Christen, A., & Voogt, J. A. (2009). Characterizing urban surface cover and structure with airborne lidar technology. *Canadian Journal of Remote Sensing, 35*(3), 297—309.

Greaves, H. E. (2017). *Applying lidar and high-resolution multispectral imagery for improved quantification and mapping of tundra vegetation structure and distribution in the Alaskan Arctic University of Idaho.*

Greaves, H. E., Vierling, L. A., Eitel, J. U., Boelman, N. T., Magney, T. S., Prager, C. M., & Griffin, K. L. (2015). Estimating aboveground biomass and leaf area of low-stature Arctic shrubs with terrestrial LiDAR. *Remote Sensing of Environment, 164*, 26–35.

Guinot, B., Roger, J.-C., Cachier, H., Pucai, W., Jianhui, B., & Tong, Y. (2006). Impact of vertical atmospheric structure on Beijing aerosol distribution. *Atmospheric Environment, 40*(27), 5167–5180.

Guo, Q., Su, Y., Hu, T., Zhao, X., Wu, F., Li, Y., Liu, J., Chen, L., Xu, G., Lin, G., Zheng, Y., Lin, Y., Mi, X., Fei, L., & Wang, X. (2017). An integrated UAV-borne lidar system for 3D habitat mapping in three forest ecosystems across China. *International Journal of Remote Sensing, 38*(8–10), 2954–2972.

Guo, Q., Wu, F., Pang, S., Zhao, X., Chen, L., Liu, J., Xue, B., Xu, G., Li, L., Jing, H., & Chu, C. (2018). Crop 3D—a LiDAR based platform for 3D high-throughput crop phenotyping. *Science China Life Sciences, 61*(3), 328–339.

ten Harkel, J., Bartholomeus, H., & Kooistra, L. (2020). Biomass and crop height estimation of different crops using UAV-based lidar. *Remote Sensing, 12*(1).

Heun, J. T., Attalah, S., French, A. N., Lehner, K. R., McKay, J. K., Mullen, J. L., Ottman, M. J., & Andrade-Sanchez, P. (2019). Deployment of lidar from a ground platform: Customizing a low-cost, information-rich and user-friendly application for field phenomics research. *Sensors, 19*(24), 5358.

Höfle, B., Hollaus, M., & Hagenauer, J. (2012). Urban vegetation detection using radiometrically calibrated small-footprint full-waveform airborne LiDAR data. *ISPRS Journal of Photogrammetry and Remote Sensing, 67*, 134–147.

Huang, M.-J., Shyue, S.-w., Lee, L.-H., & Kao, C.-C. (2008). A knowledge-based approach to urban feature classification using aerial imagery with lidar data. *Photogrammetric Engineering & Remote Sensing, 74*, 1473–1485.

Hu, X., Li, Y., Shan, J., & Zhang, J. (2014). Road centerline extraction in complex urban scenes from LiDAR data based on multiple features. *IEEE Transactions on Geoscience and Remote Sensing, 52*, 7448–7456.

Hu, S., Niu, Z., Chen, Y., Li, L., & Zhang, H. (2017). Global wetlands: Potential distribution, wetland loss, and status. *Science of the Total Environment, 586*, 318–327.

Hu, T., Wei, D., Su, Y., Wang, X., Zhang, J., Sun, X., Liu, Y., & Guo, Q. (2022). Quantifying the shape of urban street trees and evaluating its influence on their aesthetic functions based on mobile lidar data. *ISPRS Journal of Photogrammetry and Remote Sensing, 184*, 203–214.

Hu, T., Zhang, Y., Su, Y., Zheng, Y., Lin, G., & Guo, Q. (2020). Mapping the global mangrove forest aboveground biomass using multisource remote sensing data. *Remote Sensing, 12*(10), 1690.

Jansen, B. V. S., Kolden, C. A., Greaves, H. E., & Eitel, J. U. H. (2019). Lidar provides novel insights into the effect of pixel size and grazing intensity on measures of spatial heterogeneity in a native bunchgrass ecosystem. *Remote Sensing of Environment, 235*, 111432.

Jansen, V. S., Kolden, C. A., & Schmalz, H. J. (2018). The development of near real-time biomass and cover estimates for adaptive rangeland management using Landsat 7 and Landsat 8 surface reflectance products. *Remote Sensing, 10*(7), 1057.

Jimenez-Berni, J. A., Deery, D. M., Rozas-Larraondo, P., Condon, A. G., Rebetzke, G. J., James, R. A., Bovill, W. D., Furbank, R. T., & Sirault, X. R. R. (2018). High throughput determination of plant height, ground cover, and above-ground biomass in wheat with LiDAR. *Frontiers in Plant Science, 9*, 237.

Jin, S., Su, Y., Gao, S., Wu, F., Hu, T., Liu, J., Li, W., Wang, D., Chen, S., Jiang, Y., Pang, S., & Guo, Q. (2018). Deep learning: Individual maize segmentation from terrestrial lidar data using faster R-CNN and regional growth algorithms. *Frontiers in Plant Science, 9*, 866.

Jin, S., Sun, X., Wu, F., Su, Y., Li, Y., Song, S., Xu, K., Ma, Q., Baret, F., Jiang, D., Ding, Y., & Guo, Q. (2021). Lidar sheds new light on plant phenomics for plant breeding and management: Recent advances and future prospects. *ISPRS Journal of Photogrammetry and Remote Sensing, 171*, 202–223.

Jin, S., Su, Y., Song, S., Xu, K., Hu, T., Yang, Q., Wu, F., Xu, G., Ma, Q., Guan, H., Pang, S., Li, Y., & Guo, Q. (2020). Non-destructive estimation of field maize biomass using terrestrial lidar: An evaluation from plot level to individual leaf level. *Plant Methods, 16*(1), 69.

Jin, S., Su, Y., Wu, F., Pang, S., Gao, S., Hu, T., Liu, J., & Guo, Q. (2019). Stem–leaf segmentation and phenotypic trait extraction of individual maize using terrestrial LiDAR data. *IEEE Transactions on Geoscience and Remote Sensing, 57*(3), 1336–1346.

Kang, L., Han, X., Zhang, Z., & Sun, O. J. (2007). Grassland ecosystems in China: Review of current knowledge and research advancement. *Philosophical Transactions of the Royal Society of London, 362*(1482), 997–1008.

Kato, A., Moskal, L. M., Schiess, P., Swanson, M. E., Calhoun, D., & Stuetzle, W. (2009). Capturing tree crown formation through implicit surface reconstruction using airborne lidar data. *Remote Sensing of Environment, 113*(6), 1148–1162.

Kidd, C., & Chapman, L. (2012). Derivation of sky-view factors from lidar data. *International Journal of Remote Sensing, 33*, 3640–3652.

Lin, Y. (2015). LiDAR: An important tool for next-generation phenotyping technology of high potential for plant phenomics? *Computers and Electronics in Agriculture, 119*, 61–73.

Lindberg, F., & Grimmond, C. S. B. (2011). Nature of vegetation and building morphology characteristics across a city: Influence on shadow patterns and mean radiant temperatures in London. *Urban Ecosystems, 14*(4), 617–634.

Li, W., Niu, Z., Shang, R., Qin, Y., Wang, L., & Chen, H. (2020). High-resolution mapping of forest canopy height using machine learning by coupling ICESat-2 LiDAR with Sentinel-1, Sentinel-2 and Landsat-8 data. *International Journal of Applied Earth Observation and Geoinformation, 92*, 102163.

Lymburner, L., Bunting, P., Lucas, R., Scarth, P., Alam, I., Phillips, C., Ticehurst, C., & Held, A. (2020). Mapping the multi-decadal mangrove dynamics of the Australian coastline. *Remote Sensing of Environment, 238*, 111185.

MacFaden, S., O'Neil-Dunne, J., Royar, A., Lu, J., & Rundle, A. (2012). High-resolution tree canopy mapping for New York City using LIDAR and object-based image analysis. *Journal of Applied Remote Sensing, 6*, 3567.

Madec, S., Frederic, B., de Solan, B., Thomas, S., Dutartre, D., Jézéquel, S., Hemmerlé, M., Colombeau, G., & Comar, A. (2017). High-throughput phenotyping of plant height: Comparing unmanned aerial vehicles and ground LiDAR estimates. *Frontiers in Plant Science, 8*, 2002.

Malambo, L., Popescu, S. C., Horne, D. W., Pugh, N. A., & Rooney, W. L. (2019). Automated detection and measurement of individual sorghum panicles using density-based clustering of terrestrial lidar data. *ISPRS Journal of Photogrammetry and Remote Sensing, 149*, 1–13.

Mao, D., Wang, Z., Du, B., Li, L., Tian, Y., Jia, M., Zeng, Y., Song, K., Jiang, M., & Wang, Y. (2020). National wetland mapping in China: A new product resulting from object-based and hierarchical classification of Landsat 8 oli images. *ISPRS Journal of Photogrammetry and Remote Sensing, 164*, 11–25.

Marcinkowska-Ochtyra, A., Jarocińska, A., Bzdega, K., & Tokarska-Guzik, B. (2018). Classification of expansive grassland species in different growth stages based on hyperspectral and LiDAR data. *Remote Sensing, 10*(12), 2019.

Mermoz, S., Le Toan, T., Villard, L., Réjou-Méchain, M., & Seifert-Granzin, J. (2014). Biomass assessment in the Cameroon savanna using ALOS PALSAR data. *Remote Sensing of Environment, 155*, 109—119.

Mitchell, J., Glenn, N., Sankey, T., Derryberry, W. R., Anderson, M., & Hruska, R. (2011). Small-footprint lidar estimations of sagebrush canopy characteristics. *Photogrammetric Engineering and Remote Sensing, 77*, 521—530.

Moeslund, J. E., Zlinszky, A., Ejrnæs, R., Brunbjerg, A. K., Bøcher, P. K., Svenning, J.-C., & Normand, S. (2019). Light detection and ranging explains diversity of plants, fungi, lichens, and bryophytes across multiple habitats and large geographic extent. *Ecological Applications, 29*(5), e01907.

Mohamad, O., Suhaimi, O., & Abdullah, M. Z. (1994). The relationships between harvest index, grain yield and biomass in rice. *Journal of Tropical Agriculture and Food Science, 22*, 29—34.

Niemeyer, J., Wegner, J. D., Mallet, C., Rottensteiner, F., & Soergel, U. (2011). Conditional random fields for urban scene classification with full waveform LiDAR data. In U. Stilla, F. Rottensteiner, H. Mayer, B. Jutzi, & M. Butenuth (Eds.), *Photogrammetric image analysis* (pp. 233—244). Springer Berlin Heidelberg.

O'Neil-Dunne, J. P. M., MacFaden, S. W., Royar, A. R., & Pelletier, K. C. (2013). An object-based system for LiDAR data fusion and feature extraction. *Geocarto International, 28*(3), 227—242.

Ortega, S., Trujillo, A., Santana, J. M., Suárez, J. P., & Santana, J. (2019). Characterization and modeling of power line corridor elements from LiDAR point clouds. *ISPRS Journal of Photogrammetry and Remote Sensing, 152*, 24—33.

Owers, C. J., Rogers, K., & Woodroffe, C. D. (2016). Identifying spatial variability and complexity in wetland vegetation using an object-based approach. *International Journal of Remote Sensing, 37*(18), 4296—4316.

Owers, C. J., Rogers, K., & Woodroffe, C. D. (2018). Terrestrial laser scanning to quantify above-ground biomass of structurally complex coastal wetland vegetation. *Estuarine, Coastal and Shelf Science, 204*, 164—176.

Pecero-Casimiro, R., Fernández-Rodríguez, S., Tormo-Molina, R., Monroy-Colín, A., Silva-Palacios, I., Cortés-Pérez, J. P., Gonzalo-Garijo, Á., & Maya-Manzano, J. M. (2019). Urban aerobiological risk mapping of ornamental trees using a new index based on LiDAR and kriging: A case study of plane trees. *Science of the Total Environment, 693*, 133576.

Perez-Sanz, F., Navarro, P. J., & Egea-Cortines, M. (2017). Plant phenomics: An overview of image acquisition technologies and image data analysis algorithms. *GigaScience, 6*(11), gix092.

Plowright, A. A., Coops, N. C., Chance, C. M., Sheppard, S. R. J., & Aven, N. W. (2017). Multi-scale analysis of relationship between imperviousness and urban tree height using airborne remote sensing. *Remote Sensing of Environment, 194*, 391—400.

Pound, M. P., French, A. P., Atkinson, J. A., Wells, D. M., Bennett, M. J., & Pridmore, T. P. (2013). RootNav: Navigating images of complex root architectures. *Plant Physiology, 162*, 1802—1814.

Reddersen, B., Fricke, T., & Wachendorf, M. (2014). A multi-sensor approach for predicting biomass of extensively managed grassland. *Computers and Electronics in Agriculture, 109*, 247—260.

Richardson, J. J., Moskal, L. M., & Kim, S.-H. (2009). Modeling approaches to estimate effective leaf area index from aerial discrete-return LIDAR. *Agricultural and Forest Meteorology, 149*(6), 1152—1160.

Rose, L. S., & Levinson, R. (2013). Analysis of the effect of vegetation on albedo in residential areas: Case studies in suburban sacramento and los angeles, CA. *GIScience & Remote Sensing, 50*(1), 64−77.

Samadzadegan, F., Bigdeli, B., & Ramzi, P. (2010). A multiple classifier system for classification of LIDAR remote sensing data using multi-class SVM. *Proceedings of the 9th International Conference on Multiple Classifier Systems*. Cairo, Egypt.

Schaefer, M. T., & Lamb, D. W. (2016). A combination of plant NDVI and LiDAR measurements improve the estimation of pasture biomass in tall fescue (*Festuca arundinacea* var. Fletcher). *Remote Sensing, 8*(2), 109.

Schöpfer, E., Lang, S., & Blaschke, T. (2005). A Green index incorporating remote sensing and citizen's perception of green space. *International Archives of Photogramm., Remote Sensing and Spatial Information Sciences, 37*, 1−6.

Schulze-Brüninghoff, D., Hensgen, F., Wachendorf, M., & Astor, T. (2019). Methods for LiDAR-based estimation of extensive grassland biomass. *Computers and Electronics in Agriculture, 156*, 693−699.

Simard, M., Fatoyinbo, L., Smetanka, C., Rivera-Monroy, V. H., Castañeda-Moya, E., Thomas, N., & Van der Stocken, T. (2019). Mangrove canopy height globally related to precipitation, temperature and cyclone frequency. *Nature Geoscience, 12*(1), 40.

Simard, M., Zhang, K., Rivera-Monroy, V. H., Ross, M. S., Ruiz, P. L., Castañeda-Moya, E., Twilley, R. R., & Rodriguez, E. (2006). Mapping height and biomass of mangrove forests in Everglades National Park with SRTM elevation data. *Photogrammetric Engineering & Remote Sensing, 72*(3), 299−311.

Soilán, M., Riveiro, B., Martínez-Sánchez, J., & Arias, P. (2016). Traffic sign detection in MLS acquired point clouds for geometric and image-based semantic inventory. *ISPRS Journal of Photogrammetry and Remote Sensing, 114*, 92−101.

Streutker, D. R., & Glenn, N. F. (2006). LiDAR measurement of sagebrush steppe vegetation heights. *Remote Sensing of Environment, 102*(1), 135−145.

Susaki, J., Komiya, Y., & Takahashi, K. (2014). Calculation of enclosure index for assessing urban landscapes using digital surface models. *IEEE Journal of Selected Topics in Applied Earth Observations and Remote Sensing, 7*, 4038−4045.

Su, Y., Wu, F., Ao, Z., Jin, S., Qin, F., Liu, B., Pang, S., Liu, L., & Guo, Q. (2019). Evaluating maize phenotype dynamics under drought stress using terrestrial lidar. *Plant Methods, 15*(1), 11.

Su, W., Zhan, J., Zhang, M., Wu, D., & Zhang, R. (2016). Estimation method of crop leaf area index based on airborne LiDAR data. *Transactions of the Chinese Society for Agricultural Machinery, 47*(3), 272−277 (in Chinese).

Szantoi, Z., Escobedo, F., Abd-Elrahman, A., Smith, S., & Pearlstine, L. (2013). Analyzing fine-scale wetland composition using high resolution imagery and texture features. *International Journal of Applied Earth Observation and Geoinformation, 23*, 204−212.

Takahashi, T., Awaya, Y., Hirata, Y., Furuya, N., Sakai, T., & Sakai, A. (2010). Stand volume estimation by combining low laser-sampling density LiDAR data with QuickBird panchromatic imagery in closed-canopy Japanese cedar (*Cryptomeria japonica*) plantations. *International Journal of Remote Sensing, 31*(5), 1281−1301.

Tao, H., Xu, S., Tian, Y., Li, Z., Ge, Y., Zhang, J., Wang, Y., Zhou, G., Deng, X., Zhang, Z., Ding, Y., Jiang, D., Guo, Q., & Jin, S. (2022). Proximal and remote sensing in plant phenomics: Twenty years of progress, challenges and perspectives. *Plant Communications*, 100344.

Tian, Y., Zhou, W., Qian, Y., Zheng, Z., & Yan, J. (2019). The effect of urban 2D and 3D morphology on air temperature in residential neighborhoods. *Landscape Ecology, 34*(5), 1161−1178.

Tooke, T. R., van der Laan, M., & Coops, N. C. (2014). Mapping demand for residential building thermal energy services using airborne LiDAR. *Applied Energy, 127*, 125−134.

Walter, J. D. C., Edwards, J., McDonald, G., & Kuchel, H. (2019). Estimating biomass and canopy height with LiDAR for field crop breeding. *Frontiers in Plant Science, 10*, 1145.

Wang, D., Wan, B., Liu, J., Su, Y., Guo, Q., Qiu, P., & Wu, X. (2020). Estimating aboveground biomass of the mangrove forests on northeast Hainan Island in China using an upscaling method from field plots, UAV-LiDAR data and Sentinel-2 imagery. *International Journal of Applied Earth Observation and Geoinformation, 85*, 101986.

Wang, H., Zhang, W., Zhou, G., Yan, G., & Clinton, N. (2009). Image-based 3D corn reconstruction for retrieval of geometrical structural parameters. *International Journal of Remote Sensing, 30*(20), 5505−5513.

Wan, L., Lin, Y., Zhang, H., Wang, F., Liu, M., & Lin, H. (2020). GF-5 hyperspectral data for species mapping of mangrove in Mai Po, Hong Kong. *Remote Sensing, 12*(4), 656.

Wijesingha, J., Moeckel, T., Hensgen, F., & Wachendorf, M. (2019). Evaluation of 3D point cloud-based models for the prediction of grassland biomass. *International Journal of Applied Earth Observation and Geoinformation, 78*, 352−359.

Xiao, W., Vallet, B., Schindler, K., & Paparoditis, N. (2016). Street-side vehicle detection, classification and change detection using mobile laser scanning data. *ISPRS Journal of Photogrammetry and Remote Sensing, 114*, 166−178.

Xia, S., & Wang, R. (2018). Extraction of residential building instances in suburban areas from mobile LiDAR data. *ISPRS Journal of Photogrammetry and Remote Sensing, 144*, 453−468.

Xiong, X., Duan, L., Liu, L., Tu, H., Yang, P., Wu, D., Chen, G., Xiong, L., Yang, W., & Liu, Q. (2017). Panicle-SEG: A robust image segmentation method for rice panicles in the field based on deep learning and superpixel optimization. *Plant Methods, 13*(1), 104.

Xu, R., Li, C., Paterson, A. H., Jiang, Y., Sun, S., & Robertson, J. S. (2018). Aerial images and convolutional neural network for cotton bloom detection. *Frontiers in Plant Science, 8*, 2235.

Xu, K., Su, Y., Liu, J., Hu, T., Jin, S., Ma, Q., Zhai, Q., Wang, R., Zhang, J., Li, Y., Liu, H., & Guo, Q. (2020). Estimation of degraded grassland aboveground biomass using machine learning methods from terrestrial laser scanning data. *Ecological Indicators, 108*, 105747.

Xu, S., Xu, S., Ye, N., & Zhu, F. (2018). Automatic extraction of street trees' non-photosynthetic components from MLS data. *International Journal of Applied Earth Observation and Geoinformation, 69*, 64−77.

Yang, W., Duan, L., Chen, G., Xiong, L., & Liu, Q. (2013). Plant phenomics and high-throughput phenotyping: Accelerating rice functional genomics using multidisciplinary technologies. *Current Opinion in Plant Biology, 16*(2), 180−187.

Yang, J., & Kang, Z. (2018). Voxel-based extraction of transmission lines from airborne LiDAR point cloud data. *IEEE Journal of Selected Topics in Applied Earth Observations and Remote Sensing, 11*(10), 3892−3904.

Yan, W. Y., Shaker, A., & El-Ashmawy, N. (2015). Urban land cover classification using airborne LiDAR data: A review. *Remote Sensing of Environment, 158*, 295−310.

Yasrab, R., Atkinson, J. A., Wells, D. M., French, A. P., Pridmore, T. P., & Pound, M. P. (2019). RootNav 2.0: Deep learning for automatic navigation of complex plant root architectures. *GigaScience, 8*(11).

Yin, D., & Wang, L. (2019). Individual mangrove tree measurement using UAV-based LiDAR data: Possibilities and challenges. *Remote Sensing of Environment, 223*, 34−49.

Yu, B., Liu, H., Wu, J., & Lin, W.-M. (2009). Investigating impacts of urban morphology on spatio-temporal variations of solar radiation with airborne LiDAR data and a solar flux model: A case study of downtown houston. *International Journal of Remote Sensing, 30*(17), 4359−4385.

Yu, S., Yu, B., Song, W., Wu, B., Zhou, J., Huang, Y., Wu, J., Zhao, F., & Mao, W. (2016). View-based greenery: A three-dimensional assessment of city buildings' green visibility using floor green view index. *Landscape and Urban Planning, 152*, 13—26.

Zhang, K., Chen, S.-C., Whitman, D., Shyu, M.-L., Yan, J., & Zhang, C. (2003). A progressive morphological filter for removing nonground measurements from airborne LIDAR data. *IEEE Transactions on Geoscience and Remote Sensing, 41*(4), 872—882.

Zhang, C., Zhou, Y., & Qiu, F. (2015). Individual tree segmentation from LiDAR point clouds for urban forest inventory. *Remote Sensing, 7*(6), 7892—7913.

Zhao, Y., Hu, Q., Li, H., Wang, S., & Ai, M. (2018). Evaluating carbon sequestration and PM2.5 removal of urban street trees using mobile laser scanning data. *Remote Sensing, 10*(11), 1759.

Zhao, Q., Myint, S. W., Wentz, E. A., & Fan, C. (2015). Rooftop surface temperature analysis in an urban residential environment. *Remote Sensing, 7*(9), 12135—12159.

Zhao, X., Su, Y., Hu, T., Cao, M., Liu, X., Yang, Q., Guan, H., Liu, L., & Guo, Q. (2022). Analysis of UAV lidar information loss and its influence on the estimation accuracy of structural and functional traits in a meadow steppe. *Ecological Indicators, 135*, 108515.

Zhou, J., & Sun, T. (1995). Study on remote sensing model of three-dimensional green biomass and the estimation of environmental benefits of greenery. *Remote Sensing of Environment (China), 10*(3), 162—174 (in Chinese).

Zhu, Y., Liu, K., Liu, L., Wang, S., & Liu, H. (2015). Retrieval of mangrove aboveground biomass at the individual species level with worldview-2 images. *Remote Sensing, 7*(9), 12192—12214.

Zlinszky, A., Schroiff, A., Kania, A., Deák, B., Mücke, W., Vári, Á., Székely, B., & Pfeifer, N. (2014). Categorizing grassland vegetation with full-waveform airborne laser scanning: A feasibility study for detecting natura 2000 habitat types. *Remote Sensing, 6*(9), 8056—8087.

CHAPTER 15

Challenges and Opportunities for LiDAR

Contents

Since its emergence in the 1960s, light detection and ranging (LiDAR) has been widely applied in various fields, including forestry and the construction industry, because of its unique advantages over traditional optical measurement. These include high precision, spatial and temporal resolution, and resistance to interference such as clouds, as LiDAR does not rely on solar radiation. With the development of technology, new and upgraded LiDAR hardware is emerging, such as solid-state LiDAR. The availability and accumulation of LiDAR data acquired from different platforms, together with other remote sensing data, pose new challenges for data fusion, processing, and information extraction. However, LiDAR also provides numerous new opportunities in the era of big data. LiDAR applications in forest ecological studies will be promoted with hardware development and the fusion of multisource data in the future.

15.1 LiDAR Hardware Development

Light detection and ranging (LiDAR) techniques for forest applications are developing rapidly. The capacity to acquire high–accuracy three-dimensional (3-D) information on terrain and forest structures has opened up new ways to investigate forest ecosystems. Although LiDAR has broadly served forest applications over the past two decades, the costs of LiDAR systems remain high. This often prevents them from being used in

LiDAR Principles, Processing and Applications in Forest Ecology
ISBN 978-0-12-823894-3
https://doi.org/10.1016/B978-0-12-823894-3.00015-3

large-scale forest applications. Recent advances in sensor electronics are making LiDAR more affordable.

Traditional LiDAR systems are mechanical—they rely on moving parts to rotate the laser scanner to obtain a wide field of view (FoV). These moving parts limit the minimum size of a LiDAR system because making small and compact systems greatly increases the requirements for precision manufacturing, driving up the cost (Li & Ibanez-Guzman, 2020). Solid-state LiDAR is a newly developed technique that integrates the entirely LiDAR on a silicon chip. To conduct a scan, solid-state LiDAR uses a concept similar to phased-array radar to achieve an optical phased array. An optical phase modulator sends out bursts of photons in specific patterns and phases to create directional emission, the focus and size of which can be adjusted without mechanical parts (Raj et al., 2020). Therefore, solid-state LiDAR is also called optical phased-array LiDAR (Fig. 15.1). Solid-state LiDAR has many benefits because mechanical parts are eliminated. Its small size makes the production cost lower, and avoiding physical adjustments to the laser sensor makes it resistant to vibrations, giving it better scan accuracy. Beyond cost and accuracy, solid-state LiDAR scans much faster, allowing a high pulse repetition rate at the MHz level. Currently, many solid-state LiDAR developments are focused on applications in autonomous vehicles. New research has validated their performance in forest mapping. The result shows that a very low-cost unmanned aerial vehicle (UAV) LiDAR system that integrates the recently released DJI Livox Mid40 laser scanner ($\sim$600 USD) (DJI Technology, Shenzhen,

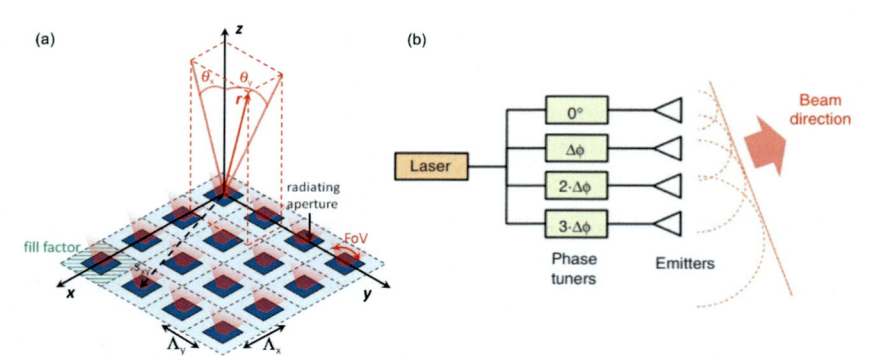

Figure 15.1 Schematic diagram of solid-state light detection and ranging (LiDAR) scanning based on waveguide grating antennas (a) and end-fire structure (b). *(From Guo, Y., Guo, Y., Li, C., Zhang, H., Zhou, X., & Zhang, L. (2021). Integrated optical phased arrays for beam forming and steering. Applied Sciences, 11(9), 4017.)*

Guangdong Province, China) can derive forest inventory attributes with an accuracy equivalent to field measurements (Hu et al., 2021). Overall, the development of solid-state LiDAR can substantially expand the use of LiDAR in forest applications by overcoming the major obstacles of traditional LiDAR systems.

Compared with traditional LiDAR, which emits a single laser beam with relatively high energy, a new technology—single-photon counting—has emerged in recent years. This uses detectors sensitive to low-energy photons and provides a new way to acquire point cloud data (Degnan, 2016). The single-photon LiDAR uses plane array detectors to emit and receive multiple photons with relatively lower energy, thereby obtaining higher point density and richer information about targeted objects. The high sensitivity of the single-photon receiver to low-intensity signals makes single-photon LiDAR much lighter and cheaper than traditional LiDAR (Pawlikowska et al., 2017). In 2016, the United States Geological Survey (USGS) evaluated the performance of single-photon LiDAR mounted on an airplane for topographic surveying. They stated that airborne single-photon LiDAR could make data collection faster (up to 30 times the data collection efficiency of traditional LiDAR) and cheaper, and the captured point cloud data could fulfill USGS requirements for point density and positioning accuracy (Fig. 15.2) (Stoker et al., 2016). In addition, single-photon LiDAR can be operated from high altitudes owing to its high receiver sensitivity to low-energy photons. For instance, ICESat-2 (Ice, Cloud, and land Elevation Satellite-2) uses a low-energy 532-nm single-photon laser to measure ice cap elevation, vegetation height, and sea surface height (Markus et al., 2017).

Another notable advance in hardware in the last decade is the introduction of multispectral and hyper-spectral LiDAR for forest observations. They combine accurate 3-D position information and spectral information at two or more wavelengths, which overcomes the weakness of current LiDAR in specific applications such as leaf—wood separation because of the lack of spectral information (Fig. 15.3) (Wallace et al., 2014). Their potential in forest applications has been demonstrated for tree species classification and pest and disease monitoring. However, mature commercial products are still not available. Current multispectral and hyperspectral LiDAR systems have been developed mostly in laboratories and face various challenges before they can be widely used. For instance, their calibration is still an issue, especially for incidence angles and range effects (Okhrimenko et al., 2019). Another challenge is the selection of wavelengths (Sun et al., 2019). Further studies are needed to choose a suitable

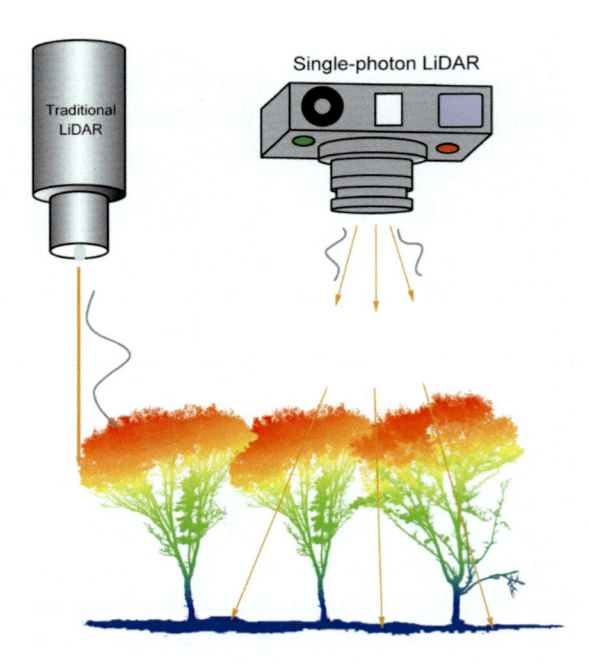

Figure 15.2 Comparison of single-photon and traditional LiDAR.

number of bands, and the corresponding central wavelength and spectral resolution is a trade-off between aims and costs.

15.2 From the Three-Dimensional to Multidimensional Big Data Era

Over the last 20 years, various LiDAR systems have been used for forest applications in multiple temporal, spatial, and spectral dimensions. Together, these dimensions of LiDAR data have fundamentally changed the way we observe and describe forest ecosystems. However, emerging data demands are increasingly expanding LiDAR-based forest applications from 3-D to multiple dimensions to compensate for the weaknesses of single-type or single-date remote sensing date (Guo, Su, et al., 2021).

15.2.1 Toward the Fusion of Structural and Spectral Information

Despite their ability to provide accurate structural information, most LiDAR systems cannot acquire the spectral information of an object. As new and frequently more complex questions emerge in forest ecology,

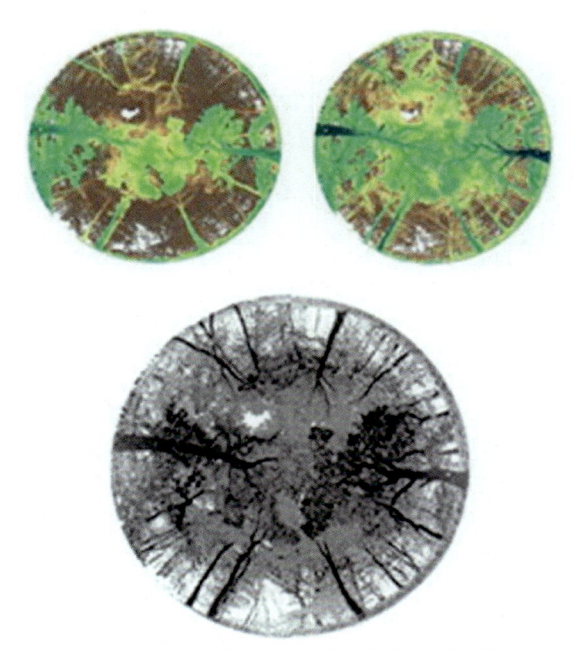

Figure 15.3 Schematic diagram of the point cloud data of a forest acquired by multispectral LiDAR. *(From Danson, F. M., Gaulton, R., Armitage, R. P., Disney, M., Gunawan, O., Lewis, P., Pearson, G., & Ramirez, A. F. (2014). Developing a dual-wavelength full-waveform terrestrial laser scanner to characterize forest canopy structure.* Agricultural and Forest Meteorology, 198, 7—14.)

researchers need to combine the structural and spectral information of forests to explore and answer these questions (Su et al., 2016). Recent developments in multispectral and hyper-spectral LiDAR have provided a potential solution to address the lack of spectral information. Yet, the absence of reliable wavelength selection and intensity calibration methods poses challenges for their practical application, as mentioned above. Fusing LiDAR data with other remote sensing data, such as optical, thermal, and fluorescence data, is another solution that has gained broad attention recently (Sankey et al., 2017). This allows the 3-D physiological properties of forests to be quantified and may revolutionize our understanding of the links between forest structure, functions, and ecosystem processes (Hosoi et al., 2019).

Another important direction for future research is how to merge the LiDAR data acquired from different platforms, because the occlusion in dense forests may result in incomplete data and thus affect the extraction of

Figure 15.4 Schematic diagram of the fusion of LiDAR data from different platforms.

vegetation structural information if using a single LiDAR system (Fig. 15.4) (Donager et al., 2021). For example, down-looking airborne and UAV LiDAR systems can provide highly accurate tree canopy information but lack tree trunk information. Meanwhile, backpack LiDAR and terrestrial laser scanning (TLS) LiDAR working under the canopy can provide detailed tree trunk information, but the limited vertical FoV and measurement range may result in the omission of upper canopy information (Guan, Su, Hu, et al., 2020). Multiplatform collaborative observations of a forest ecosystem are the only way to obtain a comprehensive and accurate quantification of forest structural features. However, different LiDAR platforms differ greatly in point density, laser range, and data accuracy (Fig. 15.5). Furthermore, forest environments have much higher complexity and irregularity than indoor and urban environments. Thus, it is hard to find robust registration features to automatically fuse multiplatform LiDAR data in forest environments. In recent years, studies have shown that tree locations are solid features to coarsely register LiDAR data from different platforms (Guan, Su, Sun, et al., 2020). However, further studies are needed on achieving high-accuracy fine registration, especially when there is only a small overlap between LiDAR data from different platforms.

15.2.2 Toward Regular Time-Series Observations

In recent years, the increased availability of LiDAR data has led to a growing number of studies using temporal LiDAR to better understand and monitor forest dynamics. For instance, multitemporal LiDAR metrics have paved the way for mapping changes in forest biomass and carbon stocks, which deepen our understanding of how forests respond to rapid global change and help develop management strategies (Srinivasan et al., 2014). Repeated TLS has provided an unprecedented opportunity in phenology

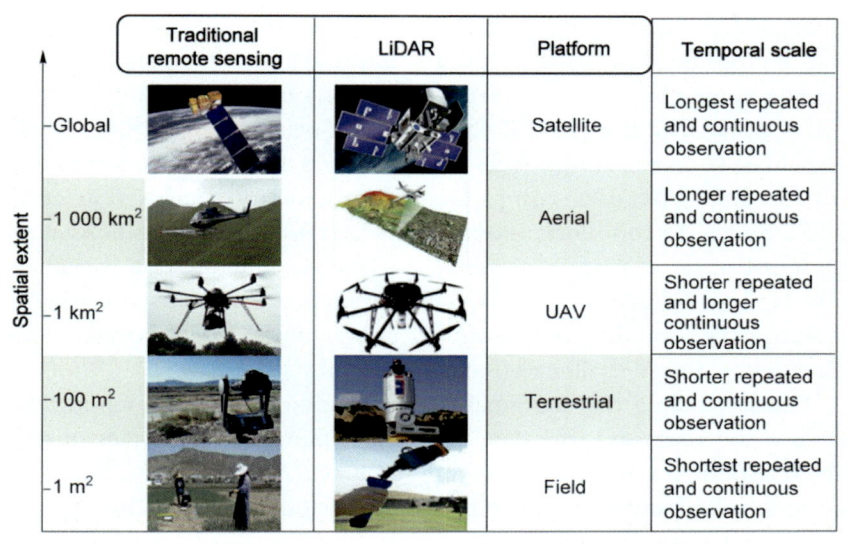

Figure 15.5 Different characteristics of multiplatform LiDAR data.

studies to explore the phenological differences between understory and canopy trees, trees with different functions, and canopy dominants and emergent trees (Calders et al., 2015). Moreover, the growing number of spaceborne LiDAR missions, e.g., the newly launched GEDI (Global Ecosystem Dynamics Investigation) and ICESat-2, can provide global-scale time-series LiDAR observations. The continuous accumulation of data will bring new opportunities for global-scale forest research.

15.2.3 Toward the Era of Big Data

With the accumulation of multiplatform and multitemporal LiDAR data, LiDAR remote sensing of forest ecosystems has entered the era of big data, facing unprecedented challenges and opportunities.

First, standardized datasets need to be established. In recent years, the building of well-labeled datasets has promoted the development of data processing algorithms. However, current open-source datasets mainly focus on image data. To both improve the accuracy and efficiency of LiDAR data processing algorithms and translate LiDAR data into ecological knowledge, there is a need for LiDAR databases that are complemented by ecological observations (e.g., phenology and carbon flux) (Liang et al., 2018). These developments can reduce the cost and difficulty associated with LiDAR-based forest studies.

Second, integrating big multisource data for ecological cognition creates new demand for data analysis tools. Deep learning has shown advantages in big data mining because it can extract spatiotemporal features automatically from massive data (Jin et al., 2020). A simple neural network has recently been used to map country-scale canopy height from massive spaceborne and UAV LiDAR data, complemented by ancillary datasets (Liu et al., 2022). Similar to this work, most current deep learning applications are in the initial stages of mapping ecological variables. The question remains of how big multisource data can be used to explore fundamental ecological questions. The coupling of data-driven and physically based process models is becoming a new challenge (Reichstein et al., 2019).

Third, handling the continually accumulating big remote sensing data needs to be addressed. Big data involves manipulating petabytes of data, which requires immense infrastructure resources (Alharthi et al., 2017). The recent development of cloud computing and cloud storage techniques can process and store massive amounts of data from a cloud service provider anywhere and anytime (e.g., Google Earth Engine), which may provide an effective solution. However, cloud-based platforms for processing and storing big LiDAR data are still in the early stage (e.g., LiEarth from GreenValley International, Inc.)

15.3 Prospects for LiDAR Applications in Forest Ecosystems

LiDAR has emerged as a robust means to measure forest structure metrics. Numerous studies have been devoted to accurately quantifying forest structures from LiDAR data at various scales (from individual trees to regional and global scales) and have revolutionized the way we consider forest structure in ecosystem studies. In addition to deriving structure metrics for forest ecosystem studies, LiDAR data can also be used to drive radiative transfer (RT) models to understand many ecological processes, such as the interaction between light and forest ecosystems and exchanges of energy, carbon, and water between the biosphere and atmosphere. RT models provide a practical and reliable approach to understanding the critical link between 3-D forest scenes and the observed remote sensing signals (Calders et al., 2018). Highly detailed, realistic 3-D forest scenes, together with 3-D RT models, can serve as a virtual laboratory to simulate the observed remote sensing signals of forests based on precisely known and controlled structural and biochemical parameters. This facilitates the testing of various assumptions about ecological processes (Widlowski et al., 2015). Normally, the 3-D scenes are constructed using a set of simple geometric primitives such as triangles, discs, cones, spheres, cylinders, and ellipsoids (Kobayashi & Iwabuchi, 2008). It is worth noting that sometimes one has to adopt these simplified geometric primitives because of the lack of necessary data to reconstruct realistic 3-D scenes, although the architecture of 3-D scenes can significantly affect the performance of RT models. The application of LiDAR provides a promising approach to reconstructing realistic scenes from point clouds and then parameterizing 3-D RT models. Approaches like the voxel-based RT model use very fine-resolution voxels derived from high-density LiDAR point cloud data to represent 3-D forest scenes without simplifying the architecture of the trees (Li et al., 2018). Up to now, RT models have provided good support for many forest ecology studies, such as testing the possible biophysical mechanisms for the green-up of forests (Morton et al., 2014), simulating the transport of energy, absorption of photosynthetically active radiation, and gross primary productivity (Van Leeuwen et al., 2015), and understanding the relationship among heterogeneous vegetation structure, radiation, and snowmelt (Ni-Meister & Gao, 2011).

Moreover, LiDAR provides a new perspective for exploring and solving traditional forest ecology problems. For example, forest canopy height is an important indicator for many ecosystem functions. However, quantitative descriptions of the determinants underlying the global pattern of forest canopy height have rarely been documented. In 2016, a study used forest canopy height derived from spaceborne LiDAR data, climate indices, and field measurements of the world's giant trees to evaluate the global patterns and determinants of forest canopy height (Tao et al., 2016). Their results showed that the canopy height exhibited a hump-shaped curve along a gradient of precipitation minus potential evapotranspiration. Moreover, based on a new forest canopy height dataset for China (Liu et al., 2022), the authors' research team conducted field surveys to search for giant trees there. With the help of backpack and UAV LiDAR systems, we found a 76.8-meter-tall *Pinus bhutanica* tree in Medog, Tibet, China, in May 2022 and collected high-precision 3-D point cloud data (Fig. 15.6). The height and structure of individual trees and the community were well captured in the point cloud data. Then an 82.6-meter-tall *Abies ernestii* var. *salouenensis* tree was found in Zayu, Tibet—the tallest known tree in China. These findings can greatly deepen our understanding of the climatic controls of

Figure 15.6 The point cloud of a 76.8-meter-tall *Pinus bhutanica* tree in Medog, Tibet, China.

the world's giant trees and help us to make forest management policies to protect these intact forests.

15.4 Chapter Summary

LiDAR provides a powerful tool for forest observations and ecological modeling. With current LiDAR hardware and processing algorithms still undergoing rapid development, LiDAR remote sensing will play an increasing role in forest applications in the future. Up to now, most studies have focused on LiDAR-derived structural information, and the full potential of LiDAR for forest applications has not yet been realized. Because remote sensing is moving into the era of big data, understanding how to manage and link the structural, temporal, and spectral information extracted from multisource data is becoming a new challenge but also opens up new opportunities to elucidate forest ecological processes.

References

Alharthi, A., Krotov, V., & Bowman, M. (2017). Addressing barriers to big data. *Business Horizons, 60*(3), 285−292.

Calders, K., Origo, N., Burt, A., Disney, M., Nightingale, J., Raumonen, P., Akerblom, M., Malhi, Y., & Lewis, P. (2018). Realistic forest stand reconstruction from terrestrial LiDAR for radiative transfer modelling. *Remote Sensing, 10*(6). Article 933.

Calders, K., Schenkels, T., Bartholomeus, H., Armston, J., Verbesselt, J., & Herold, M. (2015). Monitoring spring phenology with high temporal resolution terrestrial LiDAR measurements. *Agricultural and Forest Meteorology, 203*, 158−168.

Danson, F. M., Gaulton, R., Armitage, R. P., Disney, M., Gunawan, O., Lewis, P., Pearson, G., & Ramirez, A. F. (2014). Developing a dual-wavelength full-waveform terrestrial laser scanner to characterize forest canopy structure. *Agricultural and Forest Meteorology, 198*, 7−14.

Degnan, J. J. (2016). Scanning, multibeam, single photon lidars for rapid, large scale, high resolution, topographic and bathymetric mapping. *Remote Sensing, 8*(11). Article 958.

Donager, J. J., Meador, A. J. S., & Blackburn, R. C. (2021). Adjudicating perspectives on forest structure: How do airborne, terrestrial, and mobile lidar-derived estimates compare? *Remote Sensing, 13*(12), 2297.

Guan, H. C., Su, Y. J., Hu, T. Y., Wang, R., Ma, Q., Yang, Q. L., Sun, X. L., Li, Y. M., Jin, S. C., Zhang, J., Liu, M., Wu, F. Y., & Guo, Q. H. (2020). A novel framework to automatically fuse multiplatform LiDAR data in forest environments based on tree locations. *Ieee Transactions on Geoscience and Remote Sensing, 58*(3), 2165−2177.

Guan, H. C., Su, Y. J., Sun, X. L., Xu, G. C., Li, W. K., Ma, Q., Wu, X. Y., Wu, J., Liu, L. L., & Guo, Q. H. (2020). A marker-free method for registering multi-scan terrestrial laser scanning data in forest environments. *Isprs Journal of Photogrammetry and Remote Sensing, 166*, 82−94.

Guo, Y., Guo, Y., Li, C., Zhang, H., Zhou, X., & Zhang, L. (2021). Integrated optical phased arrays for beam forming and steering. *Applied Sciences, 11*(9).

Guo, Q. H., Su, Y. J., Hu, T. Y., Guan, H. C., Jin, S. C., Zhang, J., Zhao, X. X., Xu, K. X., Wei, D. J., Kelly, M. G., & Coops, N. C. (2021). Lidar boosts 3D ecological observations and modelings: A review and perspective. *Ieee Geoscience and Remote Sensing Magazine, 9*(1), 232−257.

Hosoi, F., Umeyama, S., & Kuo, K. T. (2019). Estimating 3D chlorophyll content distribution of trees using an image fusion method between 2D camera and 3D portable scanning lidar. *Remote Sensing, 11*(18). Article 2134.

Hu, T. Y., Sun, X. L., Su, Y. J., Guan, H. C., Sun, Q. H., Kelly, M., & Guo, Q. H. (2021). Development and performance evaluation of a very low-cost UAV-lidar system for forestry applications. *Remote Sensing, 13*(1). Article 77.

Jin, S. C., Sun, Y. J., Zhao, X. Q., Hu, T. Y., & Guo, Q. H. (2020). A point-based fully convolutional neural network for airborne LiDAR ground point filtering in forested environments. *IEEE Journal of Selected Topics in Applied Earth Observations and Remote Sensing, 13*, 3958−3974.

Kobayashi, H., & Iwabuchi, H. (2008). A coupled 1-D atmosphere and 3-D canopy radiative transfer model for canopy reflectance, light environment, and photosynthesis simulation in a heterogeneous landscape. *Remote Sensing of Environment, 112*(1), 173−185.

Liang, X. L., Hyyppa, J., Kaartinen, H., Lehtomaki, M., Pyorala, J., Pfeifer, N., Holopainen, M., Brolly, G., Pirotti, F., Hackenberg, J., Huang, H. B., Jo, H. W., Katoh, M., Liu, L. X., Mokros, M., Morel, J., Olofsson, K., Poveda-Lopez, J., Trochta, J., Wang, D., Wang, J. H., Xi, Z. X., Yang, B. S., Zheng, G., Kankare, V., Luoma, V., Yu, X. W., Chen, L., Vastaranta, M., Saarinen, N., & Wang, Y. S. (2018). International benchmarking of terrestrial laser scanning approaches for forest inventories. *ISPRS Journal of Photogrammetry and Remote Sensing, 144*, 137−179.

Li, W. K., Guo, Q. H., Tao, S. L., & Su, Y. J. (2018). Vbrt: A novel voxel-based radiative transfer model for heterogeneous three-dimensional forest scenes. *Remote Sensing of Environment, 206*, 318−335.

Li, Y., & Ibanez-Guzman, J. (2020). Lidar for autonomous driving: The principles, challenges, and trends for automotive lidar and perception systems. *IEEE Signal Processing Magazine, 37*(4), 50−61.

Liu, X. Q., Su, Y. J., Hu, T. Y., Yang, Q. L., Liu, B. B., Deng, Y. F., Tang, H., Tang, Z. Y., Fang, J. Y., & Guo, Q. H. (2022). Neural network guided interpolation for mapping canopy height of China's forests by integrating GEDI and ICESat-2 data. *Remote Sensing of Environment, 269*. Article 112844.

Markus, T., Neumann, T., Martino, A., Abdalati, W., Brunt, K., Csatho, B., Farrell, S., Fricker, H., Gardner, A., Harding, D., Jasinski, M., Kwok, R., Magruder, L., Lubin, D., Luthcke, S., Morison, J., Nelson, R., Neuenschwander, A., Palm, S., Popescu, S., Shum, C. K., Schutz, B. E., Smith, B., Yang, Y. K., & Zwally, J. (2017). The ice, cloud, and land elevation satellite-2 (ICESat-2): Science requirements, concept, and implementation. *Remote Sensing of Environment, 190,* 260–273.

Morton, D. C., Nagol, J., Carabajal, C. C., Rosette, J., Palace, M., Cook, B. D., Vermote, E. F., Harding, D. J., & North, P. R. J. (2014). Amazon forests maintain consistent canopy structure and greenness during the dry season. *Nature, 506*(7487), 221–224.

Ni-Meister, W., & Gao, H. L. (2011). Assessing the impacts of vegetation heterogeneity on energy fluxes and snowmelt in boreal forests. *Journal of Plant Ecology, 4*(1–2), 37–47.

Okhrimenko, M., Coburn, C., & Hopkinson, C. (2019). Multi-spectral lidar: Radiometric calibration, canopy spectral reflectance, and vegetation vertical SVI profiles. *Remote Sensing, 11*(13), 1556.

Pawlikowska, A. M., Halimi, A., Lamb, R. A., & Buller, G. S. (2017). Single-photon three-dimensional imaging at up to 10 kilometers range. *Optics Express, 25*(10), 11919–11931.

Raj, T., Hashim, F. H., Huddin, A. B., Ibrahim, M. F., & Hussain, A. (2020). A survey on LiDAR scanning mechanisms. *Electronics, 9*(5), 741.

Reichstein, M., Camps-Valls, G., Stevens, B., Jung, M., Denzler, J., Carvalhais, N., & Prabhat. (2019). Deep learning and process understanding for data-driven Earth system science. *Nature, 566*(7743), 195–204.

Sankey, T., Donager, J., McVay, J., & Sankey, J. B. (2017). UAV lidar and hyperspectral fusion for forest monitoring in the southwestern USA. *Remote Sensing of Environment, 195,* 30–43.

Srinivasan, S., Popescu, S. C., Eriksson, M., Sheridan, R. D., & Ku, N. W. (2014). Multi-temporal terrestrial laser scanning for modeling tree biomass change. *Forest Ecology and Management, 318,* 304–317.

Stoker, J. M., Abdullah, Q. A., Nayegandhi, A., & Winehouse, J. (2016). Evaluation of single photon and geiger mode lidar for the 3D elevation program. *Remote Sensing, 8*(9). Article 767.

Su, Y. J., Guo, Q. H., Xue, B. L., Hu, T. Y., Alvarez, O., Tao, S. L., & Fang, J. Y. (2016). Spatial distribution of forest aboveground biomass in China: Estimation through combination of spaceborne lidar, optical imagery, and forest inventory data. *Remote Sensing of Environment, 173,* 187–199.

Sun, J., Shi, S., Yang, J., Gong, W., Qiu, F., Wang, L. C., Du, L., & Chen, B. W. (2019). Wavelength selection of the multispectral lidar system for estimating leaf chlorophyll and water contents through the PROSPECT model. *Agricultural and Forest Meteorology, 266,* 43–52.

Tao, S. L., Guo, Q. H., Li, C., Wang, Z. H., & Fang, J. Y. (2016). Global patterns and determinants of forest canopy height. *Ecology, 97*(12), 3265–3270.

Van Leeuwen, M., Coops, N. C., & Black, T. A. (2015). Using stochastic ray tracing to simulate a dense time series of gross primary productivity. *Remote Sensing, 7*(12).

Wallace, A. M., McCarthy, A., Nichol, C. J., Ren, X. M., Morak, S., Martinez-Ramirez, D., Woodhouse, I. H., & Buller, G. S. (2014). Design and evaluation of multispectral LiDAR for the recovery of arboreal parameters. *Ieee Transactions on Geoscience and Remote Sensing, 52*(8), 4942–4954.

Widlowski, J. L., Mio, C., Disney, M., Adams, J., Andredakis, I., Atzberger, C., Brennan, J., Busetto, L., Chelle, M., Ceccherini, G., Colombo, R., Cote, J. F., Eenmae, A., Essery, R., Gastellu-Etchegorry, J. P., Gobron, N., Grau, E., Haverd, V., Homolova, L., Huang, H., Hunt, L., Kobayashi, H., Koetz, B., Kuusk, A., Kuusk, J., Lang, M., Lewis, P. E., Lovell, J. L., Malenovsky, Z., Meroni, M., Morsdorf, F., Mottus, M., Ni-Meister, W., Pinty, B., Rautiainen, M., Schlerf, M., Somers, B., Stuckens, J., Verstraete, M. M., Yang, W. Z., Zhao, F., & Zenone, T. (2015). The fourth phase of the radiative transfer model intercomparison (RAMI) exercise: Actual canopy scenarios and conformity testing. *Remote Sensing of Environment, 169,* 418–437.

Index

Printed in the United States
by Baker & Taylor Publisher Services